Math 265

Student Solutions Manual
to accompany
Introduction to Calculus

Custom Text

Maria Torres
Athabasca University

NELSON EDUCATION

Printed and bound in Canada
12 13 14 15 15 14 13 12

For more information contact
Nelson Education Limited
1120 Birchmount Road,
Toronto, Ontario, M1K 5G4.
Or you can visit our internet site
at http://www.nelson.com

This textbook is a Nelson custom publication. Because your instructor has chosen to produce a custom publication, you pay only for material that you will use in your course.

ISBN-13: 978-0-17-647739-4
ISBN-10: 0-17-647739-X

Consists of Selections from:

Student Solutions Manual for Single Variable Calculus, James Stewart Fifth Edition
Daniel Anderson, Jeffery Cole, Daniel Drucker
ISBN-10: 0-534-39369-1, © 2003

Contents

Selected Chapters from
Student Solutions Manual for Single Variable Calculus, James Stewart, 5th Edition

Applications of Differentiation

Integrals

Applications of Differentiation

Integrals

Applications of Integration

Integrals

Appendixes:

FUNCTIONS AND MODELS

Four Ways to Represent a Function

In exercises requiring estimations or approximations, your answers may vary slightly from the answers given here.

1. (a) The point $(-1,-2)$ is on the graph of f, so $f(-1)=-2$.

(b) When $x=2$, y is about 2.8, so $f(2)\approx 2.8$.

(c) $f(x)=2$ is equivalent to $y=2$. When $y=2$, we have $x=-3$ and $x=1$.

(d) Reasonable estimates for x when $y=0$ are $x=-2.5$ and $x=0.3$.

(e) The domain of f consists of all x-values on the graph of f. For this function, the domain is $-3\leq x\leq 3$, or $[-3,3]$. The range of f consists of all y-values on the graph of f. For this function, the range is $-2\leq y\leq 3$, or $[-2,3]$.

(f) As x increases from -1 to 3, y increases from -2 to 3. Thus, f is increasing on the interval $[-1,3]$.

3. From Figure 1 in the text, the lowest point occurs at about $(t,a)=(12,-85)$. The highest point occurs at about $(17,115)$. Thus, the range of the vertical ground acceleration is $-85\leq a\leq 115$. In Figure 11, the range of the north-south acceleration is approximately $-325\leq a\leq 485$. In Figure 12, the range of the east-west acceleration is approximately $-210\leq a\leq 200$.

5. No, the curve is not the graph of a function because a vertical line intersects the curve more than once. Hence, the curve fails the Vertical Line Test.

7. Yes, the curve is the graph of a function because it passes the Vertical Line Test. The domain is $[-3,2]$ and the range is $[-3,-2)\cup[-1,3]$.

9. The person's weight increased to about 160 pounds at age 20 and stayed fairly steady for 10 years. The person's weight dropped to about 120 pounds for the next 5 years, then increased rapidly to about 170 pounds. The next 30 years saw a gradual increase to 190 pounds. Possible reasons for the drop in weight at 30 years of age: diet, exercise, health problems.

11. The water will cool down almost to freezing as the ice melts. Then, when the ice has melted, the water will slowly warm up to room temperature.

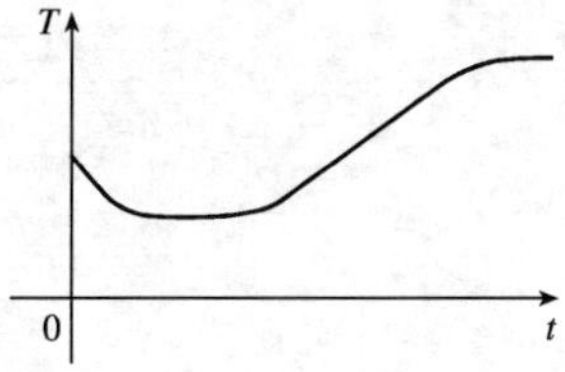

13. Of course, this graph depends strongly on the geographical location!

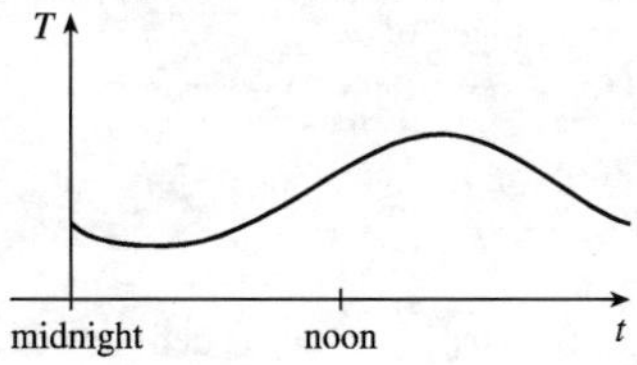

15.

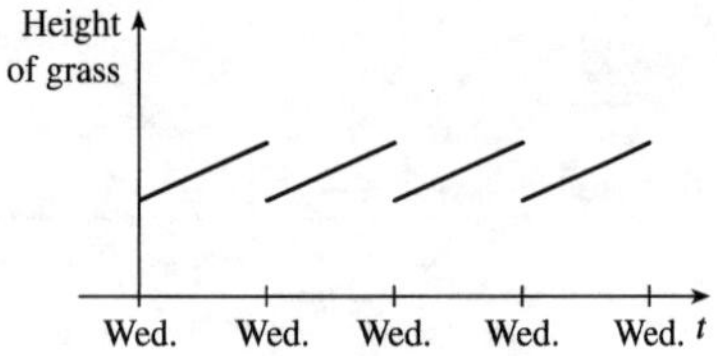

17. (a)

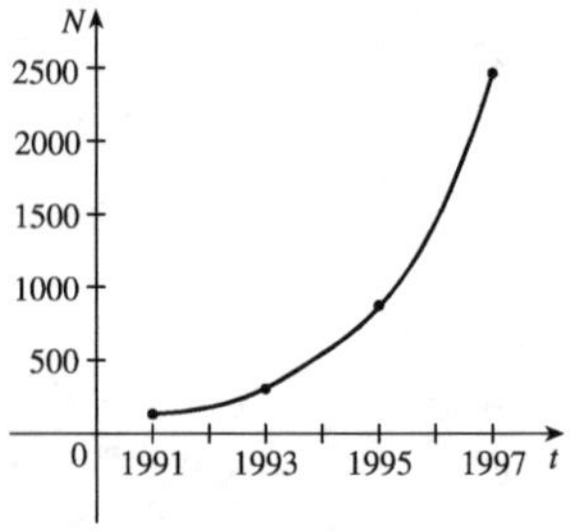

(b) From the graph, we estimate the number of cell-phone subscribers in Malaysia to be about 540 in 1994 and 1450 in 1996.

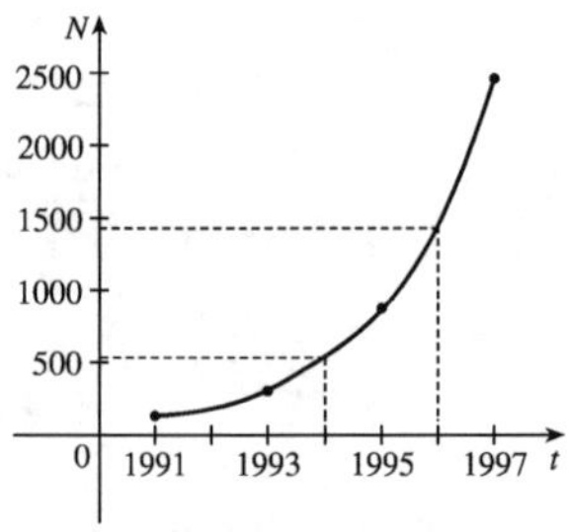

19. $f(x) = 3x^2 - x + 2.$

$f(2) = 3(2)^2 - 2 + 2 = 12 - 2 + 2 = 12.$

$f(-2) = 3(-2)^2 - (-2) + 2 = 12 + 2 + 2 = 16.$

$f(a) = 3a^2 - a + 2.$

$f(-a) = 3(-a)^2 - (-a) + 2 = 3a^2 + a + 2.$

$f(a+1) = 3(a+1)^2 - (a+1) + 2 = 3(a^2 + 2a + 1) - a - 1 + 2 = 3a^2 + 6a + 3 - a + 1 = 3a^2 + 5a + 4.$

$2f(a) = 2 \cdot f(a) = 2(3a^2 - a + 2) = 6a^2 - 2a + 4.$

$f(2a) = 3(2a)^2 - (2a) + 2 = 3(4a^2) - 2a + 2 = 12a^2 - 2a + 2.$

$f(a^2) = 3(a^2)^2 - (a^2) + 2 = 3(a^4) - a^2 + 2 = 3a^4 - a^2 + 2.$

$$[f(a)]^2 = \left[3a^2 - a + 2\right]^2 = \left(3a^2 - a + 2\right)\left(3a^2 - a + 2\right)$$
$$= 9a^4 - 3a^3 + 6a^2 - 3a^3 + a^2 - 2a + 6a^2 - 2a + 4 = 9a^4 - 6a^3 + 13a^2 - 4a + 4.$$

$f(a+h) = 3(a+h)^2 - (a+h) + 2 = 3(a^2 + 2ah + h^2) - a - h + 2 = 3a^2 + 6ah + 3h^2 - a - h + 2.$

21. $f(x) = x - x^2$, so

$f(2+h) = 2 + h - (2+h)^2 = 2 + h - (4 + 4h + h^2) = 2 + h - 4 - 4h - h^2 = -\left(h^2 + 3h + 2\right),$

$f(x+h) = x + h - (x+h)^2 = x + h - (x^2 + 2xh + h^2) = x + h - x^2 - 2xh - h^2$, and

$$\frac{f(x+h) - f(x)}{h} = \frac{x + h - x^2 - 2xh - h^2 - x + x^2}{h} = \frac{h - 2xh - h^2}{h} = \frac{h(1 - 2x - h)}{h} = 1 - 2x - h.$$

23. $f(x) = x/(3x - 1)$ is defined for all x except when $0 = 3x - 1 \quad\Leftrightarrow\quad x = \frac{1}{3}$, so the domain is $\left\{x \in \mathbb{R} \mid x \neq \frac{1}{3}\right\} = \left(-\infty, \frac{1}{3}\right) \cup \left(\frac{1}{3}, \infty\right).$

25. $f(t) = \sqrt{t} + \sqrt[3]{t}$ is defined when $t \geq 0$. These values of t give real number results for $\sqrt{t}$, whereas any value of t gives a real number result for $\sqrt[3]{t}$. The domain is $[0, \infty)$.

27. $h(x) = 1 \big/ \sqrt[4]{x^2 - 5x}$ is defined when $x^2 - 5x > 0 \quad \Leftrightarrow \quad x(x-5) > 0$. Note that $x^2 - 5x \neq 0$ since that would result in division by zero. The expression $x(x-5)$ is positive if $x < 0$ or $x > 5$. (See Appendix A for methods for solving inequalities.) Thus, the domain is $(-\infty, 0) \cup (5, \infty)$.

29. $f(x) = 5$ is defined for all real numbers, so the domain is $\mathbb{R}$, or $(-\infty, \infty)$. The graph of f is a horizontal line with y-intercept 5.

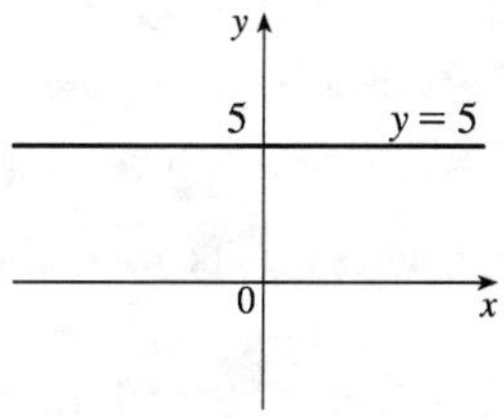

31. $f(t) = t^2 - 6t$ is defined for all real numbers, so the domain is $\mathbb{R}$, or $(-\infty, \infty)$. The graph of f is a parabola opening upward since the coefficient of t^2 is positive. To find the t-intercepts, let $y = 0$ and solve for t. $0 = t^2 - 6t = t(t-6) \quad \Rightarrow \quad t = 0$ and $t = 6$. The t-coordinate of the vertex is halfway between the t-intercepts, that is, at $t = 3$. Since $f(3) = 3^2 - 6 \cdot 3 = -9$, the vertex is $(3, -9)$.

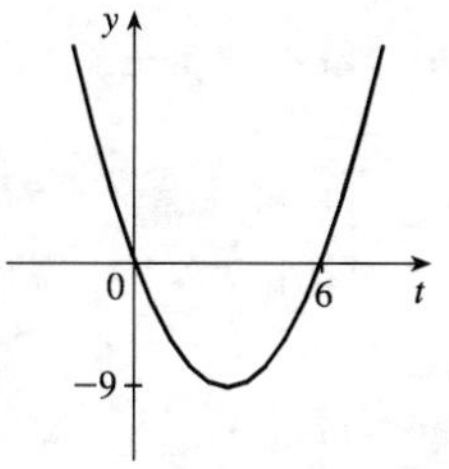

33. $g(x) = \sqrt{x - 5}$ is defined when $x - 5 \geq 0$ or $x \geq 5$, so the domain is $[5, \infty)$. Since $y = \sqrt{x-5} \quad \Rightarrow \quad y^2 = x - 5 \quad \Rightarrow \quad x = y^2 + 5$, we see that g is the top half of a parabola.

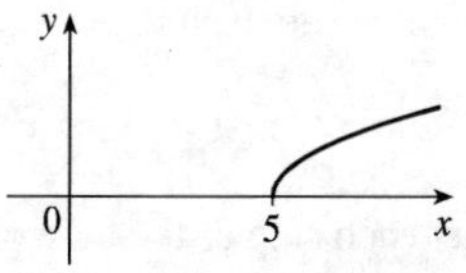

35. $G(x) = \dfrac{3x + |x|}{x}$. Since $|x| = \begin{cases} x & \text{if } x \geq 0 \\ -x & \text{if } x < 0 \end{cases}$, we have

$$G(x) = \begin{cases} \dfrac{3x + x}{x} & \text{if } x > 0 \\ \dfrac{3x - x}{x} & \text{if } x < 0 \end{cases} = \begin{cases} \dfrac{4x}{x} & \text{if } x > 0 \\ \dfrac{2x}{x} & \text{if } x < 0 \end{cases} = \begin{cases} 4 & \text{if } x > 0 \\ 2 & \text{if } x < 0 \end{cases}$$

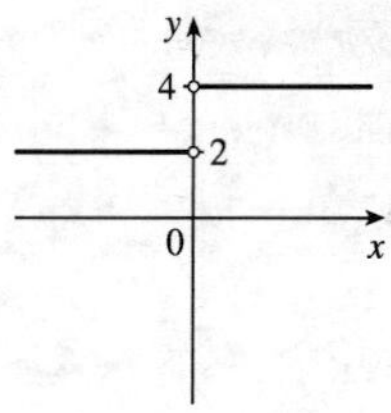

Note that G is not defined for $x = 0$. The domain is $(-\infty, 0) \cup (0, \infty)$.

37. $f(x) = \begin{cases} x & \text{if } x \leq 0 \\ x + 1 & \text{if } x > 0 \end{cases}$

Domain is $\mathbb{R}$, or $(-\infty, \infty)$.

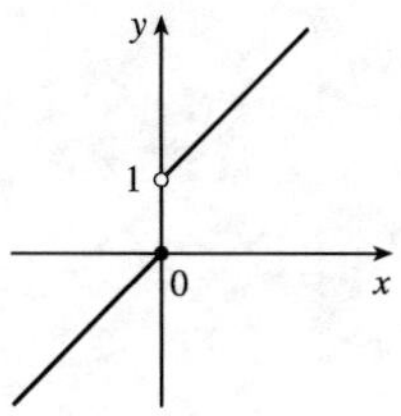

39. $f(x) = \begin{cases} x + 2 & \text{if } x \leq -1 \\ x^2 & \text{if } x > -1 \end{cases}$

Note that for $x = -1$, both $x + 2$ and x^2 are equal to 1. Domain is $\mathbb{R}$.

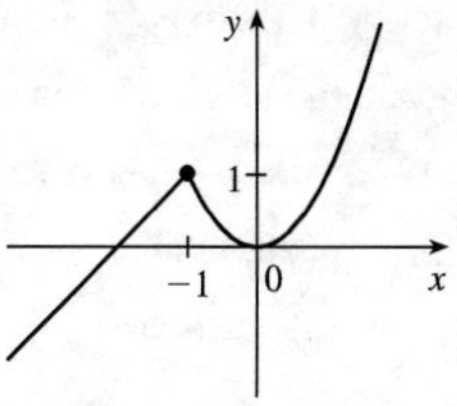

41. Recall that the slope m of a line between the two points (x_1, y_1) and (x_2, y_2) is $m = \dfrac{y_2 - y_1}{x_2 - x_1}$ and an equation of the line connecting those two points is $y - y_1 = m(x - x_1)$. The slope of this line segment is $\dfrac{-6-1}{4-(-2)} = -\dfrac{7}{6}$, so an equation is $y - 1 = -\frac{7}{6}(x+2)$. The function is $f(x) = -\frac{7}{6}x - \frac{4}{3}$, $-2 \le x \le 4$.

43. We need to solve the given equation for y. $x + (y-1)^2 = 0 \quad\Leftrightarrow\quad (y-1)^2 = -x \quad\Leftrightarrow\quad y - 1 = \pm\sqrt{-x} \quad\Leftrightarrow$ $y = 1 \pm \sqrt{-x}$. The expression with the positive radical represents the top half of the parabola, and the one with the negative radical represents the bottom half. Hence, we want $f(x) = 1 - \sqrt{-x}$. Note that the domain is $x \le 0$.

45. For $-1 \le x \le 2$, the graph is the line with slope 1 and y-intercept 1, that is, the line $y = x + 1$. For $2 < x \le 4$, the graph is the line with slope $-\frac{3}{2}$ and x-intercept 4 [which corresponds to the point $(4, 0)$], so

$y - 0 = -\frac{3}{2}(x-4) = -\frac{3}{2}x + 6$. So the function is $f(x) = \begin{cases} x+1 & \text{if } -1 \le x \le 2 \\ -\frac{3}{2}x + 6 & \text{if } 2 < x \le 4 \end{cases}$

47. Let the length and width of the rectangle be L and W. Then the perimeter is $2L + 2W = 20$ and the area is $A = LW$. Solving the first equation for W in terms of L gives $W = \dfrac{20 - 2L}{2} = 10 - L$. Thus, $A(L) = L(10 - L) = 10L - L^2$. Since lengths are positive, the domain of A is $0 < L < 10$. If we further restrict L to be larger than W, then $5 < L < 10$ would be the domain.

49. Let the length of a side of the equilateral triangle be x. Then by the Pythagorean Theorem, the height y of the triangle satisfies $y^2 + \left(\frac{1}{2}x\right)^2 = x^2$, so that $y^2 = x^2 - \frac{1}{4}x^2 = \frac{3}{4}x^2$ and $y = \frac{\sqrt{3}}{2}x$. Using the formula for the area A of a triangle, $A = \frac{1}{2}(\text{base})(\text{height})$, we obtain $A(x) = \frac{1}{2}(x)\left(\frac{\sqrt{3}}{2}x\right) = \frac{\sqrt{3}}{4}x^2$, with domain $x > 0$.

51. Let each side of the base of the box have length x, and let the height of the box be h. Since the volume is 2, we know that $2 = hx^2$, so that $h = 2/x^2$, and the surface area is $S = x^2 + 4xh$. Thus, $S(x) = x^2 + 4x(2/x^2) = x^2 + (8/x)$, with domain $x > 0$.

53. The height of the box is x and the length and width are $L = 20 - 2x$, $W = 12 - 2x$. Then $V = LWx$ and so $V(x) = (20 - 2x)(12 - 2x)(x) = 4(10 - x)(6 - x)(x) = 4x(60 - 16x + x^2) = 4x^3 - 64x^2 + 240x$. The sides L, W, and x must be positive. Thus, $L > 0 \quad\Leftrightarrow\quad 20 - 2x > 0 \quad\Leftrightarrow\quad x < 10$; $W > 0 \quad\Leftrightarrow$ $12 - 2x > 0 \quad\Leftrightarrow\quad x < 6$; and $x > 0$. Combining these restrictions gives us the domain $0 < x < 6$.

55. (a)

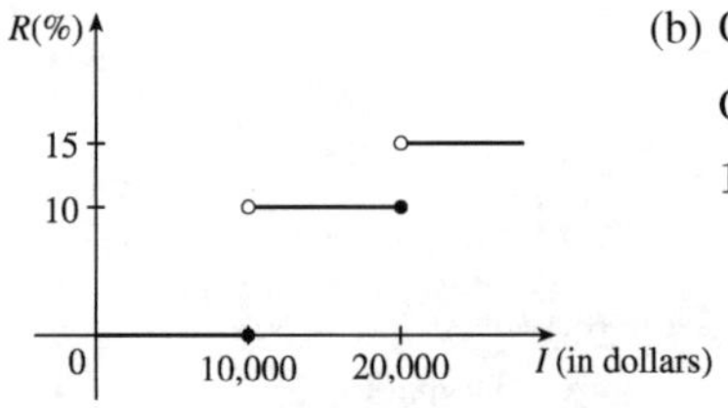

(b) On \$14,000, tax is assessed on \$4000, and 10%(\$4000) = \$400. On \$26,000, tax is assessed on \$16,000, and 10%(\$10,000) + 15%(\$6000) = \$1000 + \$900 = \$1900.

(c) As in part (b), there is \$1000 tax assessed on \$20,000 of income, so the graph of T is a line segment from (10,000, 0) to (20,000, 1000). The tax on \$30,000 is \$2500, so the graph of T for $x >$ 20,000 is the ray with initial point (20,000, 1000) that passes through (30,000, 2500).

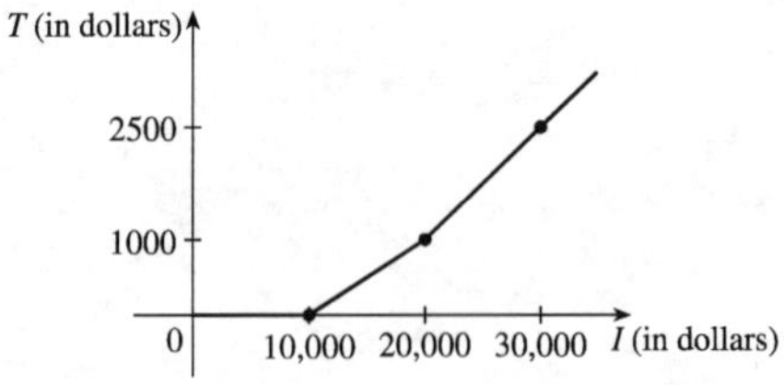

57. f is an odd function because its graph is symmetric about the origin. g is an even function because its graph is symmetric with respect to the y-axis.

59. (a) Because an even function is symmetric with respect to the y-axis, and the point $(5, 3)$ is on the graph of this even function, the point $(-5, 3)$ must also be on its graph.

(b) Because an odd function is symmetric with respect to the origin, and the point $(5, 3)$ is on the graph of this odd function, the point $(-5, -3)$ must also be on its graph.

61. $f(x) = x^{-2}$.

$$f(-x) = (-x)^{-2} = \frac{1}{(-x)^2} = \frac{1}{x^2} = x^{-2} = f(x)$$

So f is an even function.

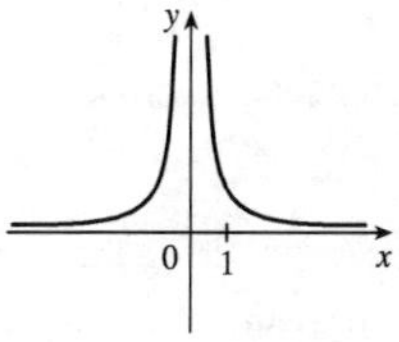

63. $f(x) = x^2 + x$, so $f(-x) = (-x)^2 + (-x) = x^2 - x$. Since this is neither $f(x)$ nor $-f(x)$, the function f is neither even nor odd.

65. $f(x) = x^3 - x$.

$$f(-x) = (-x)^3 - (-x) = -x^3 + x = -\left(x^3 - x\right) = -f(x)$$

So f is odd.

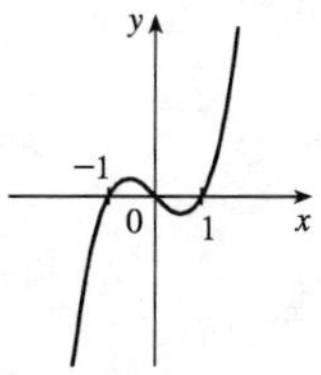

Mathematical Models: A Catalog of Essential Functions

1. (a) $f(x) = \sqrt[5]{x}$ is a root function with $n = 5$.

(b) $g(x) = \sqrt{1 - x^2}$ is an algebraic function because it is a root of a polynomial.

(c) $h(x) = x^9 + x^4$ is a polynomial of degree 9.

(d) $r(x) = \dfrac{x^2 + 1}{x^3 + x}$ is a rational function because it is a ratio of polynomials.

(e) $s(x) = \tan 2x$ is a trigonometric function.

(f) $t(x) = \log_{10} x$ is a logarithmic function.

3. We notice from the figure that g and h are even functions (symmetric with respect to the y-axis) and that f is an odd function (symmetric with respect to the origin). So (b) $\left[y = x^5\right]$ must be f. Since g is flatter than h near the origin, we must have (c) $\left[y = x^8\right]$ matched with g and (a) $\left[y = x^2\right]$ matched with h.

5. (a) An equation for the family of linear functions with slope 2 is $y = f(x) = 2x + b$, where b is the y-intercept.

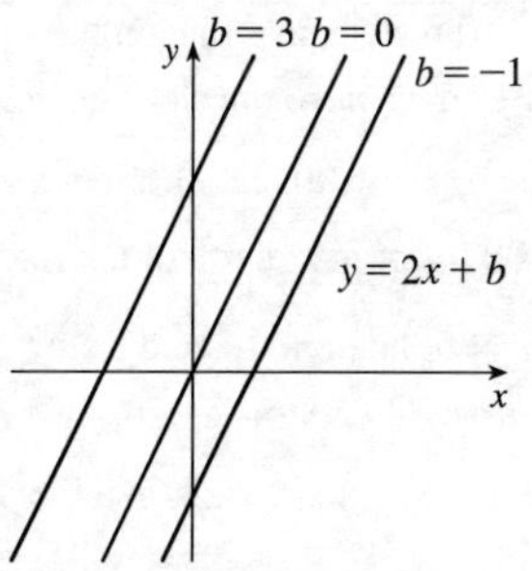

(b) $f(2) = 1$ means that the point $(2, 1)$ is on the graph of f. We can use the point-slope form of a line to obtain an equation for the family of linear functions through the point $(2, 1)$. $y - 1 = m\,(x - 2)$, which is equivalent to $y = mx + (1 - 2m)$ in slope-intercept form.

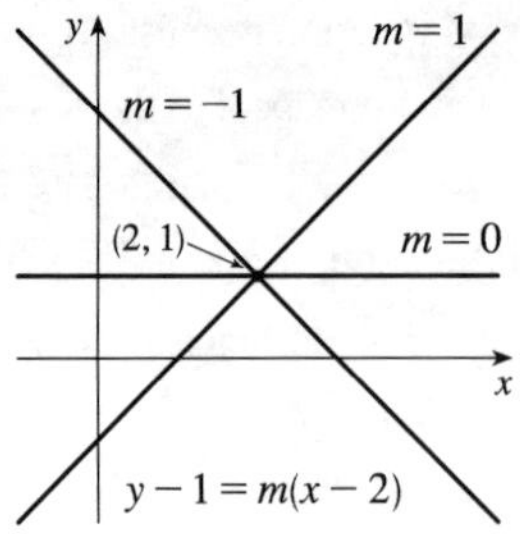

(c) To belong to both families, an equation must have slope $m = 2$, so the equation in part (b), $y = mx + (1 - 2m)$, becomes $y = 2x - 3$. It is the *only* function that belongs to both families.

7. All members of the family of linear functions $f(x) = c - x$ have graphs that are lines with slope -1. The y-intercept is c.

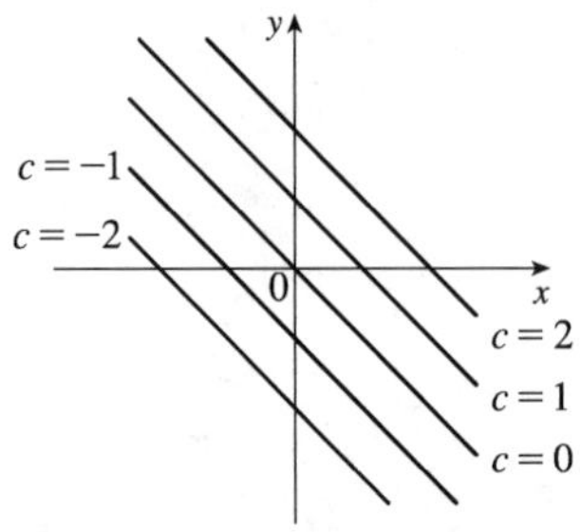

9. (a)

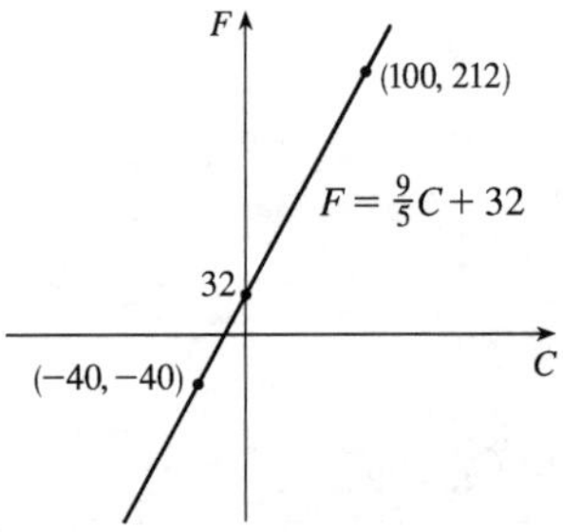

(b) The slope of $\frac{9}{5}$ means that F increases $\frac{9}{5}$ degrees for each increase of $1\,^\circ$C. (Equivalently, F increases by 9 when C increases by 5 and F decreases by 9 when C decreases by 5.) The F-intercept of 32 is the Fahrenheit temperature corresponding to a Celsius temperature of 0.

11. (a) Using N in place of x and T in place of y, we find the slope to be $\dfrac{T_2 - T_1}{N_2 - N_1} = \dfrac{80 - 70}{173 - 113} = \dfrac{10}{60} = \dfrac{1}{6}$. So a linear equation is $T - 80 = \frac{1}{6}(N - 173) \;\Leftrightarrow\; T - 80 = \frac{1}{6}N - \frac{173}{6} \;\Leftrightarrow\; T = \frac{1}{6}N + \frac{307}{6}\ \left[\frac{307}{6} = 51.1\overline{6}\right]$.

(b) The slope of $\frac{1}{6}$ means that the temperature in Fahrenheit degrees increases one-sixth as rapidly as the number of cricket chirps per minute. Said differently, each increase of 6 cricket chirps per minute corresponds to an increase of $1\,^\circ$F.

(c) When $N = 150$, the temperature is given approximately by $T = \frac{1}{6}(150) + \frac{307}{6} = 76.1\overline{6}\,^\circ\text{F} \approx 76\,^\circ\text{F}$.

13. (a) We are given $\dfrac{\text{change in pressure}}{\text{10 feet change in depth}} = \dfrac{4.34}{10} = 0.434$. Using P for pressure and d for depth with the point $(d, P) = (0, 15)$, we have the slope-intercept form of the line, $P = 0.434d + 15$.

(b) When $P = 100$, then $100 = 0.434d + 15 \;\Leftrightarrow\; 0.434d = 85 \;\Leftrightarrow\; d = \frac{85}{0.434} \approx 195.85$ feet. Thus, the pressure is 100 lb/in^2 at a depth of approximately 196 feet.

15. (a) The data appear to be periodic and a sine or cosine function would make the best model. A model of the form $f(x) = a\cos(bx) + c$ seems appropriate.

(b) The data appear to be decreasing in a linear fashion. A model of the form $f(x) = mx + b$ seems appropriate.

Some values are given to many decimal places. These are the results given by several computer algebra systems — rounding is left to the reader.

17. (a)

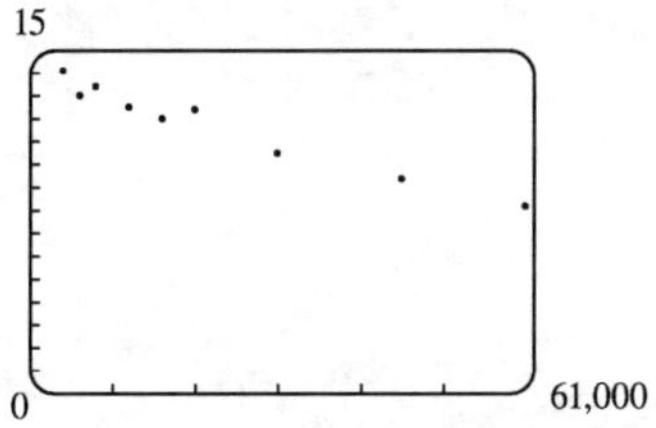

A linear model does seem appropriate.

(b) Using the points (4000, 14.1) and (60,000, 8.2), we obtain

$$y - 14.1 = \frac{8.2 - 14.1}{60{,}000 - 4000}(x - 4000) \text{ or, equivalently,}$$

$$y \approx -0.000105357x + 14.521429.$$

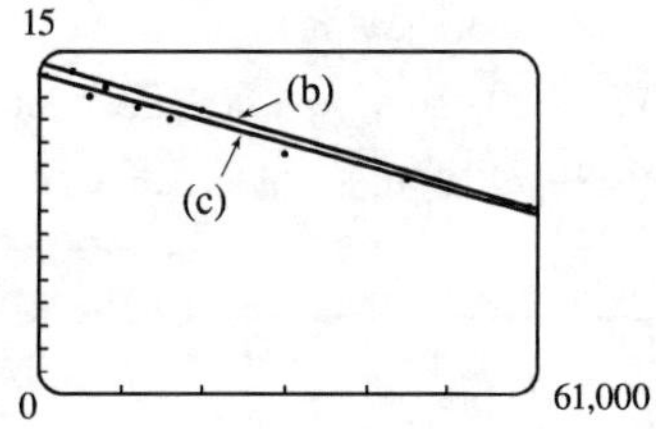

(c) Using a computing device, we obtain the least squares regression line $y = -0.0000997855x + 13.950764$. The following commands and screens illustrate how to find the least squares regression line on a TI-83 Plus. Enter the data into list one (L1) and list two (L2). Press [STAT] [1] to enter the editor.

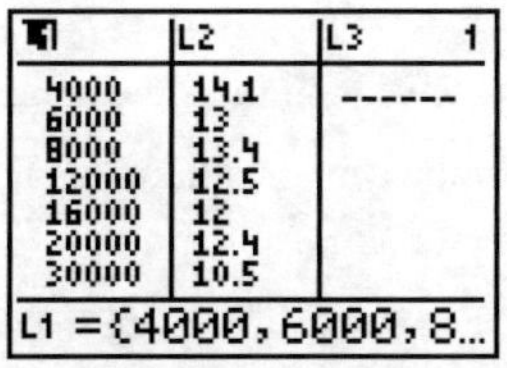

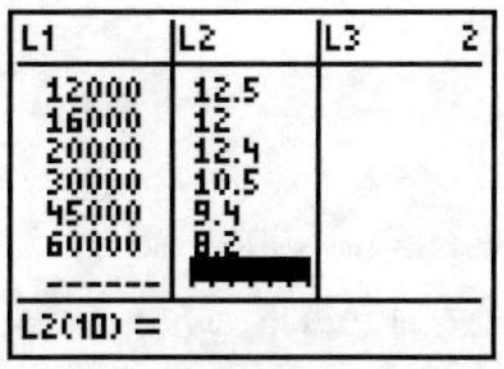

Find the regession line and store it in Y_1. Press [2nd] [QUIT] [STAT] [▸] [4] [VARS] [▸] [1] [1] [ENTER].

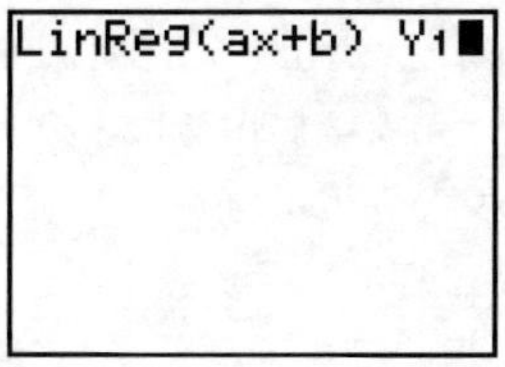

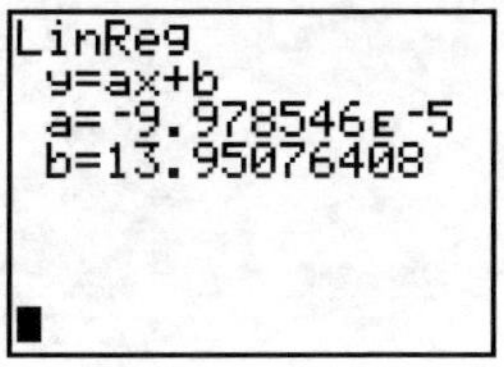

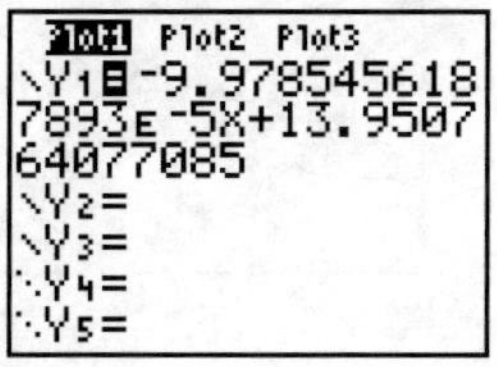

Note from the last figure that the regression line has been stored in Y_1 and that Plot1 has been turned on (Plot1 is highlighted). You can turn on Plot1 from the Y= menu by placing the cursor on Plot1 and pressing [ENTER] or by pressing [2nd] [STAT PLOT] [1] [ENTER].

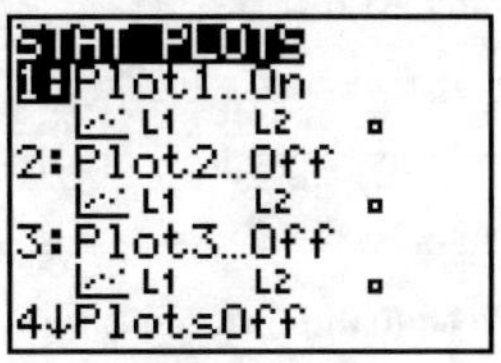

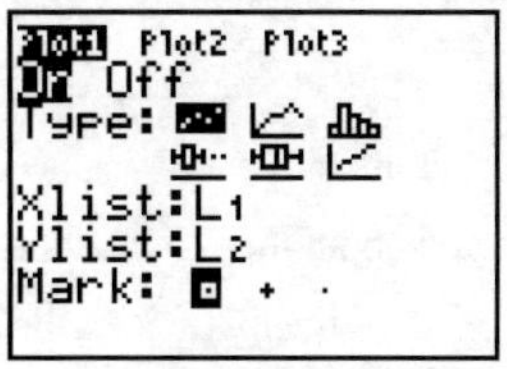

Now press ZOOM 9 to produce a graph of the data and the regression line. Note that choice 9 of the ZOOM menu automatically selects a window that displays all of the data.

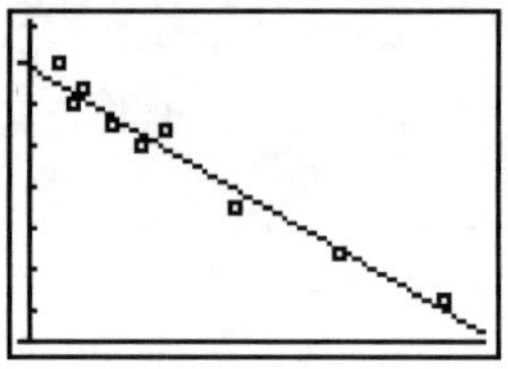

(d) When $x = 25{,}000$, $y \approx 11.456$; or about 11.5 per 100 population.

(e) When $x = 80{,}000$, $y \approx 5.968$; or about a 6% chance.

(f) When $x = 200{,}000$, y is negative, so the model does not apply.

19. (a)

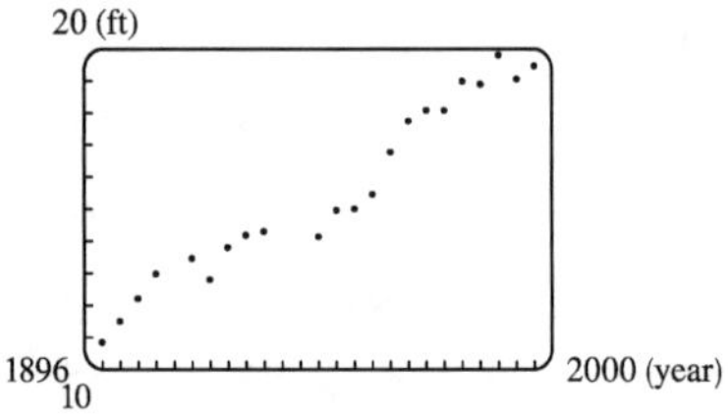

A linear model does seem appropriate.

(b)

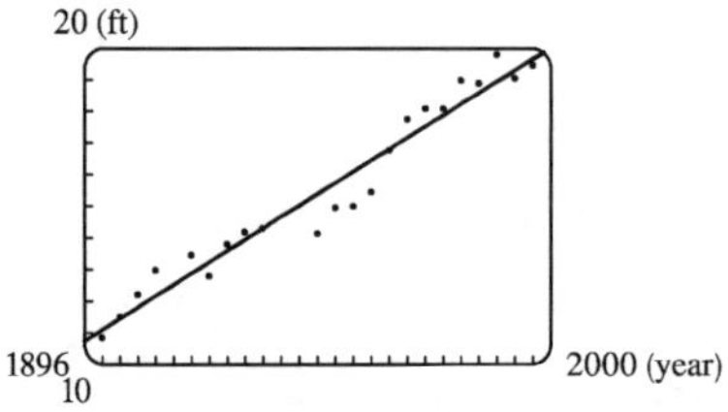

Using a computing device, we obtain the least squares regression line $y = 0.089119747x - 158.2403249$, where x is the year and y is the height in feet.

(c) When $x = 2000$, the model gives $y \approx 20.00$ ft. Note that the actual winning height for the 2000 Olympics is *less than* the winning height for 1996—so much for that prediction.

(d) When $x = 2100$, $y \approx 28.91$ ft. This would be an increase of 9.49 ft from 1996 to 2100. Even though there was an increase of 8.59 ft from 1900 to 1996, it is unlikely that a similar increase will occur over the next 100 years.

21.

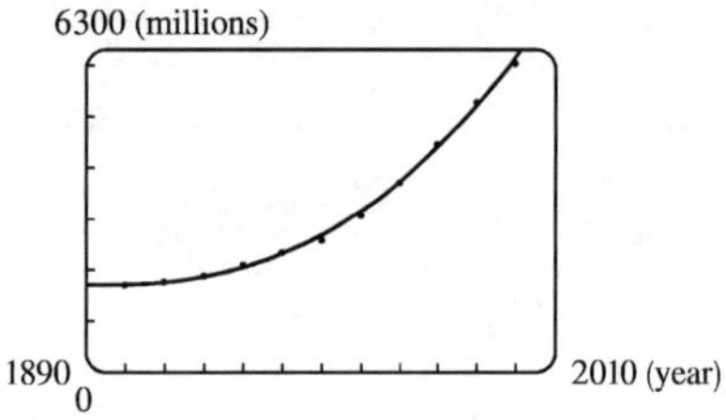

Using a computing device, we obtain the cubic function $y = ax^3 + bx^2 + cx + d$ with $a = 0.0012937$, $b = -7.06142$, $c = 12{,}823$, and $d = -7{,}743{,}770$. When $x = 1925$, $y \approx 1914$ (million).

New Functions from Old Functions

1. (a) If the graph of f is shifted 3 units upward, its equation becomes $y = f(x) + 3$.

(b) If the graph of f is shifted 3 units downward, its equation becomes $y = f(x) - 3$.

(c) If the graph of f is shifted 3 units to the right, its equation becomes $y = f(x - 3)$.

(d) If the graph of f is shifted 3 units to the left, its equation becomes $y = f(x + 3)$.

(e) If the graph of f is reflected about the x-axis, its equation becomes $y = -f(x)$.

(f) If the graph of f is reflected about the y-axis, its equation becomes $y = f(-x)$.

(g) If the graph of f is stretched vertically by a factor of 3, its equation becomes $y = 3f(x)$.

(h) If the graph of f is shrunk vertically by a factor of 3, its equation becomes $y = \frac{1}{3}f(x)$.

3. (a) (graph 3) The graph of f is shifted 4 units to the right and has equation $y = f(x-4)$.

(b) (graph 1) The graph of f is shifted 3 units upward and has equation $y = f(x) + 3$.

(c) (graph 4) The graph of f is shrunk vertically by a factor of 3 and has equation $y = \frac{1}{3}f(x)$.

(d) (graph 5) The graph of f is shifted 4 units to the left and reflected about the x-axis. Its equation is $y = -f(x+4)$.

(e) (graph 2) The graph of f is shifted 6 units to the left and stretched vertically by a factor of 2. Its equation is $y = 2f(x+6)$.

5. (a) To graph $y = f(2x)$ we shrink the graph of f horizontally by a factor of 2.

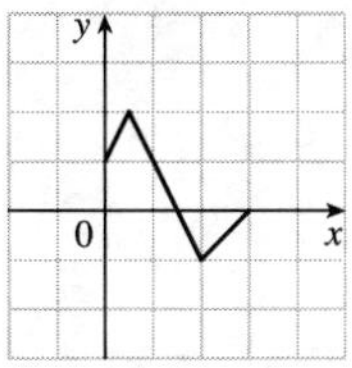

The point $(4,-1)$ on the graph of f corresponds to the point $\left(\frac{1}{2}\cdot 4,-1\right) = (2,-1)$.

(b) To graph $y = f\left(\frac{1}{2}x\right)$ we stretch the graph of f horizontally by a factor of 2.

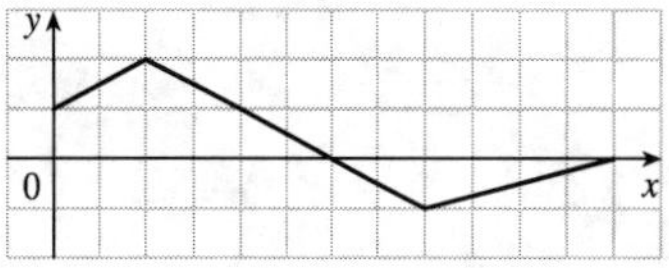

The point $(4,-1)$ on the graph of f corresponds to the point $(2\cdot 4,-1) = (8,-1)$.

(c) To graph $y = f(-x)$ we reflect the graph of f about the y-axis.

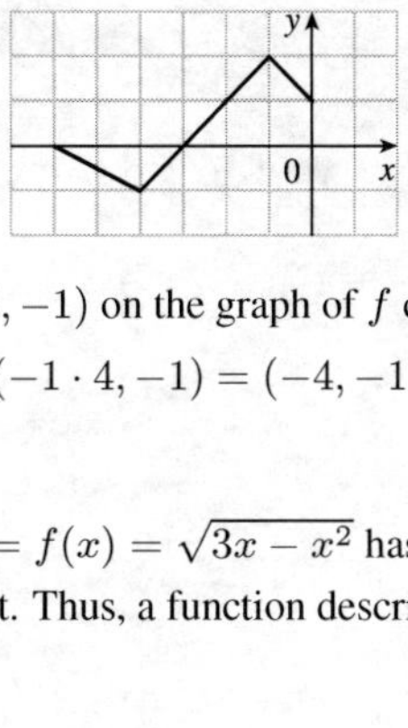

The point $(4,-1)$ on the graph of f corresponds to the point $(-1\cdot 4,-1) = (-4,-1)$.

(d) To graph $y = -f(-x)$ we reflect the graph of f about the y-axis, then about the x-axis.

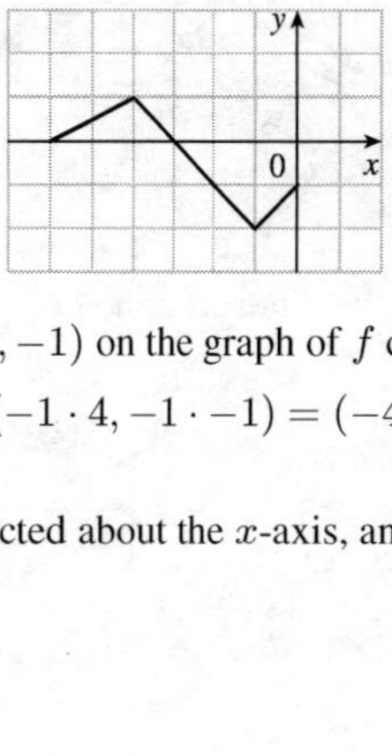

The point $(4,-1)$ on the graph of f corresponds to the point $(-1\cdot 4,-1\cdot -1) = (-4,1)$.

7. The graph of $y = f(x) = \sqrt{3x - x^2}$ has been shifted 4 units to the left, reflected about the x-axis, and shifted downward 1 unit. Thus, a function describing the graph is

$$y = \underbrace{-1\cdot}_{\substack{\text{reflect}\\\text{about}\\x\text{-axis}}}\ \underbrace{f(x+4)}_{\substack{\text{shift}\\\text{4 units}\\\text{left}}}\ \underbrace{-1}_{\substack{\text{shift}\\\text{1 unit}\\\text{down}}}$$

This function can be written as

$$\begin{aligned} y &= -f(x+4) - 1 = -\sqrt{3(x+4) - (x+4)^2} - 1 = -\sqrt{3x + 12 - (x^2 + 8x + 16)} - 1 \\ &= -\sqrt{-x^2 - 5x - 4} - 1 \end{aligned}$$

9. $y = -x^3$: Start with the graph of $y = x^3$ and reflect about the x-axis. Note: Reflecting about the y-axis gives the same result since substituting $-x$ for x gives us $y = (-x)^3 = -x^3$.

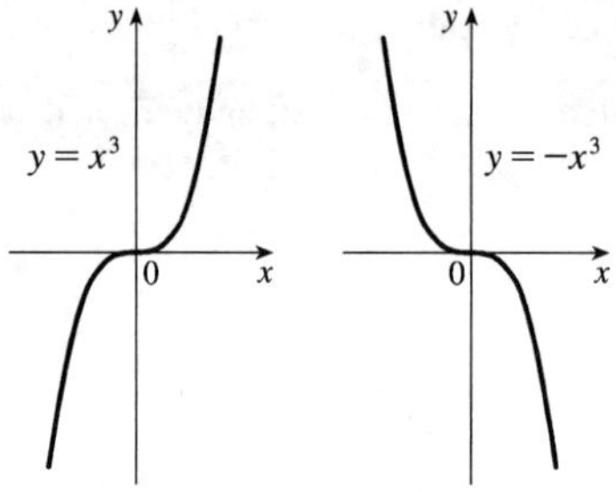

11. $y = (x+1)^2$: Start with the graph of $y = x^2$ and shift 1 unit to the left.

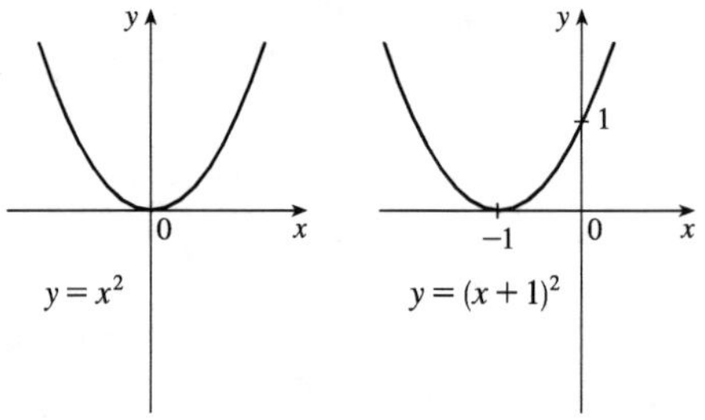

13. $y = 1 + 2\cos x$: Start with the graph of $y = \cos x$, stretch vertically by a factor of 2, and then shift 1 unit upward.

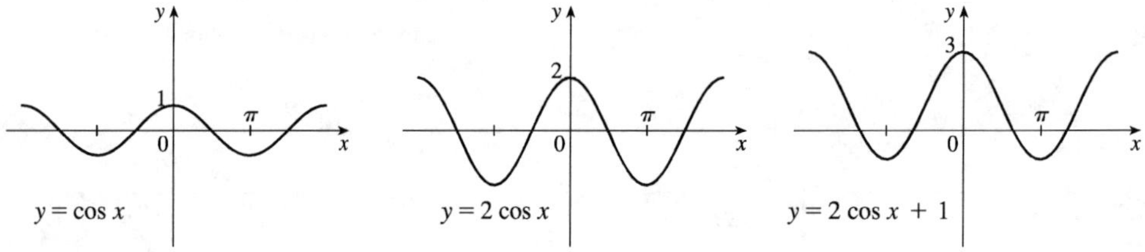

15. $y = \sin(x/2)$: Start with the graph of $y = \sin x$ and stretch horizontally by a factor of 2.

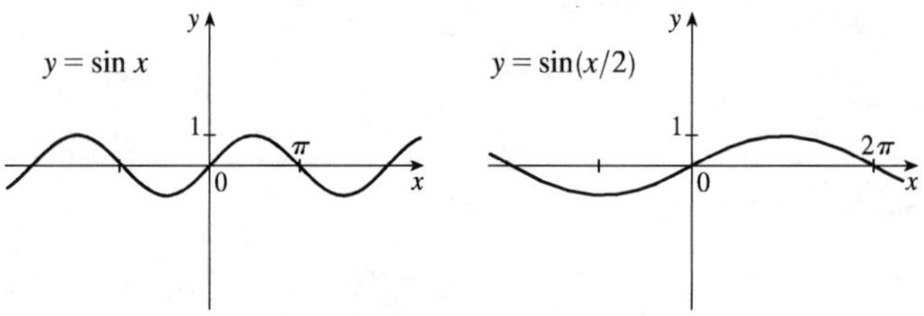

17. $y = \sqrt{x+3}$: Start with the graph of $y = \sqrt{x}$ and shift 3 units to the left.

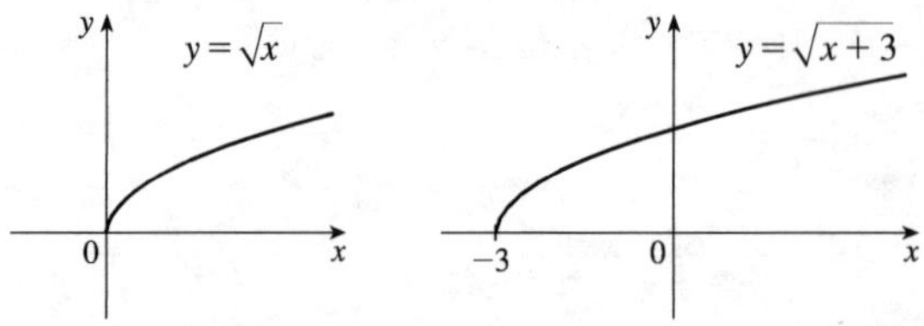

19. $y = \frac{1}{2}(x^2 + 8x) = \frac{1}{2}(x^2 + 8x + 16) - 8 = \frac{1}{2}(x+4)^2 - 8$: Start with the graph of $y = x^2$, compress vertically by a factor of 2, shift 4 units to the left, and then shift 8 units downward.

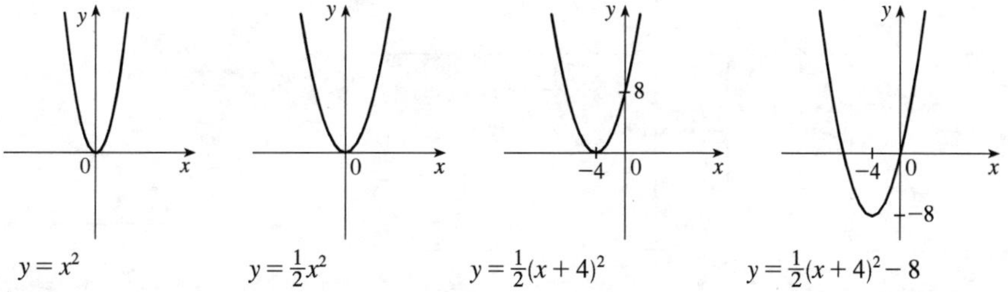

21. $y = 2/(x+1)$: Start with the graph of $y = 1/x$, shift 1 unit to the left, and then stretch vertically by a factor of 2.

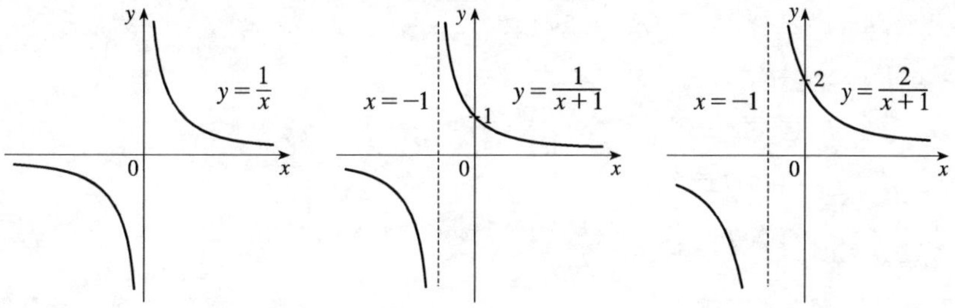

23. $y = |\sin x|$: Start with the graph of $y = \sin x$ and reflect all the parts of the graph below the x-axis about the x-axis.

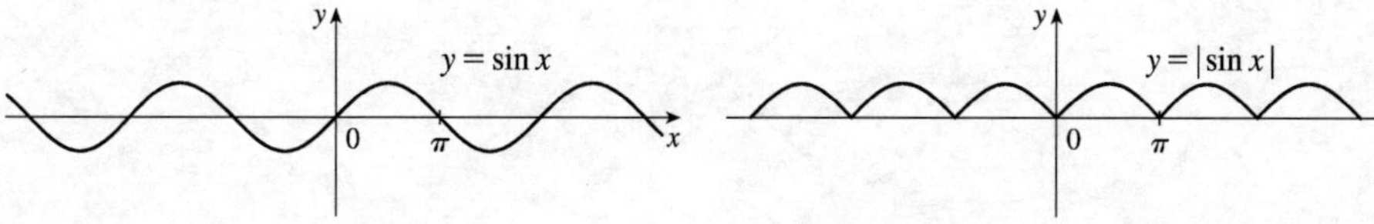

25. This is just like the solution to Example 4 except the amplitude of the curve (the 30°N curve in Figure 9 on June 21) is $14 - 12 = 2$. So the function is $L(t) = 12 + 2\sin\left[\frac{2\pi}{365}(t - 80)\right]$. March 31 is the 90th day of the year, so the model gives $L(90) \approx 12.34$ h. The daylight time (5:51 A.M. to 6:18 P.M.) is 12 hours and 27 minutes, or 12.45 h. The model value differs from the actual value by $\frac{12.45 - 12.34}{12.45} \approx 0.009$, less than 1%.

27. (a) To obtain $y = f(|x|)$, the portion of the graph of $y = f(x)$ to the right of the y-axis is reflected about the y-axis.

(b) $y = \sin|x|$

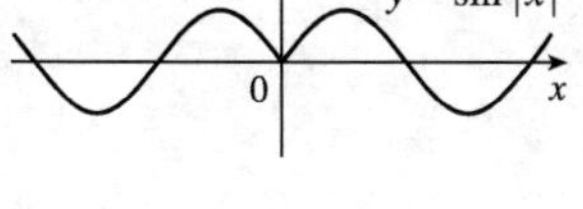

(c) $y = \sqrt{|x|}$

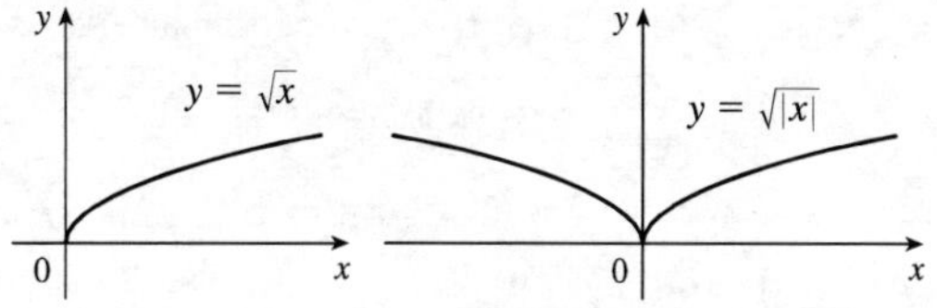

29. Assuming that successive horizontal and vertical gridlines are a unit apart, we can make a table of approximate values as follows.

x	0	1	2	3	4	5	6
$f(x)$	2	1.7	1.3	1.0	0.7	0.3	0
$g(x)$	2	2.7	3	2.8	2.4	1.7	0
$f(x)+g(x)$	4	4.4	4.3	3.8	3.1	2.0	0

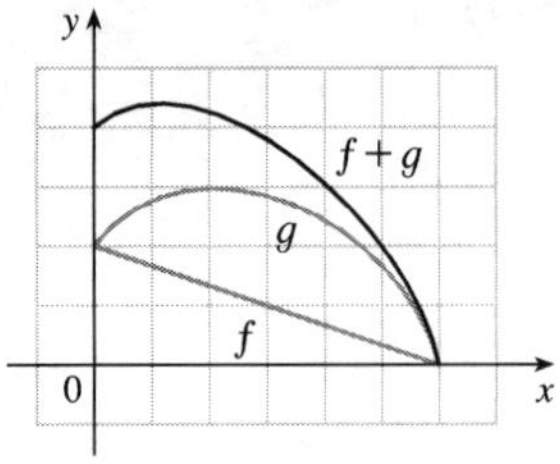

Connecting the points $(x, f(x)+g(x))$ with a smooth curve gives an approximation to the graph of $f+g$. Extra points can be plotted between those listed above if necessary.

31. $f(x) = x^3 + 2x^2$; $g(x) = 3x^2 - 1$. $D = \mathbb{R}$ for both f and g.

$(f+g)(x) = (x^3 + 2x^2) + (3x^2 - 1) = x^3 + 5x^2 - 1, \quad D = \mathbb{R}.$

$(f-g)(x) = (x^3 + 2x^2) - (3x^2 - 1) = x^3 - x^2 + 1, \quad D = \mathbb{R}.$

$(fg)(x) = (x^3 + 2x^2)(3x^2 - 1) = 3x^5 + 6x^4 - x^3 - 2x^2, \quad D = \mathbb{R}.$

$\left(\dfrac{f}{g}\right)(x) = \dfrac{x^3 + 2x^2}{3x^2 - 1}, \quad D = \left\{x \mid x \neq \pm\dfrac{1}{\sqrt{3}}\right\}$ since $3x^2 - 1 \neq 0$.

33. $f(x) = x, \quad g(x) = 1/x$

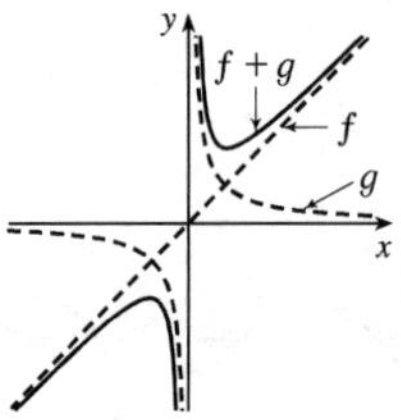

35. $f(x) = 2x^2 - x$; $g(x) = 3x + 2$. $D = \mathbb{R}$ for both f and g, and hence for their composites.

$(f \circ g)(x) = f(g(x)) = f(3x+2) = 2(3x+2)^2 - (3x+2) = 2(9x^2 + 12x + 4) - 3x - 2 = 18x^2 + 21x + 6.$

$(g \circ f)(x) = g(f(x)) = g(2x^2 - x) = 3(2x^2 - x) + 2 = 6x^2 - 3x + 2.$

$(f \circ f)(x) = f(f(x)) = f(2x^2 - x) = 2(2x^2 - x)^2 - (2x^2 - x) = 2(4x^4 - 4x^3 + x^2) - 2x^2 + x = 8x^4 - 8x^3 + x.$

$(g \circ g)(x) = g(g(x)) = g(3x+2) = 3(3x+2) + 2 = 9x + 6 + 2 = 9x + 8.$

37. $f(x) = \sin x, \quad D = \mathbb{R}; \quad g(x) = 1 - \sqrt{x}, \quad D = [0, \infty).$

$(f \circ g)(x) = f(g(x)) = f(1 - \sqrt{x}) = \sin(1 - \sqrt{x}), D = [0, \infty).$

$(g \circ f)(x) = g(f(x)) = g(\sin x) = 1 - \sqrt{\sin x}$. For $\sqrt{\sin x}$ to be defined, we must have $\sin x \geq 0 \;\Leftrightarrow\; x \in [0, \pi] \cup [2\pi, 3\pi] \cup [-2\pi, -\pi] \cup [4\pi, 5\pi] \cup [-4\pi, -3\pi] \cup \ldots$, so $D = \{x \mid x \in [2n\pi, \pi + 2n\pi]$, where n is an integer$\}$.

$(f \circ f)(x) = f(f(x)) = f(\sin x) = \sin(\sin x), D = \mathbb{R}.$

$(g \circ g)(x) = g(g(x)) = g(1 - \sqrt{x}) = 1 - \sqrt{1 - \sqrt{x}},$

$D = \{x \geq 0 \mid 1 - \sqrt{x} \geq 0\} = \{x \geq 0 \mid 1 \geq \sqrt{x}\} = \{x \geq 0 \mid \sqrt{x} \leq 1\} = [0, 1].$

39. $f(x) = x + \dfrac{1}{x}$, $D = \{x \mid x \neq 0\}$; $g(x) = \dfrac{x+1}{x+2}$, $D = \{x \mid x \neq -2\}$.

$$(f \circ g)(x) = f(g(x)) = f\left(\frac{x+1}{x+2}\right) = \frac{x+1}{x+2} + \frac{1}{\frac{x+1}{x+2}} = \frac{x+1}{x+2} + \frac{x+2}{x+1}$$
$$= \frac{(x+1)(x+1)+(x+2)(x+2)}{(x+2)(x+1)} = \frac{(x^2+2x+1)+(x^2+4x+4)}{(x+2)(x+1)} = \frac{2x^2+6x+5}{(x+2)(x+1)}$$

Since $g(x)$ is not defined for $x = -2$ and $f(g(x))$ is not defined for $x = -2$ and $x = -1$, the domain of $(f \circ g)(x)$ is $D = \{x \mid x \neq -2, -1\}$.

$$(g \circ f)(x) = g(f(x)) = g\left(x + \frac{1}{x}\right) = \frac{\left(x+\frac{1}{x}\right)+1}{\left(x+\frac{1}{x}\right)+2} = \frac{\dfrac{x^2+1+x}{x}}{\dfrac{x^2+1+2x}{x}} = \frac{x^2+x+1}{x^2+2x+1} = \frac{x^2+x+1}{(x+1)^2}.$$

Since $f(x)$ is not defined for $x = 0$ and $g(f(x))$ is not defined for $x = -1$, the domain of $(g \circ f)(x)$ is $D = \{x \mid x \neq -1, 0\}$.

$$(f \circ f)(x) = f(f(x)) = f\left(x + \frac{1}{x}\right) = \left(x + \frac{1}{x}\right) + \frac{1}{x + \frac{1}{x}} = x + \frac{1}{x} + \frac{1}{\frac{x^2+1}{x}} = x + \frac{1}{x} + \frac{x}{x^2+1}$$
$$= \frac{x(x)(x^2+1) + 1\,(x^2+1) + x(x)}{x(x^2+1)} = \frac{x^4+x^2+x^2+1+x^2}{x(x^2+1)}$$
$$= \frac{x^4+3x^2+1}{x(x^2+1)}, \quad D = \{x \mid x \neq 0\}.$$

$$(g \circ g)(x) = g(g(x)) = g\left(\frac{x+1}{x+2}\right) = \frac{\frac{x+1}{x+2}+1}{\frac{x+1}{x+2}+2} = \frac{\frac{x+1+1(x+2)}{x+2}}{\frac{x+1+2(x+2)}{x+2}} = \frac{x+1+x+2}{x+1+2x+4} = \frac{2x+3}{3x+5}.$$

Since $g(x)$ is not defined for $x = -2$ and $g(g(x))$ is not defined for $x = -\frac{5}{3}$, the domain of $(g \circ g)(x)$ is $D = \left\{x \mid x \neq -2, -\frac{5}{3}\right\}$.

41. $(f \circ g \circ h)(x) = f(g(h(x))) = f(g(x-1)) = f(2(x-1))$
$= 2(x-1) + 1 = 2x - 1$

43. $(f \circ g \circ h)(x) = f(g(h(x))) = f(g(x+3)) = f((x+3)^2 + 2)$
$= f(x^2+6x+11) = \sqrt{(x^2+6x+11) - 1} = \sqrt{x^2+6x+10}$

45. Let $g(x) = x^2 + 1$ and $f(x) = x^{10}$. Then $(f \circ g)(x) = f(g(x)) = (x^2+1)^{10} = F(x)$.

47. Let $g(x) = x^2$ and $f(x) = \dfrac{x}{x+4}$. Then $(f \circ g)(x) = f(g(x)) = \dfrac{x^2}{x^2+4} = G(x)$.

49. Let $g(t) = \cos t$ and $f(t) = \sqrt{t}$. Then $(f \circ g)(t) = f(g(t)) = \sqrt{\cos t} = u(t)$.

51. Let $h(x) = x^2$, $g(x) = 3^x$, and $f(x) = 1 - x$. Then $(f \circ g \circ h)(x) = 1 - 3^{x^2} = H(x)$.

53. Let $h(x) = \sqrt{x}$, $g(x) = \sec x$, and $f(x) = x^4$. Then $(f \circ g \circ h)(x) = (\sec \sqrt{x})^4 = \sec^4(\sqrt{x}) = H(x)$.

55. (a) $g(2) = 5$, because the point $(2, 5)$ is on the graph of g. Thus, $f(g(2)) = f(5) = 4$, because the point $(5, 4)$ is on the graph of f.

(b) $g(f(0)) = g(0) = 3$

(c) $(f \circ g)(0) = f(g(0)) = f(3) = 0$

(d) $(g \circ f)(6) = g(f(6)) = g(6)$. This value is not defined, because there is no point on the graph of g that has x-coordinate 6.

(e) $(g \circ g)(-2) = g(g(-2)) = g(1) = 4$

(f) $(f \circ f)(4) = f(f(4)) = f(2) = -2$

57. (a) Using the relationship *distance* = *rate* · *time* with the radius r as the distance, we have $r(t) = 60t$.

(b) $A = \pi r^2 \;\Rightarrow\; (A \circ r)(t) = A(r(t)) = \pi(60t)^2 = 3600\pi t^2$. This formula gives us the extent of the rippled area (in cm^2) at any time t.

59. (a)

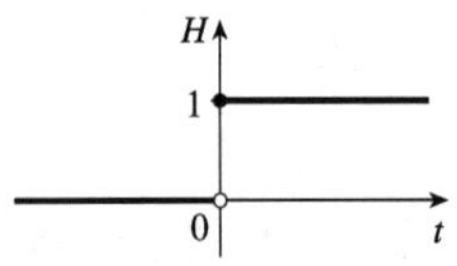

$$H(t) = \begin{cases} 0 & \text{if } t < 0 \\ 1 & \text{if } t \geq 0 \end{cases}$$

(b)

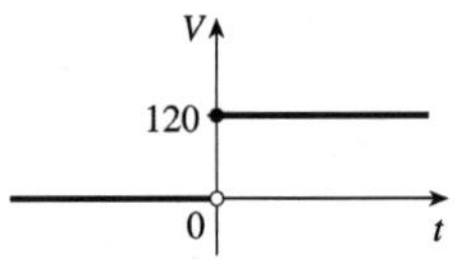

$$V(t) = \begin{cases} 0 & \text{if } t < 0 \\ 120 & \text{if } t \geq 0 \end{cases}$$ so $V(t) = 120H(t)$.

(c)

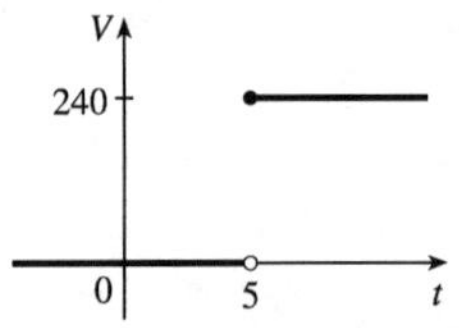

Starting with the formula in part (b), we replace 120 with 240 to reflect the different voltage. Also, because we are starting 5 units to the right of $t = 0$, we replace t with $t - 5$. Thus, the formula is $V(t) = 240H(t - 5)$.

61. (a) By examining the variable terms in g and h, we deduce that we must square g to get the terms $4x^2$ and $4x$ in h. If we let $f(x) = x^2 + c$, then $(f \circ g)(x) = f(g(x)) = f(2x + 1) = (2x + 1)^2 + c = 4x^2 + 4x + (1 + c)$. Since $h(x) = 4x^2 + 4x + 7$, we must have $1 + c = 7$. So $c = 6$ and $f(x) = x^2 + 6$.

(b) We need a function g so that $f(g(x)) = 3(g(x)) + 5 = h(x)$. But $h(x) = 3x^2 + 3x + 2 = 3(x^2 + x) + 2 = 3(x^2 + x - 1) + 5$, so we see that $g(x) = x^2 + x - 1$.

63. We need to examine $h(-x)$.

$$h(-x) = (f \circ g)(-x) = f(g(-x)) = f(g(x)) \quad [\text{because } g \text{ is even}] \quad = h(x)$$

Because $h(-x) = h(x)$, h is an even function.

Review

CONCEPT CHECK

1. (a) A **function** f is a rule that assigns to each element x in a set A exactly one element, called $f(x)$, in a set B. The set A is called the **domain** of the function. The **range** of f is the set of all possible values of $f(x)$ as x varies throughout the domain.

(b) If f is a function with domain A, then its **graph** is the set of ordered pairs $\{(x, f(x)) \mid x \in A\}$.

(c) Use the Vertical Line Test on page 17.

2. The four ways to represent a function are: verbally, numerically, visually, and algebraically. An example of each is given below.
Verbally: An assignment of students to chairs in a classroom (a description in words)
Numerically: A tax table that assigns an amount of tax to an income (a table of values)
Visually: A graphical history of the Dow Jones average (a graph)
Algebraically: A relationship between distance, rate, and time: $d = rt$ (an explicit formula)

3. (a) An **even function** f satisfies $f(-x) = f(x)$ for every number x in its domain. It is symmetric with respect to the y-axis.
   (b) An **odd function** g satisfies $g(-x) = -g(x)$ for every number x in its domain. It is symmetric with respect to the origin.

4. A function f is called **increasing** on an interval I if $f(x_1) < f(x_2)$ whenever $x_1 < x_2$ in I.

5. A **mathematical model** is a mathematical description (often by means of a function or an equation) of a real-world phenomenon.

6. (a) Linear function: $f(x) = 2x + 1$, $f(x) = ax + b$
   (b) Power function: $f(x) = x^2$, $f(x) = x^a$
   (c) Exponential function: $f(x) = 2^x$, $f(x) = a^x$
   (d) Quadratic function: $f(x) = x^2 + x + 1$, $f(x) = ax^2 + bx + c$
   (e) Polynomial of degree 5: $f(x) = x^5 + 2$
   (f) Rational function: $f(x) = \dfrac{x}{x+2}$, $f(x) = \dfrac{P(x)}{Q(x)}$ where $P(x)$ and $Q(x)$ are polynomials

7.
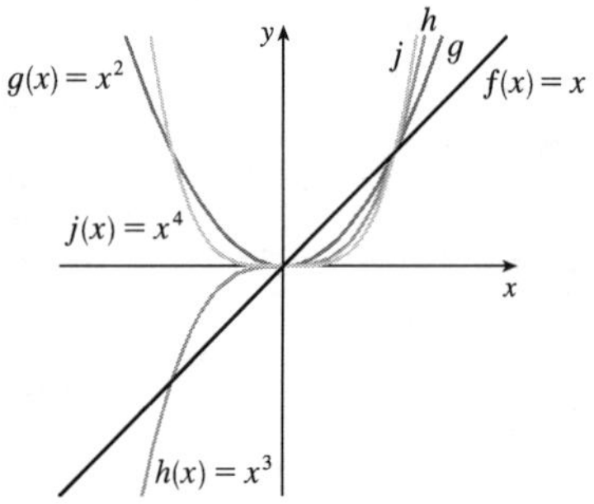

8. (a)
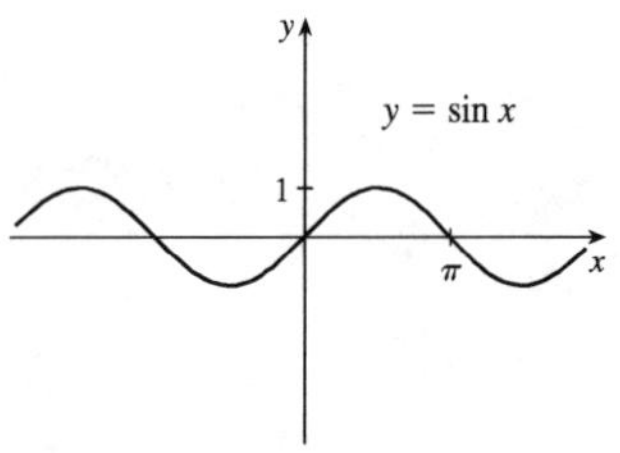

(b)
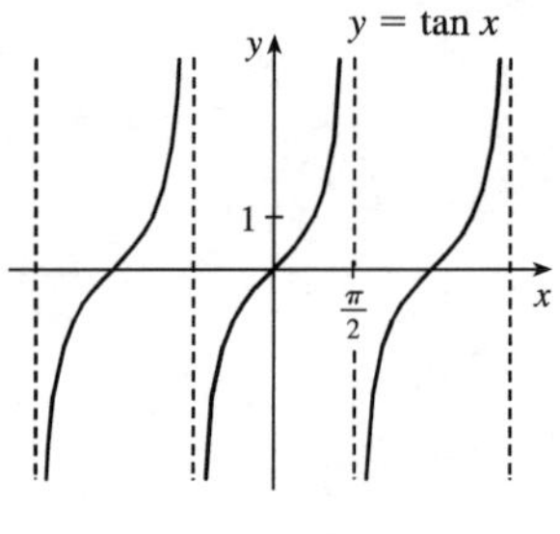

(c)
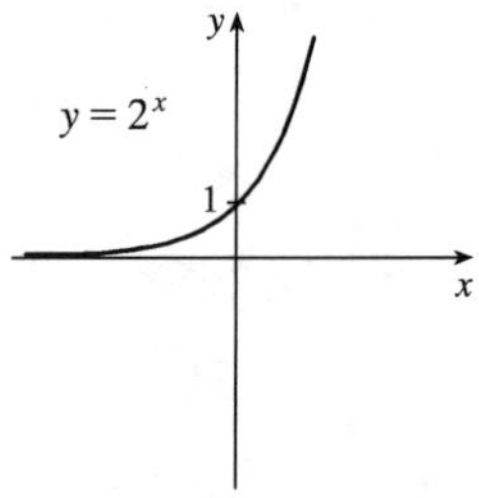

(d)
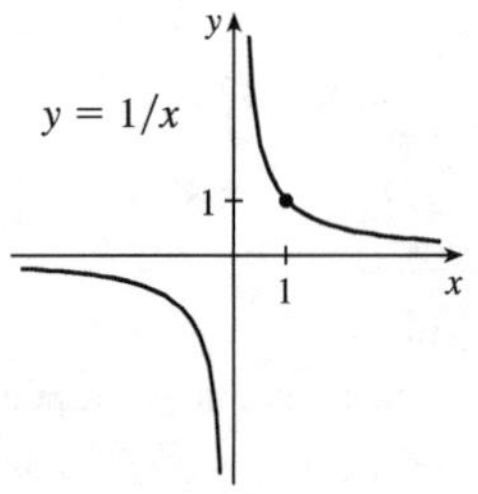

(e)
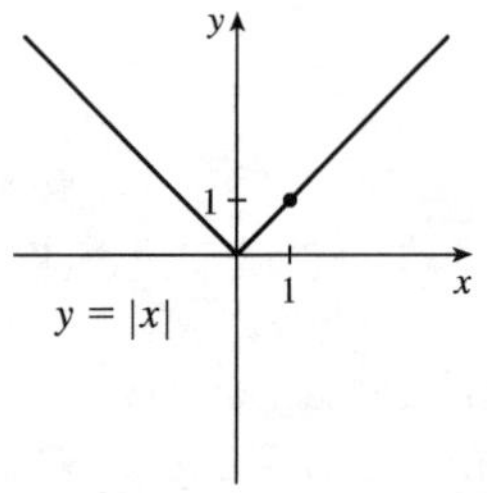

(f)
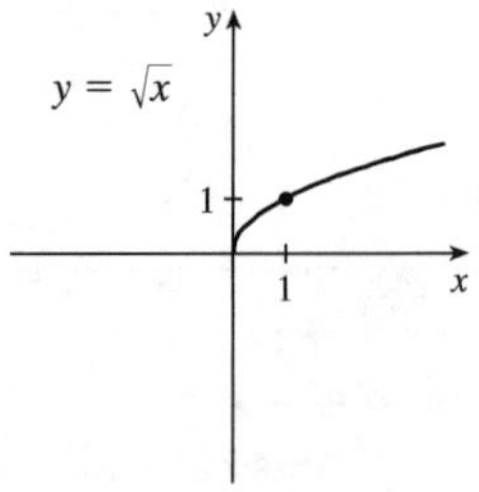

9. (a) The domain of $f + g$ is the intersection of the domain of f and the domain of g; that is, $A \cap B$.
   (b) The domain of fg is also $A \cap B$.
   (c) The domain of f/g must exclude values of x that make g equal to 0; that is, $\{x \in A \cap B \mid g(x) \neq 0\}$.

10. Given two functions f and g, the **composite** function $f \circ g$ is defined by $(f \circ g)(x) = f(g(x))$. The domain of $f \circ g$ is the set of all x in the domain of g such that $g(x)$ is in the domain of f.

11. (a) If the graph of f is shifted 2 units upward, its equation becomes $y = f(x) + 2$.

(b) If the graph of f is shifted 2 units downward, its equation becomes $y = f(x) - 2$.

(c) If the graph of f is shifted 2 units to the right, its equation becomes $y = f(x - 2)$.

(d) If the graph of f is shifted 2 units to the left, its equation becomes $y = f(x + 2)$.

(e) If the graph of f is reflected about the x-axis, its equation becomes $y = -f(x)$.

(f) If the graph of f is reflected about the y-axis, its equation becomes $y = f(-x)$.

(g) If the graph of f is stretched vertically by a factor of 2, its equation becomes $y = 2f(x)$.

(h) If the graph of f is shrunk vertically by a factor of 2, its equation becomes $y = \frac{1}{2}f(x)$.

(i) If the graph of f is stretched horizontally by a factor of 2, its equation becomes $y = f(\frac{1}{2})x$.

(j) If the graph of f is shrunk horizontally by a factor of 2, its equation becomes $y = f(2x)$.

TRUE-FALSE QUIZ

1. False. Let $f(x) = x^2$, $s = -1$, and $t = 1$. Then $f(s + t) = (-1 + 1)^2 = 0^2 = 0$, but
$f(s) + f(t) = (-1)^2 + 1^2 = 2 \neq 0 = f(s + t)$.

3. False. Let $f(x) = x^2$. Then $f(3x) = (3x)^2 = 9x^2$ and $3f(x) = 3x^2$. So $f(3x) \neq 3f(x)$.

5. True. See the Vertical Line Test.

EXERCISES

1. (a) When $x = 2$, $y \approx 2.7$. Thus, $f(2) \approx 2.7$.

(b) $f(x) = 3 \Rightarrow x \approx 2.3, 5.6$

(c) The domain of f is $-6 \leq x \leq 6$, or $[-6, 6]$.

(d) The range of f is $-4 \leq y \leq 4$, or $[-4, 4]$.

(e) f is increasing on $[-4, 4]$, that is, on $-4 \leq x \leq 4$.

(f) f is odd since its graph is symmetric about the origin.

3. (a)

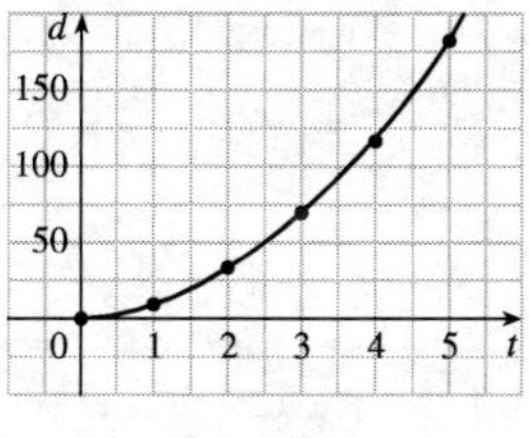

(b) From the graph, we see that the distance traveled after 4.5 seconds is slightly less than 150 feet.

5. $f(x) = \sqrt{4 - 3x^2}$. Domain: $4 - 3x^2 \geq 0 \Rightarrow 3x^2 \leq 4 \Rightarrow x^2 \leq \frac{4}{3} \Rightarrow |x| \leq \frac{2}{\sqrt{3}}$. Range: $y \geq 0$ and $y \leq \sqrt{4} \Rightarrow 0 \leq y \leq 2$.

7. $y = 1 + \sin x$. Domain: $\mathbb{R}$. Range: $-1 \leq \sin x \leq 1 \Rightarrow 0 \leq 1 + \sin x \leq 2 \Rightarrow 0 \leq y \leq 2$.

9. (a) To obtain the graph of $y = f(x) + 8$, we shift the graph of $y = f(x)$ up 8 units.

(b) To obtain the graph of $y = f(x + 8)$, we shift the graph of $y = f(x)$ left 8 units.

(c) To obtain the graph of $y = 1 + 2f(x)$, we stretch the graph of $y = f(x)$ vertically by a factor of 2, and then shift the resulting graph 1 unit upward.

(d) To obtain the graph of $y = f(x - 2) - 2$, we shift the graph of $y = f(x)$ right 2 units (for the "-2" inside the parentheses), and then shift the resulting graph 2 units downward.

(e) To obtain the graph of $y = -f(x)$, we reflect the graph of $y = f(x)$ about the x-axis.

(f) To obtain the graph of $y = 3 - f(x)$, we reflect the graph of $y = f(x)$ about the x-axis, and then shift the resulting graph 3 units upward.

11. $y = -\sin 2x$: Start with the graph of $y = \sin x$, compress horizontally by a factor of 2, and reflect about the x-axis.

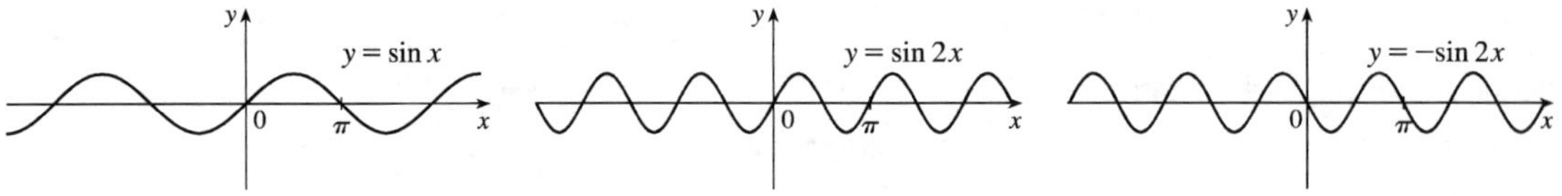

13. $y = 1 + \frac{1}{2}x^3$: Start with the graph of $y = x^3$, compress vertically by a factor of 2, and shift 1 unit upward.

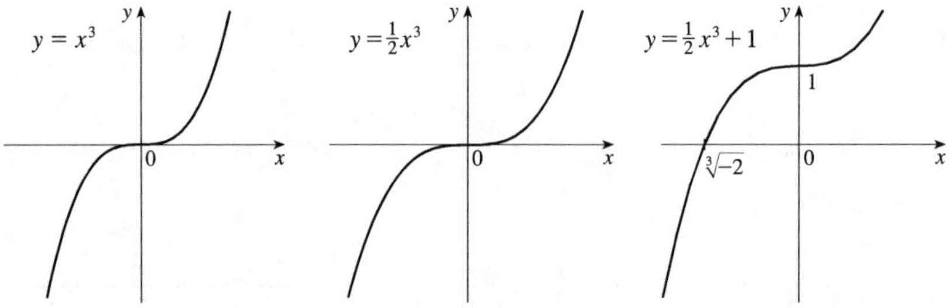

15. $f(x) = \dfrac{1}{x + 2}$: Start with the graph of $f(x) = 1/x$ and shift 2 units to the left.

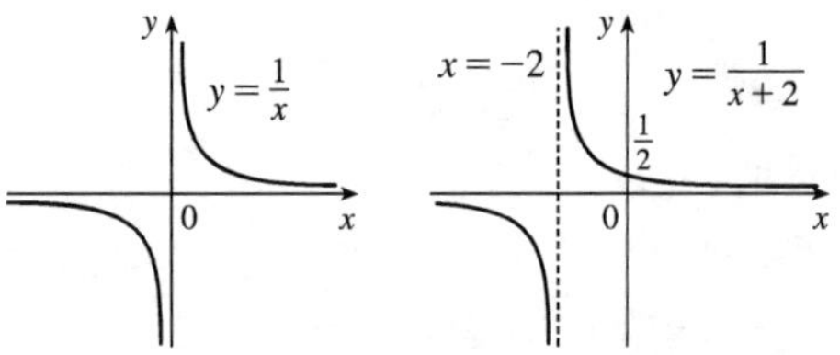

17. (a) The terms of f are a mixture of odd and even powers of x, so f is neither even nor odd.

(b) The terms of f are all odd powers of x, so f is odd.

(c) $f(-x) = \cos\left((-x)^2\right) = \cos\left(x^2\right) = f(x)$, so f is even.

(d) $f(-x) = 1 + \sin(-x) = 1 - \sin x$. Now $f(-x) \neq f(x)$ and $f(-x) \neq -f(x)$, so f is neither even nor odd.

19. $f(x) = \sqrt{x}$, $D = [0, \infty)$; $g(x) = \sin x$, $D = \mathbb{R}$.

$(f \circ g)(x) = f(g(x)) = f(\sin x) = \sqrt{\sin x}$. For $\sqrt{\sin x}$ to be defined, we must have $\sin x \geq 0 \quad \Leftrightarrow \quad x \in [0, \pi]$, $[2\pi, 3\pi]$, $[-2\pi, -\pi]$, $[4\pi, 5\pi]$, $[-4\pi, -3\pi]$, $\ldots$, so $D = \{x \mid x \in [2n\pi, \pi + 2n\pi]$, where n is an integer$\}$.

$(g \circ f)(x) = g(f(x)) = g(\sqrt{x}\,) = \sin\sqrt{x}$. x must be greater than or equal to 0 for $\sqrt{x}$ to be defined, so $D = [0, \infty)$.

$(f \circ f)(x) = f(f(x)) = f(\sqrt{x}\,) = \sqrt{\sqrt{x}} = \sqrt[4]{x}$. $D = [0, \infty)$.

$(g \circ g)(x) = g(g(x)) = g(\sin x) = \sin(\sin x)$. $D = \mathbb{R}$.

21. The graphs of $f(x) = \sin^n x$, where n is a positive integer, all have domain $\mathbb{R}$. For odd n, the range is $[-1, 1]$ and for even n, the range is $[0, 1]$. For odd n, the functions are odd and symmetric with respect to the origin. For even n, the functions are even and symmetric with respect to the y-axis. As n becomes large, the graphs become less rounded and more "spiky."

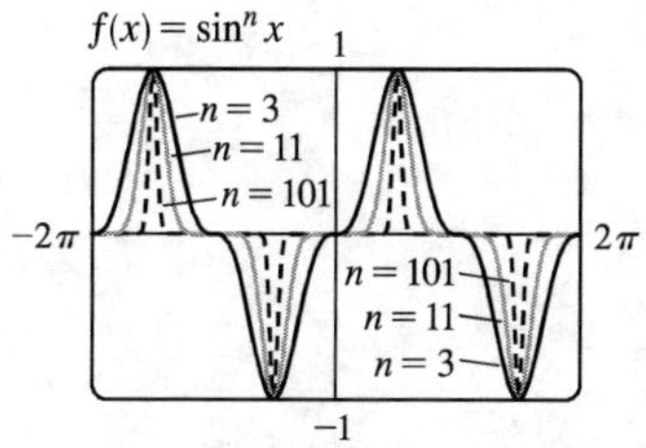

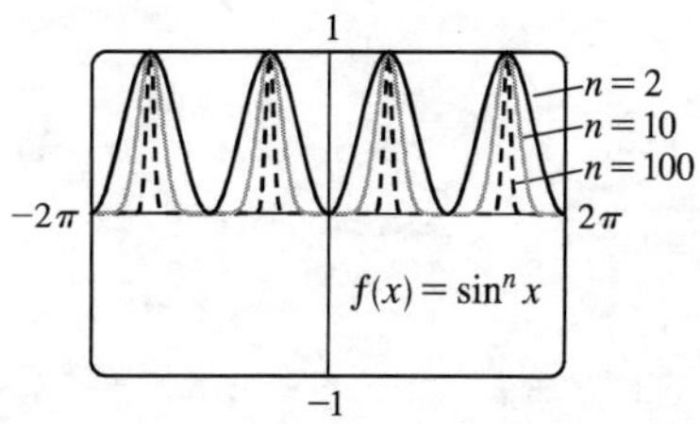

23.

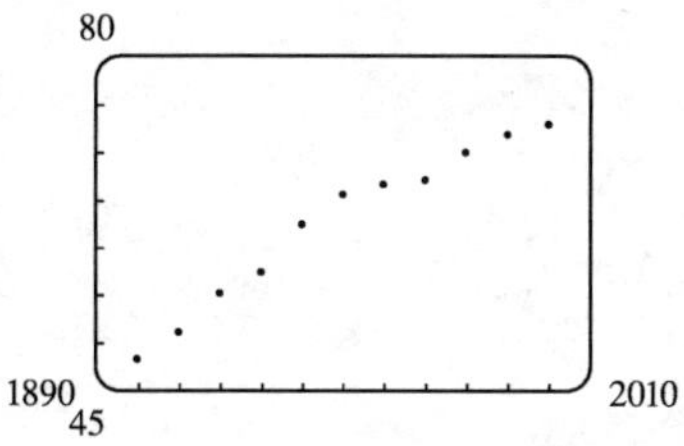

Many models appear to be plausible. Your choice depends on whether you think medical advances will keep increasing life expectancy, or if there is bound to be a natural leveling-off of life expectancy. A linear model, $y = 0.2493x - 423.4818$ gives us an estimate of 77.6 years for the year 2010.

The Limit of a Function

1. As x approaches 2, $f(x)$ approaches 5. [Or, the values of $f(x)$ can be made as close to 5 as we like by taking x sufficiently close to 2 (but $x \neq 2$).] Yes, the graph could have a hole at $(2, 5)$ and be defined such that $f(2) = 3$.

3. (a) $\lim\limits_{x \to -3} f(x) = \infty$ means that the values of $f(x)$ can be made arbitrarily large (as large as we please) by taking x sufficiently close to -3 (but not equal to -3).

(b) $\lim\limits_{x \to 4^+} f(x) = -\infty$ means that the values of $f(x)$ can be made arbitrarily large negative by taking x sufficiently close to 4 through values larger than 4.

5. (a) $f(x)$ approaches 2 as x approaches 1 from the left, so $\lim\limits_{x \to 1^-} f(x) = 2$.

(b) $f(x)$ approaches 3 as x approaches 1 from the right, so $\lim\limits_{x \to 1^+} f(x) = 3$.

(c) $\lim\limits_{x \to 1} f(x)$ does not exist because the limits in part (a) and part (b) are not equal.

(d) $f(x)$ approaches 4 as x approaches 5 from the left and from the right, so $\lim\limits_{x \to 5} f(x) = 4$.

(e) $f(5)$ is not defined, so it doesn't exist.

7. (a) $\lim\limits_{t \to 0^-} g(t) = -1$ (b) $\lim\limits_{t \to 0^+} g(t) = -2$

(c) $\lim\limits_{t \to 0} g(t)$ does not exist because the limits in part (a) and part (b) are not equal.

(d) $\lim\limits_{t \to 2^-} g(t) = 2$ (e) $\lim\limits_{t \to 2^+} g(t) = 0$

(f) $\lim\limits_{t \to 2} g(t)$ does not exist because the limits in part (d) and part (e) are not equal.

(g) $g(2) = 1$ (h) $\lim\limits_{t \to 4} g(t) = 3$

9. (a) $\lim\limits_{x \to -7} f(x) = -\infty$ (b) $\lim\limits_{x \to -3} f(x) = \infty$ (c) $\lim\limits_{x \to 0} f(x) = \infty$

(d) $\lim\limits_{x \to 6^-} f(x) = -\infty$ (e) $\lim\limits_{x \to 6^+} f(x) = \infty$

(f) The equations of the vertical asymptotes are $x = -7$, $x = -3$, $x = 0$, and $x = 6$.

11.

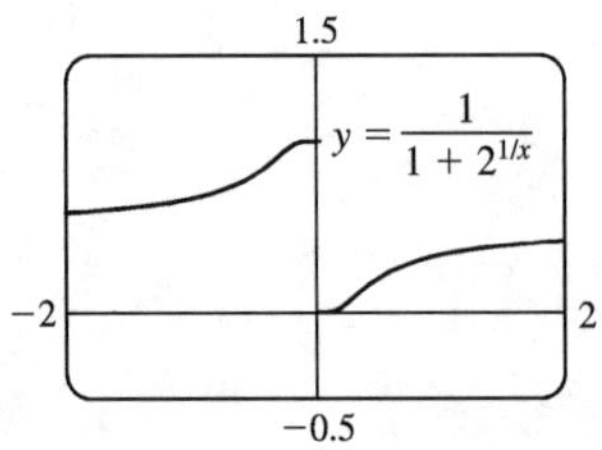

(a) $\lim\limits_{x \to 0^-} f(x) = 1$

(b) $\lim\limits_{x \to 0^+} f(x) = 0$

(c) $\lim\limits_{x \to 0} f(x)$ does not exist because the limits in part (a) and part (b) are not equal.

13. $\lim_{x\to 3^+} f(x) = 4$, $\lim_{x\to 3^-} f(x) = 2$,

$\lim_{x\to -2} f(x) = 2$, $f(3) = 3$, $f(-2) = 1$

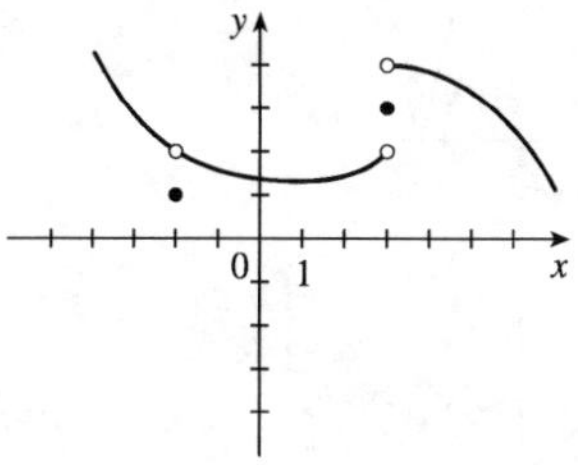

15. For $f(x) = \dfrac{x^2 - 2x}{x^2 - x - 2}$:

x	$f(x)$
2.5	0.714286
2.1	0.677419
2.05	0.672131
2.01	0.667774
2.005	0.667221
2.001	0.666778

x	$f(x)$
1.9	0.655172
1.95	0.661017
1.99	0.665552
1.995	0.666110
1.999	0.666556

It appears that $\lim_{x\to 2} \dfrac{x^2 - 2x}{x^2 - x - 2} = 0.\bar{6} = \frac{2}{3}$.

17. For $f(x) = \dfrac{\sin x}{x + \tan x}$:

x	$f(x)$
± 1	0.329033
± 0.5	0.458209
± 0.2	0.493331
± 0.1	0.498333
± 0.05	0.499583
± 0.01	0.499983

It appears that $\lim_{x\to 0} \dfrac{\sin x}{x + \tan x} = 0.5 = \dfrac{1}{2}$.

19. For $f(x) = \dfrac{\sqrt{x+4} - 2}{x}$:

x	$f(x)$
1	0.236068
0.5	0.242641
0.1	0.248457
0.05	0.249224
0.01	0.249844

x	$f(x)$
-1	0.267949
-0.5	0.258343
-0.1	0.251582
-0.05	0.250786
-0.01	0.250156

It appears that $\lim_{x\to 0} \dfrac{\sqrt{x+4} - 2}{x} = 0.25 = \frac{1}{4}$.

21. For $f(x) = \dfrac{x^6 - 1}{x^{10} - 1}$:

x	$f(x)$
0.5	0.985337
0.9	0.719397
0.95	0.660186
0.99	0.612018
0.999	0.601200

x	$f(x)$
1.5	0.183369
1.1	0.484119
1.05	0.540783
1.01	0.588022
1.001	0.598800

It appears that $\lim_{x\to 1} \dfrac{x^6 - 1}{x^{10} - 1} = 0.6 = \frac{3}{5}$.

23. $\lim_{x\to 5^+} \dfrac{6}{x - 5} = \infty$ since $(x - 5) \to 0$ as $x \to 5^+$ and $\dfrac{6}{x - 5} > 0$ for $x > 5$.

25. $\lim_{x\to 1} \dfrac{2 - x}{(x - 1)^2} = \infty$ since the numerator is positive and the denominator approaches 0 through positive values as $x \to 1$.

27. $\lim_{x \to -2^+} \frac{x-1}{x^2(x+2)} = -\infty$ since $(x+2) \to 0$ as $x \to -2^+$ and $\frac{x-1}{x^2(x+2)} < 0$ for $-2 < x < 0$.

29. $\lim_{x \to (-\pi/2)^-} \sec x = \lim_{x \to (-\pi/2)^-} (1/\cos x) = -\infty$ since $\cos x \to 0$ as $x \to (-\pi/2)^-$ and $\cos x < 0$ for $-\pi < x < -\pi/2$.

31. (a) $f(x) = 1/(x^3 - 1)$

x	$f(x)$
0.5	−1.14
0.9	−3.69
0.99	−33.7
0.999	−333.7
0.9999	−3333.7
0.99999	−33,333.7

x	$f(x)$
1.5	0.42
1.1	3.02
1.01	33.0
1.001	333.0
1.0001	3333.0
1.00001	33,333.3

From these calculations, it seems that $\lim_{x \to 1^-} f(x) = -\infty$ and $\lim_{x \to 1^+} f(x) = \infty$.

(b) If x is slightly smaller than 1, then $x^3 - 1$ will be a negative number close to 0, and the reciprocal of $x^3 - 1$, that is, $f(x)$, will be a negative number with large absolute value. So $\lim_{x \to 1^-} f(x) = -\infty$.

If x is slightly larger than 1, then $x^3 - 1$ will be a small positive number, and its reciprocal, $f(x)$, will be a large positive number. So $\lim_{x \to 1^+} f(x) = \infty$.

(c) It appears from the graph of f that $\lim_{x \to 1^-} f(x) = -\infty$ and $\lim_{x \to 1^+} f(x) = \infty$.

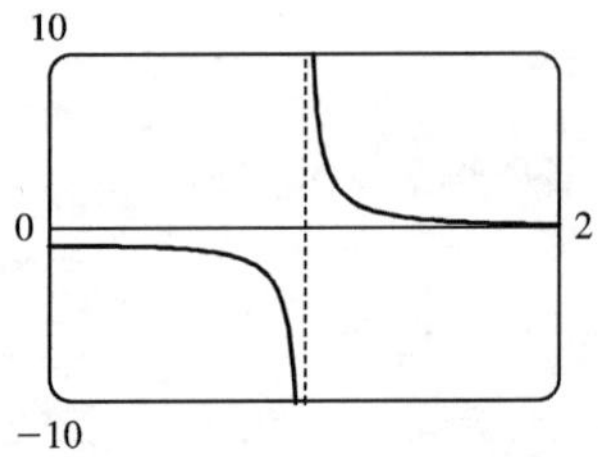

33. (a) Let $h(x) = (1+x)^{1/x}$.

x	$h(x)$
−0.001	2.71964
−0.0001	2.71842
−0.00001	2.71830
−0.000001	2.71828
0.000001	2.71828
0.00001	2.71827
0.0001	2.71815
0.001	2.71692

It appears that $\lim_{x \to 0} (1+x)^{1/x} \approx 2.71828$, which is approximately e.

In Section 7.4 we will see that the value of the limit is exactly e.

(b)

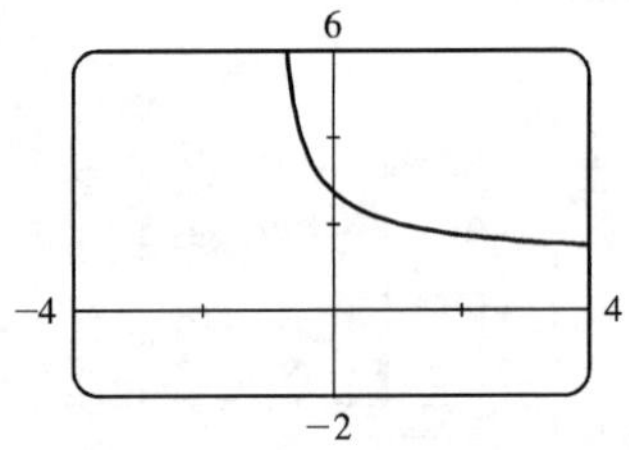

35. For $f(x) = x^2 - (2^x/1000)$:

(a)

x	$f(x)$
1	0.998000
0.8	0.638259
0.6	0.358484
0.4	0.158680
0.2	0.038851
0.1	0.008928
0.05	0.001465

It appears that $\lim\limits_{x\to 0} f(x) = 0$.

(b)

x	$f(x)$
0.04	0.000572
0.02	-0.000614
0.01	-0.000907
0.005	-0.000978
0.003	-0.000993
0.001	-0.001000

It appears that $\lim\limits_{x\to 0} f(x) = -0.001$.

37. No matter how many times we zoom in toward the origin, the graphs of $f(x) = \sin(\pi/x)$ appear to consist of almost-vertical lines. This indicates more and more frequent oscillations as $x \to 0$.

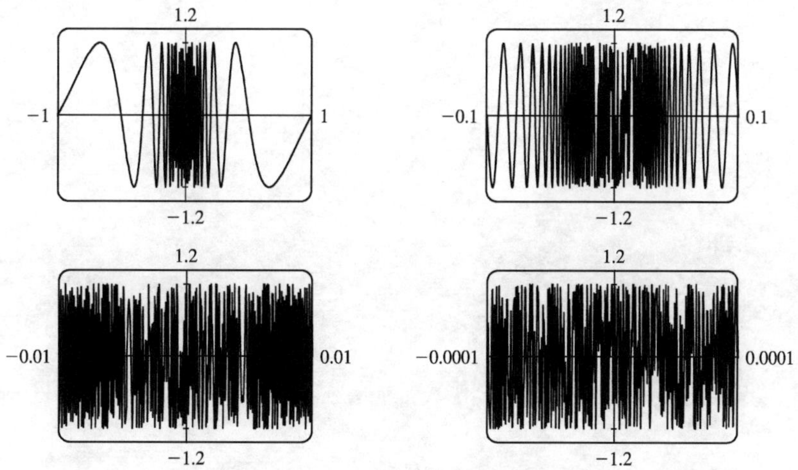

39.

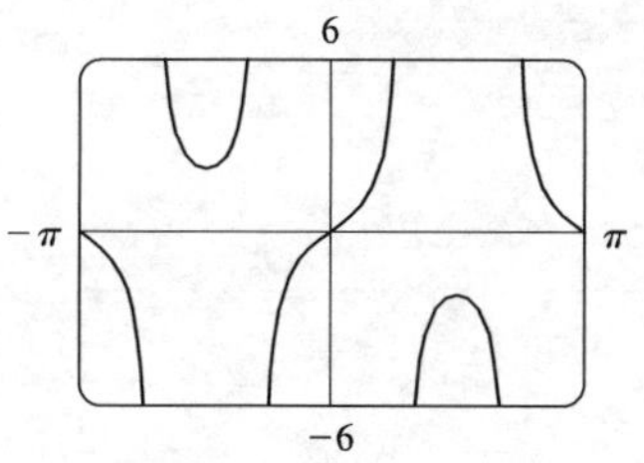

There appear to be vertical asymptotes of the curve $y = \tan(2\sin x)$ at $x \approx \pm 0.90$ and $x \approx \pm 2.24$. To find the exact equations of these asymptotes, we note that the graph of the tangent function has vertical asymptotes at $x = \frac{\pi}{2} + \pi n$. Thus, we must have $2\sin x = \frac{\pi}{2} + \pi n$, or equivalently, $\sin x = \frac{\pi}{4} + \frac{\pi}{2}n$. Since $-1 \le \sin x \le 1$, we must have $\sin x = \pm\frac{\pi}{4}$ and so $x = \pm\sin^{-1}\frac{\pi}{4}$ (corresponding to $x \approx \pm 0.90$).

Just as 150° is the reference angle for 30°, $\pi - \sin^{-1}\frac{\pi}{4}$ is the reference angle for $\sin^{-1}\frac{\pi}{4}$. So $x = \pm\left(\pi - \sin^{-1}\frac{\pi}{4}\right)$ are also equations of the vertical asymptotes (corresponding to $x \approx \pm 2.24$).

Continuity

1. From Definition 1, $\lim_{x \to 4} f(x) = f(4)$.

3. (a) The following are the numbers at which f is discontinuous and the type of discontinuity at that number: -4 (removable), -2 (jump), 2 (jump), 4 (infinite).

(b) f is continuous from the left at -2 since $\lim_{x \to -2^-} f(x) = f(-2)$. f is continuous from the right at 2 and 4 since $\lim_{x \to 2^+} f(x) = f(2)$ and $\lim_{x \to 4^+} f(x) = f(4)$. It is continuous from neither side at -4 since $f(-4)$ is undefined.

5. The graph of $y = f(x)$ must have a discontinuity at $x = 3$ and must show that $\lim_{x \to 3^-} f(x) = f(3)$.

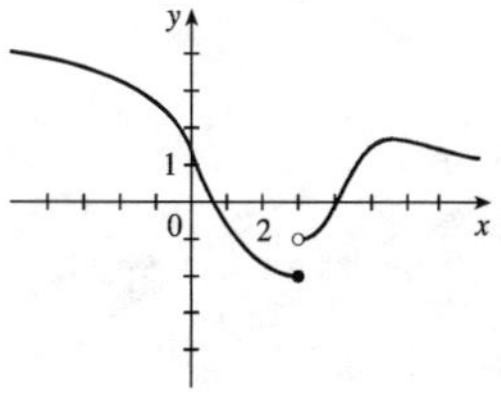

7. (a)

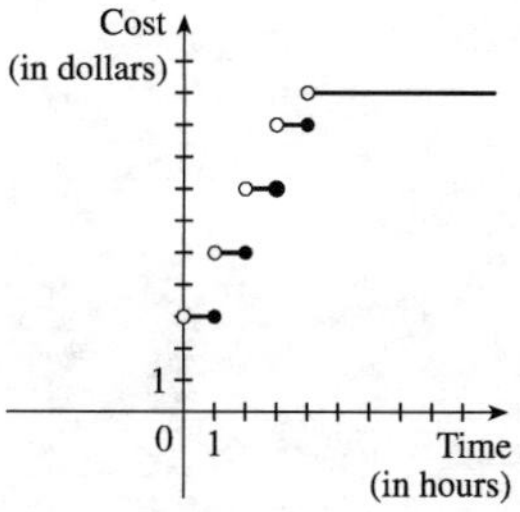

(b) There are discontinuities at times $t = 1, 2, 3,$ and 4. A person parking in the lot would want to keep in mind that the charge will jump at the beginning of each hour.

9. Since f and g are continuous functions,

$$\begin{aligned} \lim_{x \to 3} [2f(x) - g(x)] &= 2 \lim_{x \to 3} f(x) - \lim_{x \to 3} g(x) && \text{[by Limit Laws 2 and 3]} \\ &= 2f(3) - g(3) && \text{[by continuity of } f \text{ and } g \text{ at } x = 3] \\ &= 2 \cdot 5 - g(3) = 10 - g(3) \end{aligned}$$

Since it is given that $\lim_{x \to 3} [2f(x) - g(x)] = 4$, we have $10 - g(3) = 4$, so $g(3) = 6$.

11. $\lim_{x \to -1} f(x) = \lim_{x \to -1} \left(x + 2x^3\right)^4 = \left(\lim_{x \to -1} x + 2 \lim_{x \to -1} x^3\right)^4 = \left[-1 + 2(-1)^3\right]^4 = (-3)^4 = 81 = f(-1)$.

By the definition of continuity, f is continuous at $a = -1$.

13. For $a > 2$, we have $\lim_{x \to a} f(x) = \lim_{x \to a} \dfrac{2x + 3}{x - 2} = \dfrac{\lim_{x \to a} (2x + 3)}{\lim_{x \to a} (x - 2)}$ [Limit Law 5] $= \dfrac{2 \lim_{x \to a} x + \lim_{x \to a} 3}{\lim_{x \to a} x - \lim_{x \to a} 2}$

[1, 2, and 3] $= \dfrac{2a + 3}{a - 2}$ [7 and 8] $= f(a)$. Thus, f is continuous at $x = a$ for every a in $(2, \infty)$; that is, f is continuous on $(2, \infty)$.

15. $f(x) = -\dfrac{1}{(x - 1)^2}$ is discontinuous at 1 since $f(1)$ is not defined.

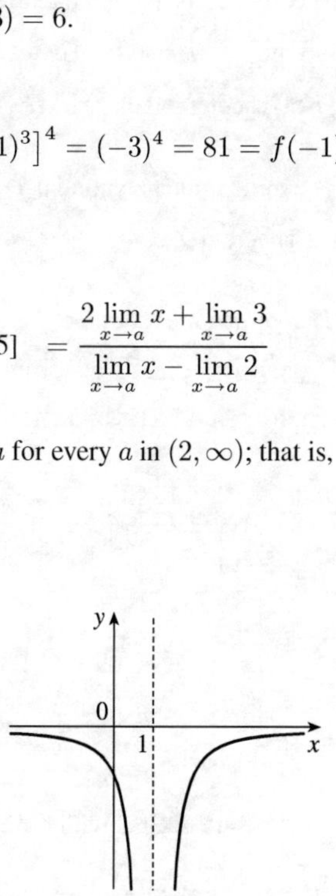

17. $f(x) = \begin{cases} 1 - x^2 & \text{if } x < 1 \\ 1/x & \text{if } x \geq 1 \end{cases}$

The left-hand limit of f at $a = 1$ is $\lim\limits_{x \to 1^-} f(x) = \lim\limits_{x \to 1^-} (1 - x^2) = 0$. The right-hand limit of f at $a = 1$ is $\lim\limits_{x \to 1^+} f(x) = \lim\limits_{x \to 1^+} (1/x) = 1$. Since these limits are not equal, $\lim\limits_{x \to 1} f(x)$ does not exist and f is discontinuous at 1.

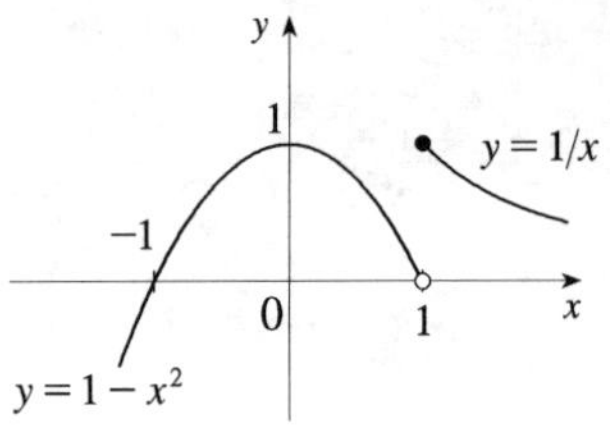

19. $f(x) = \begin{cases} \dfrac{x^2 - x - 12}{x + 3} & \text{if } x \neq -3 \\ -5 & \text{if } x = -3 \end{cases} = \begin{cases} x - 4 & \text{if } x \neq -3 \\ -5 & \text{if } x = -3 \end{cases}$

So $\lim\limits_{x \to -3} f(x) = \lim\limits_{x \to -3} (x - 4) = -7$ and $f(-3) = -5$.

Since $\lim\limits_{x \to -3} f(x) \neq f(-3)$, f is discontinuous at -3.

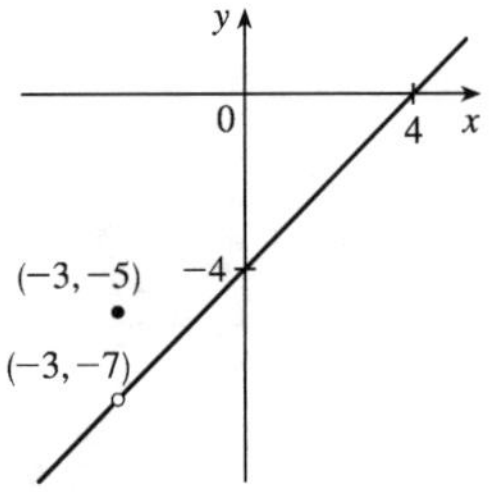

21. $F(x) = \dfrac{x}{x^2 + 5x + 6}$ is a rational function. So by Theorem 5 (or Theorem 7), F is continuous at every number in its domain, $\{x \mid x^2 + 5x + 6 \neq 0\} = \{x \mid (x + 3)(x + 2) \neq 0\} = \{x \mid x \neq -3, -2\}$ or $(-\infty, -3) \cup (-3, -2) \cup (-2, \infty)$.

23. By Theorem 5, the polynomials x^2 and $2x - 1$ are continuous on $(-\infty, \infty)$. By Theorem 7, the root function $\sqrt{x}$ is continuous on $[0, \infty)$. By Theorem 9, the composite function $\sqrt{2x - 1}$ is continuous on its domain, $[\frac{1}{2}, \infty)$. By part 1 of Theorem 4, the sum $R(x) = x^2 + \sqrt{2x - 1}$ is continuous on $[\frac{1}{2}, \infty)$.

25. By Theorem 5, the polynomial $1 - x^2$ is continuous on $(-\infty, \infty)$. By Theorem 7, cos is continuous on its domain, $\mathbb{R}$. By Theorem 9, $\cos(1 - x^2)$ is continuous on its domain, which is $\mathbb{R}$.

27. By Theorem 7, the root function $\sqrt{x}$ and the trigonometric function $\sin x$ are continuous on their domains, $[0, \infty)$ and $(-\infty, \infty)$, respectively. Thus, the product $F(x) = \sqrt{x} \sin x$ is continuous on the intersection of those domains, $[0, \infty)$, by part 4 of Theorem 4.

29.

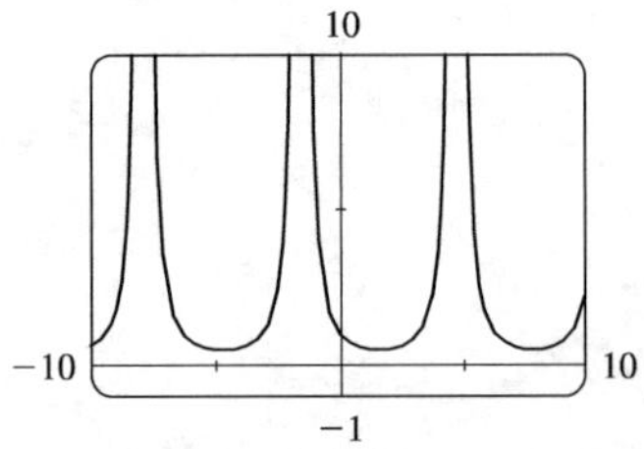

$y = \dfrac{1}{1 + \sin x}$ is undefined and hence discontinuous when $1 + \sin x = 0 \Leftrightarrow \sin x = -1 \Leftrightarrow x = -\frac{\pi}{2} + 2\pi n$, n an integer. The figure shows discontinuities for $n = -1$, 0, and 1; that is, $-\dfrac{5\pi}{2} \approx -7.85$, $-\dfrac{\pi}{2} \approx -1.57$, and $\dfrac{3\pi}{2} \approx 4.71$.

31. Because we are dealing with root functions, $5 + \sqrt{x}$ is continuous on $[0, \infty)$, $\sqrt{x + 5}$ is continuous on $[-5, \infty)$, so the quotient $f(x) = \dfrac{5 + \sqrt{x}}{\sqrt{5 + x}}$ is continuous on $[0, \infty)$. Since f is continuous at $x = 4$, $\lim\limits_{x \to 4} f(x) = f(4) = \frac{7}{3}$.

33. $f(x) = \begin{cases} x^2 & \text{if } x < 1 \\ \sqrt{x} & \text{if } x \geq 1 \end{cases}$

By Theorem 5, since $f(x)$ equals the polynomial x^2 on $(-\infty, 1)$, f is continuous on $(-\infty, 1)$. By Theorem 7, since $f(x)$ equals the root function $\sqrt{x}$ on $(1, \infty)$, f is continuous on $(1, \infty)$. At $x = 1$, $\lim\limits_{x \to 1^-} f(x) = \lim\limits_{x \to 1^-} x^2 = 1$ and $\lim\limits_{x \to 1^+} f(x) = \lim\limits_{x \to 1^+} \sqrt{x} = 1$. Thus, $\lim\limits_{x \to 1} f(x)$ exists and equals 1. Also, $f(1) = \sqrt{1} = 1$. Thus, f is continuous at $x = 1$. We conclude that f is continuous on $(-\infty, \infty)$.

35. $f(x) = \begin{cases} 1 + x^2 & \text{if } x \leq 0 \\ 2 - x & \text{if } 0 < x \leq 2 \\ (x-2)^2 & \text{if } x > 2 \end{cases}$

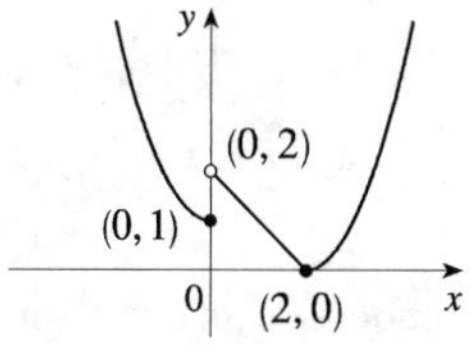

f is continuous on $(-\infty, 0)$, $(0, 2)$, and $(2, \infty)$ since it is a polynomial on each of these intervals. Now $\lim\limits_{x \to 0^-} f(x) = \lim\limits_{x \to 0^-} (1 + x^2) = 1$ and $\lim\limits_{x \to 0^+} f(x) = \lim\limits_{x \to 0^+} (2 - x) = 2$, so f is discontinuous at 0. Since $f(0) = 1$, f is continuous from the left at 0. Also, $\lim\limits_{x \to 2^-} f(x) = \lim\limits_{x \to 2^-} (2 - x) = 0$, $\lim\limits_{x \to 2^+} f(x) = \lim\limits_{x \to 2^+} (x - 2)^2 = 0$, and $f(2) = 0$, so f is continuous at 2. The only number at which f is discontinuous is 0.

37. $f(x) = \begin{cases} x + 2 & \text{if } x < 0 \\ 2x^2 & \text{if } 0 \leq x \leq 1 \\ 2 - x & \text{if } x > 1 \end{cases}$

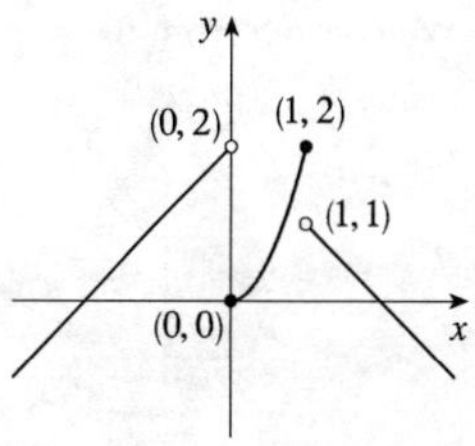

f is continuous on $(-\infty, 0)$, $(0, 1)$, and $(1, \infty)$ since on each of these intervals it is a polynomial. Now $\lim\limits_{x \to 0^-} f(x) = \lim\limits_{x \to 0^-} (x + 2) = 2$ and $\lim\limits_{x \to 0^+} f(x) = \lim\limits_{x \to 0^+} 2x^2 = 0$, so f is discontinuous at 0. Since $f(0) = 0$, f is continuous from the right at 0. Also $\lim\limits_{x \to 1^-} f(x) = \lim\limits_{x \to 1^-} 2x^2 = 2$ and $\lim\limits_{x \to 1^+} f(x) = \lim\limits_{x \to 1^+} (2 - x) = 1$, so f is discontinuous at 1. Since $f(1) = 2$, f is continuous from the left at 1.

39. f is continuous on $(-\infty, 3)$ and $(3, \infty)$. Now $\lim\limits_{x \to 3^-} f(x) = \lim\limits_{x \to 3^-} (cx + 1) = 3c + 1$ and $\lim\limits_{x \to 3^+} f(x) = \lim\limits_{x \to 3^+} (cx^2 - 1) = 9c - 1$. So f is continuous $\Leftrightarrow$ $3c + 1 = 9c - 1$ $\Leftrightarrow$ $6c = 2$ $\Leftrightarrow$ $c = \frac{1}{3}$. Thus, for f to be continuous on $(-\infty, \infty)$, $c = \frac{1}{3}$.

41. (a) $f(x) = \dfrac{x^2 - 2x - 8}{x + 2} = \dfrac{(x - 4)(x + 2)}{x + 2}$ has a removable discontinuity at -2 because $g(x) = x - 4$ is continuous on $\mathbb{R}$ and $f(x) = g(x)$ for $x \neq -2$. [The discontinuity is removed by defining $f(-2) = -6$.]

(b) $f(x) = \dfrac{x - 7}{|x - 7|}$ $\Rightarrow$ $\lim\limits_{x \to 7^-} f(x) = -1$ and $\lim\limits_{x \to 7^+} f(x) = 1$. Thus, $\lim\limits_{x \to 7} f(x)$ does not exist, so the discontinuity is not removable. (It is a jump discontinuity.)

(c) $f(x) = \dfrac{x^3 + 64}{x + 4} = \dfrac{(x + 4)(x^2 - 4x + 16)}{x + 4}$ has a removable discontinuity at -4 because $g(x) = x^2 - 4x + 16$ is continuous on $\mathbb{R}$ and $f(x) = g(x)$ for $x \neq -4$. [The discontinuity is removed by defining $f(-4) = 48$.]

(d) $f(x) = \dfrac{3-\sqrt{x}}{9-x} = \dfrac{3-\sqrt{x}}{(3-\sqrt{x})(3+\sqrt{x})}$ has a removable discontinuity at 9 because $g(x) = \dfrac{1}{3+\sqrt{x}}$ is continuous on $[0,\infty)$ and $f(x) = g(x)$ for $x \neq 9$. [The discontinuity is removed by defining $f(9) = \frac{1}{6}$.]

43. $f(x) = x^3 - x^2 + x$ is continuous on the interval $[2,3]$, $f(2) = 6$, and $f(3) = 21$. Since $6 < 10 < 21$, there is a number c in $(2,3)$ such that $f(c) = 10$ by the Intermediate Value Theorem.

45. $f(x) = x^4 + x - 3$ is continuous on the interval $[1,2]$, $f(1) = -1$, and $f(2) = 15$. Since $-1 < 0 < 15$, there is a number c in $(1,2)$ such that $f(c) = 0$ by the Intermediate Value Theorem. Thus, there is a root of the equation $x^4 + x - 3 = 0$ in the interval $(1,2)$.

47. $f(x) = \cos x - x$ is continuous on the interval $[0,1]$, $f(0) = 1$, and $f(1) = \cos 1 - 1 \approx -0.46$. Since $-0.46 < 0 < 1$, there is a number c in $(0,1)$ such that $f(c) = 0$ by the Intermediate Value Theorem. Thus, there is a root of the equation $\cos x - x = 0$, or $\cos x = x$, in the interval $(0,1)$.

49. (a) $f(x) = \sin x - 2 + x$ is continuous on $[0,2]$, $f(0) = -2$, and $f(2) = \sin 2 \approx 0.91$. Since $-2 < 0 < 0.91$, there is a number c in $(0,2)$ such that $f(c) = 0$ by the Intermediate Value Theorem. Thus, there is a root of the equation $\sin x - 2 + x = 0$, or $\sin x = 2 - x$, in the interval $(0,2)$.

(b) $f(1.10) \approx -0.009 < 0$ and $f(1.11) \approx 0.006 > 0$, so there is a root between 1.10 and 1.11.

51. (a) Let $f(x) = x^5 - x^2 - 4$. Then $f(1) = 1^5 - 1^2 - 4 = -4 < 0$ and $f(2) = 2^5 - 2^2 - 4 = 24 > 0$. So by the Intermediate Value Theorem, there is a number c in $(1,2)$ such that $f(c) = c^5 - c^2 - 4 = 0$.

(b) We can see from the graphs that, correct to three decimal places, the root is $x \approx 1.434$.

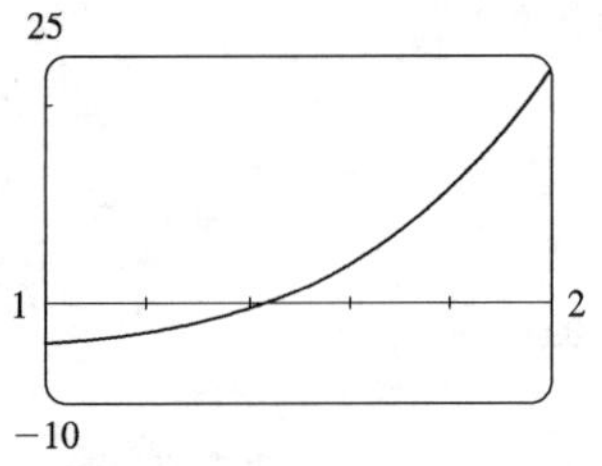

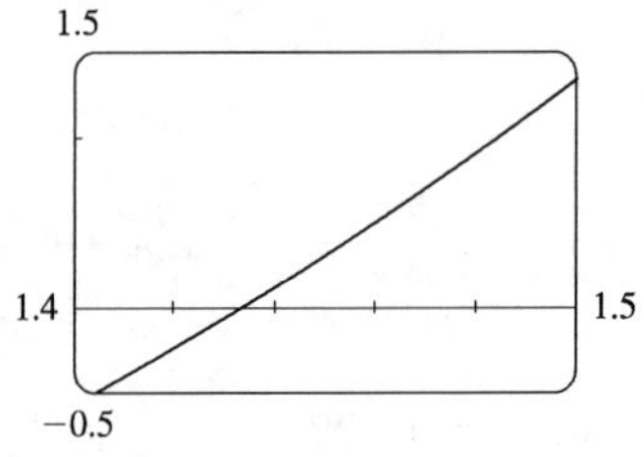

53. ($\Rightarrow$) If f is continuous at a, then by Theorem 8 with $g(h) = a + h$, we have

$$\lim_{h\to 0} f(a+h) = f\left(\lim_{h\to 0}(a+h)\right) = f(a).$$

($\Leftarrow$) Let $\varepsilon > 0$. Since $\lim_{h\to 0} f(a+h) = f(a)$, there exists $\delta > 0$ such that $0 < |h| < \delta \Rightarrow$ $|f(a+h) - f(a)| < \varepsilon$. So if $0 < |x-a| < \delta$, then $|f(x) - f(a)| = |f(a + (x-a)) - f(a)| < \varepsilon$. Thus, $\lim_{x\to a} f(x) = f(a)$ and so f is continuous at a.

55. As in the previous exercise, we must show that $\lim_{h\to 0} \cos(a+h) = \cos a$ to prove that the cosine function is continuous.

$$\begin{aligned}
\lim_{h\to 0} \cos(a+h) &= \lim_{h\to 0} (\cos a \cos h - \sin a \sin h) \\
&= \lim_{h\to 0} (\cos a \cos h) - \lim_{h\to 0} (\sin a \sin h) \\
&= \left(\lim_{h\to 0} \cos a\right)\left(\lim_{h\to 0} \cos h\right) - \left(\lim_{h\to 0} \sin a\right)\left(\lim_{h\to 0} \sin h\right) \\
&= (\cos a)(1) - (\sin a)(0) = \cos a
\end{aligned}$$

57. $f(x) = \begin{cases} 0 & \text{if } x \text{ is rational} \\ 1 & \text{if } x \text{ is irrational} \end{cases}$ is continuous nowhere. For, given any number a and any $\delta > 0$, the interval $(a - \delta, a + \delta)$ contains both infinitely many rational and infinitely many irrational numbers. Since $f(a) = 0$ or 1, there are infinitely many numbers x with $0 < |x - a| < \delta$ and $|f(x) - f(a)| = 1$. Thus, $\lim_{x \to a} f(x) \neq f(a)$. [In fact, $\lim_{x \to a} f(x)$ does not even exist.]

59. If there is such a number, it satisfies the equation $x^3 + 1 = x \iff x^3 - x + 1 = 0$. Let the left-hand side of this equation be called $f(x)$. Now $f(-2) = -5 < 0$, and $f(-1) = 1 > 0$. Note also that $f(x)$ is a polynomial, and thus continuous. So by the Intermediate Value Theorem, there is a number c between -2 and -1 such that $f(c) = 0$, so that $c = c^3 + 1$.

61. Define $u(t)$ to be the monk's distance from the monastery, as a function of time, on the first day, and define $d(t)$ to be his distance from the monastery, as a function of time, on the second day. Let D be the distance from the monastery to the top of the mountain. From the given information we know that $u(0) = 0$, $u(12) = D$, $d(0) = D$ and $d(12) = 0$. Now consider the function $u - d$, which is clearly continuous. We calculate that $(u - d)(0) = -D$ and $(u - d)(12) = D$. So by the Intermediate Value Theorem, there must be some time t_0 between 0 and 12 such that $(u - d)(t_0) = 0 \iff u(t_0) = d(t_0)$. So at time t_0 after 7:00 A.M., the monk will be at the same place on both days.

Calculating Limits Using the Limit Laws

1. (a) $\lim_{x\to a}[f(x)+h(x)] = \lim_{x\to a} f(x) + \lim_{x\to a} h(x) = -3+8=5$

(b) $\lim_{x\to a}[f(x)]^2 = \left[\lim_{x\to a} f(x)\right]^2 = (-3)^2 = 9$

(c) $\lim_{x\to a}\sqrt[3]{h(x)} = \sqrt[3]{\lim_{x\to a} h(x)} = \sqrt[3]{8} = 2$

(d) $\lim_{x\to a}\dfrac{1}{f(x)} = \dfrac{1}{\lim_{x\to a} f(x)} = \dfrac{1}{-3} = -\dfrac{1}{3}$

(e) $\lim_{x\to a}\dfrac{f(x)}{h(x)} = \dfrac{\lim_{x\to a} f(x)}{\lim_{x\to a} h(x)} = \dfrac{-3}{8} = -\dfrac{3}{8}$

(f) $\lim_{x\to a}\dfrac{g(x)}{f(x)} = \dfrac{\lim_{x\to a} g(x)}{\lim_{x\to a} f(x)} = \dfrac{0}{-3} = 0$

(g) The limit does not exist, since $\lim_{x\to a} g(x) = 0$ but $\lim_{x\to a} f(x) \neq 0$.

(h) $\lim_{x\to a}\dfrac{2f(x)}{h(x)-f(x)} = \dfrac{2\lim_{x\to a} f(x)}{\lim_{x\to a} h(x) - \lim_{x\to a} f(x)} = \dfrac{2(-3)}{8-(-3)} = -\dfrac{6}{11}$

3.
$$\begin{aligned}\lim_{x\to -2}(3x^4+2x^2-x+1) &= \lim_{x\to -2} 3x^4 + \lim_{x\to -2} 2x^2 - \lim_{x\to -2} x + \lim_{x\to -2} 1 && \text{[Limit Laws 1 and 2]}\\ &= 3\lim_{x\to -2} x^4 + 2\lim_{x\to -2} x^2 - \lim_{x\to -2} x + \lim_{x\to -2} 1 && [3]\\ &= 3(-2)^4 + 2(-2)^2 - (-2) + (1) && [9, 8, \text{and } 7]\\ &= 48+8+2+1 = 59\end{aligned}$$

5.
$$\begin{aligned}\lim_{x\to 3}(x^2-4)(x^3+5x-1) &= \lim_{x\to 3}(x^2-4)\cdot\lim_{x\to 3}(x^3+5x-1) && \text{[Limit Law 4]}\\ &= \left(\lim_{x\to 3} x^2 - \lim_{x\to 3} 4\right)\cdot\left(\lim_{x\to 3} x^3 + 5\lim_{x\to 3} x - \lim_{x\to 3} 1\right) && [2, 1, \text{and } 3]\\ &= (3^2-4)\cdot(3^3+5\cdot 3-1) && [7, 8, \text{and } 9]\\ &= 5\cdot 41 = 205\end{aligned}$$

7.
$$\begin{aligned}\lim_{x\to 1}\left(\frac{1+3x}{1+4x^2+3x^4}\right)^3 &= \left(\lim_{x\to 1}\frac{1+3x}{1+4x^2+3x^4}\right)^3 && [6]\\ &= \left[\frac{\lim_{x\to 1}(1+3x)}{\lim_{x\to 1}(1+4x^2+3x^4)}\right]^3 && [5]\\ &= \left[\frac{\lim_{x\to 1} 1 + 3\lim_{x\to 1} x}{\lim_{x\to 1} 1 + 4\lim_{x\to 1} x^2 + 3\lim_{x\to 1} x^4}\right]^3 && [2, 1, \text{and } 3]\\ &= \left[\frac{1+3(1)}{1+4(1)^2+3(1)^4}\right]^3 = \left[\frac{4}{8}\right]^3 = \left(\frac{1}{2}\right)^3 = \frac{1}{8} && [7, 8, \text{and } 9]\end{aligned}$$

9.
$$\begin{aligned}\lim_{x\to 4^-}\sqrt{16-x^2} &= \sqrt{\lim_{x\to 4^-}(16-x^2)} && [11]\\ &= \sqrt{\lim_{x\to 4^-} 16 - \lim_{x\to 4^-} x^2} && [2]\\ &= \sqrt{16-(4)^2} = 0 && [7 \text{ and } 9]\end{aligned}$$

11. $\lim_{x\to 2}\dfrac{x^2+x-6}{x-2} = \lim_{x\to 2}\dfrac{(x+3)(x-2)}{x-2} = \lim_{x\to 2}(x+3) = 2+3 = 5$

13. $\lim_{x\to 2}\dfrac{x^2-x+6}{x-2}$ does not exist since $x-2\to 0$ but $x^2-x+6\to 8$ as $x\to 2$.

15. $\lim\limits_{t\to-3} \dfrac{t^2-9}{2t^2+7t+3} = \lim\limits_{t\to-3} \dfrac{(t+3)(t-3)}{(2t+1)(t+3)} = \lim\limits_{t\to-3} \dfrac{t-3}{2t+1} = \dfrac{-3-3}{2(-3)+1} = \dfrac{-6}{-5} = \dfrac{6}{5}$

17. $\lim\limits_{h\to0} \dfrac{(4+h)^2-16}{h} = \lim\limits_{h\to0} \dfrac{(16+8h+h^2)-16}{h} = \lim\limits_{h\to0} \dfrac{8h+h^2}{h} = \lim\limits_{h\to0} \dfrac{h(8+h)}{h} = \lim\limits_{h\to0}(8+h) = 8+0 = 8$

19. $\lim\limits_{h\to0} \dfrac{(1+h)^4-1}{h} = \lim\limits_{h\to0} \dfrac{\left(1+4h+6h^2+4h^3+h^4\right)-1}{h} = \lim\limits_{h\to0} \dfrac{4h+6h^2+4h^3+h^4}{h}$

$= \lim\limits_{h\to0} \dfrac{h(4+6h+4h^2+h^3)}{h} = \lim\limits_{h\to0}\left(4+6h+4h^2+h^3\right) = 4+0+0+0 = 4$

21. $\lim\limits_{t\to9} \dfrac{9-t}{3-\sqrt{t}} = \lim\limits_{t\to9} \dfrac{\left(3+\sqrt{t}\right)\left(3-\sqrt{t}\right)}{3-\sqrt{t}} = \lim\limits_{t\to9}\left(3+\sqrt{t}\right) = 3+\sqrt{9} = 6$

23. $\lim\limits_{x\to7} \dfrac{\sqrt{x+2}-3}{x-7} = \lim\limits_{x\to7} \dfrac{\sqrt{x+2}-3}{x-7} \cdot \dfrac{\sqrt{x+2}+3}{\sqrt{x+2}+3} = \lim\limits_{x\to7} \dfrac{(x+2)-9}{(x-7)\left(\sqrt{x+2}+3\right)}$

$= \lim\limits_{x\to7} \dfrac{x-7}{(x-7)\left(\sqrt{x+2}+3\right)} = \lim\limits_{x\to7} \dfrac{1}{\sqrt{x+2}+3} = \dfrac{1}{\sqrt{9}+3} = \dfrac{1}{6}$

25. $\lim\limits_{x\to-4} \dfrac{\frac{1}{4}+\frac{1}{x}}{4+x} = \lim\limits_{x\to-4} \dfrac{\frac{x+4}{4x}}{4+x} = \lim\limits_{x\to-4} \dfrac{x+4}{4x(4+x)} = \lim\limits_{x\to-4} \dfrac{1}{4x} = \dfrac{1}{4(-4)} = -\dfrac{1}{16}$

27. $\lim\limits_{x\to9} \dfrac{x^2-81}{\sqrt{x}-3} = \lim\limits_{x\to9} \dfrac{(x-9)(x+9)}{\sqrt{x}-3} = \lim\limits_{x\to9} \dfrac{(\sqrt{x}-3)(\sqrt{x}+3)(x+9)}{\sqrt{x}-3}$ $\left[\begin{array}{c}\text{factor } x-9 \text{ as a}\\ \text{difference of squares}\end{array}\right]$

$= \lim\limits_{x\to9}\left[(\sqrt{x}+3)(x+9)\right] = \left(\sqrt{9}+3\right)(9+9) = 6\cdot18 = 108$

29. $\lim\limits_{t\to0}\left(\dfrac{1}{t\sqrt{1+t}}-\dfrac{1}{t}\right) = \lim\limits_{t\to0} \dfrac{1-\sqrt{1+t}}{t\sqrt{1+t}} = \lim\limits_{t\to0} \dfrac{\left(1-\sqrt{1+t}\right)\left(1+\sqrt{1+t}\right)}{t\sqrt{t+1}\left(1+\sqrt{1+t}\right)} = \lim\limits_{t\to0} \dfrac{-t}{t\sqrt{1+t}\left(1+\sqrt{1+t}\right)}$

$= \lim\limits_{t\to0} \dfrac{-1}{\sqrt{1+t}\left(1+\sqrt{1+t}\right)} = \dfrac{-1}{\sqrt{1+0}\left(1+\sqrt{1+0}\right)} = -\dfrac{1}{2}$

31. (a)

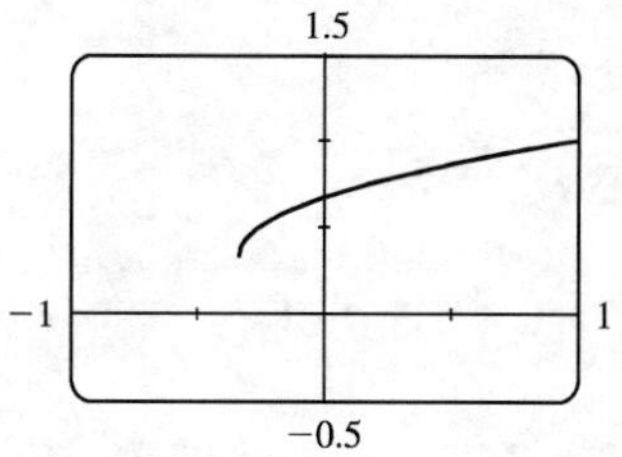

$\lim\limits_{x\to0} \dfrac{x}{\sqrt{1+3x}-1} \approx \dfrac{2}{3}$

(b)

x	$f(x)$
−0.001	0.6661663
−0.0001	0.6666167
−0.00001	0.6666617
−0.000001	0.6666662
0.000001	0.6666672
0.00001	0.6666717
0.0001	0.6667167
0.001	0.6671663

The limit appears to be $\dfrac{2}{3}$.

(c) $\lim\limits_{x\to 0}\left(\dfrac{x}{\sqrt{1+3x}-1}\cdot\dfrac{\sqrt{1+3x}+1}{\sqrt{1+3x}+1}\right)=\lim\limits_{x\to 0}\dfrac{x\left(\sqrt{1+3x}+1\right)}{(1+3x)-1}=\lim\limits_{x\to 0}\dfrac{x\left(\sqrt{1+3x}+1\right)}{3x}$

$$=\frac{1}{3}\lim_{x\to 0}\left(\sqrt{1+3x}+1\right) \qquad \text{[Limit Law 3]}$$

$$=\frac{1}{3}\left[\sqrt{\lim_{x\to 0}(1+3x)}+\lim_{x\to 0}1\right] \qquad \text{[1 and 11]}$$

$$=\frac{1}{3}\left(\sqrt{\lim_{x\to 0}1+3\lim_{x\to 0}x}+1\right) \qquad \text{[1, 3, and 7]}$$

$$=\frac{1}{3}\left(\sqrt{1+3\cdot 0}+1\right) \qquad \text{[7 and 8]}$$

$$=\frac{1}{3}(1+1)=\frac{2}{3}$$

33. Let $f(x)=-x^2$, $g(x)=x^2\cos 20\pi x$ and $h(x)=x^2$. Then $-1\le\cos 20\pi x\le 1 \;\Rightarrow\; -x^2\le x^2\cos 20\pi x\le x^2 \;\Rightarrow$ $f(x)\le g(x)\le h(x)$. So since $\lim\limits_{x\to 0}f(x)=\lim\limits_{x\to 0}h(x)=0$, by the Squeeze Theorem we have $\lim\limits_{x\to 0}g(x)=0$.

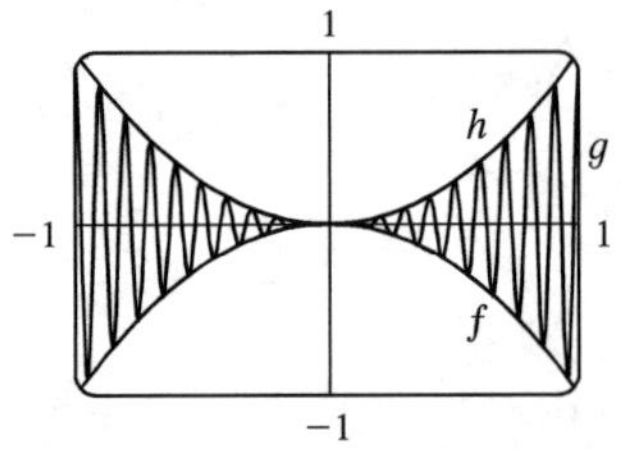

35. $1\le f(x)\le x^2+2x+2$ for all x. Now $\lim\limits_{x\to -1}1=1$ and $\lim\limits_{x\to -1}\left(x^2+2x+2\right)=\lim\limits_{x\to -1}x^2+2\lim\limits_{x\to -1}x+\lim\limits_{x\to -1}2=(-1)^2+2(-1)+2=1$. Therefore, by the Squeeze Theorem, $\lim\limits_{x\to -1}f(x)=1$.

37. $-1\le\cos(2/x)\le 1 \;\Rightarrow\; -x^4\le x^4\cos(2/x)\le x^4$. Since $\lim\limits_{x\to 0}\left(-x^4\right)=0$ and $\lim\limits_{x\to 0}x^4=0$, we have $\lim\limits_{x\to 0}\left[x^4\cos(2/x)\right]=0$ by the Squeeze Theorem.

39. If $x>-4$, then $|x+4|=x+4$, so $\lim\limits_{x\to -4^+}|x+4|=\lim\limits_{x\to -4^+}(x+4)=-4+4=0$.
If $x<-4$, then $|x+4|=-(x+4)$, so $\lim\limits_{x\to -4^-}|x+4|=\lim\limits_{x\to -4^-}-(x+4)=-(-4+4)=0$.
Since the right and left limits are equal, $\lim\limits_{x\to -4}|x+4|=0$.

41. If $x>2$, then $|x-2|=x-2$, so $\lim\limits_{x\to 2^+}\dfrac{|x-2|}{x-2}=\lim\limits_{x\to 2^+}\dfrac{x-2}{x-2}=\lim\limits_{x\to 2^+}1=1$. If $x<2$, then $|x-2|=-(x-2)$, so $\lim\limits_{x\to 2^-}\dfrac{|x-2|}{x-2}=\lim\limits_{x\to 2^-}\dfrac{-(x-2)}{x-2}=\lim\limits_{x\to 2^-}-1=-1$. The right and left limits are different, so $\lim\limits_{x\to 2}\dfrac{|x-2|}{x-2}$ does not exist.

43. Since $|x|=-x$ for $x<0$, we have $\lim\limits_{x\to 0^-}\left(\dfrac{1}{x}-\dfrac{1}{|x|}\right)=\lim\limits_{x\to 0^-}\left(\dfrac{1}{x}-\dfrac{1}{-x}\right)=\lim\limits_{x\to 0^-}\dfrac{2}{x}$, which does not exist since the denominator approaches 0 and the numerator does not.

45. (a) 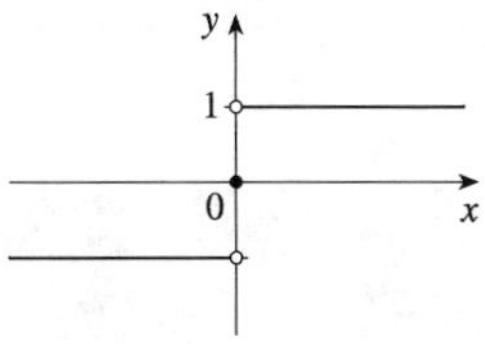

(b) (i) Since $\operatorname{sgn} x = 1$ for $x > 0$, $\lim_{x\to 0^+} \operatorname{sgn} x = \lim_{x\to 0^+} 1 = 1$.

(ii) Since $\operatorname{sgn} x = -1$ for $x < 0$, $\lim_{x\to 0^-} \operatorname{sgn} x = \lim_{x\to 0^-} -1 = -1$.

(iii) Since $\lim_{x\to 0^-} \operatorname{sgn} x \neq \lim_{x\to 0^+} \operatorname{sgn} x$, $\lim_{x\to 0} \operatorname{sgn} x$ does not exist.

(iv) Since $|\operatorname{sgn} x| = 1$ for $x \neq 0$, $\lim_{x\to 0} |\operatorname{sgn} x| = \lim_{x\to 0} 1 = 1$.

47. (a) (i) $\lim_{x\to 1^+} \dfrac{x^2-1}{|x-1|} = \lim_{x\to 1^+} \dfrac{x^2-1}{x-1} = \lim_{x\to 1^+} (x+1) = 2$

(ii) $\lim_{x\to 1^-} \dfrac{x^2-1}{|x-1|} = \lim_{x\to 1^-} \dfrac{x^2-1}{-(x-1)} = \lim_{x\to 1^-} -(x+1) = -2$

(b) No, $\lim_{x\to 1} F(x)$ does not exist since $\lim_{x\to 1^+} F(x) \neq \lim_{x\to 1^-} F(x)$.

(c) 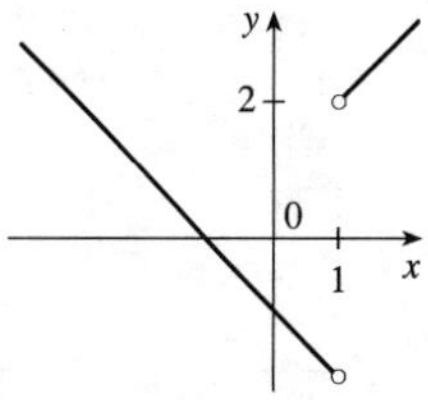

49. (a) (i) $[\![x]\!] = -2$ for $-2 \le x < -1$, so $\lim_{x\to -2^+} [\![x]\!] = \lim_{x\to -2^+} (-2) = -2$

(ii) $[\![x]\!] = -3$ for $-3 \le x < -2$, so $\lim_{x\to -2^-} [\![x]\!] = \lim_{x\to -2^-} (-3) = -3$. The right and left limits are different, so $\lim_{x\to -2} [\![x]\!]$ does not exist.

(iii) $[\![x]\!] = -3$ for $-3 \le x < -2$, so $\lim_{x\to -2.4} [\![x]\!] = \lim_{x\to -2.4} (-3) = -3$.

(b) (i) $[\![x]\!] = n-1$ for $n-1 \le x < n$, so $\lim_{x\to n^-} [\![x]\!] = \lim_{x\to n^-} (n-1) = n-1$.

(ii) $[\![x]\!] = n$ for $n \le x < n+1$, so $\lim_{x\to n^+} [\![x]\!] = \lim_{x\to n^+} n = n$.

(c) $\lim_{x\to a} [\![x]\!]$ exists $\Leftrightarrow$ a is not an integer.

51. The graph of $f(x) = [\![x]\!] + [\![-x]\!]$ is the same as the graph of $g(x) = -1$ with holes at each integer, since $f(a) = 0$ for any integer a. Thus, $\lim_{x\to 2^-} f(x) = -1$ and $\lim_{x\to 2^+} f(x) = -1$, so $\lim_{x\to 2} f(x) = -1$. However, $f(2) = [\![2]\!] + [\![-2]\!] = 2 + (-2) = 0$, so $\lim_{x\to 2} f(x) \neq f(2)$.

53. Since $p(x)$ is a polynomial, $p(x) = a_0 + a_1x + a_2x^2 + \cdots + a_nx^n$. Thus, by the Limit Laws,

$$\begin{aligned}\lim_{x\to a} p(x) &= \lim_{x\to a} \left(a_0 + a_1x + a_2x^2 + \cdots + a_nx^n\right)\\ &= a_0 + a_1 \lim_{x\to a} x + a_2 \lim_{x\to a} x^2 + \cdots + a_n \lim_{x\to a} x^n\\ &= a_0 + a_1a + a_2a^2 + \cdots + a_na^n = p(a)\end{aligned}$$

Thus, for any polynomial p, $\lim_{x\to a} p(x) = p(a)$.

55. Observe that $0 \le f(x) \le x^2$ for all x, and $\lim_{x\to 0} 0 = 0 = \lim_{x\to 0} x^2$. So, by the Squeeze Theorem, $\lim_{x\to 0} f(x) = 0$.

57. Let $f(x) = H(x)$ and $g(x) = 1 - H(x)$, where H is the Heaviside function defined in Exercise 1.3.59. Thus, either f or g is 0 for any value of x. Then $\lim_{x\to 0} f(x)$ and $\lim_{x\to 0} g(x)$ do not exist, but $\lim_{x\to 0} [f(x)g(x)] = \lim_{x\to 0} 0 = 0$.

59. Since the denominator approaches 0 as $x \to -2$, the limit will exist only if the numerator also approaches 0 as $x \to -2$. In order for this to happen, we need $\lim\limits_{x \to -2} \left(3x^2 + ax + a + 3\right) = 0 \iff$ $3(-2)^2 + a(-2) + a + 3 = 0 \iff 12 - 2a + a + 3 = 0 \iff a = 15$. With $a = 15$, the limit becomes

$$\lim_{x \to -2} \frac{3x^2 + 15x + 18}{x^2 + x - 2} = \lim_{x \to -2} \frac{3(x+2)(x+3)}{(x-1)(x+2)} = \lim_{x \to -2} \frac{3(x+3)}{x-1} = \frac{3(-2+3)}{-2-1} = \frac{3}{-3} = -1.$$

Limits at Infinity; Horizontal Asymptotes

1. (a) As x becomes large, the values of $f(x)$ approach 5.

(b) As x becomes large negative, the values of $f(x)$ approach 3.

3. (a) $\lim\limits_{x\to 2} f(x) = \infty$

(b) $\lim\limits_{x\to -1^-} f(x) = \infty$

(c) $\lim\limits_{x\to -1^+} f(x) = -\infty$

(d) $\lim\limits_{x\to \infty} f(x) = 1$

(e) $\lim\limits_{x\to -\infty} f(x) = 2$

(f) Vertical: $x = -1$, $x = 2$; Horizontal: $y = 1$, $y = 2$

5. If $f(x) = x^2/2^x$, then a calculator gives $f(0) = 0$, $f(1) = 0.5$, $f(2) = 1$, $f(3) = 1.125$, $f(4) = 1$,

$f(5) = 0.78125$, $f(6) = 0.5625$, $f(7) = 0.3828125$, $f(8) = 0.25$, $f(9) = 0.158203125$, $f(10) = 0.09765625$,

$f(20) \approx 0.00038147$, $f(50) \approx 2.2204 \times 10^{-12}$, $f(100) \approx 7.8886 \times 10^{-27}$.

It appears that $\lim\limits_{x\to\infty} \left(x^2/2^x\right) = 0$.

7. $$\lim_{x\to\infty} \frac{3x^2 - x + 4}{2x^2 + 5x - 8} = \lim_{x\to\infty} \frac{(3x^2 - x + 4)/x^2}{(2x^2 + 5x - 8)/x^2}$$ [divide both the numerator and denominator by x^2 (the highest power of x that appears in the denominator)]

$$= \frac{\lim\limits_{x\to\infty} (3 - 1/x + 4/x^2)}{\lim\limits_{x\to\infty} (2 + 5/x - 8/x^2)}$$ [Limit Law 5]

$$= \frac{\lim\limits_{x\to\infty} 3 - \lim\limits_{x\to\infty}(1/x) + \lim\limits_{x\to\infty}(4/x^2)}{\lim\limits_{x\to\infty} 2 + \lim\limits_{x\to\infty}(5/x) - \lim\limits_{x\to\infty}(8/x^2)} \quad \text{[Limit Laws 1 and 2]}$$

$$= \frac{3 - \lim\limits_{x\to\infty}(1/x) + 4\lim\limits_{x\to\infty}(1/x^2)}{2 + 5\lim\limits_{x\to\infty}(1/x) - 8\lim\limits_{x\to\infty}(1/x^2)} \quad \text{[Limit Laws 7 and 3]}$$

$$= \frac{3 - 0 + 4(0)}{2 + 5(0) - 8(0)} \quad \text{[Theorem 4]}$$

$$= \frac{3}{2}$$

9. $$\lim_{x\to\infty} \frac{1}{2x+3} = \lim_{x\to\infty} \frac{1/x}{(2x+3)/x} = \frac{\lim\limits_{x\to\infty}(1/x)}{\lim\limits_{x\to\infty}(2+3/x)} = \frac{\lim\limits_{x\to\infty}(1/x)}{\lim\limits_{x\to\infty} 2 + 3\lim\limits_{x\to\infty}(1/x)} = \frac{0}{2+3(0)} = \frac{0}{2} = 0$$

11. $$\lim_{x\to-\infty} \frac{1-x-x^2}{2x^2-7} = \lim_{x\to-\infty} \frac{(1-x-x^2)/x^2}{(2x^2-7)/x^2} = \frac{\lim\limits_{x\to-\infty}(1/x^2 - 1/x - 1)}{\lim\limits_{x\to-\infty}(2 - 7/x^2)}$$

$$= \frac{\lim\limits_{x\to-\infty}(1/x^2) - \lim\limits_{x\to-\infty}(1/x) - \lim\limits_{x\to-\infty} 1}{\lim\limits_{x\to-\infty} 2 - 7\lim\limits_{x\to-\infty}(1/x^2)} = \frac{0-0-1}{2-7(0)} = -\frac{1}{2}$$

13. Divide both the numerator and denominator by x^3 (the highest power of x that occurs in the denominator).

$$\lim_{x\to\infty} \frac{x^3+5x}{2x^3-x^2+4} = \lim_{x\to\infty} \frac{\dfrac{x^3+5x}{x^3}}{\dfrac{2x^3-x^2+4}{x^3}} = \lim_{x\to\infty} \frac{1+\dfrac{5}{x^2}}{2-\dfrac{1}{x}+\dfrac{4}{x^3}} = \frac{\lim\limits_{x\to\infty}\left(1+\dfrac{5}{x^2}\right)}{\lim\limits_{x\to\infty}\left(2-\dfrac{1}{x}+\dfrac{4}{x^3}\right)}$$

$$= \frac{\lim\limits_{x\to\infty} 1 + 5\lim\limits_{x\to\infty}\dfrac{1}{x^2}}{\lim\limits_{x\to\infty} 2 - \lim\limits_{x\to\infty}\dfrac{1}{x} + 4\lim\limits_{x\to\infty}\dfrac{1}{x^3}} = \frac{1+5(0)}{2-0+4(0)} = \frac{1}{2}$$

15. First, multiply the factors in the denominator. Then divide both the numerator and denominator by u^4.

$$\lim_{u\to\infty} \frac{4u^4+5}{(u^2-2)(2u^2-1)} = \lim_{u\to\infty} \frac{4u^4+5}{2u^4-5u^2+2} = \lim_{u\to\infty} \frac{\dfrac{4u^4+5}{u^4}}{\dfrac{2u^4-5u^2+2}{u^4}} = \lim_{u\to\infty} \frac{4+\dfrac{5}{u^4}}{2-\dfrac{5}{u^2}+\dfrac{2}{u^4}}$$

$$= \frac{\lim\limits_{u\to\infty}\left(4+\dfrac{5}{u^4}\right)}{\lim\limits_{u\to\infty}\left(2-\dfrac{5}{u^2}+\dfrac{2}{u^4}\right)} = \frac{\lim\limits_{u\to\infty} 4 + 5\lim\limits_{u\to\infty}\dfrac{1}{u^4}}{\lim\limits_{u\to\infty} 2 - 5\lim\limits_{u\to\infty}\dfrac{1}{u^2} + 2\lim\limits_{u\to\infty}\dfrac{1}{u^4}} = \frac{4+5(0)}{2-5(0)+2(0)}$$

$$= \frac{4}{2} = 2$$

17. $$\lim_{x\to\infty} \frac{\sqrt{9x^6-x}}{x^3+1} = \lim_{x\to\infty} \frac{\sqrt{9x^6-x}/x^3}{(x^3+1)/x^3} = \frac{\lim\limits_{x\to\infty}\sqrt{(9x^6-x)/x^6}}{\lim\limits_{x\to\infty}(1+1/x^3)} \quad \left[\text{since } x^3 = \sqrt{x^6} \text{ for } x > 0\right]$$

$$= \frac{\lim\limits_{x\to\infty}\sqrt{9-1/x^5}}{\lim\limits_{x\to\infty} 1 + \lim\limits_{x\to\infty}(1/x^3)} = \frac{\sqrt{\lim\limits_{x\to\infty} 9 - \lim\limits_{x\to\infty}(1/x^5)}}{1+0}$$

$$= \sqrt{9-0} = 3$$

19. $\lim\limits_{x\to-\infty} \dfrac{x}{\sqrt{x^2+4}} = \lim\limits_{x\to-\infty} \dfrac{x/\sqrt{x^2}}{\sqrt{x^2+4}/\sqrt{x^2}} = \lim\limits_{x\to-\infty} \dfrac{-1}{\sqrt{1+4/x^2}}$ [since $-x = |x| = \sqrt{x^2}$ for $x < 0$] $= -1$

21. $\lim\limits_{x\to\infty} \left(\sqrt{9x^2+x} - 3x\right) = \lim\limits_{x\to\infty} \dfrac{\left(\sqrt{9x^2+x} - 3x\right)\left(\sqrt{9x^2+x} + 3x\right)}{\sqrt{9x^2+x} + 3x} = \lim\limits_{x\to\infty} \dfrac{\left(\sqrt{9x^2+x}\right)^2 - (3x)^2}{\sqrt{9x^2+x} + 3x}$

$$= \lim_{x\to\infty} \frac{(9x^2+x) - 9x^2}{\sqrt{9x^2+x}+3x} = \lim_{x\to\infty} \frac{x}{\sqrt{9x^2+x}+3x} \cdot \frac{1/x}{1/x}$$

$$= \lim_{x\to\infty} \frac{x/x}{\sqrt{9x^2/x^2 + x/x^2} + 3x/x} = \lim_{x\to\infty} \frac{1}{\sqrt{9+1/x}+3} = \frac{1}{\sqrt{9}+3} = \frac{1}{3+3} = \frac{1}{6}$$

23. $\lim\limits_{x\to\infty} \left(\sqrt{x^2+ax} - \sqrt{x^2+bx}\right) = \lim\limits_{x\to\infty} \dfrac{\left(\sqrt{x^2+ax} - \sqrt{x^2+bx}\right)\left(\sqrt{x^2+ax} + \sqrt{x^2+bx}\right)}{\sqrt{x^2+ax} + \sqrt{x^2+bx}}$

$$= \lim_{x\to\infty} \frac{(x^2+ax)-(x^2+bx)}{\sqrt{x^2+ax}+\sqrt{x^2+bx}} = \lim_{x\to\infty} \frac{[(a-b)x]/x}{\left(\sqrt{x^2+ax}+\sqrt{x^2+bx}\right)/\sqrt{x^2}}$$

$$= \lim_{x\to\infty} \frac{a-b}{\sqrt{1+a/x}+\sqrt{1+b/x}} = \frac{a-b}{\sqrt{1+0}+\sqrt{1+0}} = \frac{a-b}{2}$$

25. $\sqrt{x}$ is large when x is large, so $\lim\limits_{x\to\infty} \sqrt{x} = \infty$.

27. $\lim\limits_{x\to\infty} (x - \sqrt{x}) = \lim\limits_{x\to\infty} \sqrt{x}\,(\sqrt{x} - 1) = \infty$ since $\sqrt{x} \to \infty$ and $\sqrt{x} - 1 \to \infty$ as $x \to \infty$.

29. $\lim\limits_{x\to-\infty} (x^4 + x^5) = \lim\limits_{x\to-\infty} x^5(\frac{1}{x} + 1)$ [factor out the largest power of x] $= -\infty$ because $x^5 \to -\infty$ and $1/x + 1 \to 1$ as $x \to -\infty$.

31. $\lim\limits_{x\to\infty} \dfrac{x + x^3 + x^5}{1 - x^2 + x^4} = \lim\limits_{x\to\infty} \dfrac{(x + x^3 + x^5)/x^4}{(1 - x^2 + x^4)/x^4}$ [divide by the highest power of x in the denominator]

$$= \lim_{x\to\infty} \frac{1/x^3 + 1/x + x}{1/x^4 - 1/x^2 + 1} = \infty$$

because $(1/x^3 + 1/x + x) \to \infty$ and $(1/x^4 - 1/x^2 + 1) \to 1$ as $x \to \infty$.

33. (a)

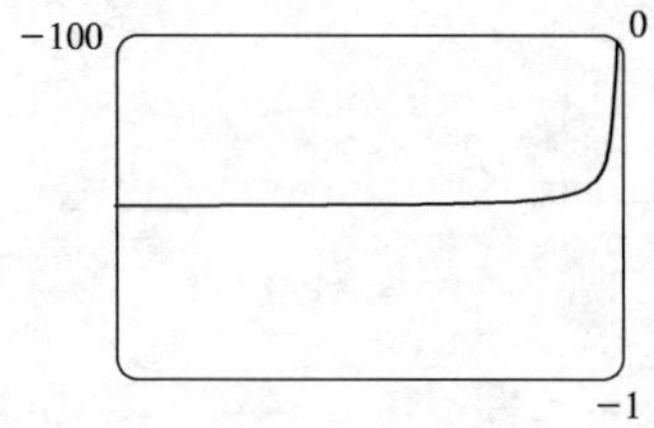

From the graph of $f(x) = \sqrt{x^2+x+1} + x$, we estimate the value of $\lim\limits_{x\to-\infty} f(x)$ to be -0.5.

(b)

x	$f(x)$
−10,000	−0.4999625
−100,000	−0.4999962
−1,000,000	−0.4999996

From the table, we estimate the limit to be -0.5.

(c) $\lim\limits_{x\to-\infty}\left(\sqrt{x^2+x+1}+x\right)=\lim\limits_{x\to-\infty}\left(\sqrt{x^2+x+1}+x\right)\left[\dfrac{\sqrt{x^2+x+1}-x}{\sqrt{x^2+x+1}-x}\right]=\lim\limits_{x\to-\infty}\dfrac{(x^2+x+1)-x^2}{\sqrt{x^2+x+1}-x}$

$$=\lim_{x\to-\infty}\frac{(x+1)(1/x)}{\left(\sqrt{x^2+x+1}-x\right)(1/x)}=\lim_{x\to-\infty}\frac{1+(1/x)}{-\sqrt{1+(1/x)+(1/x^2)}-1}$$

$$=\frac{1+0}{-\sqrt{1+0+0}-1}=-\frac{1}{2}$$

Note that for $x<0$, we have $\sqrt{x^2}=|x|=-x$, so when we divide the radical by x, with $x<0$, we get

$$\frac{1}{x}\sqrt{x^2+x+1}=-\frac{1}{\sqrt{x^2}}\sqrt{x^2+x+1}=-\sqrt{1+(1/x)+(1/x^2)}.$$

35. $\lim\limits_{x\to\pm\infty}\dfrac{x}{x+4}=\lim\limits_{x\to\pm\infty}\dfrac{1}{1+4/x}=\dfrac{1}{1+0}=1$, so $y=1$ is a horizontal asymptote. $\lim\limits_{x\to-4^-}\dfrac{x}{x+4}=\infty$ and $\lim\limits_{x\to-4^+}\dfrac{x}{x+4}=-\infty$, so $x=-4$ is a vertical asymptote. The graph confirms these calculations.

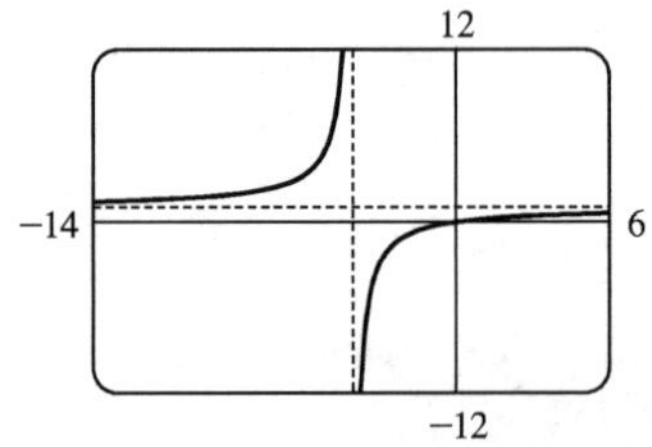

37. $\lim\limits_{x\to\pm\infty}\dfrac{x^3}{x^2+3x-10}=\lim\limits_{x\to\pm\infty}\dfrac{x}{1+(3/x)-(10/x^2)}=\pm\infty$, so there is no horizontal asymptote.

$\lim\limits_{x\to2^+}\dfrac{x^3}{x^2+3x-10}=\lim\limits_{x\to2^+}\dfrac{x^3}{(x+5)(x-2)}=\infty$, since

$\dfrac{x^3}{(x+5)(x-2)}>0$ for $x>2$. Similarly, $\lim\limits_{x\to2^-}\dfrac{x^3}{x^2+3x-10}=-\infty$,

$\lim\limits_{x\to-5^-}\dfrac{x^3}{x^2+3x-10}=-\infty$, and $\lim\limits_{x\to-5^+}\dfrac{x^3}{x^2+3x-10}=\infty$, so

$x=2$ and $x=-5$ are vertical asymptotes. The graph confirms these calculations.

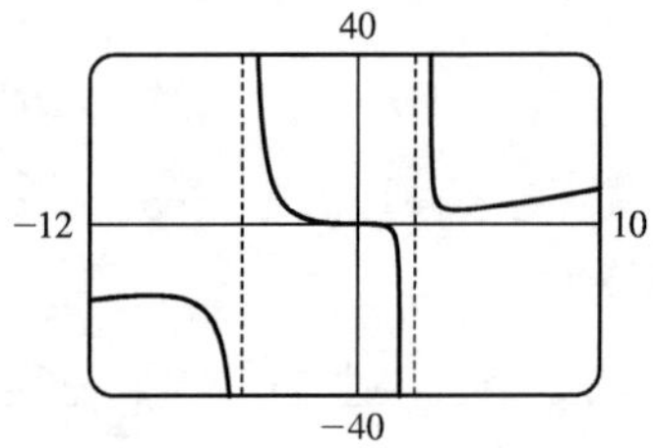

39. $\lim\limits_{x\to\infty}\dfrac{x}{\sqrt[4]{x^4+1}}\cdot\dfrac{1/x}{1/\sqrt[4]{x^4}}=\lim\limits_{x\to\infty}\dfrac{1}{\sqrt[4]{1+\dfrac{1}{x^4}}}=\dfrac{1}{\sqrt[4]{1+0}}=1$ and

$\lim\limits_{x\to-\infty}\dfrac{x}{\sqrt[4]{x^4+1}}\cdot\dfrac{1/x}{-1/\sqrt[4]{x^4}}=\lim\limits_{x\to-\infty}\dfrac{1}{-\sqrt[4]{1+\dfrac{1}{x^4}}}=\dfrac{1}{-\sqrt[4]{1+0}}=-1$,

so $y=\pm1$ are horizontal asymptotes. There is no vertical asymptote.

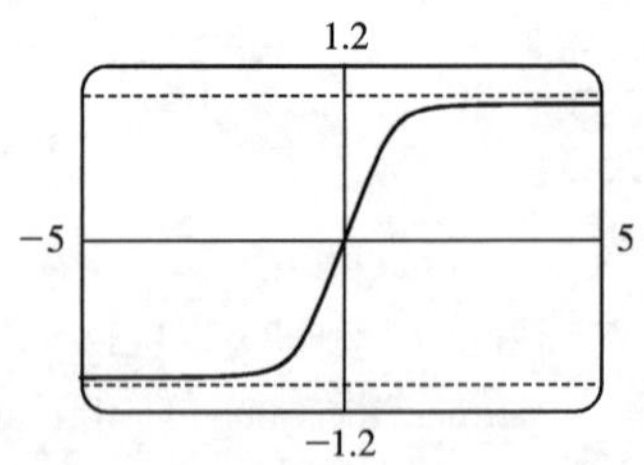

41. Let's look for a rational function.

(1) $\lim_{x\to\pm\infty} f(x) = 0 \;\Rightarrow\;$ degree of numerator < degree of denominator

(2) $\lim_{x\to 0} f(x) = -\infty \;\Rightarrow\;$ there is a factor of x^2 in the denominator (not just x, since that would produce a sign change at $x = 0$), and the function is negative near $x = 0$.

(3) $\lim_{x\to 3^-} f(x) = \infty$ and $\lim_{x\to 3^+} f(x) = -\infty \;\Rightarrow\;$ vertical asymptote at $x = 3$; there is a factor of $(x - 3)$ in the denominator.

(4) $f(2) = 0 \;\Rightarrow\;$ 2 is an x-intercept; there is at least one factor of $(x - 2)$ in the numerator.

Combining all of this information and putting in a negative sign to give us the desired left- and right-hand limits gives us $f(x) = \dfrac{2 - x}{x^2(x - 3)}$ as one possibility.

43. $y = \dfrac{1 - x}{1 + x}$ has domain $(-\infty, -1) \cup (-1, \infty)$.

$\lim_{x\to\pm\infty} \dfrac{1 - x}{1 + x} = \lim_{x\to\pm\infty} \dfrac{1/x - 1}{1/x + 1} = \dfrac{0 - 1}{0 + 1} = -1$, so $y = -1$ is a HA.

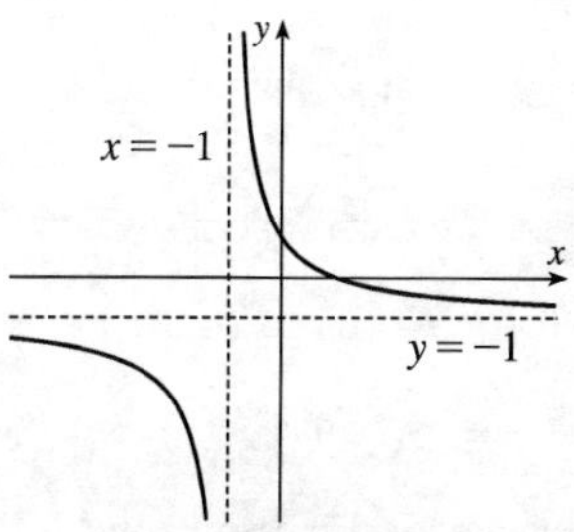

The line $x = -1$ is a VA.

$y' = \dfrac{(1 + x)(-1) - (1 - x)(1)}{(1 + x)^2} = \dfrac{-2}{(1 + x)^2} < 0$ for $x \neq 1$. Thus,

$(-\infty, -1)$ and $(-1, \infty)$ are intervals of decrease.

$y'' = -2 \cdot \dfrac{-2(1 + x)}{[(1 + x)^2]^2} = \dfrac{4}{(1 + x)^3} < 0$ for $x < -1$ and $y'' > 0$ for $x > -1$, so the curve is CD on $(-\infty, -1)$ and CU on $(-1, \infty)$. Since $x = -1$ is not in the domain, there is no IP.

45. $\lim_{x\to\pm\infty} \dfrac{x}{x^2 + 1} = \lim_{x\to\pm\infty} \dfrac{1/x}{1 + 1/x^2} = \dfrac{0}{1 + 0} = 0$, so $y = 0$ is a horizontal asymptote.

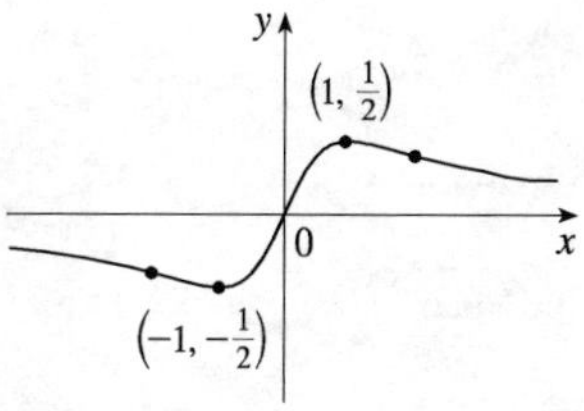

$y' = \dfrac{x^2 + 1 - x(2x)}{(x^2 + 1)^2} = \dfrac{1 - x^2}{(x^2 + 1)^2} = 0$ when $x = \pm 1$ and $y' > 0 \;\Leftrightarrow$ $x^2 < 1 \;\Leftrightarrow\; -1 < x < 1$, so y is increasing on $(-1, 1)$ and decreasing on $(-\infty, -1)$ and $(1, \infty)$.

$y'' = \dfrac{(1 + x^2)^2(-2x) - (1 - x^2)2(x^2 + 1)2x}{(1 + x^2)^4} = \dfrac{2x(x^2 - 3)}{(1 + x^2)^3} > 0 \;\Leftrightarrow\; x > \sqrt{3}$ or $-\sqrt{3} < x < 0$, so y is CU on $(\sqrt{3}, \infty)$ and $(-\sqrt{3}, 0)$ and CD on $(-\infty, -\sqrt{3})$ and $(0, \sqrt{3})$.

47. $y = f(x) = x^2(x-2)(1-x)$. The y-intercept is $f(0) = 0$, and the x-intercepts occur when $y = 0 \Rightarrow x = 0$, 1, and 2. Notice that, since x^2 is always positive, the graph does not cross the x-axis at 0, but does cross the x-axis at 1 and 2.

$\lim\limits_{x\to\infty} x^2(x-2)(1-x) = -\infty$, since the first two factors are large positive and the third large negative when x is large positive. $\lim\limits_{x\to-\infty} x^2(x-2)(1-x) = -\infty$ because the first and third factors are large positive and the second large negative as $x \to -\infty$.

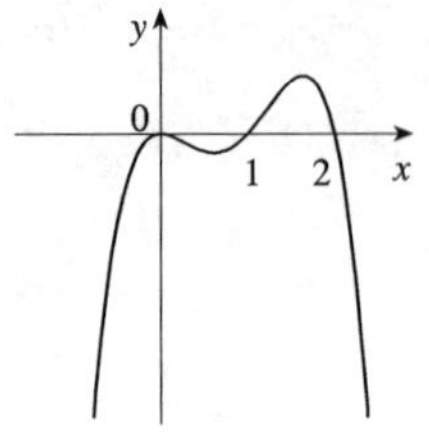

49. $y = f(x) = (x+4)^5(x-3)^4$. The y-intercept is $f(0) = 4^5(-3)^4 = 82{,}944$. The x-intercepts occur when $y = 0 \Rightarrow x = -4, 3$. Notice that the graph does not cross the x-axis at 3 because $(x-3)^4$ is always positive, but does cross the x-axis at -4. $\lim\limits_{x\to\infty}(x+4)^5(x-3)^4 = \infty$ since both factors are large positive when x is large positive. $\lim\limits_{x\to-\infty}(x+4)^5(x-3)^4 = -\infty$ since the first factor is large negative and the second factor is large positive when x is large negative.

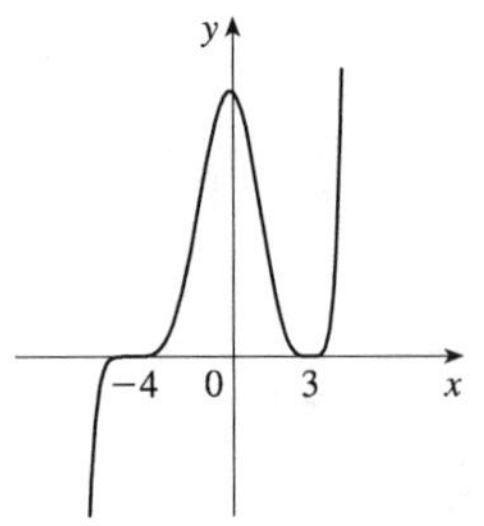

51. First we plot the points which are known to be on the graph: $(2,-1)$ and $(0,0)$. We can also draw a short line segment of slope 0 at $x = 2$, since we are given that $f'(2) = 0$. Now we know that $f'(x) < 0$ (that is, the function is decreasing) on $(0,2)$, and that $f''(x) < 0$ on $(0,1)$ and $f''(x) > 0$ on $(1,2)$. So we must join the points $(0,0)$ and $(2,-1)$ in such a way that the curve is concave down on $(0,1)$ and concave up on $(1,2)$. The curve must be concave up and increasing on $(2,4)$ and concave down and increasing toward $y = 1$ on $(4,\infty)$. Now we just need to reflect the curve in the y-axis, since we are given that f is an even function [the condition that $f(-x) = f(x)$ for all x].

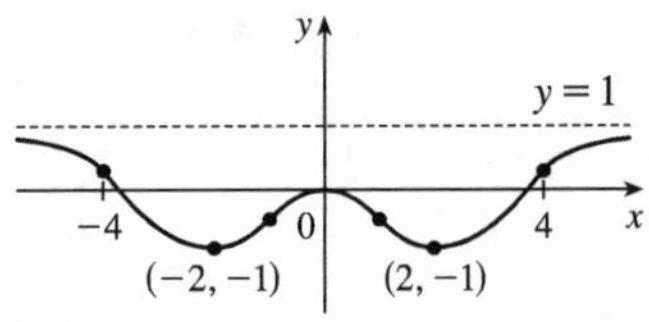

53. We are given that $f(1) = f'(1) = 0$. So we can draw a short horizontal line at the point $(1,0)$ to represent this situation. We are given that $x = 0$ and $x = 2$ are vertical asymptotes, with $\lim\limits_{x\to0} f(x) = -\infty$, $\lim\limits_{x\to2^+} f(x) = \infty$ and $\lim\limits_{x\to2^-} f(x) = -\infty$, so we can draw the parts of the curve which approach these asymptotes.

On the interval $(-\infty,0)$, the graph is concave down, and $f(x) \to \infty$ as $x \to -\infty$. Between the asymptotes the graph is concave down. On the interval $(2,\infty)$ the graph is concave up, and $f(x) \to 0$ as $x \to \infty$, so $y = 0$ is a horizontal asymptote. The diagram shows one possible function satisfying all of the given conditions.

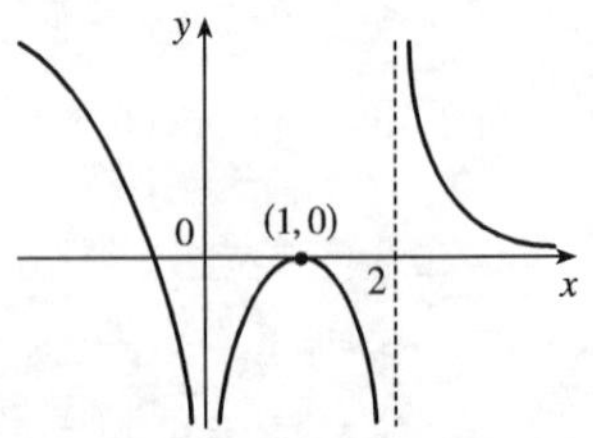

55. (a) Since $-1 \le \sin x \le 1$ for all x, $-\dfrac{1}{x} \le \dfrac{\sin x}{x} \le \dfrac{1}{x}$ for $x > 0$. As $x \to \infty$, $-1/x \to 0$ and $1/x \to 0$, so by the Squeeze Theorem, $(\sin x)/x \to 0$. Thus, $\lim\limits_{x\to\infty} \dfrac{\sin x}{x} = 0$.

(b) From part (a), the horizontal asymptote is $y = 0$. The function $y = (\sin x)/x$ crosses the horizontal asymptote whenever $\sin x = 0$; that is, at $x = \pi n$ for every integer n. Thus, the graph crosses the asymptote *an infinite number of times.*

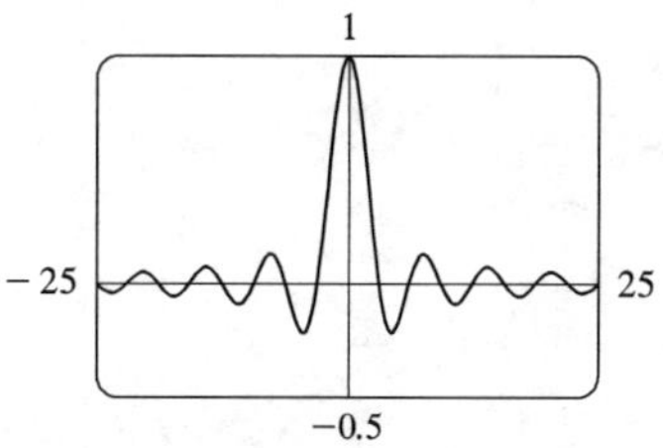

57. Divide the numerator and the denominator by the highest power of x in $Q(x)$.

(a) If $\deg P < \deg Q$, then the numerator $\to 0$ but the denominator doesn't. So $\lim\limits_{x\to\infty} [P(x)/Q(x)] = 0$.

(b) If $\deg P > \deg Q$, then the numerator $\to \pm\infty$ but the denominator doesn't, so $\lim\limits_{x\to\infty} [P(x)/Q(x)] = \pm\infty$ (depending on the ratio of the leading coefficients of P and Q).

59. $\lim\limits_{x\to\infty} \dfrac{4x-1}{x} = \lim\limits_{x\to\infty} \left(4 - \dfrac{1}{x}\right) = 4$, and $\lim\limits_{x\to\infty} \dfrac{4x^2+3x}{x^2} = \lim\limits_{x\to\infty} \left(4 + \dfrac{3}{x}\right) = 4$. Therefore, by the Squeeze Theorem, $\lim\limits_{x\to\infty} f(x) = 4$.

61. $\left|\dfrac{6x^2+5x-3}{2x^2-1} - 3\right| < 0.2 \quad\Leftrightarrow\quad 2.8 < \dfrac{6x^2+5x-3}{2x^2-1} < 3.2$. So we graph the three parts of this inequality on the same screen, and find that the curve $y = \dfrac{6x^2+5x-3}{2x^2-1}$ seems to lie between the lines $y = 2.8$ and $y = 3.2$ whenever $x > 12.8$. So we can choose $N = 13$ (or any larger number) so that the inequality holds whenever $x \ge N$.

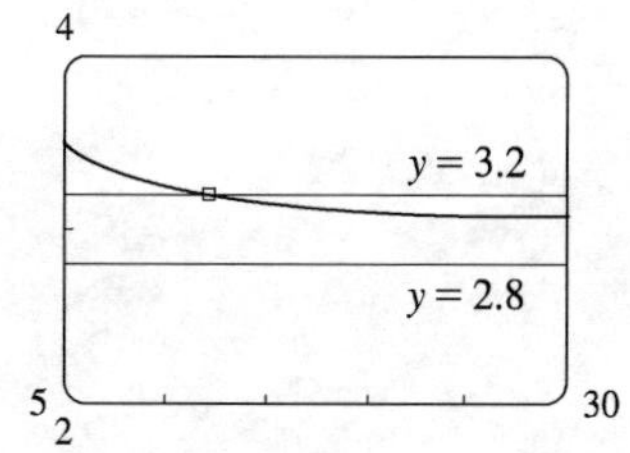

63. For $\varepsilon = 0.5$, we need to find N such that $\left|\dfrac{\sqrt{4x^2+1}}{x+1} - (-2)\right| < 0.5$ $\Leftrightarrow\quad -2.5 < \dfrac{\sqrt{4x^2+1}}{x+1} < -1.5$ whenever $x \le N$. We graph the three parts of this inequality on the same screen, and see that the inequality holds for $x \le -6$. So we choose $N = -6$ (or any smaller number).

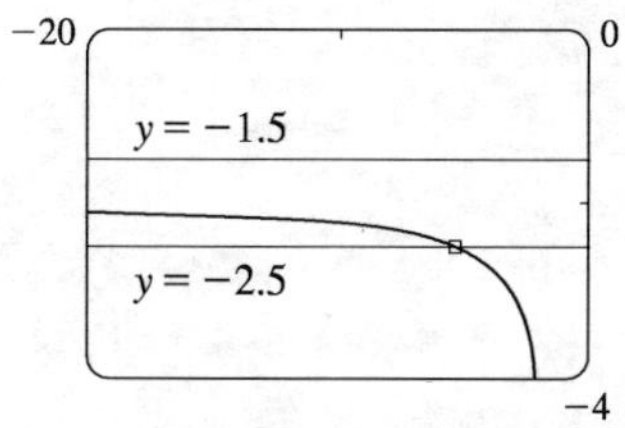

For $\varepsilon = 0.1$, we need $-2.1 < \dfrac{\sqrt{4x^2+1}}{x+1} < -1.9$ whenever $x \le N$. From the graph, it seems that this inequality holds for $x \le -22$. So we choose $N = -22$ (or any smaller number).

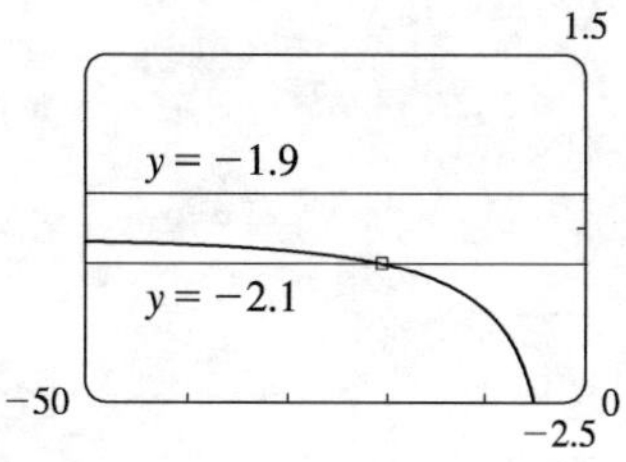

65. (a) $1/x^2 < 0.0001 \quad \Leftrightarrow \quad x^2 > 1/0.0001 = 10{,}000 \quad \Leftrightarrow \quad x > 100 \quad (x > 0)$

(b) If $\varepsilon > 0$ is given, then $1/x^2 < \varepsilon \quad \Leftrightarrow \quad x^2 > 1/\varepsilon \quad \Leftrightarrow \quad x > 1/\sqrt{\varepsilon}$. Let $N = 1/\sqrt{\varepsilon}$.

Then $x > N \quad \Rightarrow \quad x > \dfrac{1}{\sqrt{\varepsilon}} \quad \Rightarrow \quad \left|\dfrac{1}{x^2} - 0\right| = \dfrac{1}{x^2} < \varepsilon$, so $\displaystyle\lim_{x\to\infty} \frac{1}{x^2} = 0$.

67. For $x < 0$, $|1/x - 0| = -1/x$. If $\varepsilon > 0$ is given, then $-1/x < \varepsilon \quad \Leftrightarrow \quad x < -1/\varepsilon$.

Take $N = -1/\varepsilon$. Then $x < N \quad \Rightarrow \quad x < -1/\varepsilon \quad \Rightarrow \quad |(1/x) - 0| = -1/x < \varepsilon$, so $\displaystyle\lim_{x\to-\infty} (1/x) = 0$.

69. Suppose that $\displaystyle\lim_{x\to\infty} f(x) = L$. Then for every $\varepsilon > 0$ there is a corresponding positive number N such that $|f(x) - L| < \varepsilon$ whenever $x > N$. If $t = 1/x$, then $x > N \quad \Leftrightarrow \quad 0 < 1/x < 1/N \quad \Leftrightarrow \quad 0 < t < 1/N$. Thus, for every $\varepsilon > 0$ there is a corresponding $\delta > 0$ (namely $1/N$) such that $|f(1/t) - L| < \varepsilon$ whenever $0 < t < \delta$. This proves that $\displaystyle\lim_{t\to0^+} f(1/t) = L = \lim_{x\to\infty} f(x)$.

Now suppose that $\displaystyle\lim_{x\to-\infty} f(x) = L$. Then for every $\varepsilon > 0$ there is a corresponding negative number N such that $|f(x) - L| < \varepsilon$ whenever $x < N$. If $t = 1/x$, then $x < N \quad \Leftrightarrow \quad 1/N < 1/x < 0 \quad \Leftrightarrow \quad 1/N < t < 0$. Thus, for every $\varepsilon > 0$ there is a corresponding $\delta > 0$ (namely $-1/N$) such that $|f(1/t) - L| < \varepsilon$ whenever $-\delta < t < 0$. This proves that $\displaystyle\lim_{t\to0^-} f(1/t) = L = \lim_{x\to-\infty} f(x)$.

Review

CONCEPT CHECK

1. (a) $\lim_{x \to a} f(x) = L$: See Definition 2.2.1 and Figures 1 and 2 in Section 2.2.

(b) $\lim_{x \to a^+} f(x) = L$: See the paragraph after Definition 2.2.2 and Figure 9(b) in Section 2.2.

(c) $\lim_{x \to a^-} f(x) = L$: See Definition 2.2.2 and Figure 9(a) in Section 2.2.

(d) $\lim_{x \to a} f(x) = \infty$: See Definition 2.2.4 and Figure 12 in Section 2.2.

(e) $\lim_{x \to a} f(x) = -\infty$: See Definition 2.2.5 and Figure 13 in Section 2.2.

2. In general, the limit of a function fails to exist when the function does not approach a fixed number. For each of the following functions, the limit fails to exist at $x = 2$.

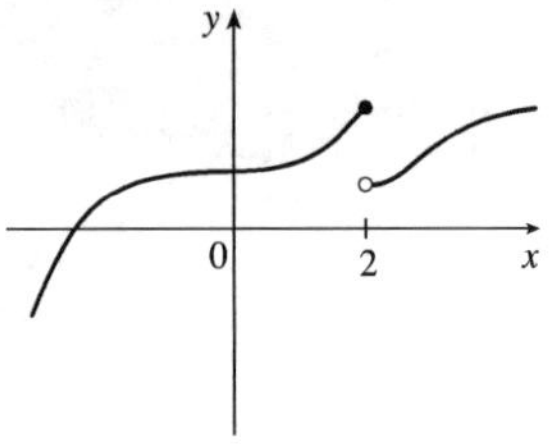

The left- and right-hand limits are not equal.

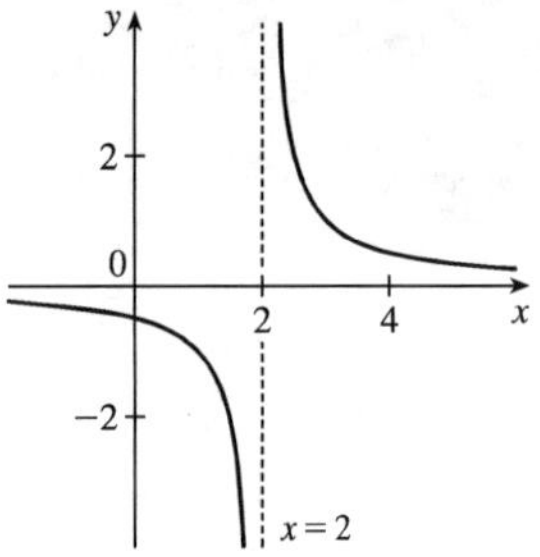

There is an infinite discontinuity.

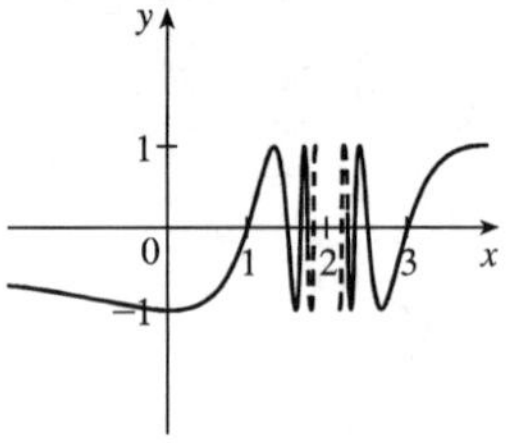

There are an infinite number of oscillations.

3. See Definition 2.2.6 and Figure 14 in Section 2.2.

4. (a)–(g) See the statements of Limit Laws 1–6 and 11 in Section 2.3.

5. See Theorem 3 in Section 2.3.

6. (a) A function f is continuous at a number a if $f(x)$ approaches $f(a)$ as x approaches a; that is, $\lim_{x \to a} f(x) = f(a)$.

(b) A function f is continuous on the interval $(-\infty, \infty)$ if f is continuous at every real number a. The graph of such a function has no breaks and every vertical line crosses it.

7. See Theorem 2.5.10.

8. See Definition 2.6.1.

9. See the paragraph containing Formula 3 in Section 2.6.

10. (a) The average rate of change of y with respect to x over the interval $[x_1, x_2]$ is $\dfrac{f(x_2) - f(x_1)}{x_2 - x_1}$.

(b) The instantaneous rate of change of y with respect to x at $x = x_1$ is $\lim_{x_2 \to x_1} \dfrac{f(x_2) - f(x_1)}{x_2 - x_1}$.

TRUE-FALSE QUIZ

1. False. Limit Law 2 applies only if the individual limits exist (these don't).

3. True. Limit Law 5 applies.

5. False. Consider $\lim_{x \to 5} \dfrac{x(x-5)}{x-5}$ or $\lim_{x \to 5} \dfrac{\sin(x-5)}{x-5}$. The first limit exists and is equal to 5. By Example 3 in Section 2.2, we know that the latter limit exists (and it is equal to 1).

7. True. A polynomial is continuous everywhere, so $\lim_{x \to b} p(x)$ exists and is equal to $p(b)$.

9. False. Consider $f(x) = \begin{cases} 1/(x-1) & \text{if } x \neq 1 \\ 2 & \text{if } x = 1 \end{cases}$

11. True. Use Theorem 2.5.8 with $a = 2$, $b = 5$, and $g(x) = 4x^2 - 11$. Note that $f(4) = 3$ is not needed.

13. True, by the definition of a limit with $\varepsilon = 1$.

EXERCISES

1. (a) (i) $\lim_{x \to 2^+} f(x) = 3$ (ii) $\lim_{x \to -3^+} f(x) = 0$

(iii) $\lim_{x \to -3} f(x)$ does not exist since the left and right limits are not equal. (The left limit is -2.) (iv) $\lim_{x \to 4} f(x) = 2$

(v) $\lim_{x \to 0} f(x) = \infty$ (vi) $\lim_{x \to 2^-} f(x) = -\infty$

(b) The equations of the vertical asymptotes are $x = 0$ and $x = 2$.

(c) f is discontinuous at $x = -3$, 0, 2, and 4. The discontinuities are jump, infinite, infinite, and removable, respectively.

3. $\lim_{x \to 0} \cos(x + \sin x) = \cos\left[\lim_{x \to 0}(x + \sin x)\right]$ [by Theorem 2.5.8] $= \cos 0 = 1$

5. $$\lim_{x \to -3} \frac{x^2 - 9}{x^2 + 2x - 3} = \lim_{x \to -3} \frac{(x+3)(x-3)}{(x+3)(x-1)} = \lim_{x \to -3} \frac{x-3}{x-1} = \frac{-3-3}{-3-1} = \frac{-6}{-4} = \frac{3}{2}$$

7. $$\lim_{h \to 0} \frac{(h-1)^3 + 1}{h} = \lim_{h \to 0} \frac{(h^3 - 3h^2 + 3h - 1) + 1}{h} = \lim_{h \to 0} \frac{h^3 - 3h^2 + 3h}{h} = \lim_{h \to 0} \left(h^2 - 3h + 3\right) = 3$$

Another solution: Factor the numerator as a sum of two cubes and then simplify.

$$\lim_{h \to 0} \frac{(h-1)^3 + 1}{h} = \lim_{h \to 0} \frac{(h-1)^3 + 1^3}{h} = \lim_{h \to 0} \frac{[(h-1) + 1]\left[(h-1)^2 - 1(h-1) + 1^2\right]}{h}$$
$$= \lim_{h \to 0} \left[(h-1)^2 - h + 2\right] = 1 - 0 + 2 = 3$$

9. $\lim_{r \to 9} \dfrac{\sqrt{r}}{(r-9)^4} = \infty$ since $(r-9)^4 \to 0$ as $r \to 9$ and $\dfrac{\sqrt{r}}{(r-9)^4} > 0$ for $r \neq 9$.

11. $$\lim_{s \to 16} \frac{4 - \sqrt{s}}{s - 16} = \lim_{s \to 16} \frac{4 - \sqrt{s}}{(\sqrt{s} + 4)(\sqrt{s} - 4)} = \lim_{s \to 16} \frac{-1}{\sqrt{s} + 4} = \frac{-1}{\sqrt{16} + 4} = -\frac{1}{8}$$

13. $$\frac{|x-8|}{x-8} = \begin{cases} \dfrac{x-8}{x-8} & \text{if } x - 8 > 0 \\ \dfrac{-(x-8)}{x-8} & \text{if } x - 8 < 0 \end{cases} = \begin{cases} 1 & \text{if } x > 8 \\ -1 & \text{if } x < 8 \end{cases}$$

Thus, $\lim_{x \to 8^-} \dfrac{|x-8|}{x-8} = \lim_{x \to 8^-} (-1) = -1$.

15. $$\lim_{x \to 0} \frac{1 - \sqrt{1 - x^2}}{x} \cdot \frac{1 + \sqrt{1 - x^2}}{1 + \sqrt{1 - x^2}} = \lim_{x \to 0} \frac{1 - (1 - x^2)}{x\left(1 + \sqrt{1 - x^2}\right)} = \lim_{x \to 0} \frac{x^2}{x\left(1 + \sqrt{1 - x^2}\right)} = \lim_{x \to 0} \frac{x}{1 + \sqrt{1 - x^2}} = 0$$

17. Since $2x - 1 \leq f(x) \leq x^2$ for $0 < x < 3$ and $\lim_{x \to 1} (2x - 1) = 1 = \lim_{x \to 1} x^2$, we have $\lim_{x \to 1} f(x) = 1$ by the Squeeze Theorem.

19. Given $\varepsilon > 0$, we need $\delta > 0$ so that if $0 < |x - 5| < \delta$, then $|(7x - 27) - 8| < \varepsilon \Leftrightarrow |7x - 35| < \varepsilon \Leftrightarrow |x - 5| < \varepsilon/7$. So take $\delta = \varepsilon/7$. Then $0 < |x - 5| < \delta \Rightarrow |(7x - 27) - 8| < \varepsilon$. Thus, $\lim_{x \to 5}(7x - 27) = 8$ by the definition of a limit.

21. Given $\varepsilon > 0$, we need $\delta > 0$ so that if $0 < |x - 2| < \delta$, then $\left|x^2 - 3x - (-2)\right| < \varepsilon$. First, note that if $|x - 2| < 1$, then $-1 < x - 2 < 1$, so $0 < x - 1 < 2 \;\Rightarrow\; |x - 1| < 2$. Now let $\delta = \min\{\varepsilon/2, 1\}$. Then $0 < |x - 2| < \delta$ $\Rightarrow\; \left|x^2 - 3x - (-2)\right| = |(x - 2)(x - 1)| = |x - 2|\,|x - 1| < (\varepsilon/2)(2) = \varepsilon$.

Thus, $\lim\limits_{x\to 2}(x^2 - 3x) = -2$ by the definition of a limit.

23. (a) $f(x) = \sqrt{-x}$ if $x < 0$, $f(x) = 3 - x$ if $0 \le x < 3$, $f(x) = (x - 3)^2$ if $x > 3$.

(i) $\lim\limits_{x\to 0^+} f(x) = \lim\limits_{x\to 0^+}(3 - x) = 3$

(ii) $\lim\limits_{x\to 0^-} f(x) = \lim\limits_{x\to 0^-}\sqrt{-x} = 0$

(iii) Because of (i) and (ii), $\lim\limits_{x\to 0} f(x)$ does not exist.

(iv) $\lim\limits_{x\to 3^-} f(x) = \lim\limits_{x\to 3^-}(3 - x) = 0$

(v) $\lim\limits_{x\to 3^+} f(x) = \lim\limits_{x\to 3^+}(x - 3)^2 = 0$

(vi) Because of (iv) and (v), $\lim\limits_{x\to 3} f(x) = 0$.

(b) f is discontinuous at 0 since $\lim\limits_{x\to 0} f(x)$ does not exist.

f is discontinuous at 3 since $f(3)$ does not exist.

(c)

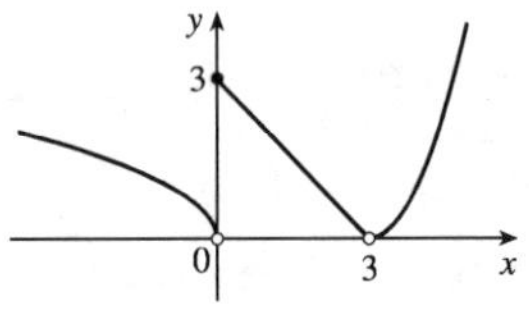

25. x^3 is continuous on $\mathbb{R}$ since it is a polynomial and $\cos x$ is also continuous on $\mathbb{R}$, so the product $x^3\cos x$ is continuous on $\mathbb{R}$. The root function $\sqrt[4]{x}$ is continuous on its domain, $[0, \infty)$, and so the sum, $h(x) = \sqrt[4]{x} + x^3\cos x$, is continuous on its domain, $[0, \infty)$.

27. $f(x) = 2x^3 + x^2 + 2$ is a polynomial, so it is continuous on $[-2, -1]$ and $f(-2) = -10 < 0 < 1 = f(-1)$. So by the Intermediate Value Theorem there is a number c in $(-2, -1)$ such that $f(c) = 0$, that is, the equation $2x^3 + x^2 + 2 = 0$ has a root in $(-2, -1)$.

29. (a) The slope of the tangent line at $(2, 1)$ is

$$\lim_{x\to 2}\frac{f(x) - f(2)}{x - 2} = \lim_{x\to 2}\frac{9 - 2x^2 - 1}{x - 2} = \lim_{x\to 2}\frac{8 - 2x^2}{x - 2} = \lim_{x\to 2}\frac{-2(x^2 - 4)}{x - 2} = \lim_{x\to 2}\frac{-2(x - 2)(x + 2)}{x - 2}$$
$$= \lim_{x\to 2}[-2(x + 2)] = -2\cdot 4 = -8$$

(b) An equation of this tangent line is $y - 1 = -8(x - 2)$ or $y = -8x + 17$.

31. (a) $s = s(t) = 1 + 2t + t^2/4$. The average velocity over the time interval $[1, 1 + h]$ is

$$v_{\text{ave}} = \frac{s(1 + h) - s(1)}{(1 + h) - 1} = \frac{1 + 2(1 + h) + (1 + h)^2/4 - 13/4}{h} = \frac{10h + h^2}{4h} = \frac{10 + h}{4}.$$

So for the following intervals the average velocities are:

(i) $[1, 3]$: $h = 2$, $v_{\text{ave}} = (10 + 2)/4 = 3$ m/s

(ii) $[1, 2]$: $h = 1$, $v_{\text{ave}} = (10 + 1)/4 = 2.75$ m/s

(iii) $[1, 1.5]$: $h = 0.5$, $v_{\text{ave}} = (10 + 0.5)/4 = 2.625$ m/s

(iv) $[1, 1.1]$: $h = 0.1$, $v_{\text{ave}} = (10 + 0.1)/4 = 2.525$ m/s

(b) When $t = 1$, the instantaneous velocity is $\lim\limits_{h\to 0}\dfrac{s(1 + h) - s(1)}{h} = \lim\limits_{h\to 0}\dfrac{10 + h}{4} = \dfrac{10}{4} = 2.5$ m/s.

33. $\left|\dfrac{x+1}{x-1}-3\right|<0.2 \;\Rightarrow\; -0.2<\dfrac{x+1}{x-1}-3<0.2 \;\Rightarrow$

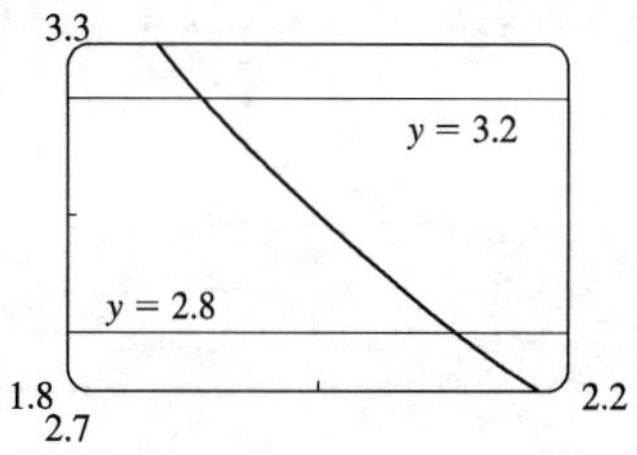

$2.8<\dfrac{x+1}{x-1}<3.2$. Graphing the functions $y=2.8$, $y=(x+1)/(x-1)$, and $y=3.2$ on the interval $[1.8, 2.2]$, we see that the inequality holds whenever $1.91<x<2.11$ (approximately). So since $|2-1.91|=0.09$ and $|2-2.15|=0.15$, any positive $\delta\leq 0.09$ will do.

35. $|f(x)|\leq g(x) \;\Leftrightarrow\; -g(x)\leq f(x)\leq g(x)$ and $\lim\limits_{x\to a} g(x)=0=\lim\limits_{x\to a} -g(x)$. Thus, by the Squeeze Theorem, $\lim\limits_{x\to a} f(x)=0$.

□ LIMITS AND RATES OF CHANGE

The Tangent and Velocity Problems

1. (a) Using $P(15, 250)$, we construct the following table:

t	Q	slope $= m_{PQ}$
5	$(5, 694)$	$\frac{694-250}{5-15} = -\frac{444}{10} = -44.4$
10	$(10, 444)$	$\frac{444-250}{10-15} = -\frac{194}{5} = -38.8$
20	$(20, 111)$	$\frac{111-250}{20-15} = -\frac{139}{5} = -27.8$
25	$(25, 28)$	$\frac{28-250}{25-15} = -\frac{222}{10} = -22.2$
30	$(30, 0)$	$\frac{0-250}{30-15} = -\frac{250}{15} = -16.\overline{6}$

(b) Using the values of t that correspond to the points closest to P ($t = 10$ and $t = 20$), we have

$$\frac{-38.8 + (-27.8)}{2} = -33.3$$

(c) From the graph, we can estimate the slope of the tangent line at P to be $\frac{-300}{9} = -33.\overline{3}$.

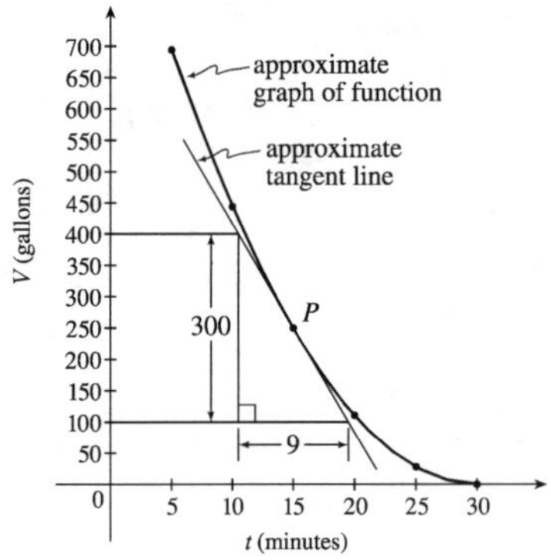

3. For the curve $y = x/(1 + x)$ and the point $P\left(1, \frac{1}{2}\right)$:

(a)

	x	Q	m_{PQ}
(i)	0.5	$(0.5, 0.333333)$	0.333333
(ii)	0.9	$(0.9, 0.473684)$	0.263158
(iii)	0.99	$(0.99, 0.497487)$	0.251256
(iv)	0.999	$(0.999, 0.499750)$	0.250125
(v)	1.5	$(1.5, 0.6)$	0.2
(vi)	1.1	$(1.1, 0.523810)$	0.238095
(vii)	1.01	$(1.01, 0.502488)$	0.248756
(viii)	1.001	$(1.001, 0.500250)$	0.249875

(b) The slope appears to be $\frac{1}{4}$.

(c) $y - \frac{1}{2} = \frac{1}{4}(x - 1)$ or $y = \frac{1}{4}x + \frac{1}{4}$.

5. (a) $y = y(t) = 40t - 16t^2$. At $t = 2$, $y = 40(2) - 16(2)^2 = 16$. The average velocity between times 2 and $2 + h$ is $v_{\text{ave}} = \dfrac{y(2+h) - y(2)}{(2+h) - 2} = \dfrac{\left[40(2+h) - 16(2+h)^2\right] - 16}{h} = \dfrac{-24h - 16h^2}{h} = -24 - 16h$, if $h \neq 0$.

(i) $[2, 2.5]$: $h = 0.5$, $v_{\text{ave}} = -32$ ft/s (ii) $[2, 2.1]$: $h = 0.1$, $v_{\text{ave}} = -25.6$ ft/s

(iii) $[2, 2.05]$: $h = 0.05$, $v_{\text{ave}} = -24.8$ ft/s (iv) $[2, 2.01]$: $h = 0.01$, $v_{\text{ave}} = -24.16$ ft/s

(b) The instantaneous velocity when $t = 2$ (h approaches 0) is -24 ft/s.

7. $s = s(t) = t^3/6$. Average velocity between times 1 and $1 + h$ is

$$v_{\text{ave}} = \frac{s(1+h) - s(1)}{(1+h) - 1} = \frac{(1+h)^3/6 - 1/6}{h} = \frac{h^3 + 3h^2 + 3h}{6h} = \frac{h^2 + 3h + 3}{6} \text{ if } h \neq 0.$$

(a) (i) $[1, 3]$: $h = 2$, $v_{\text{ave}} = \frac{13}{6}$ ft/s (ii) $[1, 2]$: $h = 1$, $v_{\text{ave}} = \frac{7}{6}$ ft/s

(iii) $[1, 1.5]$: $h = 0.5$, $v_{\text{ave}} = \frac{19}{24}$ ft/s (iv) $[1, 1.1]$: $h = 0.1$, $v_{\text{ave}} = \frac{331}{600}$ ft/s

(b) As h approaches 0, the velocity approaches $\frac{3}{6} = \frac{1}{2}$ ft/s.

(c)

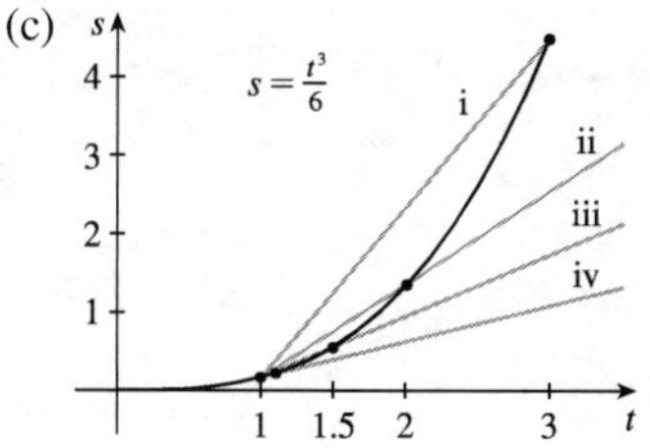

(d)

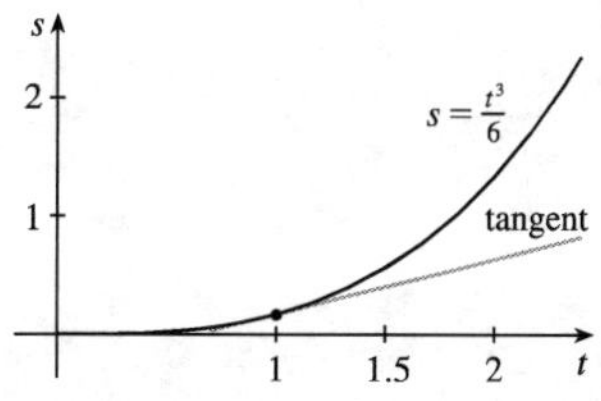

9. For the curve $y = \sin(10\pi/x)$ and the point $P(1, 0)$:

(a)

x	Q	m_{PQ}
2	$(2, 0)$	0
1.5	$(1.5, 0.8660)$	1.7321
1.4	$(1.4, -0.4339)$	-1.0847
1.3	$(1.3, -0.8230)$	-2.7433
1.2	$(1.2, 0.8660)$	4.3301
1.1	$(1.1, -0.2817)$	-2.8173

x	Q	m_{PQ}
0.5	$(0.5, 0)$	0
0.6	$(0.6, 0.8660)$	-2.1651
0.7	$(0.7, 0.7818)$	-2.6061
0.8	$(0.8, 1)$	-5
0.9	$(0.9, -0.3420)$	3.4202

As x approaches 1, the slopes do not appear to be approaching any particular value.

(b)

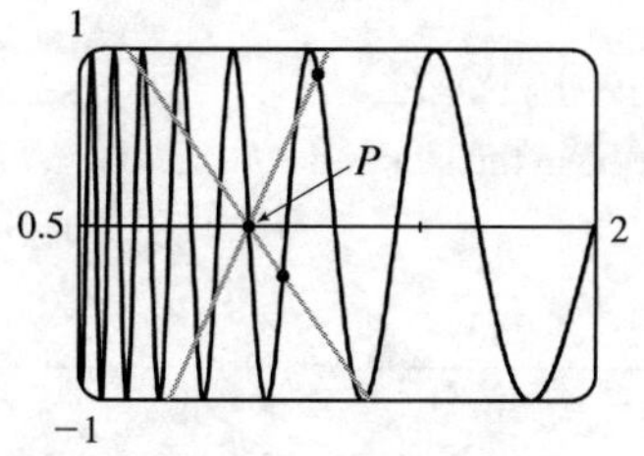

We see that problems with estimation are caused by the frequent oscillations of the graph. The tangent is so steep at P that we need to take x-values much closer to 1 in order to get accurate estimates of its slope.

(c) If we choose $x = 1.001$, then the point Q is $(1.001, -0.0314)$ and $m_{PQ} \approx -31.3794$. If $x = 0.999$, then Q is $(0.999, 0.0314)$ and $m_{PQ} = -31.4422$. The average of these slopes is -31.4108. So we estimate that the slope of the tangent line at P is about -31.4.

Tangents, Velocities, and Other Rates of Change

1. (a) This is just the slope of the line through two points: $m_{PQ} = \dfrac{\Delta y}{\Delta x} = \dfrac{f(x) - f(3)}{x - 3}$.

(b) This is the limit of the slope of the secant line PQ as Q approaches P: $m = \lim\limits_{x \to 3} \dfrac{f(x) - f(3)}{x - 3}$.

3. The slope at D is the largest positive slope, followed by the positive slope at E. The slope at C is zero. The slope at B is steeper than at A (both are negative). In decreasing order, we have the slopes at: D, E, C, A, and B.

5. (a) (i) Using Definition 1,

$$m = \lim_{x \to a} \frac{f(x) - f(a)}{x - a} \quad \lim_{x \to -3} \frac{f(x) - f(-3)}{x - (-3)} = \lim_{x \to -3} \frac{(x^2 + 2x) - (3)}{x - (-3)} = \lim_{x \to -3} \frac{(x + 3)(x - 1)}{x + 3}$$

$$= \lim_{x \to -3} (x - 1) = -4$$

(ii) Using Equation 2,

$$m = \lim_{h \to 0} \frac{f(a + h) - f(a)}{h} = \lim_{h \to 0} \frac{f(-3 + h) - f(-3)}{h} = \lim_{h \to 0} \frac{[(-3 + h)^2 + 2(-3 + h)] - (3)}{h}$$

$$= \lim_{h \to 0} \frac{9 - 6h + h^2 - 6 + 2h - 3}{h} = \lim_{h \to 0} \frac{h(h - 4)}{h} = \lim_{h \to 0} (h - 4) = -4$$

(b) Using the point-slope form of the equation of a line, an equation of the tangent line is $y - 3 = -4(x + 3)$. Solving for y gives us $y = -4x - 9$, which is the slope-intercept form of the equation of the tangent line.

(c)

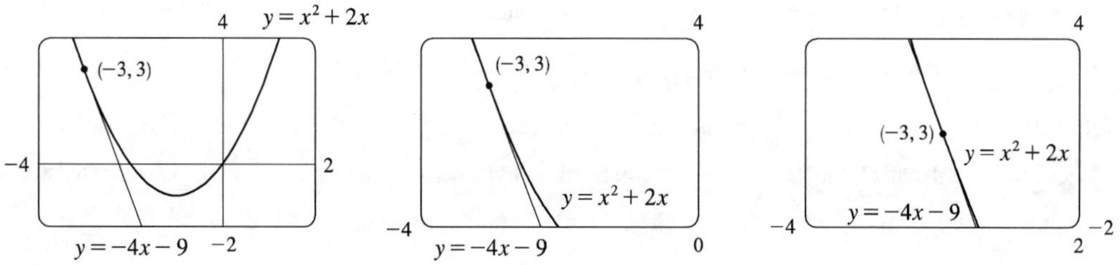

7. Using (2) with $f(x) = 1 + 2x - x^3$ and $P(1, 2)$,

$$m = \lim_{h\to 0} \frac{f(a+h) - f(a)}{h} = \lim_{h\to 0} \frac{f(1+h) - f(1)}{h} = \lim_{h\to 0} \frac{\left[1 + 2(1+h) - (1+h)^3\right] - 2}{h}$$

$$= \lim_{h\to 0} \frac{1 + 2 + 2h - (1 + 3h + 3h^2 + h^3) - 2}{h} = \lim_{h\to 0} \frac{-h^3 - 3h^2 - h}{h}$$

$$= \lim_{h\to 0} \frac{h(-h^2 - 3h - 1)}{h} = \lim_{h\to 0}(-h^2 - 3h - 1) = -1$$

Tangent line: $y - 2 = -1(x - 1) \quad\Leftrightarrow\quad y - 2 = -x + 1 \quad\Leftrightarrow\quad y = -x + 3$

9. Using (1) with $f(x) = \dfrac{x-1}{x-2}$ and $P(3, 2)$,

$$m = \lim_{x\to a} \frac{f(x) - f(a)}{x - a} = \lim_{x\to 3} \frac{\dfrac{x-1}{x-2} - 2}{x - 3} = \lim_{x\to 3} \frac{\dfrac{x - 1 - 2(x-2)}{x-2}}{x - 3} = \lim_{x\to 3} \frac{3 - x}{(x-2)(x-3)}$$

$$= \lim_{x\to 3} \frac{-1}{x-2} = \frac{-1}{1} = -1.$$

Tangent line: $y - 2 = -1(x - 3) \quad\Leftrightarrow\quad y - 2 = -x + 3 \quad\Leftrightarrow\quad y = -x + 5$

11. (a) $m = \displaystyle\lim_{x\to a} \frac{f(x) - f(a)}{x - a} = \lim_{x\to a} \frac{2/(x+3) - 2/(a+3)}{x - a} = \lim_{x\to a} \frac{2(a+3) - 2(x+3)}{(x-a)(x+3)(a+3)}$

$$= \lim_{x\to a} \frac{2(a - x)}{(x-a)(x+3)(a+3)} = \lim_{x\to a} \frac{-2}{(x+3)(a+3)} = \frac{-2}{(a+3)^2}$$

(b) (i) $a = -1 \;\Rightarrow\; m = \dfrac{-2}{(-1+3)^2} = -\dfrac{1}{2}$ (ii) $a = 0 \;\Rightarrow\; m = \dfrac{-2}{(0+3)^2} = -\dfrac{2}{9}$

(iii) $a = 1 \;\Rightarrow\; m = \dfrac{-2}{(1+3)^2} = -\dfrac{1}{8}$

13. (a) Using (1),

$$m = \lim_{x\to a} \frac{(x^3 - 4x + 1) - (a^3 - 4a + 1)}{x - a} = \lim_{x\to a} \frac{(x^3 - a^3) - 4(x - a)}{x - a}$$

$$= \lim_{x\to a} \frac{(x-a)(x^2 + ax + a^2) - 4(x - a)}{x - a} = \lim_{x\to a}(x^2 + ax + a^2 - 4) = 3a^2 - 4$$

(b) At $(1, -2)$: $m = 3(1)^2 - 4 = -1$, so an equation of the tangent line is $y - (-2) = -1(x - 1) \quad\Leftrightarrow\quad y = -x - 1$. At $(2, 1)$: $m = 3(2)^2 - 4 = 8$, so an equation of the tangent line is $y - 1 = 8(x - 2) \quad\Leftrightarrow\quad y = 8x - 15$.

(c)

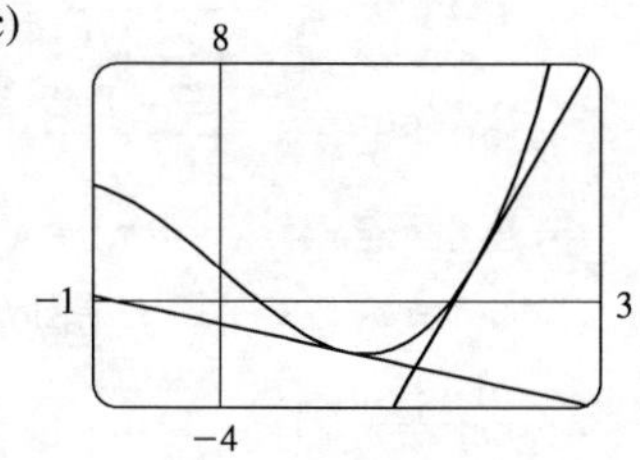

15. (a) Since the slope of the tangent at $t = 0$ is 0, the car's initial velocity was 0.

(b) The slope of the tangent is greater at C than at B, so the car was going faster at C.

(c) Near A, the tangent lines are becoming steeper as x increases, so the velocity was increasing, so the car was speeding up. Near B, the tangent lines are becoming less steep, so the car was slowing down. The steepest tangent near C is the one at C, so at C the car had just finished speeding up, and was about to start slowing down.

(d) Between D and E, the slope of the tangent is 0, so the car did not move during that time.

17. Let $s(t) = 40t - 16t^2$.

$$v(2) = \lim_{t\to 2} \frac{s(t) - s(2)}{t - 2} = \lim_{t\to 2} \frac{(40t - 16t^2) - 16}{t - 2} = \lim_{t\to 2} \frac{-16t^2 + 40t - 16}{t - 2} = \lim_{t\to 2} \frac{-8(2t^2 - 5t + 2)}{t - 2}$$

$$= \lim_{t\to 2} \frac{-8(t - 2)(2t - 1)}{t - 2} = -8 \lim_{t\to 2}(2t - 1) = -8(3) = -24$$

Thus, the instantaneous velocity when $t = 2$ is -24 ft/s.

19. $$v(a) = \lim_{h\to 0} \frac{s(a + h) - s(a)}{h} = \lim_{h\to 0} \frac{4(a + h)^3 + 6(a + h) + 2 - (4a^3 + 6a + 2)}{h}$$

$$= \lim_{h\to 0} \frac{4a^3 + 12a^2h + 12ah^2 + 4h^3 + 6a + 6h + 2 - 4a^3 - 6a - 2}{h}$$

$$= \lim_{h\to 0} \frac{12a^2h + 12ah^2 + 4h^3 + 6h}{h} = \lim_{h\to 0} \left(12a^2 + 12ah + 4h^2 + 6\right) = \left(12a^2 + 6\right) \text{ m/s}$$

So $v(1) = 12(1)^2 + 6 = 18$ m/s, $v(2) = 12(2)^2 + 6 = 54$ m/s, and $v(3) = 12(3)^2 + 6 = 114$ m/s.

21. The sketch shows the graph for a room temperature of 72° and a refrigerator temperature of 38°. The initial rate of change is greater in magnitude than the rate of change after an hour.

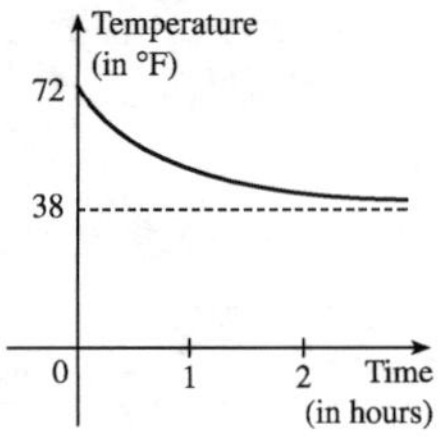

23. (a) (i) $[20, 23]$: $\dfrac{7.9 - 11.5}{23 - 20} = -1.2$ °C/h

(ii) $[20, 22]$: $\dfrac{9.0 - 11.5}{22 - 20} = -1.25$ °C/h

(iii) $[20, 21]$: $\dfrac{10.2 - 11.5}{21 - 20} = -1.3$ °C/h

(b) In the figure, we estimate A to be $(18, 15.5)$ and B as $(23, 6)$. So the slope is $\dfrac{6 - 15.5}{23 - 18} = -1.9$ °C/h at 8:00 P.M.

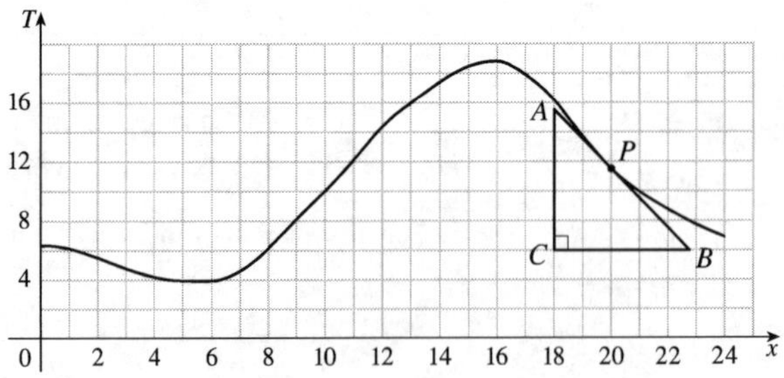

25. (a) (i) $[1995, 1997]$: $\dfrac{N(1997) - N(1995)}{1997 - 1995} = \dfrac{2461 - 873}{2} = \dfrac{1588}{2} = 794$ thousand/year

(ii) $[1995, 1996]$: $\dfrac{N(1996) - N(1995)}{1996 - 1995} = \dfrac{1513 - 873}{1} = 640$ thousand/year

(iii) $[1994, 1995]$: $\dfrac{N(1995) - N(1994)}{1995 - 1994} = \dfrac{873 - 572}{1} = 301$ thousand/year

(b) Using the values from (ii) and (iii), we have $\dfrac{640 + 301}{2} = \dfrac{941}{2} = 470.5$ thousand/year.

(c) Estimating A as $(1994, 420)$ and B as $(1996, 1275)$, the slope at 1995 is $\dfrac{1275 - 420}{1996 - 1994} = \dfrac{855}{2} = 427.5$ thousand/year

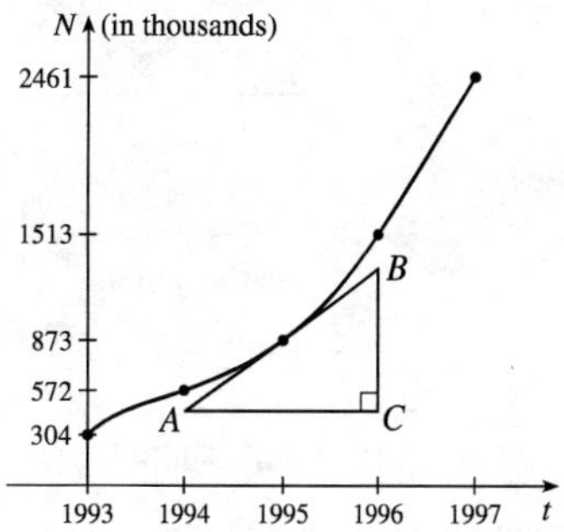

27. (a) (i) $\dfrac{\Delta C}{\Delta x} = \dfrac{C(105) - C(100)}{105 - 100} = \dfrac{6601.25 - 6500}{5} = \$20.25/\text{unit}.$

(ii) $\dfrac{\Delta C}{\Delta x} = \dfrac{C(101) - C(100)}{101 - 100} = \dfrac{6520.05 - 6500}{1} = \$20.05/\text{unit}.$

(b) $\dfrac{C(100 + h) - C(100)}{h} = \dfrac{\left[5000 + 10(100 + h) + 0.05(100 + h)^2\right] - 6500}{h} = \dfrac{20h + 0.05h^2}{h}$

$= 20 + 0.05h,\ h \neq 0$

So the instantaneous rate of change is $\lim\limits_{h \to 0} \dfrac{C(100 + h) - C(100)}{h} = \lim\limits_{h \to 0} (20 + 0.05h) = \$20/\text{unit}.$

□ DERIVATIVES

Derivatives

1. 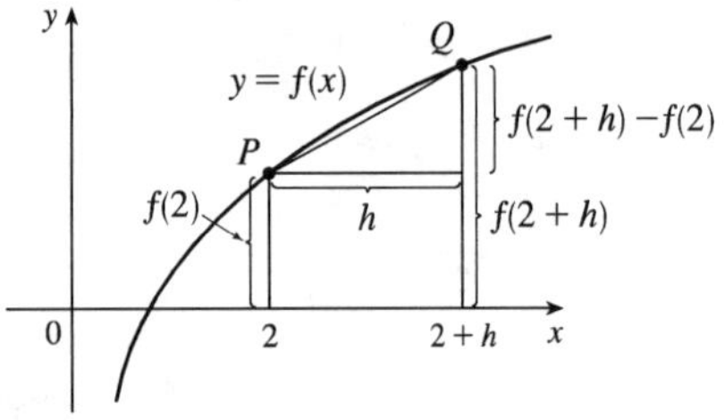

The line from $P(2, f(2))$ to $Q(2+h, f(2+h))$ is the line that has slope $\dfrac{f(2+h)-f(2)}{h}$.

3. $g'(0)$ is the only negative value. The slope at $x = 4$ is smaller than the slope at $x = 2$ and both are smaller than the slope at $x = -2$. Thus, $g'(0) < 0 < g'(4) < g'(2) < g'(-2)$.

5. We begin by drawing a curve through the origin at a slope of 3 to satisfy $f(0) = 0$ and $f'(0) = 3$. Since $f'(1) = 0$, we will round off our figure so that there is a horizontal tangent directly over $x = 1$. Lastly, we make sure that the curve has a slope of -1 as we pass over $x = 2$. Two of the many possibilities are shown.

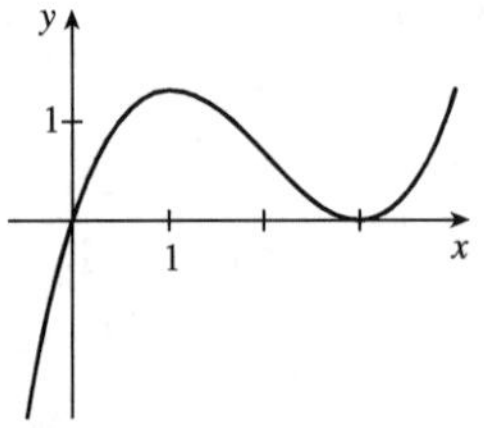

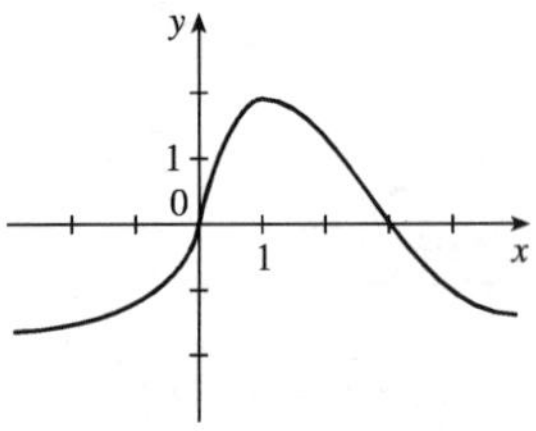

7. Using Definition 2 with $f(x) = 3x^2 - 5x$ and the point $(2, 2)$, we have

$$f'(2) = \lim_{h\to 0}\frac{f(2+h)-f(2)}{h} = \lim_{h\to 0}\frac{\left[3(2+h)^2 - 5(2+h)\right] - 2}{h}$$
$$= \lim_{h\to 0}\frac{(12 + 12h + 3h^2 - 10 - 5h) - 2}{h} = \lim_{h\to 0}\frac{3h^2 + 7h}{h} = \lim_{h\to 0}(3h+7) = 7.$$

So an equation of the tangent line at $(2, 2)$ is $y - 2 = 7(x - 2)$ or $y = 7x - 12$.

9. (a) Using Definition 2 with $F(x) = x^3 - 5x + 1$ and the point $(1, -3)$, we have

$$F'(1) = \lim_{h\to 0}\frac{F(1+h)-F(1)}{h} = \lim_{h\to 0}\frac{\left[(1+h)^3 - 5(1+h) + 1\right] - (-3)}{h}$$
$$= \lim_{h\to 0}\frac{(1 + 3h + 3h^2 + h^3 - 5 - 5h + 1) + 3}{h} = \lim_{h\to 0}\frac{h^3 + 3h^2 - 2h}{h}$$
$$= \lim_{h\to 0}\frac{h(h^2 + 3h - 2)}{h} = \lim_{h\to 0}(h^2 + 3h - 2) = -2$$

So an equation of the tangent line at $(1, -3)$ is $y - (-3) = -2(x - 1) \;\Leftrightarrow\; y = -2x - 1$.

(b)

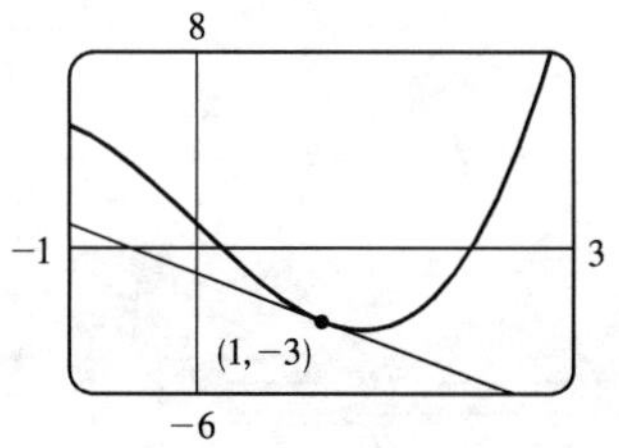

11. (a) $f'(1) = \lim_{h \to 0} \frac{f(1+h) - f(1)}{h} = \lim_{h \to 0} \frac{3^{1+h} - 3^1}{h}$.

So let $F(h) = \frac{3^{1+h} - 3}{h}$. We calculate:

h	$F(h)$	h	$F(h)$
0.1	3.484	−0.1	3.121
0.01	3.314	−0.01	3.278
0.001	3.298	−0.001	3.294
0.0001	3.296	−0.0001	3.296

We estimate that $f'(1) \approx 3.296$.

(b)

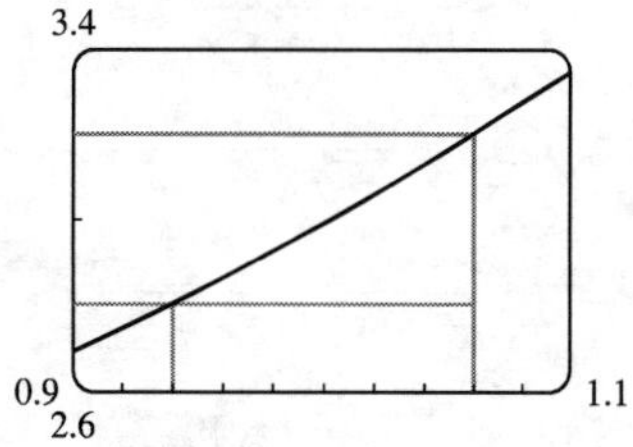

From the graph, we estimate that the slope of the tangent is about

$$\frac{3.2 - 2.8}{1.06 - 0.94} = \frac{0.4}{0.12} \approx 3.3.$$

13. Use Definition 2 with $f(x) = 3 - 2x + 4x^2$.

$$f'(a) = \lim_{h \to 0} \frac{f(a+h) - f(a)}{h} = \lim_{h \to 0} \frac{\left[3 - 2(a+h) + 4(a+h)^2\right] - (3 - 2a + 4a^2)}{h}$$

$$= \lim_{h \to 0} \frac{(3 - 2a - 2h + 4a^2 + 8ah + 4h^2) - (3 - 2a + 4a^2)}{h}$$

$$= \lim_{h \to 0} \frac{-2h + 8ah + 4h^2}{h} = \lim_{h \to 0} \frac{h(-2 + 8a + 4h)}{h} = \lim_{h \to 0} (-2 + 8a + 4h) = -2 + 8a$$

15. Use Definition 2 with $f(t) = (2t + 1)/(t + 3)$.

$$f'(a) = \lim_{h \to 0} \frac{f(a+h) - f(a)}{h} = \lim_{h \to 0} \frac{\dfrac{2(a+h) + 1}{(a+h) + 3} - \dfrac{2a + 1}{a + 3}}{h}$$

$$= \lim_{h \to 0} \frac{(2a + 2h + 1)(a + 3) - (2a + 1)(a + h + 3)}{h(a + h + 3)(a + 3)}$$

$$= \lim_{h \to 0} \frac{(2a^2 + 6a + 2ah + 6h + a + 3) - (2a^2 + 2ah + 6a + a + h + 3)}{h(a + h + 3)(a + 3)}$$

$$= \lim_{h \to 0} \frac{5h}{h(a + h + 3)(a + 3)} = \lim_{h \to 0} \frac{5}{(a + h + 3)(a + 3)} = \frac{5}{(a + 3)^2}$$

17. Use Definition 2 with $f(x) = 1/\sqrt{x+2}$.

$$f'(a) = \lim_{h\to 0} \frac{f(a+h)-f(a)}{h} = \lim_{h\to 0} \frac{\dfrac{1}{\sqrt{(a+h)+2}} - \dfrac{1}{\sqrt{a+2}}}{h}$$

$$= \lim_{h\to 0} \frac{\dfrac{\sqrt{a+2}-\sqrt{a+h+2}}{\sqrt{a+h+2}\,\sqrt{a+2}}}{h} = \lim_{h\to 0} \left[\frac{\sqrt{a+2}-\sqrt{a+h+2}}{h\sqrt{a+h+2}\,\sqrt{a+2}} \cdot \frac{\sqrt{a+2}+\sqrt{a+h+2}}{\sqrt{a+2}+\sqrt{a+h+2}}\right]$$

$$= \lim_{h\to 0} \frac{(a+2)-(a+h+2)}{h\sqrt{a+h+2}\,\sqrt{a+2}\left(\sqrt{a+2}+\sqrt{a+h+2}\right)}$$

$$= \lim_{h\to 0} \frac{-h}{h\sqrt{a+h+2}\,\sqrt{a+2}\left(\sqrt{a+2}+\sqrt{a+h+2}\right)}$$

$$= \lim_{h\to 0} \frac{-1}{\sqrt{a+h+2}\,\sqrt{a+2}\left(\sqrt{a+2}+\sqrt{a+h+2}\right)}$$

$$= \frac{-1}{\left(\sqrt{a+2}\right)^2\left(2\sqrt{a+2}\right)} = -\frac{1}{2\,(a+2)^{3/2}}$$

Note that the answers to Exercises 19–24 are not unique.

19. By Definition 2, $\displaystyle\lim_{h\to 0} \frac{(1+h)^{10}-1}{h} = f'(1)$, where $f(x) = x^{10}$ and $a = 1$.

Or: By Definition 2, $\displaystyle\lim_{h\to 0} \frac{(1+h)^{10}-1}{h} = f'(0)$, where $f(x) = (1+x)^{10}$ and $a = 0$.

21. By Equation 3, $\displaystyle\lim_{x\to 5} \frac{2^x-32}{x-5} = f'(5)$, where $f(x) = 2^x$ and $a = 5$.

23. By Definition 2, $\displaystyle\lim_{h\to 0} \frac{\cos(\pi+h)+1}{h} = f'(\pi)$, where $f(x) = \cos x$ and $a = \pi$.

Or: By Definition 2, $\displaystyle\lim_{h\to 0} \frac{\cos(\pi+h)+1}{h} = f'(0)$, where $f(x) = \cos(\pi+x)$ and $a = 0$.

25. $$v(2) = f'(2) = \lim_{h\to 0} \frac{f(2+h)-f(2)}{h} = \lim_{h\to 0} \frac{\left[(2+h)^2-6(2+h)-5\right]-\left[2^2-6(2)-5\right]}{h}$$

$$= \lim_{h\to 0} \frac{(4+4h+h^2-12-6h-5)-(-13)}{h} = \lim_{h\to 0} \frac{h^2-2h}{h} = \lim_{h\to 0}(h-2) = -2 \text{ m/s}$$

27. (a) $f'(x)$ is the rate of change of the production cost with respect to the number of ounces of gold produced. Its units are dollars per ounce.

(b) After 800 ounces of gold have been produced, the rate at which the production cost is increasing is \$17/ounce. So the cost of producing the 800th (or 801st) ounce is about \$17.

(c) In the short term, the values of $f'(x)$ will decrease because more efficient use is made of start-up costs as x increases. But eventually $f'(x)$ might increase due to large-scale operations.

29. (a) $f'(v)$ is the rate at which the fuel consumption is changing with respect to the speed. Its units are $(\text{gal/h})/(\text{mi/h})$.

(b) The fuel consumption is decreasing by $0.05\ (\text{gal/h})/(\text{mi/h})$ as the car's speed reaches 20 mi/h. So if you increase your speed to 21 mi/h, you could expect to decrease your fuel consumption by about $0.05\ (\text{gal/h})/(\text{mi/h})$.

31. $T'(10)$ is the rate at which the temperature is changing at 10:00 A.M. To estimate the value of $T'(10)$, we will average the difference quotients obtained using the times $t = 8$ and $t = 12$. Let

$A = \dfrac{T(8) - T(10)}{8 - 10} = \dfrac{72 - 81}{-2} = 4.5$ and $B = \dfrac{T(12) - T(10)}{12 - 10} = \dfrac{88 - 81}{2} = 3.5$. Then

$T'(10) = \lim\limits_{t \to 10} \dfrac{T(t) - T(10)}{t - 10} \approx \dfrac{A + B}{2} = \dfrac{4.5 + 3.5}{2} = 4°\text{F/h}$.

33. (a) $S'(T)$ is the rate at which the oxygen solubility changes with respect to the water temperature. Its units are $(\text{mg/L})/°\text{C}$.

(b) For $T = 16°\text{C}$, it appears that the tangent line to the curve goes through the points $(0, 14)$ and $(32, 6)$. So $S'(16) \approx \dfrac{6 - 14}{32 - 0} = -\dfrac{8}{32} = -0.25\ (\text{mg/L})/°\text{C}$. This means that as the temperature increases past $16°\text{C}$, the oxygen solubility is decreasing at a rate of $0.25\ (\text{mg/L})/°\text{C}$.

35. Since $f(x) = x\sin(1/x)$ when $x \neq 0$ and $f(0) = 0$, we have

$f'(0) = \lim\limits_{h \to 0} \dfrac{f(0 + h) - f(0)}{h} = \lim\limits_{h \to 0} \dfrac{h\sin(1/h) - 0}{h} = \lim\limits_{h \to 0} \sin(1/h)$. This limit does not exist since $\sin(1/h)$ takes the values -1 and 1 on any interval containing 0. (Compare with Example 4 in Section 2.2.)

The Derivative as a Function

1. *Note:* Your answers may vary depending on your estimates. By estimating the slopes of tangent lines on the graph of f, it appears that

(a) $f'(1) \approx -2$ (b) $f'(2) \approx 0.8$

(c) $f'(3) \approx -1$ (d) $f'(4) \approx -0.5$

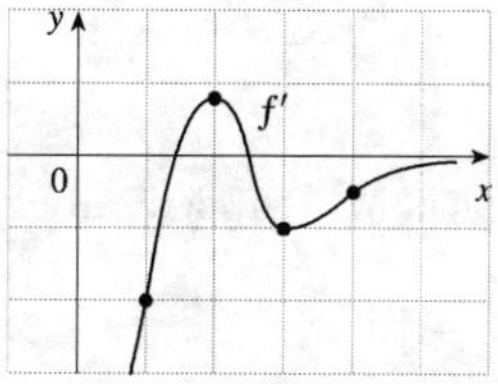

3. It appears that f is an odd function, so f' will be an even function—that is, $f'(-a) = f'(a)$.

(a) $f'(-3) \approx 1.5$ (b) $f'(-2) \approx 1$

(c) $f'(-1) \approx 0$ (d) $f'(0) \approx -4$

(e) $f'(1) \approx 0$ (f) $f'(2) \approx 1$

(g) $f'(3) \approx 1.5$

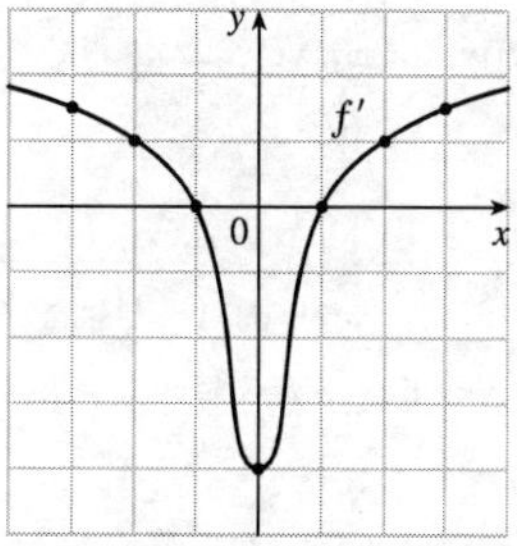

Hints for Exercises 5–13: First plot x-intercepts on the graph of f' for any horizontal tangents on the graph of f. Look for any corners on the graph of f — there will be a discontinuity on the graph of f'. On any interval where f has a tangent with positive (or negative) slope, the graph of f' will be positive (or negative). If the graph of the function is linear, the graph of f' will be a horizontal line.

5.

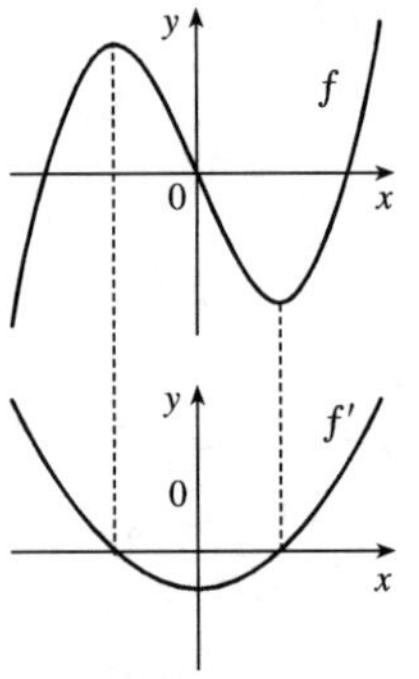

7.

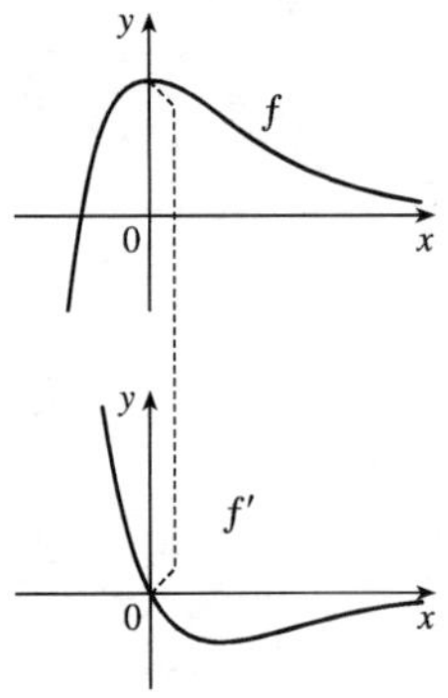

9.

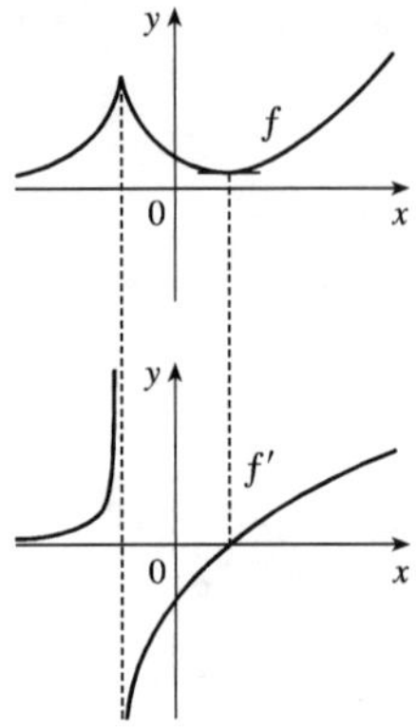

11.

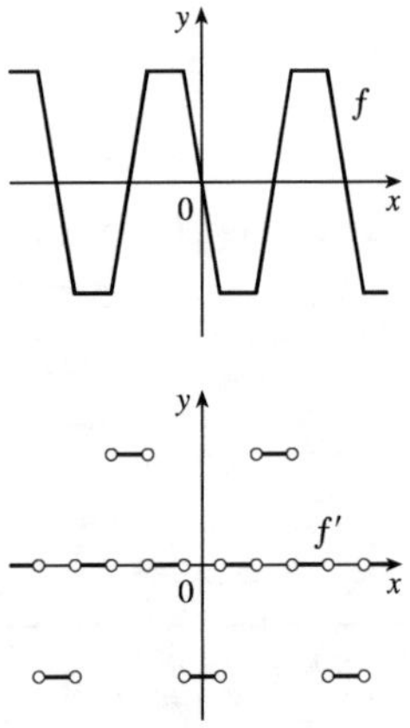

13.

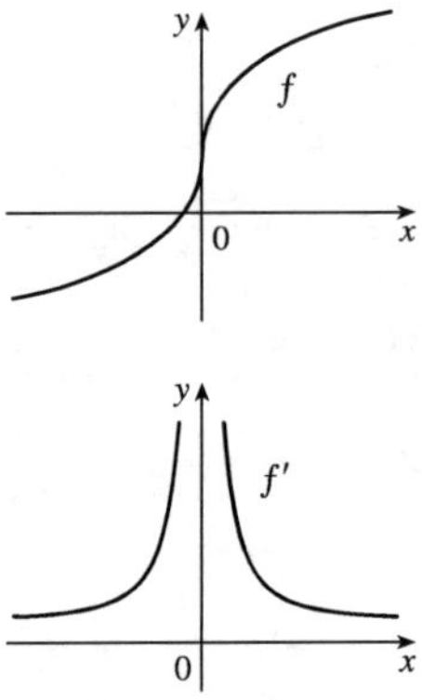

15. It appears that there are horizontal tangents on the graph of M for $t = 1963$ and $t = 1971$. Thus, there are zeros for those values of t on the graph of M'. The derivative is negative for the years 1963 to 1971.

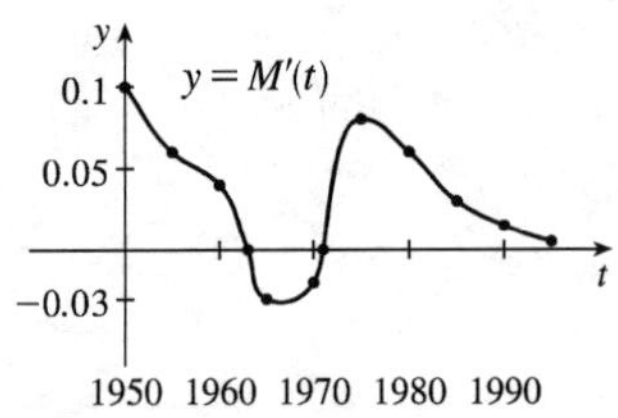

17. (a) By zooming in, we estimate that $f'(0) = 0$, $f'\left(\frac{1}{2}\right) = 1$, $f'(1) = 2$, and $f'(2) = 4$.

(b) By symmetry, $f'(-x) = -f'(x)$. So $f'\left(-\frac{1}{2}\right) = -1$, $f'(-1) = -2$, and $f'(-2) = -4$.

(c) It appears that $f'(x)$ is twice the value of x, so we guess that $f'(x) = 2x$.

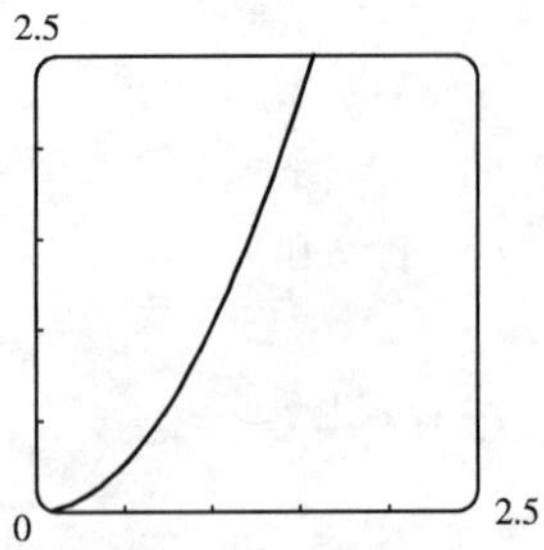

(d) $f'(x) = \lim\limits_{h \to 0} \dfrac{f(x+h) - f(x)}{h} = \lim\limits_{h \to 0} \dfrac{(x+h)^2 - x^2}{h}$

$= \lim\limits_{h \to 0} \dfrac{(x^2 + 2hx + h^2) - x^2}{h} = \lim\limits_{h \to 0} \dfrac{2hx + h^2}{h} = \lim\limits_{h \to 0} \dfrac{h(2x+h)}{h} = \lim\limits_{h \to 0} (2x + h) = 2x$

19. $f'(x) = \lim\limits_{h \to 0} \dfrac{f(x+h) - f(x)}{h} = \lim\limits_{h \to 0} \dfrac{37 - 37}{h} = \lim\limits_{h \to 0} \dfrac{0}{h} = \lim\limits_{h \to 0} 0 = 0$

Domain of f = domain of $f' = \mathbb{R}$.

21. $f'(x) = \lim\limits_{h \to 0} \dfrac{f(x+h) - f(x)}{h} = \lim\limits_{h \to 0} \dfrac{[1 - 3(x+h)^2] - (1 - 3x^2)}{h}$

$= \lim\limits_{h \to 0} \dfrac{[1 - 3(x^2 + 2xh + h^2)] - (1 - 3x^2)}{h} = \lim\limits_{h \to 0} \dfrac{1 - 3x^2 - 6xh - 3h^2 - 1 + 3x^2}{h}$

$= \lim\limits_{h \to 0} \dfrac{-6xh - 3h^2}{h} = \lim\limits_{h \to 0} \dfrac{h(-6x - 3h)}{h} = \lim\limits_{h \to 0} (-6x - 3h) = -6x$

Domain of f = domain of $f' = \mathbb{R}$.

23. $f'(x) = \lim\limits_{h \to 0} \dfrac{f(x+h) - f(x)}{h} = \lim\limits_{h \to 0} \dfrac{\left[(x+h)^3 - 3(x+h) + 5\right] - (x^3 - 3x + 5)}{h}$

$= \lim\limits_{h \to 0} \dfrac{(x^3 + 3x^2h + 3xh^2 + h^3 - 3x - 3h + 5) - (x^3 - 3x + 5)}{h}$

$= \lim\limits_{h \to 0} \dfrac{3x^2h + 3xh^2 + h^3 - 3h}{h} = \lim\limits_{h \to 0} \dfrac{h(3x^2 + 3xh + h^2 - 3)}{h}$

$= \lim\limits_{h \to 0} (3x^2 + 3xh + h^2 - 3) = 3x^2 - 3$

Domain of f = domain of $f' = \mathbb{R}$.

25. $g'(x) = \lim\limits_{h \to 0} \dfrac{g(x+h) - g(x)}{h} = \lim\limits_{h \to 0} \dfrac{\sqrt{1 + 2(x+h)} - \sqrt{1 + 2x}}{h} \left[\dfrac{\sqrt{1 + 2(x+h)} + \sqrt{1 + 2x}}{\sqrt{1 + 2(x+h)} + \sqrt{1 + 2x}}\right]$

$= \lim\limits_{h \to 0} \dfrac{(1 + 2x + 2h) - (1 + 2x)}{h\left[\sqrt{1 + 2(x+h)} + \sqrt{1 + 2x}\right]} = \lim\limits_{h \to 0} \dfrac{2}{\sqrt{1 + 2x + 2h} + \sqrt{1 + 2x}} = \dfrac{2}{2\sqrt{1 + 2x}} = \dfrac{1}{\sqrt{1 + 2x}}$

Domain of $g = \left[-\frac{1}{2}, \infty\right)$, domain of $g' = \left(-\frac{1}{2}, \infty\right)$.

27. $G'(t) = \lim\limits_{h \to 0} \dfrac{G(t+h) - G(t)}{h} = \lim\limits_{h \to 0} \dfrac{\dfrac{4(t+h)}{(t+h)+1} - \dfrac{4t}{t+1}}{h} = \lim\limits_{h \to 0} \dfrac{\dfrac{4(t+h)(t+1) - 4t(t+h+1)}{(t+h+1)(t+1)}}{h}$

$= \lim\limits_{h \to 0} \dfrac{(4t^2 + 4ht + 4t + 4h) - (4t^2 + 4ht + 4t)}{h(t+h+1)(t+1)}$

$= \lim\limits_{h \to 0} \dfrac{4h}{h(t+h+1)(t+1)} = \lim\limits_{h \to 0} \dfrac{4}{(t+h+1)(t+1)} = \dfrac{4}{(t+1)^2}$

Domain of G = domain of $G' = (-\infty, -1) \cup (-1, \infty)$.

29. $f'(x) = \lim\limits_{h \to 0} \dfrac{f(x+h) - f(x)}{h} = \lim\limits_{h \to 0} \dfrac{(x+h)^4 - x^4}{h} = \lim\limits_{h \to 0} \dfrac{(x^4 + 4x^3h + 6x^2h^2 + 4xh^3 + h^4) - x^4}{h}$

$= \lim\limits_{h \to 0} \dfrac{4x^3h + 6x^2h^2 + 4xh^3 + h^4}{h} = \lim\limits_{h \to 0} (4x^3 + 6x^2h + 4xh^2 + h^3) = 4x^3$

Domain of f = domain of $f' = \mathbb{R}$.

31. (a) $f'(x) = \lim\limits_{h \to 0} \dfrac{f(x+h) - f(x)}{h} = \lim\limits_{h \to 0} \dfrac{\left[x + h - \left(\dfrac{2}{x+h}\right)\right] - \left[x - \left(\dfrac{2}{x}\right)\right]}{h}$

$$= \lim_{h \to 0} \left[\frac{h - \dfrac{2}{(x+h)} + \dfrac{2}{x}}{h}\right] = \lim_{h \to 0} \left[1 + \frac{-2x + 2(x+h)}{hx(x+h)}\right] = \lim_{h \to 0} \left[1 + \frac{2h}{hx(x+h)}\right]$$

$$= \lim_{h \to 0} \left[1 + \frac{2}{x(x+h)}\right] = 1 + \frac{2}{x^2}$$

(b) Notice that when f has steep tangent lines, $f'(x)$ is very large. When f is flatter, $f'(x)$ is smaller.

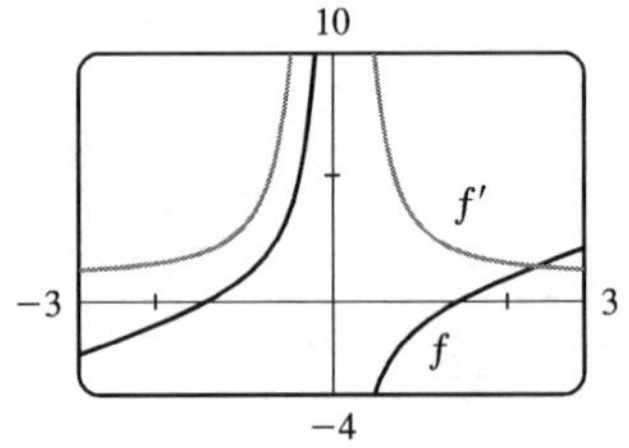

33. (a) $U'(t)$ is the rate at which the unemployment rate is changing with respect to time. Its units are percent per year.

(b) To find $U'(t)$, we use $\lim\limits_{h \to 0} \dfrac{U(t+h) - U(t)}{h} \approx \dfrac{U(t+h) - U(t)}{h}$ for small values of h.

For 1991: $U'(1991) = \dfrac{U(1992) - U(1991)}{1992 - 1991} = \dfrac{7.5 - 6.8}{1} = 0.70$

For 1992: We estimate $U'(1992)$ by using $h = -1$ and $h = 1$, and then average the two results to obtain a final estimate.

$h = -1 \;\Rightarrow\; U'(1992) \approx \dfrac{U(1991) - U(1992)}{1991 - 1992} = \dfrac{6.8 - 7.5}{-1} = 0.70;$

$h = 1 \;\Rightarrow\; U'(1992) \approx \dfrac{U(1993) - U(1992)}{1993 - 1992} = \dfrac{6.9 - 7.5}{1} = -0.60.$

So we estimate that $U'(1992) \approx \frac{1}{2}[0.70 + (-0.60)] = 0.05$.

t	1991	1992	1993	1994	1995	1996	1997	1998	1999	2000
$U'(t)$	0.70	0.05	−0.70	−0.65	−0.35	−0.35	−0.45	−0.35	−0.25	−0.20

35. f is not differentiable at $x = -1$ or at $x = 11$ because the graph has vertical tangents at those points; at $x = 4$, because there is a discontinuity there; and at $x = 8$, because the graph has a corner there.

37. As we zoom in toward $(-1, 0)$, the curve appears more and more like a straight line, so $f(x) = x + \sqrt{|x|}$ is differentiable at $x = -1$. But no matter how much we zoom in toward the origin, the curve doesn't straighten out—we can't eliminate the sharp point (a cusp). So f is not differentiable at $x = 0$.

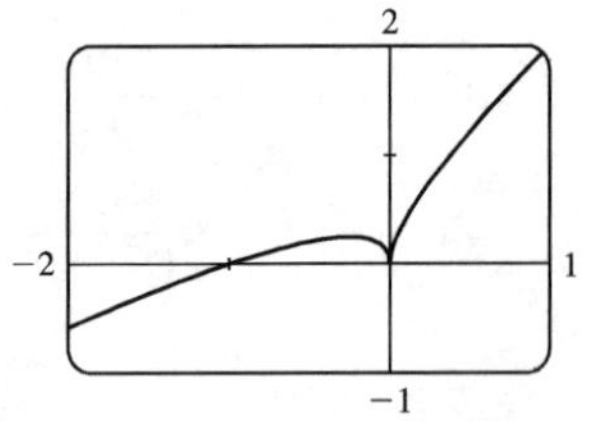

39. (a) Note that we have factored $x - a$ as the difference of two cubes in the third step.

$$f'(a) = \lim_{x \to a} \frac{f(x) - f(a)}{x - a} = \lim_{x \to a} \frac{x^{1/3} - a^{1/3}}{x - a} = \lim_{x \to a} \frac{x^{1/3} - a^{1/3}}{(x^{1/3} - a^{1/3})(x^{2/3} + x^{1/3}a^{1/3} + a^{2/3})}$$

$$= \lim_{x \to a} \frac{1}{x^{2/3} + x^{1/3}a^{1/3} + a^{2/3}} = \frac{1}{3a^{2/3}} \text{ or } \tfrac{1}{3}a^{-2/3}$$

(b) $f'(0) = \lim\limits_{h\to 0} \dfrac{f(0+h)-f(0)}{h} = \lim\limits_{h\to 0} \dfrac{\sqrt[3]{h}-0}{h} = \lim\limits_{h\to 0} \dfrac{1}{h^{2/3}}$. This function increases without bound, so the limit does not exist, and therefore $f'(0)$ does not exist.

(c) $\lim\limits_{x\to 0} |f'(x)| = \lim\limits_{x\to 0} \dfrac{1}{3x^{2/3}} = \infty$ and f is continuous at $x = 0$ (root function), so f has a vertical tangent at $x = 0$.

41. $f(x) = |x-6| = \begin{cases} -(x-6) & \text{if } x < 6 \\ x-6 & \text{if } x \geq 6 \end{cases} = \begin{cases} 6-x & \text{if } x < 6 \\ x-6 & \text{if } x \geq 6 \end{cases}$

$$\lim_{x\to 6^+} \frac{f(x)-f(6)}{x-6} = \lim_{x\to 6^+} \frac{|x-6|-0}{x-6} = \lim_{x\to 6^+} \frac{x-6}{x-6} = \lim_{x\to 6^+} 1 = 1.$$

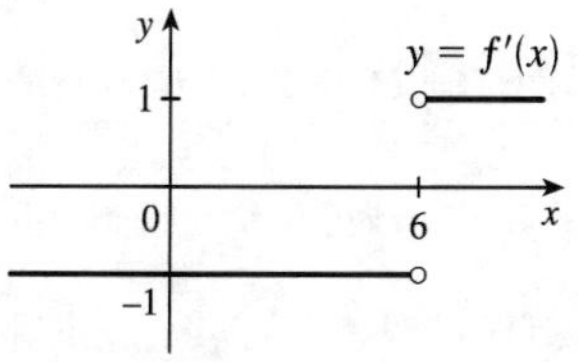

But $\lim\limits_{x\to 6^-} \dfrac{f(x)-f(6)}{x-6} = \lim\limits_{x\to 6^-} \dfrac{|x-6|-0}{x-6} = \lim\limits_{x\to 6^-} \dfrac{6-x}{x-6}$

$= \lim\limits_{x\to 6^-} (-1) = -1$

So $f'(6) = \lim\limits_{x\to 6} \dfrac{f(x)-f(6)}{x-6}$ does not exist. However, $f'(x) = \begin{cases} -1 & \text{if } x < 6 \\ 1 & \text{if } x > 6 \end{cases}$

Another way of writing the answer is $f'(x) = \dfrac{x-6}{|x-6|}$.

43. (a) $f(x) = x\,|x| = \begin{cases} x^2 & \text{if } x \geq 0 \\ -x^2 & \text{if } x < 0 \end{cases}$

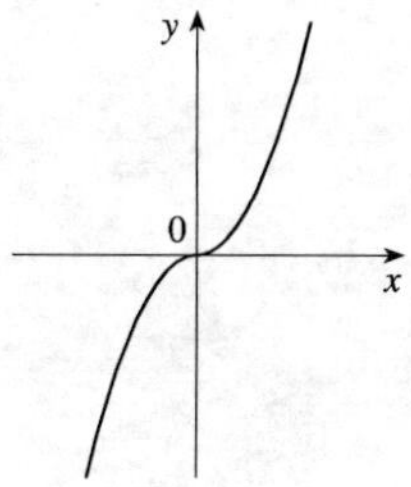

(b) Since $f(x) = x^2$ for $x \geq 0$, we have $f'(x) = 2x$ for $x > 0$. [See Exercise 3.2.17(d).] Similarly, since $f(x) = -x^2$ for $x < 0$, we have $f'(x) = -2x$ for $x < 0$. At $x = 0$, we have

$f'(0) = \lim\limits_{x\to 0} \dfrac{f(x)-f(0)}{x-0} = \lim\limits_{x\to 0} \dfrac{x\,|x|}{x} = \lim\limits_{x\to 0} |x| = 0.$

So f is differentiable at 0. Thus, f is differentiable for all x.

(c) From part (b), we have $f'(x) = \left\{ \begin{array}{ll} 2x & \text{if } x \geq 0 \\ -2x & \text{if } x < 0 \end{array} \right\} = 2\,|x|$.

45. (a) If f is even, then

$$f'(-x) = \lim_{h\to 0} \frac{f(-x+h)-f(-x)}{h} = \lim_{h\to 0} \frac{f[-(x-h)]-f(-x)}{h} = \lim_{h\to 0} \frac{f(x-h)-f(x)}{h}$$

$$= -\lim_{h\to 0} \frac{f(x-h)-f(x)}{-h} \quad [\text{let } \Delta x = -h] \quad = -\lim_{\Delta x\to 0} \frac{f(x+\Delta x)-f(x)}{\Delta x} = -f'(x)$$

Therefore, f' is odd.

(b) If f is odd, then

$$f'(-x) = \lim_{h\to 0} \frac{f(-x+h)-f(-x)}{h} = \lim_{h\to 0} \frac{f[-(x-h)]-f(-x)}{h} = \lim_{h\to 0} \frac{-f(x-h)+f(x)}{h}$$

$$= \lim_{h\to 0} \frac{f(x-h)-f(x)}{-h} \quad [\text{let } \Delta x = -h] \quad = \lim_{\Delta x\to 0} \frac{f(x+\Delta x)-f(x)}{\Delta x} = f'(x)$$

Therefore, f' is even.

47.

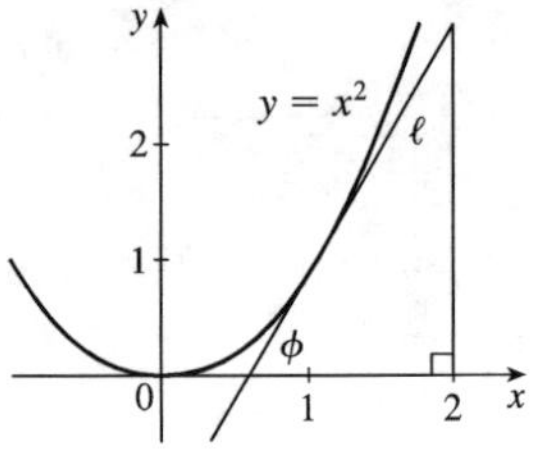

In the right triangle in the diagram, let Δy be the side opposite angle ϕ and Δx the side adjacent angle ϕ. Then the slope of the tangent line ℓ is $m = \Delta y/\Delta x = \tan\phi$. Note that $0 < \phi < \frac{\pi}{2}$. We know (see Exercise 17) that the derivative of $f(x) = x^2$ is $f'(x) = 2x$. So the slope of the tangent to the curve at the point $(1, 1)$ is 2. Thus, ϕ is the angle between 0 and $\frac{\pi}{2}$ whose tangent is 2; that is, $\phi = \tan^{-1} 2 \approx 63^\circ$.

Differentiation Formulas

1. $f(x) = 186.5$ is a constant function, so its derivative is 0, that is, $f'(x) = 0$.

3. $f(x) = 5x - 1 \;\Rightarrow\; f'(x) = 5 - 0 = 5$

5. $f(x) = x^2 + 3x - 4 \;\Rightarrow\; f'(x) = 2x^{2-1} + 3 - 0 = 2x + 3$

7. $f(t) = \frac{1}{4}(t^4 + 8) \;\Rightarrow\; f'(t) = \frac{1}{4}(t^4 + 8)' = \frac{1}{4}(4t^{4-1} + 0) = t^3$

9. $V(r) = \frac{4}{3}\pi r^3 \;\Rightarrow\; V'(r) = \frac{4}{3}\pi(3r^2) = 4\pi r^2$

11. $Y(t) = 6t^{-9} \;\Rightarrow\; Y'(t) = 6(-9)t^{-10} = -54t^{-10}$

13. $F(x) = (\frac{1}{2}x)^5 = \left(\frac{1}{2}\right)^5 x^5 = \frac{1}{32}x^5 \;\Rightarrow\; F'(x) = \frac{1}{32}(5x^4) = \frac{5}{32}x^4$

15. $y = x^{-2/5} \;\Rightarrow\; y' = -\frac{2}{5}x^{(-2/5)-1} = -\frac{2}{5}x^{-7/5} = -\dfrac{2}{5x^{7/5}}$

17. $y = 4\pi^2 \;\Rightarrow\; y' = 0$ since $4\pi^2$ is a constant.

19. $v = t^2 - \dfrac{1}{\sqrt[4]{t^3}} = t^2 - t^{-3/4} \;\Rightarrow\; v' = 2t - (-\frac{3}{4})t^{-7/4} = 2t + \dfrac{3}{4t^{7/4}} = 2t + \dfrac{3}{4t\,\sqrt[4]{t^3}}$

21. Product Rule: $y = (x^2 + 1)(x^3 + 1) \;\Rightarrow$

$$y' = (x^2 + 1)(3x^2) + (x^3 + 1)(2x) = 3x^4 + 3x^2 + 2x^4 + 2x = 5x^4 + 3x^2 + 2x.$$

Multiplying first: $y = (x^2 + 1)(x^3 + 1) = x^5 + x^3 + x^2 + 1 \;\Rightarrow\; y' = 5x^4 + 3x^2 + 2x$ (equivalent).

The notations $\overset{\text{PR}}{\Rightarrow}$ and $\overset{\text{QR}}{\Rightarrow}$ indicate use of the Product and Quotient Rules, respectively.

23. $V(x) = (2x^3 + 3)(x^4 - 2x) \;\overset{\text{PR}}{\Rightarrow}$

$V'(x) = (2x^3 + 3)(4x^3 - 2) + (x^4 - 2x)(6x^2) = (8x^6 + 8x^3 - 6) + (6x^6 - 12x^3) = 14x^6 - 4x^3 - 6$

25. $F(y) = \left(\dfrac{1}{y^2} - \dfrac{3}{y^4}\right)(y + 5y^3) = (y^{-2} - 3y^{-4})(y + 5y^3) \;\overset{\text{PR}}{\Rightarrow}$

$$\begin{aligned} F'(y) &= (y^{-2} - 3y^{-4})(1 + 15y^2) + (y + 5y^3)(-2y^{-3} + 12y^{-5}) \\ &= (y^{-2} + 15 - 3y^{-4} - 45y^{-2}) + (-2y^{-2} + 12y^{-4} - 10 + 60y^{-2}) \\ &= 5 + 14y^{-2} + 9y^{-4} \text{ or } 5 + 14/y^2 + 9/y^4 \end{aligned}$$

27. $g(x) = \dfrac{3x - 1}{2x + 1} \;\overset{\text{QR}}{\Rightarrow}\; g'(x) = \dfrac{(2x + 1)(3) - (3x - 1)(2)}{(2x + 1)^2} = \dfrac{6x + 3 - 6x + 2}{(2x + 1)^2} = \dfrac{5}{(2x + 1)^2}$

29. $y = \dfrac{t^2}{3t^2 - 2t + 1} \overset{\text{QR}}{\Rightarrow}$

$$y' = \frac{(3t^2 - 2t + 1)(2t) - t^2(6t - 2)}{(3t^2 - 2t + 1)^2} = \frac{2t\left[3t^2 - 2t + 1 - t(3t - 1)\right]}{(3t^2 - 2t + 1)^2}$$
$$= \frac{2t(3t^2 - 2t + 1 - 3t^2 + t)}{(3t^2 - 2t + 1)^2} = \frac{2t(1 - t)}{(3t^2 - 2t + 1)^2}$$

31. $y = \dfrac{v^3 - 2v\sqrt{v}}{v} = v^2 - 2\sqrt{v} = v^2 - 2v^{1/2} \Rightarrow y' = 2v - 2\left(\frac{1}{2}\right)v^{-1/2} = 2v - v^{-1/2}$.

We can change the form of the answer as follows: $2v - v^{-1/2} = 2v - \dfrac{1}{\sqrt{v}} = \dfrac{2v\sqrt{v} - 1}{\sqrt{v}} = \dfrac{2v^{3/2} - 1}{\sqrt{v}}$

33. $y = \dfrac{1}{x^4 + x^2 + 1} \Rightarrow y' = \dfrac{(x^4 + x^2 + 1)(0) - 1\,(4x^3 + 2x)}{(x^4 + x^2 + 1)^2} = -\dfrac{2x(2x^2 + 1)}{(x^4 + x^2 + 1)^2}$

35. $y = ax^2 + bx + c \Rightarrow y' = 2ax + b$

37. $y = \dfrac{r^2}{1 + \sqrt{r}} \Rightarrow$

$$y' = \frac{(1 + \sqrt{r})(2r) - r^2\left(\frac{1}{2}r^{-1/2}\right)}{(1 + \sqrt{r})^2} = \frac{2r + 2r^{3/2} - \frac{1}{2}r^{3/2}}{(1 + \sqrt{r})^2} = \frac{2r + \frac{3}{2}r^{3/2}}{(1 + \sqrt{r})^2} = \frac{\frac{1}{2}r(4 + 3r^{1/2})}{(1 + \sqrt{r})^2} = \frac{r(4 + 3\sqrt{r})}{2(1 + \sqrt{r})^2}$$

39. $y = \sqrt[3]{t}\,(t^2 + t + t^{-1}) = t^{1/3}(t^2 + t + t^{-1}) = t^{7/3} + t^{4/3} + t^{-2/3} \Rightarrow$
$y' = \frac{7}{3}t^{4/3} + \frac{4}{3}t^{1/3} - \frac{2}{3}t^{-5/3} = \frac{1}{3}t^{-5/3}(7t^{9/3} + 4t^{6/3} - 2) = (7t^3 + 4t^2 - 2)/(3t^{5/3})$

41. $f(x) = \dfrac{x}{x + c/x} \Rightarrow$

$$f'(x) = \frac{(x + c/x)(1) - x(1 - c/x^2)}{\left(x + \frac{c}{x}\right)^2} = \frac{x + c/x - x + c/x}{\left(\frac{x^2 + c}{x}\right)^2} = \frac{2c/x}{\frac{(x^2 + c)^2}{x^2}} \cdot \frac{x^2}{x^2} = \frac{2cx}{(x^2 + c)^2}$$

43. $P(x) = a_n x^n + a_{n-1}x^{n-1} + \cdots + a_2 x^2 + a_1 x + a_0 \Rightarrow$
$P'(x) = na_n x^{n-1} + (n - 1)a_{n-1}x^{n-2} + \cdots + 2a_2 x + a_1$

45. $f(x) = 3x^{15} - 5x^3 + 3 \Rightarrow f'(x) = 45x^{14} - 15x^2$.

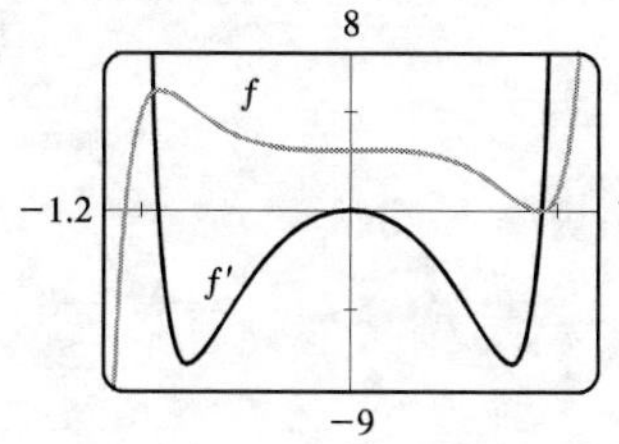

Notice that $f'(x) = 0$ when f has a horizontal tangent, f' is positive when f is increasing, and f' is negative when f is decreasing.

47. To graphically estimate the value of $f'(1)$ for $f(x) = 3x^2 - x^3$, we'll graph f in the viewing rectangle $[1 - 0.1, 1 + 0.1]$ by $[f(0.9), f(1.1)]$, as shown in the figure. [When assigning values to the window variables, it is convenient to use $Y_1(0.9)$ for $Y_{\min}$ and $Y_1(1.1)$ for $Y_{\max}$.] If we have sufficiently zoomed in on the graph of f, we should obtain a graph that looks like a diagonal line; if not, graph again with $1 - 0.01$ and $1 + 0.01$, etc.

Estimated value:

$$f'(1) \approx \frac{2.299 - 1.701}{1.1 - 0.9} = \frac{0.589}{0.2} = 2.99.$$

Exact value: $f(x) = 3x^2 - x^3 \quad \Rightarrow \quad f'(x) = 6x - 3x^2$,

so $f'(1) = 6 - 3 = 3$.

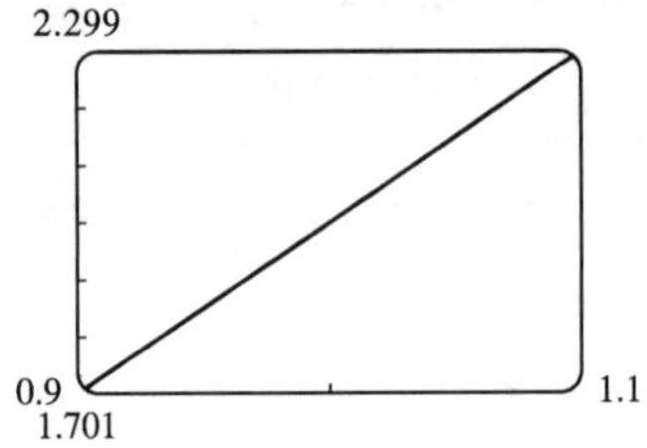

49. (a)

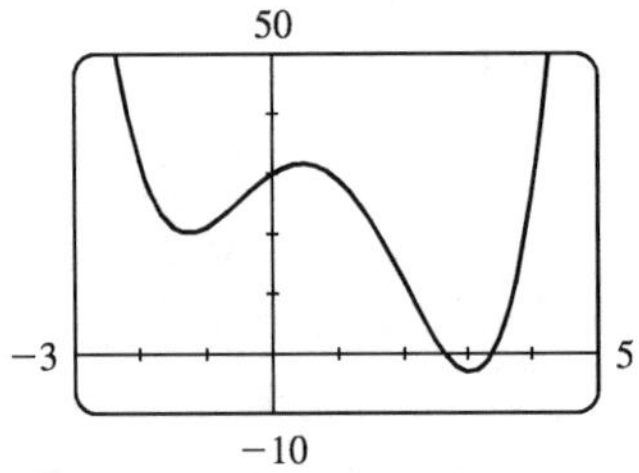

(b)

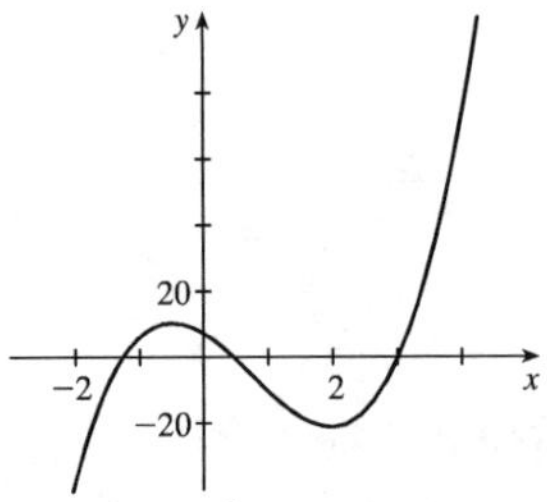

From the graph in part (a), it appears that f' is zero at $x_1 \approx -1.25$, $x_2 \approx 0.5$, and $x_3 \approx 3$. The slopes are negative (so f' is negative) on $(-\infty, x_1)$ and (x_2, x_3). The slopes are positive (so f' is positive) on (x_1, x_2) and (x_3, ∞).

(c) $f(x) = x^4 - 3x^3 - 6x^2 + 7x + 30 \quad \Rightarrow$

$f'(x) = 4x^3 - 9x^2 - 12x + 7$

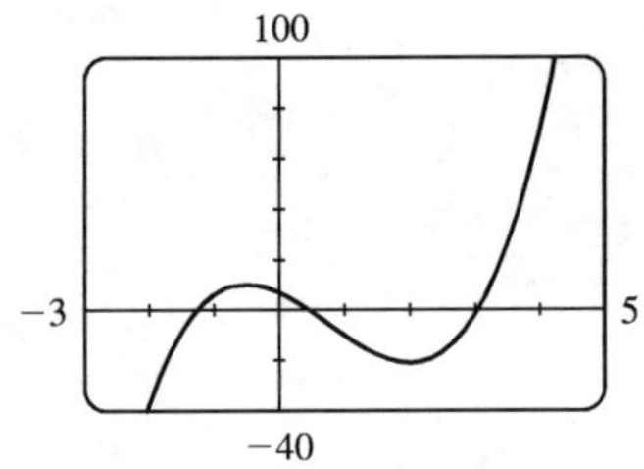

51. $y = \dfrac{2x}{x+1} \quad \Rightarrow \quad y' = \dfrac{(x+1)(2) - (2x)(1)}{(x+1)^2} = \dfrac{2}{(x+1)^2}$. At $(1, 1)$, $y' = \frac{1}{2}$, and an equation of the tangent line is $y - 1 = \frac{1}{2}(x - 1)$, or $y = \frac{1}{2}x + \frac{1}{2}$.

53. $y = f(x) = x + \sqrt{x} \quad \Rightarrow \quad f'(x) = 1 + \frac{1}{2}x^{-1/2}$. So the slope of the tangent line at $(1, 2)$ is $f'(1) = 1 + \frac{1}{2}(1) = \frac{3}{2}$ and its equation is $y - 2 = \frac{3}{2}(x - 1)$ or $y = \frac{3}{2}x + \frac{1}{2}$.

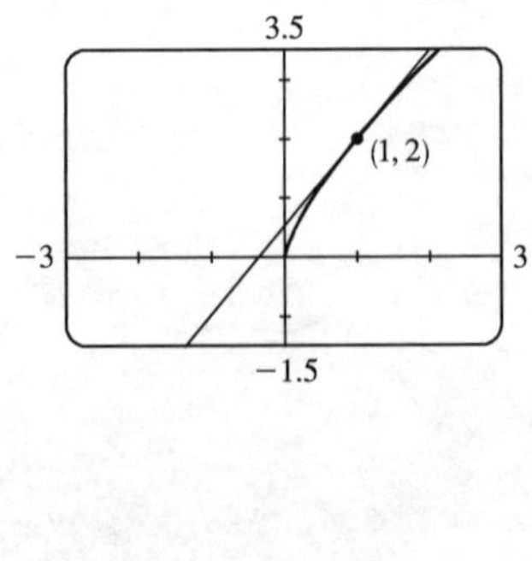

55. (a) $y = f(x) = \dfrac{1}{1+x^2} \quad \Rightarrow$

$f'(x) = \dfrac{(1+x^2)(0) - 1(2x)}{(1+x^2)^2} = \dfrac{-2x}{(1+x^2)^2}$. So the slope of the

tangent line at the point $\left(-1, \frac{1}{2}\right)$ is $f'(-1) = \dfrac{2}{2^2} = \frac{1}{2}$ and its

equation is $y - \frac{1}{2} = \frac{1}{2}(x+1)$ or $y = \frac{1}{2}x + 1$.

(b)

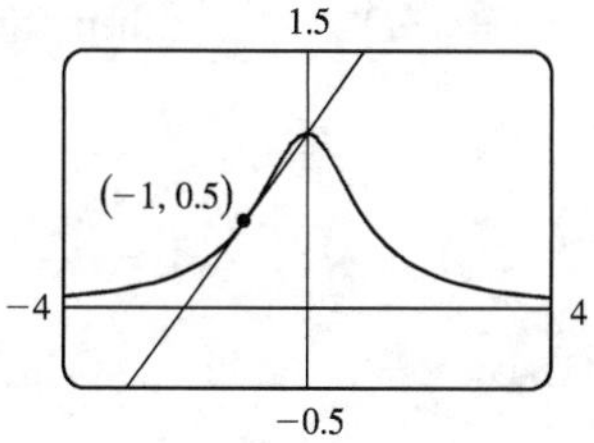

57. We are given that $f(5) = 1$, $f'(5) = 6$, $g(5) = -3$, and $g'(5) = 2$.

(a) $(fg)'(5) = f(5)g'(5) + g(5)f'(5) = (1)(2) + (-3)(6) = 2 - 18 = -16$

(b) $\left(\dfrac{f}{g}\right)'(5) = \dfrac{g(5)f'(5) - f(5)g'(5)}{[g(5)]^2} = \dfrac{(-3)(6) - (1)(2)}{(-3)^2} = -\dfrac{20}{9}$

(c) $\left(\dfrac{g}{f}\right)'(5) = \dfrac{f(5)g'(5) - g(5)f'(5)}{[f(5)]^2} = \dfrac{(1)(2) - (-3)(6)}{(1)^2} = 20$

59. $f(x) = \sqrt{x}\,g(x) \quad \Rightarrow \quad f'(x) = \sqrt{x}\,g'(x) + g(x) \cdot \frac{1}{2}x^{-1/2}$, so
$f'(4) = \sqrt{4}\,g'(4) + g(4) \cdot \frac{1}{2\sqrt{4}} = 2 \cdot 7 + 8 \cdot \frac{1}{4} = 16$.

61. (a) From the graphs of f and g, we obtain the following values: $f(1) = 2$ since the point $(1, 2)$ is on the graph of f; $g(1) = 1$ since the point $(1, 1)$ is on the graph of g; $f'(1) = 2$ since the slope of the line segment between $(0, 0)$ and $(2, 4)$ is $\dfrac{4-0}{2-0} = 2$; $g'(1) = -1$ since the slope of the line segment between $(-2, 4)$ and $(2, 0)$ is $\dfrac{0-4}{2-(-2)} = -1$. Now $u(x) = f(x)g(x)$, so $u'(1) = f(1)g'(1) + g(1)\,f'(1) = 2 \cdot (-1) + 1 \cdot 2 = 0$.

(b) $v(x) = f(x)/g(x)$, so $v'(5) = \dfrac{g(5)f'(5) - f(5)g'(5)}{[g(5)]^2} = \dfrac{2\left(-\frac{1}{3}\right) - 3 \cdot \frac{2}{3}}{2^2} = \dfrac{-\frac{8}{3}}{4} = -\dfrac{2}{3}$

63. (a) $y = xg(x) \quad \Rightarrow \quad y' = xg'(x) + g(x) \cdot 1 = xg'(x) + g(x)$

(b) $y = \dfrac{x}{g(x)} \quad \Rightarrow \quad y' = \dfrac{g(x) \cdot 1 - xg'(x)}{[g(x)]^2} = \dfrac{g(x) - xg'(x)}{[g(x)]^2}$

(c) $y = \dfrac{g(x)}{x} \quad \Rightarrow \quad y' = \dfrac{xg'(x) - g(x) \cdot 1}{(x)^2} = \dfrac{xg'(x) - g(x)}{x^2}$

65. $y = f(x) = 1 - x^2 \quad \Rightarrow \quad f'(x) = -2x$, so the tangent line at $(2, -3)$ has slope $f'(2) = -4$. The normal line has slope $-\frac{1}{-4} = \frac{1}{4}$ and equation $y + 3 = \frac{1}{4}(x - 2)$ or $y = \frac{1}{4}x - \frac{7}{2}$.

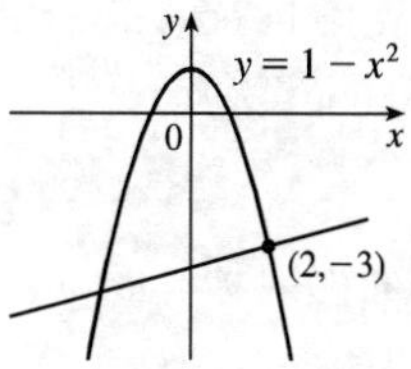

67. $y = x^3 - x^2 - x + 1$ has a horizontal tangent when $y' = 3x^2 - 2x - 1 = 0$. $(3x+1)(x-1) = 0 \quad \Leftrightarrow$ $x = 1$ or $-\frac{1}{3}$. Therefore, the points are $(1, 0)$ and $\left(-\frac{1}{3}, \frac{32}{27}\right)$.

69. If $y = f(x) = \dfrac{x}{x+1}$, then $f'(x) = \dfrac{(x+1)(1) - x(1)}{(x+1)^2} = \dfrac{1}{(x+1)^2}$. When $x = a$, the equation of the tangent line is $y - \dfrac{a}{a+1} = \dfrac{1}{(a+1)^2}(x-a)$. This line passes through $(1,2)$ when $2 - \dfrac{a}{a+1} = \dfrac{1}{(a+1)^2}(1-a)$ $\Leftrightarrow$ $2(a+1)^2 - a(a+1) = 1 - a$ $\Leftrightarrow$ $2a^2 + 4a + 2 - a^2 - a - 1 + a = 0$ $\Leftrightarrow$ $a^2 + 4a + 1 = 0$.

The quadratic formula gives the roots of this equation as $a = \dfrac{-4 \pm \sqrt{4^2 - 4(1)(1)}}{2(1)} = \dfrac{-4 \pm \sqrt{12}}{2} = -2 \pm \sqrt{3}$, so there are two such tangent lines. Since

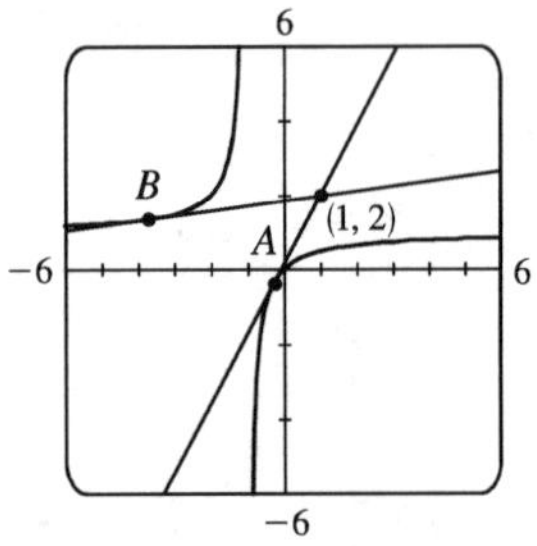

$$f\left(-2 \pm \sqrt{3}\right) = \frac{-2 \pm \sqrt{3}}{-2 \pm \sqrt{3} + 1} = \frac{-2 \pm \sqrt{3}}{-1 \pm \sqrt{3}} \cdot \frac{-1 \mp \sqrt{3}}{-1 \mp \sqrt{3}}$$
$$= \frac{2 \pm 2\sqrt{3} \mp \sqrt{3} - 3}{1 - 3} = \frac{-1 \pm \sqrt{3}}{-2} = \frac{1 \mp \sqrt{3}}{2},$$

the lines touch the curve at $A\left(-2 + \sqrt{3}, \frac{1 - \sqrt{3}}{2}\right) \approx (-0.27, -0.37)$ and $B\left(-2 - \sqrt{3}, \frac{1 + \sqrt{3}}{2}\right) \approx (-3.73, 1.37)$.

71. $y = 6x^3 + 5x - 3$ $\Rightarrow$ $m = y' = 18x^2 + 5$, but $x^2 \geq 0$ for all x, so $m \geq 5$ for all x.

73. (a) $(fgh)' = [(fg)h]' = (fg)'h + (fg)h' = (f'g + fg')h + (fg)h' = f'gh + fg'h + fgh'$

(b) $y = \sqrt{x}\,(x^4 + x + 1)(2x - 3)$. Using part (a), we have

$$y' = \frac{1}{2\sqrt{x}}\,(x^4 + x + 1)(2x - 3) + \sqrt{x}\,(4x^3 + 1)(2x - 3) + \sqrt{x}\,(x^4 + x + 1)(2)$$
$$= (x^4 + x + 1)\,\frac{2x - 3}{2\sqrt{x}} + \sqrt{x}\left[(4x^3 + 1)(2x - 3) + 2(x^4 + x + 1)\right]$$

75. $y = f(x) = ax^3 + bx^2 + cx + d$ $\Rightarrow$ $f'(x) = 3ax^2 + 2bx + c$. The point $(-2, 6)$ is on f, so $f(-2) = 6$ $\Rightarrow$ $-8a + 4b - 2c + d = 6$ **(1)**. The point $(2, 0)$ is on f, so $f(2) = 0$ $\Rightarrow$ $8a + 4b + 2c + d = 0$ **(2)**. Since there are horizontal tangents at $(-2, 6)$ and $(2, 0)$, $f'(\pm 2) = 0$. $f'(-2) = 0$ $\Rightarrow$ $12a - 4b + c = 0$ **(3)** and $f'(2) = 0$ $\Rightarrow$ $12a + 4b + c = 0$ **(4)**. Subtracting equation **(3)** from **(4)** gives $8b = 0$ $\Rightarrow$ $b = 0$. Adding **(1)** and **(2)** gives $8b + 2d = 6$, so $d = 3$ since $b = 0$. From **(3)** we have $c = -12a$, so **(2)** becomes $8a + 4(0) + 2(-12a) + 3 = 0$ $\Rightarrow$ $3 = 16a$ $\Rightarrow$ $a = \frac{3}{16}$. Now $c = -12a = -12\left(\frac{3}{16}\right) = -\frac{9}{4}$ and the desired cubic function is $y = \frac{3}{16}x^3 - \frac{9}{4}x + 3$.

77. If $P(t)$ denotes the population at time t and $A(t)$ the average annual income, then $T(t) = P(t)A(t)$ is the total personal income. The rate at which $T(t)$ is rising is given by $T'(t) = P(t)A'(t) + A(t)P'(t)$ $\Rightarrow$

$$T'(1999) = P(1999)A'(1999) + A(1999)P'(1999) = (961{,}400)(\$1400/\text{yr}) + (\$30{,}593)(9200/\text{yr})$$
$$= \$1{,}345{,}960{,}000/\text{yr} + \$281{,}455{,}600/\text{yr} = \$1{,}627{,}415{,}600/\text{yr}$$

So the total personal income was rising by about \$1.627 billion per year in 1999.

The term $P(t)A'(t) \approx \$1.346$ billion represents the portion of the rate of change of total income due to the existing population's increasing income. The term $A(t)P'(t) \approx \$281$ million represents the portion of the rate of change of total income due to increasing population.

79. $f(x) = 2 - x$ if $x \le 1$ and $f(x) = x^2 - 2x + 2$ if $x > 1$. Now we compute the right- and left-hand derivatives defined in Exercise 3.2.44:

$$f'_-(1) = \lim_{h\to 0^-} \frac{f(1+h) - f(1)}{h} = \lim_{h\to 0^-} \frac{2-(1+h)-1}{h} = \lim_{h\to 0^-} \frac{-h}{h} = \lim_{h\to 0^-} -1 = -1 \text{ and}$$

$$f'_+(1) = \lim_{h\to 0^+} \frac{f(1+h) - f(1)}{h} = \lim_{h\to 0^+} \frac{(1+h)^2 - 2(1+h) + 2 - 1}{h} = \lim_{h\to 0^+} \frac{h^2}{h} = \lim_{h\to 0^+} h = 0.$$

Thus, $f'(1)$ does not exist since $f'_-(1) \neq f'_+(1)$, so f is not differentiable at 1. But $f'(x) = -1$ for $x < 1$ and $f'(x) = 2x - 2$ if $x > 1$.

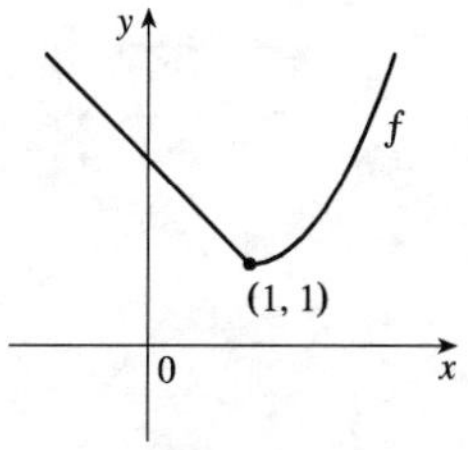

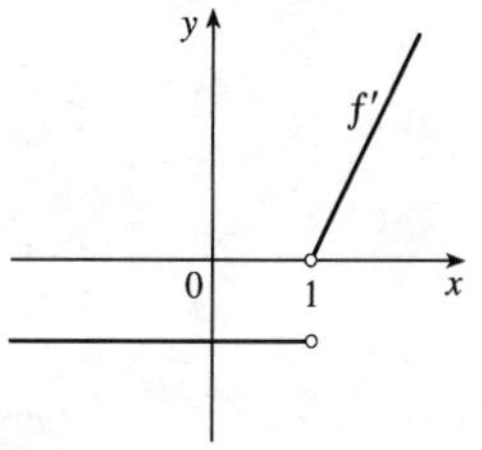

81. (a) Note that $x^2 - 9 < 0$ for $x^2 < 9 \Leftrightarrow |x| < 3 \Leftrightarrow -3 < x < 3$. So

$$f(x) = \begin{cases} x^2 - 9 & \text{if } x \le -3 \\ -x^2 + 9 & \text{if } -3 < x < 3 \\ x^2 - 9 & \text{if } x \ge 3 \end{cases} \Rightarrow$$

$$f'(x) = \begin{cases} 2x & \text{if } x < -3 \\ -2x & \text{if } -3 < x < 3 \\ 2x & \text{if } x > 3 \end{cases} = \begin{cases} 2x & \text{if } |x| > 3 \\ -2x & \text{if } |x| < 3 \end{cases}$$

To show that $f'(3)$ does not exist we investigate $\lim_{h\to 0} \frac{f(3+h) - f(3)}{h}$ by computing the left- and right-hand derivatives defined in Exercise 3.2.44.

$$f'_-(3) = \lim_{h\to 0^-} \frac{f(3+h) - f(3)}{h} = \lim_{h\to 0^-} \frac{[-(3+h)^2 + 9] - 0}{h} = \lim_{h\to 0^-} (-6 - h) = -6 \quad \text{and}$$

$$f'_+(3) = \lim_{h\to 0^+} \frac{f(3+h) - f(3)}{h} = \lim_{h\to 0^+} \frac{[(3+h)^2 - 9] - 0}{h} = \lim_{h\to 0^+} \frac{6h + h^2}{h} = \lim_{h\to 0^+} (6 + h) = 6.$$

Since the left and right limits are different, $\lim_{h\to 0} \frac{f(3+h) - f(3)}{h}$ does not exist, that is, $f'(3)$ does not exist. Similarly, $f'(-3)$ does not exist. Therefore, f is not differentiable at 3 or at -3.

(b)

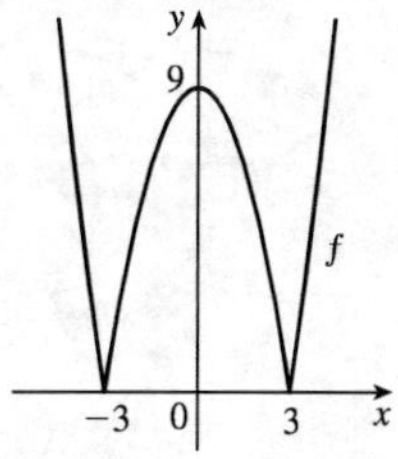

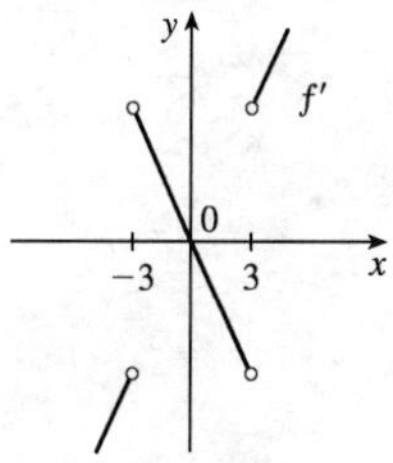

83. $y = f(x) = ax^2 \Rightarrow f'(x) = 2ax$. So the slope of the tangent to the parabola at $x = 2$ is $m = 2a(2) = 4a$. The slope of the given line, $2x + y = b \Leftrightarrow y = -2x + b$, is seen to be -2, so we must have $4a = -2 \Leftrightarrow a = -\frac{1}{2}$. So when $x = 2$, the point in question has y-coordinate $-\frac{1}{2} \cdot 2^2 = -2$. Now we simply require that the given line, whose equation is $2x + y = b$, pass through the point $(2, -2)$: $2(2) + (-2) = b \Leftrightarrow b = 2$. So we must have $a = -\frac{1}{2}$ and $b = 2$.

85. $F = f/g \Rightarrow f = Fg \Rightarrow f' = F'g + Fg' \Rightarrow F' = \dfrac{f' - Fg'}{g} = \dfrac{f' - (f/g)g'}{g} = \dfrac{f'g - fg'}{g^2}$

87. *Solution 1:* Let $f(x) = x^{1000}$. Then, by the definition of a derivative,

$f'(1) = \lim\limits_{x \to 1} \dfrac{f(x) - f(1)}{x - 1} = \lim\limits_{x \to 1} \dfrac{x^{1000} - 1}{x - 1}$. But this is just the limit we want to find, and we know (from the Power Rule) that $f'(x) = 1000x^{999}$, so $f'(1) = 1000(1)^{999} = 1000$. So $\lim\limits_{x \to 1} \dfrac{x^{1000} - 1}{x - 1} = 1000$.

Solution 2: Note that $\left(x^{1000} - 1\right) = (x - 1)\left(x^{999} + x^{998} + x^{997} + \cdots + x^2 + x + 1\right)$. So

$$\begin{aligned}
\lim_{x \to 1} \frac{x^{1000} - 1}{x - 1} &= \lim_{x \to 1} \frac{(x - 1)\left(x^{999} + x^{998} + x^{997} + \cdots + x^2 + x + 1\right)}{x - 1} \\
&= \lim_{x \to 1} \left(x^{999} + x^{998} + x^{997} + \cdots + x^2 + x + 1\right) = \underbrace{1 + 1 + 1 + \cdots + 1 + 1 + 1}_{\text{1000 ones}} \\
&= 1000, \text{ as above.}
\end{aligned}$$

Derivatives of Trigonometric Functions

1. $f(x) = x - 3\sin x \;\Rightarrow\; f'(x) = 1 - 3\cos x$

3. $y = \sin x + 10\tan x \;\Rightarrow\; y' = \cos x + 10\sec^2 x$

5. $g(t) = t^3\cos t \;\Rightarrow\; g'(t) = t^3(-\sin t) + (\cos t)\cdot 3t^2 = 3t^2\cos t - t^3\sin t$ or $t^2(3\cos t - t\sin t)$

7. $h(\theta) = \theta\csc\theta - \cot\theta \;\Rightarrow\; h'(\theta) = \theta(-\csc\theta\cot\theta) + (\csc\theta)\cdot 1 - \left(-\csc^2\theta\right) = \csc\theta - \theta\csc\theta\cot\theta + \csc^2\theta$

9. $y = \dfrac{x}{\cos x} \;\Rightarrow\; y' = \dfrac{(\cos x)(1) - (x)(-\sin x)}{(\cos x)^2} = \dfrac{\cos x + x\sin x}{\cos^2 x}$

11. $f(\theta) = \dfrac{\sec\theta}{1 + \sec\theta} \;\Rightarrow$

$$f'(\theta) = \frac{(1+\sec\theta)(\sec\theta\tan\theta) - (\sec\theta)(\sec\theta\tan\theta)}{(1+\sec\theta)^2} = \frac{(\sec\theta\tan\theta)[(1+\sec\theta) - \sec\theta]}{(1+\sec\theta)^2} = \frac{\sec\theta\tan\theta}{(1+\sec\theta)^2}$$

13. $y = \dfrac{\sin x}{x^2} \;\Rightarrow\; y' = \dfrac{x^2\cos x - (\sin x)(2x)}{(x^2)^2} = \dfrac{x(x\cos x - 2\sin x)}{x^4} = \dfrac{x\cos x - 2\sin x}{x^3}$

15. $y = \sec\theta\,\tan\theta \;\Rightarrow\; y' = \sec\theta\left(\sec^2\theta\right) + \tan\theta\,(\sec\theta\tan\theta) = \sec\theta\left(\sec^2\theta + \tan^2\theta\right)$

Using the identity $1 + \tan^2\theta = \sec^2\theta$, we can write alternative forms of the answer as

$$\sec\theta\left(1 + 2\tan^2\theta\right) \qquad \text{or} \qquad \sec\theta\left(2\sec^2\theta - 1\right)$$

17. $\dfrac{d}{dx}(\csc x) = \dfrac{d}{dx}\left(\dfrac{1}{\sin x}\right) = \dfrac{(\sin x)(0) - 1(\cos x)}{\sin^2 x} = \dfrac{-\cos x}{\sin^2 x} = -\dfrac{1}{\sin x}\cdot\dfrac{\cos x}{\sin x} = -\csc x\cot x$

19. $\dfrac{d}{dx}(\cot x) = \dfrac{d}{dx}\left(\dfrac{\cos x}{\sin x}\right) = \dfrac{(\sin x)(-\sin x) - (\cos x)(\cos x)}{\sin^2 x} = -\dfrac{\sin^2 x + \cos^2 x}{\sin^2 x} = -\dfrac{1}{\sin^2 x} = -\csc^2 x$

21. $y = \tan x \;\Rightarrow\; y' = \sec^2 x \;\Rightarrow$ the slope of the tangent line at $\left(\frac{\pi}{4}, 1\right)$ is $\sec^2\frac{\pi}{4} = \left(\sqrt{2}\right)^2 = 2$ and an equation of the tangent line is $y - 1 = 2\left(x - \frac{\pi}{4}\right)$ or $y = 2x + 1 - \frac{\pi}{2}$.

23. $y = x + \cos x \;\Rightarrow\; y' = 1 - \sin x$. At $(0, 1)$, $y' = 1$, and an equation of the tangent line is $y - 1 = 1(x - 0)$, or $y = x + 1$.

25. (a) $y = x\cos x \quad \Rightarrow \quad y' = x(-\sin x) + \cos x(1) = \cos x - x\sin x$.
So the slope of the tangent at the point $(\pi, -\pi)$ is
$\cos\pi - \pi\sin\pi = -1 - \pi(0) = -1$, and an equation is
$y + \pi = -(x - \pi)$ or $y = -x$.

(b)

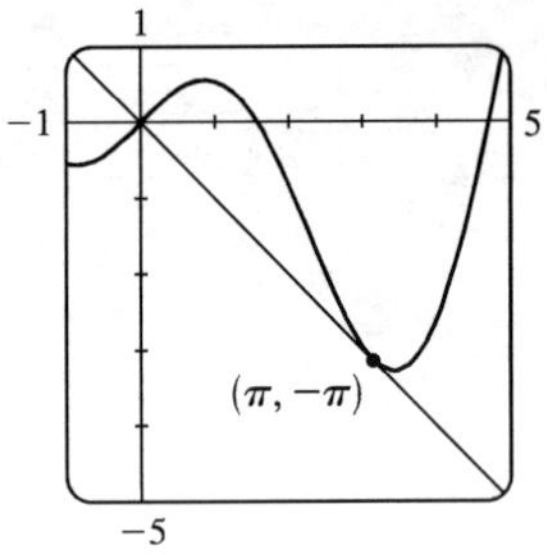

27. (a) $f(x) = 2x + \cot x \quad \Rightarrow \quad f'(x) = 2 - \csc^2 x$

(b)

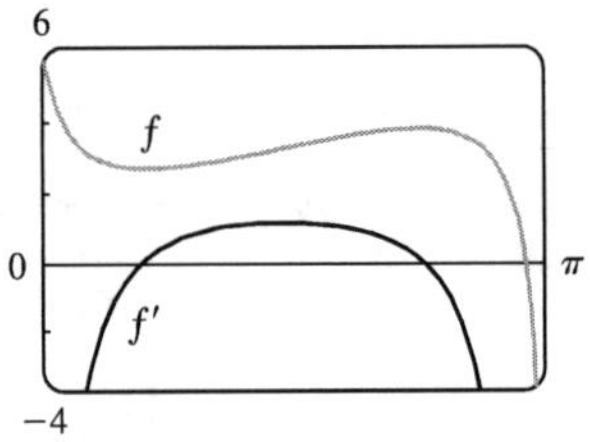

Notice that $f'(x) = 0$ when f has a horizontal tangent. f' is positive when f is increasing and f' is negative when f is decreasing. Also, $f'(x)$ is large negative when the graph of f is steep.

29. $f(x) = x + 2\sin x$ has a horizontal tangent when $f'(x) = 0 \quad \Leftrightarrow \quad 1 + 2\cos x = 0 \quad \Leftrightarrow \quad \cos x = -\frac{1}{2} \quad \Leftrightarrow$ $x = \frac{2\pi}{3} + 2\pi n$ or $\frac{4\pi}{3} + 2\pi n$, where n is an integer. Note that $\frac{4\pi}{3}$ and $\frac{2\pi}{3}$ are $\pm\frac{\pi}{3}$ units from π. This allows us to write the solutions in the more compact equivalent form $(2n + 1)\pi \pm \frac{\pi}{3}$, n an integer.

31. (a) $x(t) = 8\sin t \quad \Rightarrow \quad v(t) = x'(t) = 8\cos t$

(b) The mass at time $t = \frac{2\pi}{3}$ has position $x\left(\frac{2\pi}{3}\right) = 8\sin\frac{2\pi}{3} = 8\left(\frac{\sqrt{3}}{2}\right) = 4\sqrt{3}$ and velocity
$v\left(\frac{2\pi}{3}\right) = 8\cos\frac{2\pi}{3} = 8\left(-\frac{1}{2}\right) = -4$. Since $v\left(\frac{2\pi}{3}\right) < 0$, the particle is moving to the left.

33.

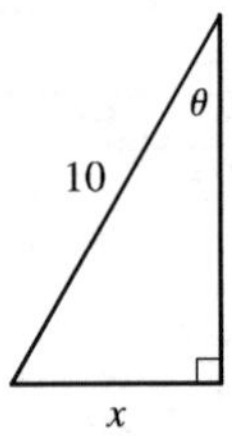

From the diagram we can see that $\sin\theta = x/10 \quad \Leftrightarrow \quad x = 10\sin\theta$. We want to find the rate of change of x with respect to θ; that is, $dx/d\theta$. Taking the derivative of the above expression, $dx/d\theta = 10(\cos\theta)$. So when $\theta = \frac{\pi}{3}$,

$$dx/d\theta = 10\cos\tfrac{\pi}{3} = 10\left(\tfrac{1}{2}\right) = 5 \text{ ft/rad}$$

35.
$$\begin{aligned}
\lim_{x\to 0}\frac{\sin 3x}{x} &= \lim_{x\to 0}\frac{3\sin 3x}{3x} && \text{[multiply numerator and denominator by 3]}\\
&= 3\lim_{3x\to 0}\frac{\sin 3x}{3x} && [\text{as } x \to 0,\ 3x \to 0]\\
&= 3\lim_{\theta\to 0}\frac{\sin\theta}{\theta} && [\text{let } \theta = 3x]\\
&= 3(1) && \text{[Equation 2]}\\
&= 3
\end{aligned}$$

37.
$$\lim_{t\to 0}\frac{\tan 6t}{\sin 2t} = \lim_{t\to 0}\left(\frac{\sin 6t}{t}\cdot\frac{1}{\cos 6t}\cdot\frac{t}{\sin 2t}\right) = \lim_{t\to 0}\frac{6\sin 6t}{6t}\cdot\lim_{t\to 0}\frac{1}{\cos 6t}\cdot\lim_{t\to 0}\frac{2t}{2\sin 2t}$$
$$= 6\lim_{t\to 0}\frac{\sin 6t}{6t}\cdot\lim_{t\to 0}\frac{1}{\cos 6t}\cdot\frac{1}{2}\lim_{t\to 0}\frac{2t}{\sin 2t} = 6(1)\cdot\frac{1}{1}\cdot\frac{1}{2}(1) = 3$$

39. $\lim\limits_{\theta\to0}\dfrac{\sin(\cos\theta)}{\sec\theta}=\dfrac{\sin\left(\lim\limits_{\theta\to0}\cos\theta\right)}{\lim\limits_{\theta\to0}\sec\theta}=\dfrac{\sin 1}{1}=\sin 1$

41. $\lim\limits_{x\to0}\dfrac{\cot 2x}{\csc x}=\lim\limits_{x\to0}\dfrac{\cos 2x\sin x}{\sin 2x}=\lim\limits_{x\to0}\cos 2x\left[\dfrac{(\sin x)/x}{(\sin 2x)/x}\right]=\lim\limits_{x\to0}\cos 2x\left[\dfrac{\lim\limits_{x\to0}[(\sin x)/x]}{2\lim\limits_{x\to0}[(\sin 2x)/2x]}\right]$

$=1\cdot\dfrac{1}{2\cdot1}=\dfrac{1}{2}$

43. Divide numerator and denominator by θ. ($\sin\theta$ also works.)

$$\lim_{\theta\to0}\frac{\sin\theta}{\theta+\tan\theta}=\lim_{\theta\to0}\frac{\dfrac{\sin\theta}{\theta}}{1+\dfrac{\sin\theta}{\theta}\cdot\dfrac{1}{\cos\theta}}=\frac{\lim\limits_{\theta\to0}\dfrac{\sin\theta}{\theta}}{1+\lim\limits_{\theta\to0}\dfrac{\sin\theta}{\theta}\lim\limits_{\theta\to0}\dfrac{1}{\cos\theta}}=\frac{1}{1+1\cdot1}=\frac{1}{2}$$

45. (a) $\dfrac{d}{dx}\tan x=\dfrac{d}{dx}\dfrac{\sin x}{\cos x}\ \Rightarrow\ \sec^2 x=\dfrac{\cos x\cos x-\sin x(-\sin x)}{\cos^2 x}=\dfrac{\cos^2 x+\sin^2 x}{\cos^2 x}$. So $\sec^2 x=\dfrac{1}{\cos^2 x}$.

(b) $\dfrac{d}{dx}\sec x=\dfrac{d}{dx}\dfrac{1}{\cos x}\ \Rightarrow\ \sec x\tan x=\dfrac{(\cos x)(0)-1(-\sin x)}{\cos^2 x}$. So $\sec x\tan x=\dfrac{\sin x}{\cos^2 x}$.

(c) $\dfrac{d}{dx}(\sin x+\cos x)=\dfrac{d}{dx}\dfrac{1+\cot x}{\csc x}\ \Rightarrow$

$\cos x-\sin x=\dfrac{\csc x\left(-\csc^2 x\right)-(1+\cot x)(-\csc x\cot x)}{\csc^2 x}=\dfrac{\csc x\left[-\csc^2 x+(1+\cot x)\cot x\right]}{\csc^2 x}$

$=\dfrac{-\csc^2 x+\cot^2 x+\cot x}{\csc x}=\dfrac{-1+\cot x}{\csc x}$

So $\cos x-\sin x=\dfrac{\cot x-1}{\csc x}$.

47. By the definition of radian measure, $s=r\theta$, where r is the radius of the circle.

By drawing the bisector of the angle θ, we can see that $\sin\dfrac{\theta}{2}=\dfrac{d/2}{r}\ \Rightarrow\ d=2r\sin\dfrac{\theta}{2}$.

So $\lim\limits_{\theta\to0^+}\dfrac{s}{d}=\lim\limits_{\theta\to0^+}\dfrac{r\theta}{2r\sin(\theta/2)}=\lim\limits_{\theta\to0^+}\dfrac{2\cdot(\theta/2)}{2\sin(\theta/2)}=\lim\limits_{\theta\to0}\dfrac{\theta/2}{\sin(\theta/2)}=1$. [This is just the reciprocal of the limit $\lim\limits_{x\to0}\frac{\sin x}{x}=1$ combined with the fact that as $\theta\to0$, $\frac{\theta}{2}\to0$ also.]

The Chain Rule

1. Let $u=g(x)=4x$ and $y=f(u)=\sin u$. Then $\dfrac{dy}{dx}=\dfrac{dy}{du}\dfrac{du}{dx}=(\cos u)(4)=4\cos 4x$.

3. Let $u=g(x)=1-x^2$ and $y=f(u)=u^{10}$. Then $\dfrac{dy}{dx}=\dfrac{dy}{du}\dfrac{du}{dx}=(10u^9)(-2x)=-20x(1-x^2)^9$.

5. Let $u=g(x)=\sin x$ and $y=f(u)=\sqrt{u}$. Then $\dfrac{dy}{dx}=\dfrac{dy}{du}\dfrac{du}{dx}=\frac{1}{2}u^{-1/2}\cos x=\dfrac{\cos x}{2\sqrt{u}}=\dfrac{\cos x}{2\sqrt{\sin x}}$.

7. $F(x)=(x^3+4x)^7\ \Rightarrow\ F'(x)=7(x^3+4x)^6(3x^2+4)$ [or $7x^6(x^2+4)^6(3x^2+4)$]

9. $F(x) = \sqrt[4]{1+2x+x^3} = (1+2x+x^3)^{1/4} \Rightarrow$

$$F'(x) = \tfrac{1}{4}(1+2x+x^3)^{-3/4} \cdot \frac{d}{dx}(1+2x+x^3) = \frac{1}{4(1+2x+x^3)^{3/4}} \cdot (2+3x^2)$$

$$= \frac{2+3x^2}{4(1+2x+x^3)^{3/4}} = \frac{2+3x^2}{4\sqrt[4]{(1+2x+x^3)^3}}$$

11. $g(t) = \dfrac{1}{(t^4+1)^3} = (t^4+1)^{-3} \Rightarrow g'(t) = -3(t^4+1)^{-4}(4t^3) = -12t^3(t^4+1)^{-4} = \dfrac{-12t^3}{(t^4+1)^4}$

13. $y = \cos(a^3+x^3) \Rightarrow y' = -\sin(a^3+x^3) \cdot 3x^2$ [a^3 is just a constant] $= -3x^2\sin(a^3+x^3)$

15. $y = \cot(x/2) \Rightarrow y' = -\csc^2(x/2) \cdot \frac{1}{2} = -\frac{1}{2}\csc^2(x/2)$

17. $g(x) = (1+4x)^5(3+x-x^2)^8 \Rightarrow$

$$\begin{aligned} g'(x) &= (1+4x)^5 \cdot 8(3+x-x^2)^7(1-2x) + (3+x-x^2)^8 \cdot 5(1+4x)^4 \cdot 4 \\ &= 4(1+4x)^4(3+x-x^2)^7\left[2(1+4x)(1-2x) + 5(3+x-x^2)\right] \\ &= 4(1+4x)^4(3+x-x^2)^7\left[(2+4x-16x^2) + (15+5x-5x^2)\right] \\ &= 4(1+4x)^4(3+x-x^2)^7(17+9x-21x^2) \end{aligned}$$

19. $y = (2x-5)^4(8x^2-5)^{-3} \Rightarrow$

$$\begin{aligned} y' &= 4(2x-5)^3(2)(8x^2-5)^{-3} + (2x-5)^4(-3)(8x^2-5)^{-4}(16x) \\ &= 8(2x-5)^3(8x^2-5)^{-3} - 48x(2x-5)^4(8x^2-5)^{-4} \end{aligned}$$

[This simplifies to $8(2x-5)^3(8x^2-5)^{-4}(-4x^2+30x-5)$.]

21. $y = x^3\cos nx \Rightarrow y' = x^3(-\sin nx)(n) + \cos nx\,(3x^2) = x^2(3\cos nx - nx\sin nx)$

23. $y = \sin(x\cos x) \Rightarrow y' = \cos(x\cos x) \cdot [x(-\sin x) + \cos x \cdot 1] = (\cos x - x\sin x)\cos(x\cos x)$

25. $F(z) = \sqrt{\dfrac{z-1}{z+1}} = \left(\dfrac{z-1}{z+1}\right)^{1/2} \Rightarrow$

$$F'(z) = \frac{1}{2}\left(\frac{z-1}{z+1}\right)^{-1/2} \cdot \frac{d}{dz}\left(\frac{z-1}{z+1}\right) = \frac{1}{2}\left(\frac{z+1}{z-1}\right)^{1/2} \cdot \frac{(z+1)(1)-(z-1)(1)}{(z+1)^2}$$

$$= \frac{1}{2}\frac{(z+1)^{1/2}}{(z-1)^{1/2}} \cdot \frac{z+1-z+1}{(z+1)^2} = \frac{1}{2}\frac{(z+1)^{1/2}}{(z-1)^{1/2}} \cdot \frac{2}{(z+1)^2} = \frac{1}{(z-1)^{1/2}(z+1)^{3/2}}$$

27. $y = \dfrac{r}{\sqrt{r^2+1}} \Rightarrow$

$$y' = \frac{\sqrt{r^2+1}\,(1) - r \cdot \frac{1}{2}(r^2+1)^{-1/2}(2r)}{\left(\sqrt{r^2+1}\right)^2} = \frac{\sqrt{r^2+1} - \dfrac{r^2}{\sqrt{r^2+1}}}{\left(\sqrt{r^2+1}\right)^2} = \frac{\dfrac{\sqrt{r^2+1}\,\sqrt{r^2+1} - r^2}{\sqrt{r^2+1}}}{\left(\sqrt{r^2+1}\right)^2}$$

$$= \frac{(r^2+1) - r^2}{\left(\sqrt{r^2+1}\right)^3} = \frac{1}{(r^2+1)^{3/2}} \text{ or } (r^2+1)^{-3/2}$$

Another solution: Write y as a product and make use of the Product Rule. $y = r(r^2+1)^{-1/2} \Rightarrow$

$$\begin{aligned} y' &= r \cdot -\tfrac{1}{2}(r^2+1)^{-3/2}(2r) + (r^2+1)^{-1/2} \cdot 1 \\ &= (r^2+1)^{-3/2}\left[-r^2 + (r^2+1)^1\right] = (r^2+1)^{-3/2}(1) = (r^2+1)^{-3/2} \end{aligned}$$

The step that students usually have trouble with is factoring out $(r^2+1)^{-3/2}$. But this is no different than factoring out x^2 from $x^2 + x^5$; that is, we are just factoring out a factor with the *smallest* exponent that appears on it. In this case, $-\frac{3}{2}$ is smaller than $-\frac{1}{2}$.

29. $y = \tan(\cos x) \Rightarrow y' = \sec^2(\cos x) \cdot (-\sin x) = -\sin x \sec^2(\cos x)$

31. $y = \sin\sqrt{1+x^2} \Rightarrow y' = \cos\sqrt{1+x^2} \cdot \frac{1}{2}(1+x^2)^{-1/2} \cdot 2x = \left(x \cos\sqrt{1+x^2}\right)/\sqrt{1+x^2}$

33. $y = (1+\cos^2 x)^6 \Rightarrow y' = 6(1+\cos^2 x)^5 \cdot 2\cos x\,(-\sin x) = -12\cos x \sin x\,(1+\cos^2 x)^5$

35. $y = \sec^2 x + \tan^2 x = (\sec x)^2 + (\tan x)^2 \Rightarrow$

$y' = 2(\sec x)(\sec x \tan x) + 2(\tan x)(\sec^2 x) = 2\sec^2 x \tan x + 2\sec^2 x \tan x = 4\sec^2 x \tan x$

37. $y = \cot^2(\sin\theta) = [\cot(\sin\theta)]^2 \Rightarrow$

$y' = 2\,[\cot(\sin\theta)] \cdot \dfrac{d}{d\theta}[\cot(\sin\theta)] = 2\cot(\sin\theta) \cdot [-\csc^2(\sin\theta) \cdot \cos\theta] = -2\cos\theta\,\cot(\sin\theta)\,\csc^2(\sin\theta)$

39. $y = \sqrt{x+\sqrt{x}} \Rightarrow y' = \frac{1}{2}\left(x+\sqrt{x}\right)^{-1/2}\left(1+\frac{1}{2}x^{-1/2}\right) = \dfrac{1}{2\sqrt{x+\sqrt{x}}}\left(1+\dfrac{1}{2\sqrt{x}}\right)$

41. $y = \sin\left(\tan\sqrt{\sin x}\right) \Rightarrow$

$$\begin{aligned} y' &= \cos\left(\tan\sqrt{\sin x}\right) \cdot \frac{d}{dx}\left(\tan\sqrt{\sin x}\right) = \cos\left(\tan\sqrt{\sin x}\right)\sec^2\sqrt{\sin x} \cdot \frac{d}{dx}(\sin x)^{1/2} \\ &= \cos\left(\tan\sqrt{\sin x}\right)\sec^2\sqrt{\sin x} \cdot \tfrac{1}{2}(\sin x)^{-1/2} \cdot \cos x \\ &= \cos\left(\tan\sqrt{\sin x}\right)\left(\sec^2\sqrt{\sin x}\right)\left(\frac{1}{2\sqrt{\sin x}}\right)(\cos x) \end{aligned}$$

43. $y = (1+2x)^{10} \Rightarrow y' = 10(1+2x)^9 \cdot 2 = 20(1+2x)^9$. At $(0,1)$, $y' = 20(1+0)^9 = 20$, and an equation of the tangent line is $y - 1 = 20(x-0)$, or $y = 20x+1$.

45. $y = \sin(\sin x) \Rightarrow y' = \cos(\sin x) \cdot \cos x$. At $(\pi, 0)$, $y' = \cos(\sin\pi) \cdot \cos\pi = \cos(0) \cdot (-1) = 1(-1) = -1$, and an equation of the tangent line is $y - 0 = -1(x-\pi)$, or $y = -x+\pi$.

47. (a) $y = f(x) = \tan\left(\frac{\pi}{4}x^2\right) \Rightarrow f'(x) = \sec^2\left(\frac{\pi}{4}x^2\right)\left(2 \cdot \frac{\pi}{4}x\right)$.

The slope of the tangent at $(1,1)$ is thus

$f'(1) = \sec^2\frac{\pi}{4}\left(\frac{\pi}{2}\right) = 2 \cdot \frac{\pi}{2} = \pi$, and its equation is

$y - 1 = \pi(x-1)$ or $y = \pi x - \pi + 1$.

(b)

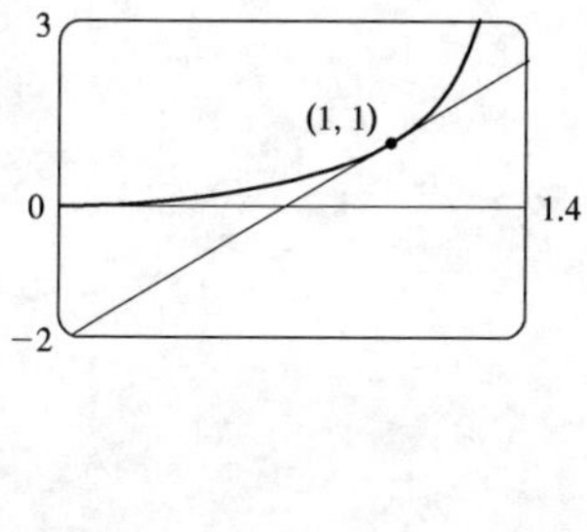

49. (a) $f(x) = \dfrac{\sqrt{1-x^2}}{x} \Rightarrow$

$$f'(x) = \frac{x \cdot \frac{1}{2}\left(1-x^2\right)^{-1/2}(-2x) - \sqrt{1-x^2}\,(1)}{x^2} \cdot \frac{\sqrt{1-x^2}}{\sqrt{1-x^2}}$$

$$= \frac{-x^2 - \left(1-x^2\right)}{x^2\sqrt{1-x^2}} = \frac{-1}{x^2\sqrt{1-x^2}}$$

(b)

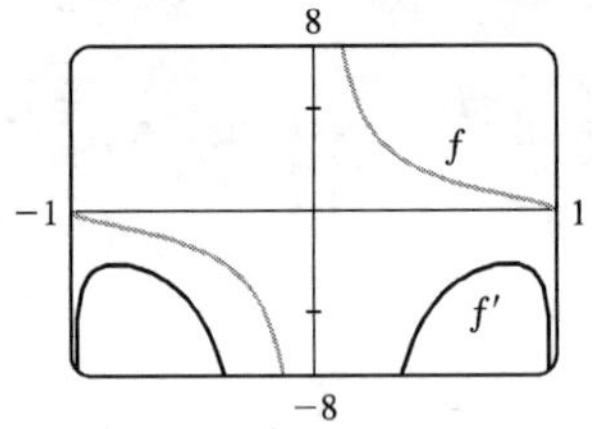

Notice that all tangents to the graph of f have negative slopes and $f'(x) < 0$ always.

51. For the tangent line to be horizontal, $f'(x) = 0$. $f(x) = 2\sin x + \sin^2 x \Rightarrow$ $f'(x) = 2\cos x + 2\sin x \cos x = 0 \Leftrightarrow 2\cos x\,(1 + \sin x) = 0 \Leftrightarrow \cos x = 0$ or $\sin x = -1$, so $x = \frac{\pi}{2} + 2n\pi$ or $\frac{3\pi}{2} + 2n\pi$, where n is any integer. Now $f\left(\frac{\pi}{2}\right) = 3$ and $f\left(\frac{3\pi}{2}\right) = -1$, so the points on the curve with a horizontal tangent are $\left(\frac{\pi}{2} + 2n\pi, 3\right)$ and $\left(\frac{3\pi}{2} + 2n\pi, -1\right)$, where n is any integer.

53. $F(x) = f(g(x)) \Rightarrow F'(x) = f'(g(x)) \cdot g'(x)$, so $F'(3) = f'(g(3)) \cdot g'(3) = f'(6) \cdot g'(3) = 7 \cdot 4 = 28$. Notice that we did not use $f'(3) = 2$.

55. (a) $h(x) = f(g(x)) \Rightarrow h'(x) = f'(g(x)) \cdot g'(x)$, so $h'(1) = f'(g(1)) \cdot g'(1) = f'(2) \cdot 6 = 5 \cdot 6 = 30$.

(b) $H(x) = g(f(x)) \Rightarrow H'(x) = g'(f(x)) \cdot f'(x)$, so $H'(1) = g'(f(1)) \cdot f'(1) = g'(3) \cdot 4 = 9 \cdot 4 = 36$.

57. (a) $u(x) = f(g(x)) \Rightarrow u'(x) = f'(g(x))g'(x)$. So $u'(1) = f'(g(1))g'(1) = f'(3)g'(1)$. To find $f'(3)$, note that f is linear from $(2,4)$ to $(6,3)$, so its slope is $\dfrac{3-4}{6-2} = -\dfrac{1}{4}$. To find $g'(1)$, note that g is linear from $(0,6)$ to $(2,0)$, so its slope is $\dfrac{0-6}{2-0} = -3$. Thus, $f'(3)g'(1) = \left(-\frac{1}{4}\right)(-3) = \frac{3}{4}$.

(b) $v(x) = g(f(x)) \Rightarrow v'(x) = g'(f(x))f'(x)$. So $v'(1) = g'(f(1))f'(1) = g'(2)f'(1)$, which does not exist since $g'(2)$ does not exist.

(c) $w(x) = g(g(x)) \Rightarrow w'(x) = g'(g(x))g'(x)$. So $w'(1) = g'(g(1))g'(1) = g'(3)g'(1)$. To find $g'(3)$, note that g is linear from $(2,0)$ to $(5,2)$, so its slope is $\dfrac{2-0}{5-2} = \dfrac{2}{3}$. Thus, $g'(3)g'(1) = \left(\frac{2}{3}\right)(-3) = -2$.

59. $h(x) = f(g(x)) \Rightarrow h'(x) = f'(g(x))g'(x)$. So $h'(0.5) = f'(g(0.5))g'(0.5) = f'(0.1)g'(0.5)$. We can estimate the derivatives by taking the average of two secant slopes.

For $f'(0.1)$: $m_1 = \dfrac{14.8 - 12.6}{0.1 - 0} = 22$, $m_2 = \dfrac{18.4 - 14.8}{0.2 - 0.1} = 36$. So $f'(0.1) \approx \dfrac{m_1 + m_2}{2} = \dfrac{22 + 36}{2} = 29$.

For $g'(0.5)$: $m_1 = \dfrac{0.10 - 0.17}{0.5 - 0.4} = -0.7$, $m_2 = \dfrac{0.05 - 0.10}{0.6 - 0.5} = -0.5$. So $g'(0.5) \approx \dfrac{m_1 + m_2}{2} = -0.6$.

Hence, $h'(0.5) = f'(0.1)g'(0.5) \approx (29)(-0.6) = -17.4$.

61. (a) $F(x) = f(\cos x) \Rightarrow F'(x) = f'(\cos x)\dfrac{d}{dx}(\cos x) = -\sin x\, f'(\cos x)$

(b) $G(x) = \cos(f(x)) \Rightarrow G'(x) = -\sin(f(x))\, f'(x)$

63. (a) $f(x) = L(x^4) \Rightarrow f'(x) = L'(x^4) \cdot 4x^3 = (1/x^4) \cdot 4x^3 = 4/x$ for $x > 0$.

(b) $g(x) = L(4x) \Rightarrow g'(x) = L'(4x) \cdot 4 = (1/(4x)) \cdot 4 = 1/x$ for $x > 0$.

(c) $F(x) = [L(x)]^4 \Rightarrow F'(x) = 4[L(x)]^3 \cdot L'(x) = 4[L(x)]^3 \cdot (1/x) = 4[L(x)]^3/x$

(d) $G(x) = L(1/x) \Rightarrow G'(x) = L'(1/x) \cdot (-1/x^2) = (1/(1/x)) \cdot (-1/x^2) = x \cdot (-1/x^2) = -1/x$ for $x > 0$.

65. $s(t) = 10 + \frac{1}{4}\sin(10\pi t) \Rightarrow$ the velocity after t seconds is

$v(t) = s'(t) = \frac{1}{4}\cos(10\pi t)(10\pi) = \frac{5\pi}{2}\cos(10\pi t)$ cm/s.

67. (a) $B(t) = 4.0 + 0.35\sin\dfrac{2\pi t}{5.4} \Rightarrow \dfrac{dB}{dt} = \left(0.35\cos\dfrac{2\pi t}{5.4}\right)\left(\dfrac{2\pi}{5.4}\right) = \dfrac{0.7\pi}{5.4}\cos\dfrac{2\pi t}{5.4} = \dfrac{7\pi}{54}\cos\dfrac{2\pi t}{5.4}$

(b) At $t = 1$, $\dfrac{dB}{dt} = \dfrac{7\pi}{54}\cos\dfrac{2\pi}{5.4} \approx 0.16$.

69. (a) Derive gives $g'(t) = \dfrac{45(t-2)^8}{(2t+1)^{10}}$ without simplifying. With either Maple or Mathematica, we first get

$g'(t) = 9\dfrac{(t-2)^8}{(2t+1)^9} - 18\dfrac{(t-2)^9}{(2t+1)^{10}}$, and the simplification command results in the above expression.

(b) Derive gives $y' = 2(x^3 - x + 1)^3(2x+1)^4(17x^3 + 6x^2 - 9x + 3)$ without simplifying.

With either Maple or Mathematica, we first get

$y' = 10(2x+1)^4(x^3 - x + 1)^4 + 4(2x+1)^5(x^3 - x + 1)^3(3x^2 - 1)$. If we use Mathematica's `Factor` or `Simplify`, or Maple's `factor`, we get the above expression, but Maple's `simplify` gives the polynomial expansion instead. For locating horizontal tangents, the factored form is the most helpful.

71. (a) If f is even, then $f(x) = f(-x)$. Using the Chain Rule to differentiate this equation, we get

$f'(x) = f'(-x)\dfrac{d}{dx}(-x) = -f'(-x)$. Thus, $f'(-x) = -f'(x)$, so f' is odd.

(b) If f is odd, then $f(x) = -f(-x)$. Differentiating this equation, we get $f'(x) = -f'(-x)(-1) = f'(-x)$, so f' is even.

73. (a)

$$\begin{aligned}\frac{d}{dx}(\sin^n x\cos nx) &= n\sin^{n-1}x\cos x\cos nx + \sin^n x(-n\sin nx) && \text{[Product Rule]}\\ &= n\sin^{n-1}x(\cos nx\cos x - \sin nx\sin x) && [\text{factor out } n\sin^{n-1}x]\\ &= n\sin^{n-1}x\cos(nx + x) && \text{[Addition Formula for cosine]}\\ &= n\sin^{n-1}x\cos[(n+1)x] && [\text{factor out } x]\end{aligned}$$

(b)

$$\begin{aligned}\frac{d}{dx}(\cos^n x\cos nx) &= n\cos^{n-1}x(-\sin x)\cos nx + \cos^n x(-n\sin nx) && \text{[Product Rule]}\\ &= -n\cos^{n-1}x(\cos nx\sin x + \sin nx\cos x) && [\text{factor out } -n\cos^{n-1}x]\\ &= -n\cos^{n-1}x\sin(nx + x) && \text{[Addition Formula for sine]}\\ &= -n\cos^{n-1}x\sin[(n+1)x] && [\text{factor out } x]\end{aligned}$$

75. Since $\theta^\circ = \left(\frac{\pi}{180}\right)\theta$ rad, we have $\dfrac{d}{d\theta}(\sin\theta^\circ) = \dfrac{d}{d\theta}\left(\sin\frac{\pi}{180}\theta\right) = \frac{\pi}{180}\cos\frac{\pi}{180}\theta = \frac{\pi}{180}\cos\theta^\circ$.

77. First note that products and differences of polynomials are polynomials and that the derivative of a polynomial is also a polynomial. When $n = 1$, $f^{(1)}(x) = \left(\dfrac{P(x)}{Q(x)}\right)' = \dfrac{Q(x)P'(x) - P(x)Q'(x)}{[Q(x)]^2} = \dfrac{A_1(x)}{[Q(x)]^{1+1}}$, where

$A_1(x) = Q(x)P'(x) - P(x)Q'(x)$. Suppose the result is true for $n = k$, where $k \geq 1$. Then

$f^{(k)}(x) = \dfrac{A_k(x)}{[Q(x)]^{k+1}}$, so

$$\begin{aligned} f^{(k+1)}(x) &= \left(\frac{A_k(x)}{[Q(x)]^{k+1}}\right)' = \frac{[Q(x)]^{k+1}A_k'(x) - A_k(x)\cdot(k+1)[Q(x)]^k\cdot Q'(x)}{\{[Q(x)]^{k+1}\}^2} \\ &= \frac{[Q(x)]^{k+1}A_k'(x) - (k+1)A_k(x)[Q(x)]^kQ'(x)}{[Q(x)]^{2k+2}} \\ &= \frac{[Q(x)]^k\{[Q(x)]^1A_k'(x) - (k+1)A_k(x)Q'(x)\}}{[Q(x)]^k[Q(x)]^{k+2}} = \frac{Q(x)A_k'(x) - (k+1)A_k(x)Q'(x)}{[Q(x)]^{k+2}} \\ &= A_{k+1}(x)/[Q(x)]^{k+2}, \text{ where } A_{k+1}(x) = Q(x)A_k'(x) - (k+1)A_k(x)Q'(x). \end{aligned}$$

We have shown that the formula holds for $n = 1$, and that when it holds for $n = k$ it also holds for $n = k + 1$. Thus, by mathematical induction, the formula holds for all positive integers n.

Linear Approximations and Differentials

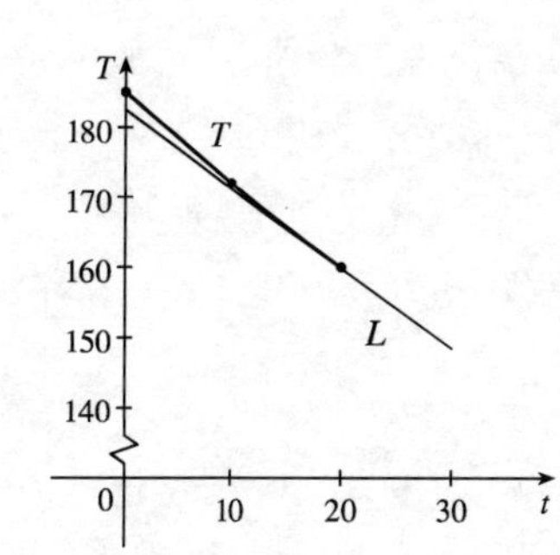

1. As in Example 1, $T(0) = 185$, $T(10) = 172$, $T(20) = 160$, and

$$T'(20) \approx \frac{T(10) - T(20)}{10 - 20} = \frac{172 - 160}{-10} = -1.2\ ^\circ\text{F/min.}$$

$T(30) \approx T(20) + T'(20)(30 - 20) \approx 160 - 1.2(10) = 148\ ^\circ$F.

We would expect the temperature of the turkey to get closer to $75\ ^\circ$F as time increases. Since the temperature decreased $13\ ^\circ$F in the first 10 minutes and $12\ ^\circ$F in the second 10 minutes, we can assume that the slopes of the tangent line are increasing through negative values: $-1.3, -1.2, \ldots$. Hence, the tangent lines are under the curve and $148\ ^\circ$F is an underestimate. From the figure, we estimate the slope of the tangent line at $t = 20$ to be $\frac{184 - 147}{0 - 30} = -\frac{37}{30}$. Then the linear approximation becomes $T(30) \approx T(20) + T'(20) \cdot 10 \approx 160 - \frac{37}{30}(10) = 147\frac{2}{3} \approx 147.7$.

3. Extend the tangent line at the point $(2030, 21)$ to the t-axis. Answers will vary based on this approximation—we'll use $t = 1900$ as our t-intercept. The linearization is then

$$P(t) \approx P(2030) + P'(2030)(t - 2030)$$
$$\approx 21 + \tfrac{21}{130}(t - 2030)$$

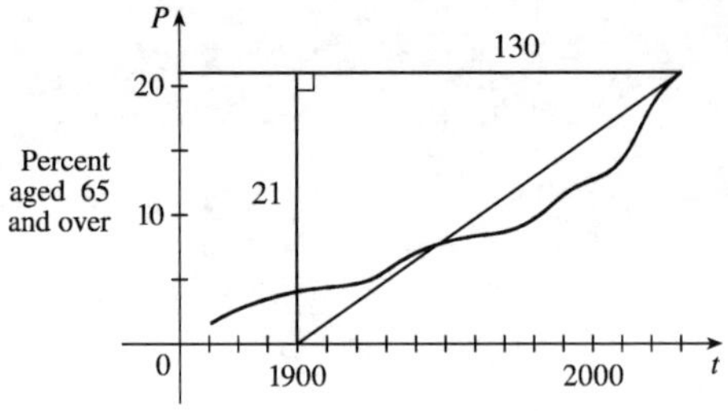

$P(2040) = 21 + \frac{21}{130}(2040 - 2030) \approx 22.6\%$

$P(2050) = 21 + \frac{21}{130}(2050 - 2030) \approx 24.2\%$

These predictions are probably too high since the tangent line lies above the graph at $t = 2030$.

5. $f(x) = x^3 \;\Rightarrow\; f'(x) = 3x^2$, so $f(1) = 1$ and $f'(1) = 3$. With $a = 1$, $L(x) = f(a) + f'(a)(x - a)$ becomes $L(x) = f(1) + f'(1)(x - 1) = 1 + 3(x - 1) = 3x - 2$.

7. $f(x) = \cos x \;\Rightarrow\; f'(x) = -\sin x$, so $f\left(\frac{\pi}{2}\right) = 0$ and $f'\left(\frac{\pi}{2}\right) = -1$. Thus, $L(x) = f\left(\frac{\pi}{2}\right) + f'\left(\frac{\pi}{2}\right)\left(x - \frac{\pi}{2}\right) = 0 - 1\left(x - \frac{\pi}{2}\right) = -x + \frac{\pi}{2}$.

9. $f(x) = \sqrt{1-x} \;\Rightarrow\; f'(x) = \dfrac{-1}{2\sqrt{1-x}}$, so $f(0) = 1$ and $f'(0) = -\frac{1}{2}$. Therefore,

$$\sqrt{1-x} = f(x) \approx f(0) + f'(0)(x - 0)$$
$$= 1 + \left(-\tfrac{1}{2}\right)(x - 0) = 1 - \tfrac{1}{2}x$$

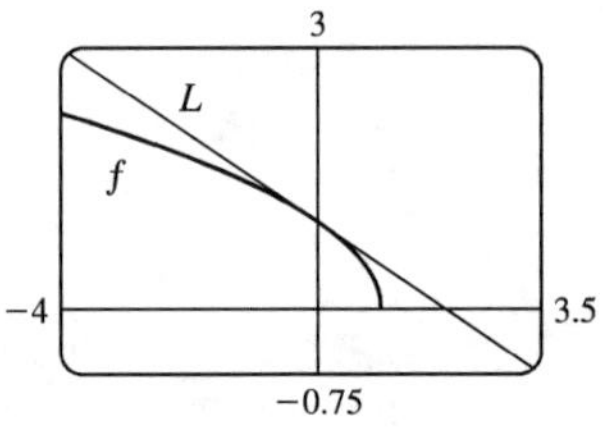

So $\sqrt{0.9} = \sqrt{1 - 0.1} \approx 1 - \frac{1}{2}(0.1) = 0.95$ and $\sqrt{0.99} = \sqrt{1 - 0.01} \approx 1 - \frac{1}{2}(0.01) = 0.995$.

11. $f(x) = \sqrt[3]{1-x} = (1-x)^{1/3} \;\Rightarrow\; f'(x) = -\frac{1}{3}(1-x)^{-2/3}$, so $f(0) = 1$ and $f'(0) = -\frac{1}{3}$. Thus, $f(x) \approx f(0) + f'(0)(x - 0) = 1 - \frac{1}{3}x$. We need $\sqrt[3]{1-x} - 0.1 < 1 - \frac{1}{3}x < \sqrt[3]{1-x} + 0.1$, which is true when $-1.204 < x < 0.706$.

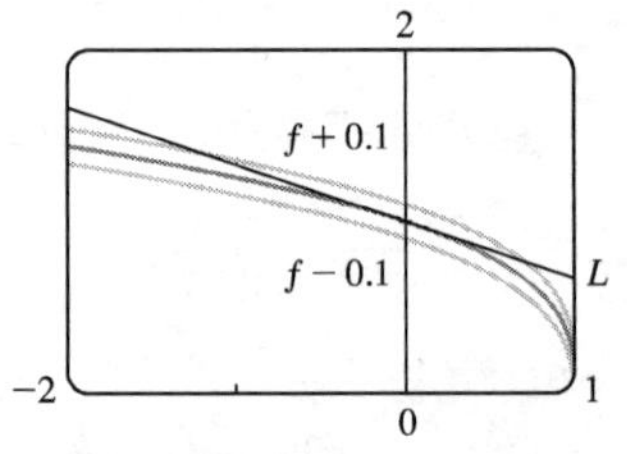

13. $f(x) = \dfrac{1}{(1+2x)^4} = (1+2x)^{-4} \;\Rightarrow$

$f'(x) = -4(1+2x)^{-5}(2) = \dfrac{-8}{(1+2x)^5}$, so $f(0) = 1$ and $f'(0) = -8$.

Thus, $f(x) \approx f(0) + f'(0)(x - 0) = 1 + (-8)(x - 0) = 1 - 8x$.

We need $1/(1+2x)^4 - 0.1 < 1 - 8x < 1/(1+2x)^4 + 0.1$, which is true when $-0.045 < x < 0.055$.

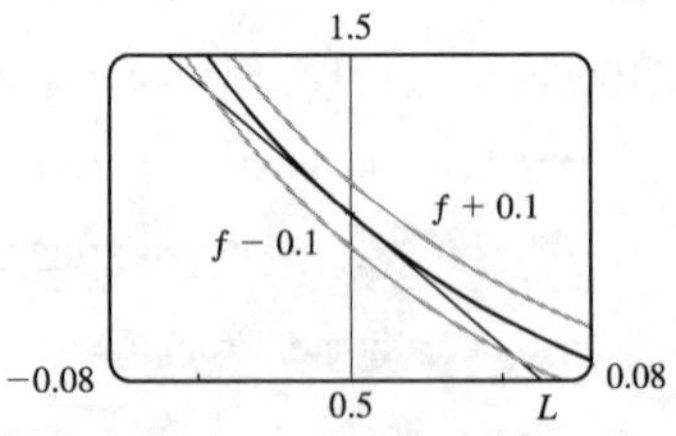

15. If $y = f(x)$, then the differential dy is equal to $f'(x)\,dx$. $y = x^4 + 5x \;\Rightarrow\; dy = \left(4x^3 + 5\right) dx$.

17. $y = x^2 \tan x \;\Rightarrow\; dy = \left(x^2 \sec^2 x + \tan x \cdot 2x\right) dx = \left(x^2 \sec^2 x + 2x \tan x\right) dx$

19. $y = \dfrac{u+1}{u-1} \;\Rightarrow\; dy = \dfrac{(u-1)(1) - (u+1)(1)}{(u-1)^2}\,du = \dfrac{-2}{(u-1)^2}\,du$

21. (a) $y = x^2 + 2x \;\Rightarrow\; dy = (2x + 2)\,dx$

(b) When $x = 3$ and $dx = \frac{1}{2}$, $dy = [2(3) + 2]\left(\frac{1}{2}\right) = 4$.

23. (a) $y = \sqrt{4 + 5x} \;\Rightarrow\; dy = \frac{1}{2}(4 + 5x)^{-1/2} \cdot 5\,dx = \dfrac{5}{2\sqrt{4 + 5x}}\,dx$

(b) When $x = 0$ and $dx = 0.04$, $dy = \dfrac{5}{2\sqrt{4}}(0.04) = \frac{5}{4} \cdot \frac{1}{25} = \frac{1}{20} = 0.05$.

25. (a) $y = \tan x \;\Rightarrow\; dy = \sec^2 x\,dx$

(b) When $x = \pi/4$ and $dx = -0.1$, $dy = [\sec(\pi/4)]^2\,(-0.1) = \left(\sqrt{2}\right)^2(-0.1) = -0.2$.

27. $y = x^2$, $x = 1$, $\Delta x = 0.5 \;\Rightarrow$

$\Delta y = (1.5)^2 - 1^2 = 1.25$.

$dy = 2x\,dx = 2(1)(0.5) = 1$

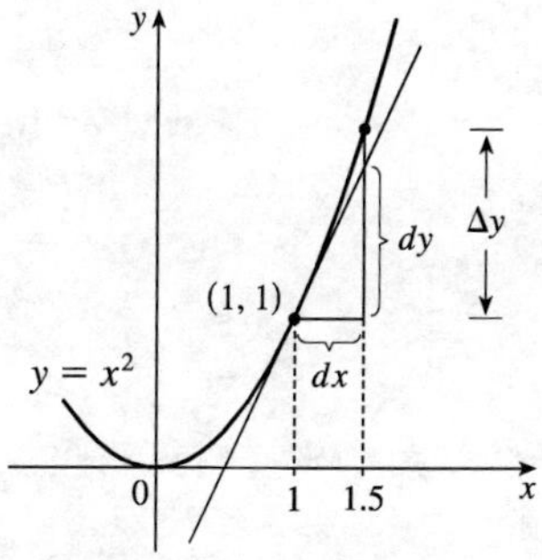

29. $y = 6 - x^2$, $x = -2$, $\Delta x = 0.4 \;\Rightarrow$

$\Delta y = \left(6 - (-1.6)^2\right) - \left(6 - (-2)^2\right) = 1.44$

$dy = -2x\,dx = -2(-2)(0.4) = 1.6$

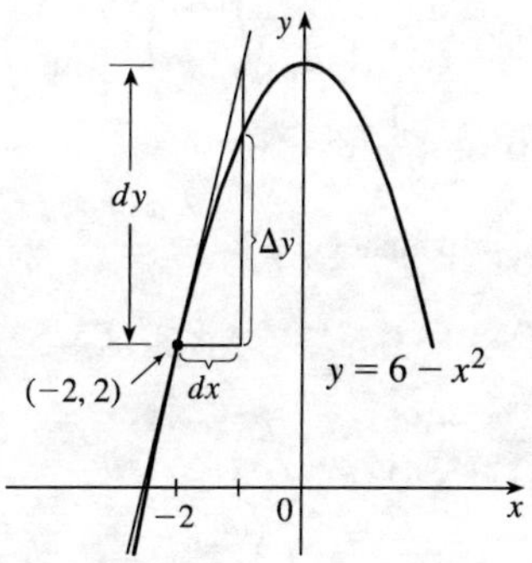

31. $y = f(x) = x^5 \;\Rightarrow\; dy = 5x^4\,dx$. When $x = 2$ and $dx = 0.001$, $dy = 5(2)^4(0.001) = 0.08$, so $(2.001)^5 = f(2.001) \approx f(2) + dy = 32 + 0.08 = 32.08$.

33. $y = f(x) = x^{2/3} \;\Rightarrow\; dy = \dfrac{2}{3\sqrt[3]{x}}\,dx$. When $x = 8$ and $dx = 0.06$, $dy = \dfrac{2}{3\sqrt[3]{8}}(0.06) = 0.02$, so

$(8.06)^{2/3} = f(8.06) \approx f(8) + dy = 4 + 0.02 = 4.02$.

35. $y = f(x) = \tan x \;\Rightarrow\; dy = \sec^2 x\,dx$. When $x = 45^\circ$ and $dx = -1^\circ$,

$dy = \sec^2 45^\circ(-\pi/180) = \left(\sqrt{2}\right)^2(-\pi/180) = -\pi/90$, so

$\tan 44^\circ = f(44^\circ) \approx f(45^\circ) + dy = 1 - \pi/90 \approx 0.965$.

37. $y = f(x) = \sec x \;\Rightarrow\; f'(x) = \sec x\,\tan x$, so $f(0) = 1$ and $f'(0) = 1 \cdot 0 = 0$. The linear approximation of f at 0 is $f(0) + f'(0)(x - 0) = 1 + 0(x) = 1$. Since 0.08 is close to 0, approximating $\sec 0.08$ with 1 is reasonable.

39. (a) If x is the edge length, then $V = x^3 \Rightarrow dV = 3x^2\,dx$. When $x = 30$ and $dx = 0.1$, $dV = 3(30)^2(0.1) = 270$, so the maximum possible error in computing the volume of the cube is about 270 cm^3. The relative error is calculated by dividing the change in V, ΔV, by V. We approximate ΔV with dV.

Relative error $= \dfrac{\Delta V}{V} \approx \dfrac{dV}{V} = \dfrac{3x^2\,dx}{x^3} = 3\dfrac{dx}{x} = 3\left(\dfrac{0.1}{30}\right) = 0.01.$

Percentage error $=$ relative error $\times\ 100\% = 0.01 \times 100\% = 1\%$.

(b) $S = 6x^2 \Rightarrow dS = 12x\,dx$. When $x = 30$ and $dx = 0.1$, $dS = 12(30)(0.1) = 36$, so the maximum possible error in computing the surface area of the cube is about 36 cm^2.

Relative error $= \dfrac{\Delta S}{S} \approx \dfrac{dS}{S} = \dfrac{12x\,dx}{6x^2} = 2\dfrac{dx}{x} = 2\left(\dfrac{0.1}{30}\right) = 0.00\overline{6}.$

Percentage error $=$ relative error $\times\ 100\% = 0.00\overline{6} \times 100\% = 0.\overline{6}\%$.

41. (a) For a sphere of radius r, the circumference is $C = 2\pi r$ and the surface area is $S = 4\pi r^2$, so $r = C/(2\pi) \Rightarrow S = 4\pi(C/2\pi)^2 = C^2/\pi \Rightarrow dS = (2/\pi)C\,dC$. When $C = 84$ and $dC = 0.5$, $dS = \dfrac{2}{\pi}(84)(0.5) = \dfrac{84}{\pi}$, so the maximum error is about $\dfrac{84}{\pi} \approx 27\text{ cm}^2$. Relative error $\approx \dfrac{dS}{S} = \dfrac{84/\pi}{84^2/\pi} = \dfrac{1}{84} \approx 0.012$

(b) $V = \dfrac{4}{3}\pi r^3 = \dfrac{4}{3}\pi\left(\dfrac{C}{2\pi}\right)^3 = \dfrac{C^3}{6\pi^2} \Rightarrow dV = \dfrac{1}{2\pi^2}C^2\,dC$. When $C = 84$ and $dC = 0.5$, $dV = \dfrac{1}{2\pi^2}(84)^2(0.5) = \dfrac{1764}{\pi^2}$, so the maximum error is about $\dfrac{1764}{\pi^2} \approx 179\text{ cm}^3$. The relative error is approximately $\dfrac{dV}{V} = \dfrac{1764/\pi^2}{(84)^3/(6\pi^2)} = \dfrac{1}{56} \approx 0.018.$

43. (a) $V = \pi r^2 h \Rightarrow \Delta V \approx dV = 2\pi rh\,dr = 2\pi rh\,\Delta r$

(b) The error is

$$\begin{aligned}\Delta V - dV &= [\pi(r+\Delta r)^2 h - \pi r^2 h] - 2\pi rh\,\Delta r = \pi r^2 h + 2\pi rh\,\Delta r + \pi(\Delta r)^2 h - \pi r^2 h - 2\pi rh\,\Delta r\\ &= \pi(\Delta r)^2 h\end{aligned}$$

45. (a) $dc = \dfrac{dc}{dx}\,dx = 0\,dx = 0$

(b) $d(cu) = \dfrac{d}{dx}(cu)\,dx = c\dfrac{du}{dx}\,dx = c\,du$

(c) $d(u+v) = \dfrac{d}{dx}(u+v)\,dx = \left(\dfrac{du}{dx} + \dfrac{dv}{dx}\right)dx = \dfrac{du}{dx}\,dx + \dfrac{dv}{dx}\,dx = du + dv$

(d) $d(uv) = \dfrac{d}{dx}(uv)\,dx = \left(u\dfrac{dv}{dx} + v\dfrac{du}{dx}\right)dx = u\dfrac{dv}{dx}\,dx + v\dfrac{du}{dx}\,dx = u\,dv + v\,du$

(e) $d\left(\dfrac{u}{v}\right) = \dfrac{d}{dx}\left(\dfrac{u}{v}\right)dx = \dfrac{v\dfrac{du}{dx} - u\dfrac{dv}{dx}}{v^2}\,dx = \dfrac{v\dfrac{du}{dx}\,dx - u\dfrac{dv}{dx}\,dx}{v^2} = \dfrac{v\,du - u\,dv}{v^2}$

(f) $d\,(x^n) = \dfrac{d}{dx}(x^n)\,dx = nx^{n-1}\,dx$

47. (a) The graph shows that $f'(1) = 2$, so $L(x) = f(1) + f'(1)(x-1) = 5 + 2(x-1) = 2x + 3$. $f(0.9) \approx L(0.9) = 4.8$ and $f(1.1) \approx L(1.1) = 5.2$.

(b) From the graph, we see that $f'(x)$ is positive and decreasing. This means that the slopes of the tangent lines are positive, but the tangents are becoming less steep. So the tangent lines lie *above* the curve. Thus, the estimates in part (a) are too large.

Rates of Change in the Natural and Social Sciences

1. (a) $s = f(t) = t^2 - 10t + 12 \quad \Rightarrow \quad v(t) = f'(t) = 2t - 10$

(b) $v(3) = 2(3) - 10 = -4$ ft/s

(c) The particle is at rest when $v(t) = 0 \quad \Leftrightarrow \quad 2t - 10 = 0 \quad \Leftrightarrow \quad t = 5$ s.

(d) The particle is moving in the positive direction when $v(t) > 0 \quad \Leftrightarrow \quad 2t - 10 > 0 \quad \Leftrightarrow \quad 2t > 10 \quad \Leftrightarrow \quad t > 5$.

(e) Since the particle is moving in the positive direction and in the negative direction, we need to calculate the distance traveled in the intervals $[0, 5]$ and $[5, 8]$ separately. $|f(5) - f(0)| = |-13 - 12| = 25$ ft and $|f(8) - f(5)| = |-4 - (-13)| = 9$ ft. The total distance traveled during the first 8 s is $25 + 9 = 34$ ft.

(f)

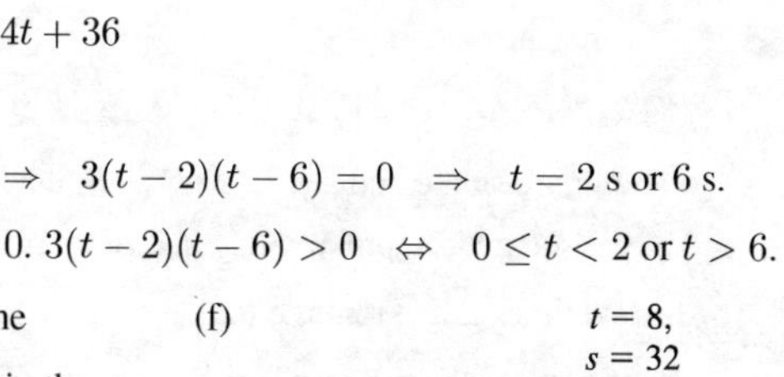

3. (a) $s = f(t) = t^3 - 12t^2 + 36t \quad \Rightarrow \quad v(t) = f'(t) = 3t^2 - 24t + 36$

(b) $v(3) = 27 - 72 + 36 = -9$ ft/s

(c) The particle is at rest when $v(t) = 0$. $3t^2 - 24t + 36 = 0 \quad \Rightarrow \quad 3(t - 2)(t - 6) = 0 \quad \Rightarrow \quad t = 2$ s or 6 s.

(d) The particle is moving in the positive direction when $v(t) > 0$. $3(t - 2)(t - 6) > 0 \quad \Leftrightarrow \quad 0 \leq t < 2$ or $t > 6$.

(e) Since the particle is moving in the positive direction and in the negative direction, we need to calculate the distance traveled in the intervals $[0, 2]$, $[2, 6]$, and $[6, 8]$ separately.

$|f(2) - f(0)| = |32 - 0| = 32.$

$|f(6) - f(2)| = |0 - 32| = 32.$

$|f(8) - f(6)| = |32 - 0| = 32.$

The total distance is $32 + 32 + 32 = 96$ ft.

(f)

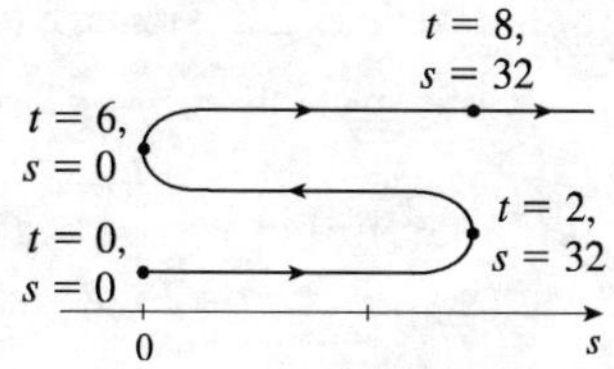

5. (a) $s = \dfrac{t}{t^2+1} \Rightarrow v(t) = s'(t) = \dfrac{(t^2+1)(1) - t(2t)}{(t^2+1)^2} = \dfrac{1-t^2}{(t^2+1)^2}$

(b) $v(3) = \dfrac{1-(3)^2}{(3^2+1)^2} = \dfrac{1-9}{10^2} = \dfrac{-8}{100} = -\dfrac{2}{25}$ ft/s

(c) It is at rest when $v = 0 \Leftrightarrow 1 - t^2 = 0 \Leftrightarrow t = 1$ s $[t \neq -1$ since $t \geq 0]$.

(d) It moves in the positive direction when $v > 0 \Leftrightarrow 1 - t^2 > 0 \Leftrightarrow t^2 < 1 \Leftrightarrow 0 \leq t < 1$.

(e) Distance in positive direction $= |s(1) - s(0)| = \left|\frac{1}{2} - 0\right| = \frac{1}{2}$ ft

Distance in negative direction $= |s(8) - s(1)| = \left|\frac{8}{65} - \frac{1}{2}\right| = \frac{49}{130}$ ft

Total distance traveled $= \frac{1}{2} + \frac{49}{130} = \frac{57}{65}$ ft

(f)

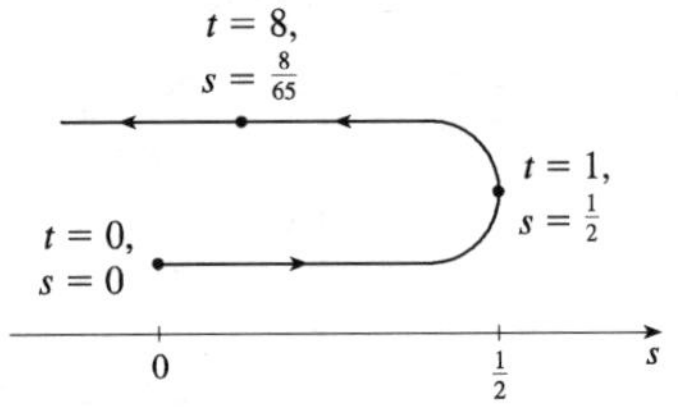

7. $s(t) = t^3 - 4.5t^2 - 7t \Rightarrow v(t) = s'(t) = 3t^2 - 9t - 7 = 5 \Leftrightarrow 3t^2 - 9t - 12 = 0 \Leftrightarrow$ $3(t-4)(t+1) = 0 \Leftrightarrow t = 4$ or -1. Since $t \geq 0$, the particle reaches a velocity of 5 m/s at $t = 4$ s.

9. (a) $h = 10t - 0.83t^2 \Rightarrow v(t) = \dfrac{dh}{dt} = 10 - 1.66t$, so $v(3) = 10 - 1.66(3) = 5.02$ m/s.

(b) $h = 25 \Rightarrow 10t - 0.83t^2 = 25 \Rightarrow 0.83t^2 - 10t + 25 = 0 \Rightarrow t = \dfrac{10 \pm \sqrt{17}}{1.66} \approx 3.54$ or 8.51.

The value $t_1 = \left(10 - \sqrt{17}\right)/1.66$ corresponds to the time it takes for the stone to rise 25 m and $t_2 = \left(10 + \sqrt{17}\right)/1.66$ corresponds to the time when the stone is 25 m high on the way down. Thus, $v(t_1) = 10 - 1.66\left[\left(10 - \sqrt{17}\right)/1.66\right] = \sqrt{17} \approx 4.12$ m/s.

11. (a) $A(x) = x^2 \Rightarrow A'(x) = 2x$. $A'(15) = 30\ \text{mm}^2/\text{mm}$ is the rate at which the area is increasing with respect to the side length as x reaches 15 mm.

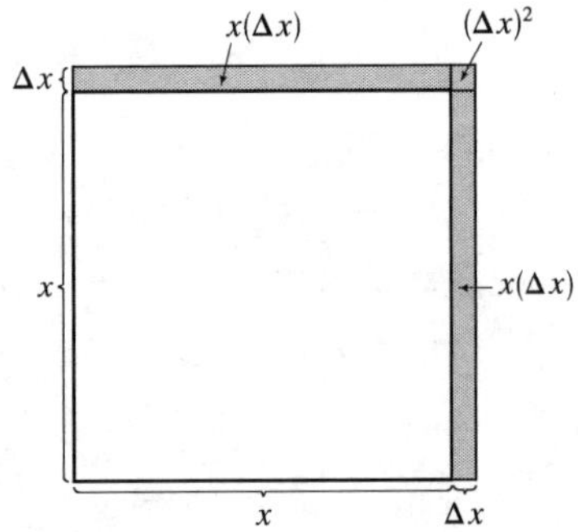

(b) The perimeter is $P(x) = 4x$, so $A'(x) = 2x = \frac{1}{2}(4x) = \frac{1}{2}P(x)$. The figure suggests that if Δx is small, then the change in the area of the square is approximately half of its perimeter (2 of the 4 sides) times Δx. From the figure, $\Delta A = 2x(\Delta x) + (\Delta x)^2$. If Δx is small, then $\Delta A \approx 2x(\Delta x)$ and so $\Delta A/\Delta x \approx 2x$.

13. (a) Using $A(r) = \pi r^2$, we find that the average rate of change is:

(i) $\dfrac{A(3) - A(2)}{3-2} = \dfrac{9\pi - 4\pi}{1} = 5\pi$ (ii) $\dfrac{A(2.5) - A(2)}{2.5-2} = \dfrac{6.25\pi - 4\pi}{0.5} = 4.5\pi$

(iii) $\dfrac{A(2.1) - A(2)}{2.1-2} = \dfrac{4.41\pi - 4\pi}{0.1} = 4.1\pi$

(b) $A(r) = \pi r^2 \Rightarrow A'(r) = 2\pi r$, so $A'(2) = 4\pi$.

(c) The circumference is $C(r) = 2\pi r = A'(r)$. The figure suggests that if Δr is small, then the change in the area of the circle (a ring around the outside) is approximately equal to its circumference times Δr. Straightening out this ring gives us a shape that is approximately rectangular with length $2\pi r$ and width Δr, so $\Delta A \approx 2\pi r(\Delta r)$. Algebraically,

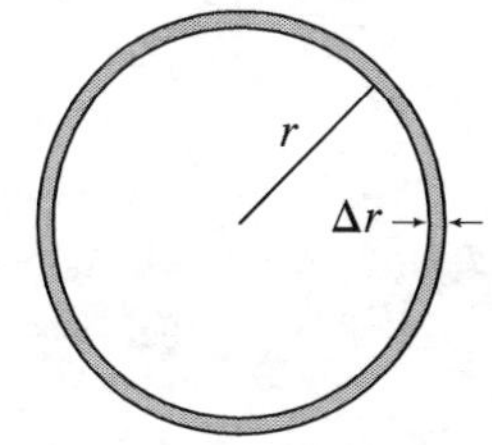

$\Delta A = A(r + \Delta r) - A(r) = \pi(r + \Delta r)^2 - \pi r^2 = 2\pi r(\Delta r) + \pi(\Delta r)^2$.

So we see that if Δr is small, then $\Delta A \approx 2\pi r(\Delta r)$ and therefore, $\Delta A/\Delta r \approx 2\pi r$.

15. $S(r) = 4\pi r^2 \Rightarrow S'(r) = 8\pi r \Rightarrow$

(a) $S'(1) = 8\pi\ \text{ft}^2/\text{ft}$ (b) $S'(2) = 16\pi\ \text{ft}^2/\text{ft}$ (c) $S'(3) = 24\pi\ \text{ft}^2/\text{ft}$

As the radius increases, the surface area grows at an increasing rate. In fact, the rate of change is linear with respect to the radius.

17. The mass is $f(x) = 3x^2$, so the linear density at x is $\rho(x) = f'(x) = 6x$.

(a) $\rho(1) = 6$ kg/m (b) $\rho(2) = 12$ kg/m (c) $\rho(3) = 18$ kg/m

Since ρ is an increasing function, the density will be the highest at the right end of the rod and lowest at the left end.

19. The quantity of charge is $Q(t) = t^3 - 2t^2 + 6t + 2$, so the current is $Q'(t) = 3t^2 - 4t + 6$.

(a) $Q'(0.5) = 3(0.5)^2 - 4(0.5) + 6 = 4.75$ A (b) $Q'(1) = 3(1)^2 - 4(1) + 6 = 5$ A

The current is lowest when Q' has a minimum. $Q''(t) = 6t - 4 < 0$ when $t < \frac{2}{3}$. So the current decreases when $t < \frac{2}{3}$ and increases when $t > \frac{2}{3}$. Thus, the current is lowest at $t = \frac{2}{3}$ s.

21. (a) To find the rate of change of volume with respect to pressure, we first solve for V in terms of P.

$$PV = C \Rightarrow V = \frac{C}{P} \Rightarrow \frac{dV}{dP} = -\frac{C}{P^2}.$$

(b) From the formula for dV/dP in part (a), we see that as P increases, the absolute value of dV/dP decreases. Thus, the volume is decreasing more rapidly at the beginning.

(c) $\beta = -\dfrac{1}{V}\dfrac{dV}{dP} = -\dfrac{1}{V}\left(-\dfrac{C}{P^2}\right) = \dfrac{C}{(PV)P} = \dfrac{C}{CP} = \dfrac{1}{P}$

23. (a) **1920:** $m_1 = \dfrac{1860 - 1750}{1920 - 1910} = \frac{110}{10} = 11$, $m_2 = \dfrac{2070 - 1860}{1930 - 1920} = \frac{210}{10} = 21$,

$(m_1 + m_2)/2 = (11 + 21)/2 = 16$ million/year

1980: $m_1 = \dfrac{4450 - 3710}{1980 - 1970} = \frac{740}{10} = 74$, $m_2 = \dfrac{5280 - 4450}{1990 - 1980} = \frac{830}{10} = 83$,

$(m_1 + m_2)/2 = (74 + 83)/2 = 78.5$ million/year

(b) $P(t) = at^3 + bt^2 + ct + d$ (in millions of people), where $a \approx 0.0012937063$, $b \approx -7.061421911$, $c \approx 12{,}822.97902$, and $d \approx -7{,}743{,}770.396$.

(c) $P(t) = at^3 + bt^2 + ct + d \Rightarrow P'(t) = 3at^2 + 2bt + c$ (in millions of people per year)

(d) $P'(1920) = 3(0.0012937063)(1920)^2 + 2(-7.061421911)(1920) + 12{,}822.97902$

≈ 14.48 million/year [smaller than the answer in part (a), but close to it]

$P'(1980) \approx 75.29$ million/year (smaller, but close)

(e) $P'(1985) \approx 81.62$ million/year, so the rate of growth in 1985 was about 81.62 million/year.

25. (a) $[\mathrm{C}] = \dfrac{a^2kt}{akt+1} \quad \Rightarrow$

rate of reaction $= \dfrac{d\,[\mathrm{C}]}{dt} = \dfrac{(akt+1)(a^2k) - (a^2kt)(ak)}{(akt+1)^2} = \dfrac{a^2k(akt+1-akt)}{(akt+1)^2} = \dfrac{a^2k}{(akt+1)^2}$

(b) If $x = [\mathrm{C}]$, then $a - x = a - \dfrac{a^2kt}{akt+1} = \dfrac{a^2kt + a - a^2kt}{akt+1} = \dfrac{a}{akt+1}$.

So $k(a-x)^2 = k\left(\dfrac{a}{akt+1}\right)^2 = \dfrac{a^2k}{(akt+1)^2} = \dfrac{d\,[\mathrm{C}]}{dt}$ [from part (a)] $= \dfrac{dx}{dt}$.

27. (a) Using $v = \dfrac{P}{4\eta l}\left(R^2 - r^2\right)$ with $R = 0.01$, $l = 3$, $P = 3000$, and $\eta = 0.027$, we have v as a function of r:

$v(r) = \dfrac{3000}{4(0.027)3}\left(0.01^2 - r^2\right)$. $v(0) = 0.\overline{925}$ cm/s, $v(0.005) = 0.69\overline{4}$ cm/s, $v(0.01) = 0$.

(b) $v(r) = \dfrac{P}{4\eta l}\left(R^2 - r^2\right) \quad \Rightarrow \quad v'(r) = \dfrac{P}{4\eta l}(-2r) = -\dfrac{Pr}{2\eta l}$. When $l = 3$, $P = 3000$, and $\eta = 0.027$, we have

$v'(r) = -\dfrac{3000r}{2(0.027)3}$. $v'(0) = 0$, $v'(0.005) = -92.\overline{592}$ (cm/s)/cm, and $v'(0.01) = -185.\overline{185}$ (cm/s)/cm.

(c) The velocity is greatest where $r = 0$ (at the center) and the velocity is changing most where $r = R = 0.01$ cm (at the edge).

29. (a) $C(x) = 2000 + 3x + 0.01x^2 + 0.0002x^3 \quad \Rightarrow \quad C'(x) = 3 + 0.02x + 0.0006x^2$

(b) $C'(100) = 3 + 0.02(100) + 0.0006(10{,}000) = 3 + 2 + 6 = \$11$/pair. $C'(100)$ is the rate at which the cost is increasing as the 100th pair of jeans is produced. It predicts the cost of the 101st pair.

(c) The cost of manufacturing the 101st pair of jeans is

$$\begin{aligned} C(101) - C(100) &= (2000 + 303 + 102.01 + 206.0602) - (2000 + 300 + 100 + 200) \\ &= 11.0702 \approx \$11.07 \end{aligned}$$

31. (a) $A(x) = \dfrac{p(x)}{x} \quad \Rightarrow \quad A'(x) = \dfrac{xp'(x) - p(x)\cdot 1}{x^2} = \dfrac{xp'(x) - p(x)}{x^2}$. $A'(x) > 0 \quad \Rightarrow \quad A(x)$ is increasing; that is, the average productivity increases as the size of the workforce increases.

(b) $p'(x)$ is greater than the average productivity $\Rightarrow \quad p'(x) > A(x) \quad \Rightarrow \quad p'(x) > \dfrac{p(x)}{x} \quad \Rightarrow$

$xp'(x) > p(x) \quad \Rightarrow \quad xp'(x) - p(x) > 0 \quad \Rightarrow \quad \dfrac{xp'(x) - p(x)}{x^2} > 0 \quad \Rightarrow \quad A'(x) > 0.$

33. $PV = nRT \quad \Rightarrow \quad T = \dfrac{PV}{nR} = \dfrac{PV}{(10)(0.0821)} = \dfrac{1}{0.821}(PV)$. Using the Product Rule, we have

$\dfrac{dT}{dt} = \dfrac{1}{0.821}\left[P(t)V'(t) + V(t)P'(t)\right] = \dfrac{1}{0.821}\left[(8)(-0.15) + (10)(0.10)\right] \approx -0.2436$ K/min.

35. (a) If the populations are stable, then the growth rates are neither positive nor negative; that is, $\dfrac{dC}{dt} = 0$ and $\dfrac{dW}{dt} = 0$.

(b) "The caribou go extinct" means that the population is zero, or mathematically, $C = 0$.

(c) We have the equations $\dfrac{dC}{dt} = aC - bCW$ and $\dfrac{dW}{dt} = -cW + dCW$. Let $dC/dt = dW/dt = 0$, $a = 0.05$, $b = 0.001$, $c = 0.05$, and $d = 0.0001$ to obtain $0.05C - 0.001CW = 0$ **(1)** and $-0.05W + 0.0001CW = 0$ **(2)**. Adding 10 times **(2)** to **(1)** eliminates the CW-terms and gives us $0.05C - 0.5W = 0 \quad \Rightarrow \quad C = 10W$. Substituting $C = 10W$ into **(1)** results in $0.05(10W) - 0.001(10W)W = 0 \quad \Leftrightarrow \quad 0.5W - 0.01W^2 = 0 \quad \Leftrightarrow \quad 50W - W^2 = 0 \quad \Leftrightarrow$ $W(50 - W) = 0 \quad \Leftrightarrow \quad W = 0$ or 50. Since $C = 10W$, $C = 0$ or 500. Thus, the population pairs (C, W) that lead to stable populations are $(0, 0)$ and $(500, 50)$. So it is possible for the two species to live in harmony.

Implicit Differentiation

1. (a) $\dfrac{d}{dx}\left(xy + 2x + 3x^2\right) = \dfrac{d}{dx}(4) \;\Rightarrow\; (x \cdot y' + y \cdot 1) + 2 + 6x = 0 \;\Rightarrow\; xy' = -y - 2 - 6x \;\Rightarrow$

$y' = \dfrac{-y - 2 - 6x}{x}$ or $y' = -6 - \dfrac{y+2}{x}$.

(b) $xy + 2x + 3x^2 = 4 \;\Rightarrow\; xy = 4 - 2x - 3x^2 \;\Rightarrow\; y = \dfrac{4 - 2x - 3x^2}{x} = \dfrac{4}{x} - 2 - 3x$, so $y' = -\dfrac{4}{x^2} - 3$.

(c) From part (a), $y' = \dfrac{-y - 2 - 6x}{x} = \dfrac{-(4/x - 2 - 3x) - 2 - 6x}{x} = \dfrac{-4/x - 3x}{x} = -\dfrac{4}{x^2} - 3$.

3. (a) $\dfrac{d}{dx}\left(\dfrac{1}{x} + \dfrac{1}{y}\right) = \dfrac{d}{dx}(1) \;\Rightarrow\; -\dfrac{1}{x^2} - \dfrac{1}{y^2}y' = 0 \;\Rightarrow\; -\dfrac{1}{y^2}y' = \dfrac{1}{x^2} \;\Rightarrow\; y' = -\dfrac{y^2}{x^2}$

(b) $\dfrac{1}{x} + \dfrac{1}{y} = 1 \;\Rightarrow\; \dfrac{1}{y} = 1 - \dfrac{1}{x} = \dfrac{x-1}{x} \;\Rightarrow\; y = \dfrac{x}{x-1}$, so $y' = \dfrac{(x-1)(1) - (x)(1)}{(x-1)^2} = \dfrac{-1}{(x-1)^2}$.

(c) $y' = -\dfrac{y^2}{x^2} = -\dfrac{[x/(x-1)]^2}{x^2} = -\dfrac{x^2}{x^2(x-1)^2} = -\dfrac{1}{(x-1)^2}$

5. $\dfrac{d}{dx}\left(x^2+y^2\right)=\dfrac{d}{dx}(1) \;\Rightarrow\; 2x+2yy'=0 \;\Rightarrow\; 2yy'=-2x \;\Rightarrow\; y'=-\dfrac{x}{y}$

7. $\dfrac{d}{dx}\left(x^3+x^2y+4y^2\right)=\dfrac{d}{dx}(6) \;\Rightarrow\; 3x^2+\left(x^2y'+y\cdot 2x\right)+8yy'=0 \;\Rightarrow\; x^2y'+8yy'=-3x^2-2xy$

$\Rightarrow\; \left(x^2+8y\right)y'=-3x^2-2xy \;\Rightarrow\; y'=-\dfrac{3x^2+2xy}{x^2+8y}=-\dfrac{x(3x+2y)}{x^2+8y}$

9. $\dfrac{d}{dx}\left(x^2y+xy^2\right)=\dfrac{d}{dx}(3x) \;\Rightarrow\; \left(x^2y'+y\cdot 2x\right)+\left(x\cdot 2yy'+y^2\cdot 1\right)=3 \;\Rightarrow$

$x^2y'+2xyy'=3-2xy-y^2 \;\Rightarrow\; y'\left(x^2+2xy\right)=3-2xy-y^2 \;\Rightarrow\; y'=\dfrac{3-2xy-y^2}{x^2+2xy}$

11. $\dfrac{d}{dx}\left(x^2y^2+x\sin y\right)=\dfrac{d}{dx}(4) \;\Rightarrow\; x^2\cdot 2yy'+y^2\cdot 2x+x\cos y\cdot y'+\sin y\cdot 1=0 \;\Rightarrow$

$2x^2yy'+x\cos y\cdot y'=-2xy^2-\sin y \;\Rightarrow\; (2x^2y+x\cos y)y'=-2xy^2-\sin y \;\Rightarrow\; y'=\dfrac{-2xy^2-\sin y}{2x^2y+x\cos y}$

13. $\dfrac{d}{dx}(4\cos x\sin y)=\dfrac{d}{dx}(1) \;\Rightarrow\; 4\left[\cos x\cdot\cos y\cdot y'+\sin y\cdot(-\sin x)\right]=0 \;\Rightarrow$

$y'(4\cos x\cos y)=4\sin x\sin y \;\Rightarrow\; y'=\dfrac{4\sin x\sin y}{4\cos x\cos y}=\tan x\tan y$

15. $\dfrac{d}{dx}\left[\tan(x/y)\right]=\dfrac{d}{dx}(x+y) \;\Rightarrow\; \sec^2(x/y)\cdot\dfrac{y\cdot 1-x\cdot y'}{y^2}=1+y' \;\Rightarrow$

$y\sec^2(x/y)-x\sec^2(x/y)\cdot y'=y^2+y^2y' \;\Rightarrow\; y\sec^2(x/y)-y^2=y^2y'+x\sec^2(x/y) \;\Rightarrow$

$y\sec^2(x/y)-y^2=\left[y^2+x\sec^2(x/y)\right]\cdot y' \;\Rightarrow\; y'=\dfrac{y\sec^2(x/y)-y^2}{y^2+x\sec^2(x/y)}$

17. $\sqrt{xy}=1+x^2y \;\Rightarrow\; \frac{1}{2}(xy)^{-1/2}(xy'+y\cdot 1)=0+x^2y'+y\cdot 2x \;\Rightarrow\; \dfrac{x}{2\sqrt{xy}}y'+\dfrac{y}{2\sqrt{xy}}=x^2y'+2xy$

$\Rightarrow\; y'\left(\dfrac{x}{2\sqrt{xy}}-x^2\right)=2xy-\dfrac{y}{2\sqrt{xy}} \;\Rightarrow\; y'\left(\dfrac{x-2x^2\sqrt{xy}}{2\sqrt{xy}}\right)=\dfrac{4xy\sqrt{xy}-y}{2\sqrt{xy}} \;\Rightarrow\; y'=\dfrac{4xy\sqrt{xy}-y}{x-2x^2\sqrt{xy}}$

19. $xy=\cot(xy) \;\Rightarrow\; y+xy'=-\csc^2(xy)(y+xy') \;\Rightarrow\; (y+xy')\left[1+\csc^2(xy)\right]=0 \;\Rightarrow$

$y+xy'=0$ [since $1+\csc^2(xy)>0$] $\Rightarrow\; y'=-y/x$

21. $\frac{d}{dx}\{1 + f(x) + x^2[f(x)]^3\} = \frac{d}{dx}(0) \Rightarrow f'(x) + x^2 \cdot 3[f(x)]^2 \cdot f'(x) + [f(x)]^3 \cdot 2x = 0$. If $x = 1$, we have

$f'(1) + 1^2 \cdot 3[f(1)]^2 \cdot f'(1) + [f(1)]^3 \cdot 2(1) = 0 \Rightarrow f'(1) + 1 \cdot 3 \cdot 2^2 \cdot f'(1) + 2^3 \cdot 2 = 0 \Rightarrow$

$f'(1) + 12f'(1) = -16 \Rightarrow 13f'(1) = -16 \Rightarrow f'(1) = -\frac{16}{13}$.

23. $y^4 + x^2y^2 + yx^4 = y + 1 \Rightarrow 4y^3 + \left(x^2 \cdot 2y + y^2 \cdot 2x\frac{dx}{dy}\right) + \left(y \cdot 4x^3\frac{dx}{dy} + x^4 \cdot 1\right) = 1 \Rightarrow$

$2xy^2\frac{dx}{dy} + 4x^3y\frac{dx}{dy} = 1 - 4y^3 - 2x^2y - x^4 \Rightarrow \frac{dx}{dy} = \frac{1 - 4y^3 - 2x^2y - x^4}{2xy^2 + 4x^3y}$

25. $x^2 + xy + y^2 = 3 \Rightarrow 2x + xy' + y \cdot 1 + 2yy' = 0 \Rightarrow xy' + 2yy' = -2x - y \Rightarrow$

$y'(x + 2y) = -2x - y \Rightarrow y' = \frac{-2x - y}{x + 2y}$. When $x = 1$ and $y = 1$, we have $y' = \frac{-2 - 1}{1 + 2 \cdot 1} = \frac{-3}{3} = -1$, so

an equation of the tangent line is $y - 1 = -1(x - 1)$ or $y = -x + 2$.

27. $x^2 + y^2 = (2x^2 + 2y^2 - x)^2 \Rightarrow 2x + 2yy' = 2(2x^2 + 2y^2 - x)(4x + 4yy' - 1)$. When $x = 0$ and $y = \frac{1}{2}$, we

have $0 + y' = 2(\frac{1}{2})(2y' - 1) \Rightarrow y' = 2y' - 1 \Rightarrow y' = 1$, so an equation of the tangent line is

$y - \frac{1}{2} = 1(x - 0)$ or $y = x + \frac{1}{2}$.

29. $2(x^2 + y^2)^2 = 25(x^2 - y^2) \Rightarrow 4(x^2 + y^2)(2x + 2yy') = 25(2x - 2yy') \Rightarrow$

$4(x + yy')(x^2 + y^2) = 25(x - yy') \Rightarrow 4yy'(x^2 + y^2) + 25yy' = 25x - 4x(x^2 + y^2) \Rightarrow$

$y' = \frac{25x - 4x(x^2 + y^2)}{25y + 4y(x^2 + y^2)}$. When $x = 3$ and $y = 1$, we have $y' = \frac{75 - 120}{25 + 40} = -\frac{45}{65} = -\frac{9}{13}$, so an equation of the

tangent line is $y - 1 = -\frac{9}{13}(x - 3)$ or $y = -\frac{9}{13}x + \frac{40}{13}$.

31. (a) $y^2 = 5x^4 - x^2 \Rightarrow 2yy' = 5(4x^3) - 2x \Rightarrow y' = \frac{10x^3 - x}{y}$.

So at the point $(1, 2)$ we have $y' = \frac{10(1)^3 - 1}{2} = \frac{9}{2}$, and an

equation of the tangent line is $y - 2 = \frac{9}{2}(x - 1)$ or $y = \frac{9}{2}x - \frac{5}{2}$.

(b)

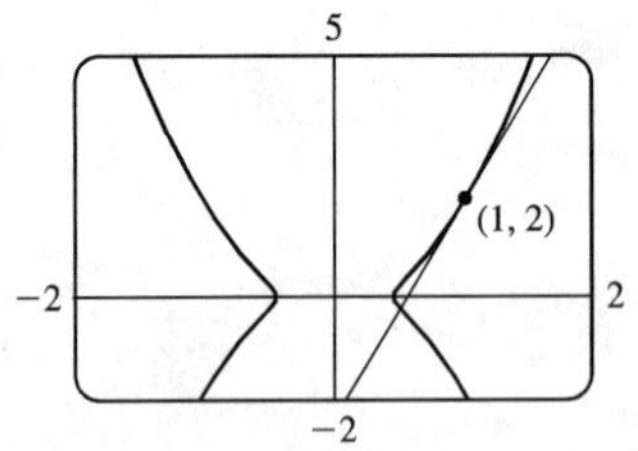

33. (a)

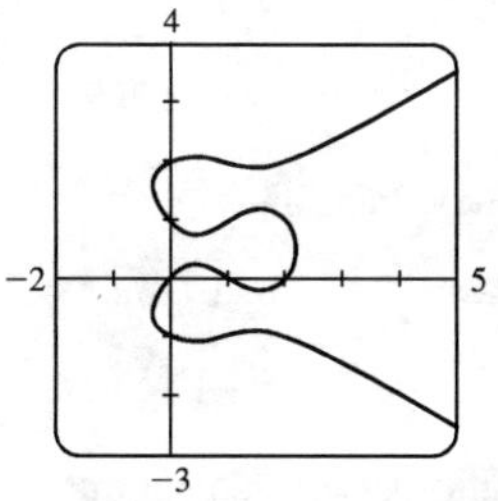

There are eight points with horizontal tangents: four at $x \approx 1.57735$ and four at $x \approx 0.42265$.

(b) $y' = \dfrac{3x^2 - 6x + 2}{2(2y^3 - 3y^2 - y + 1)} \Rightarrow y' = -1$ at $(0, 1)$ and $y' = \frac{1}{3}$ at $(0, 2)$.

Equations of the tangent lines are $y = -x + 1$ and $y = \frac{1}{3}x + 2$.

(c) $y' = 0 \Rightarrow 3x^2 - 6x + 2 = 0 \Rightarrow x = 1 \pm \frac{1}{3}\sqrt{3}$

(d) By multiplying the right side of the equation by $x - 3$, we obtain the first graph.

By modifying the equation in other ways, we can generate the other graphs.

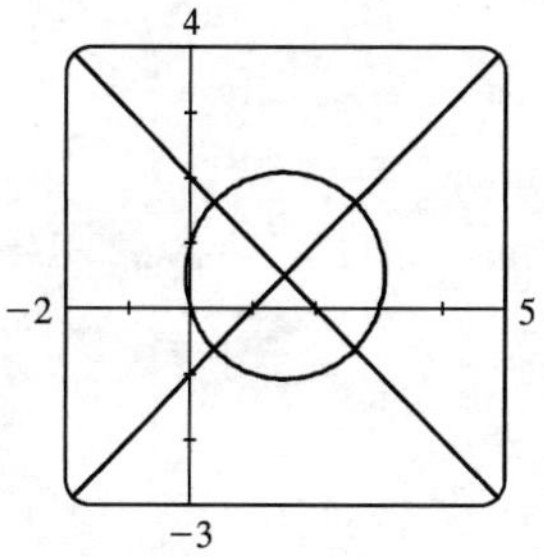

$y(y^2 - 1)(y - 2)$
$= x(x - 1)(x - 2)(x - 3)$

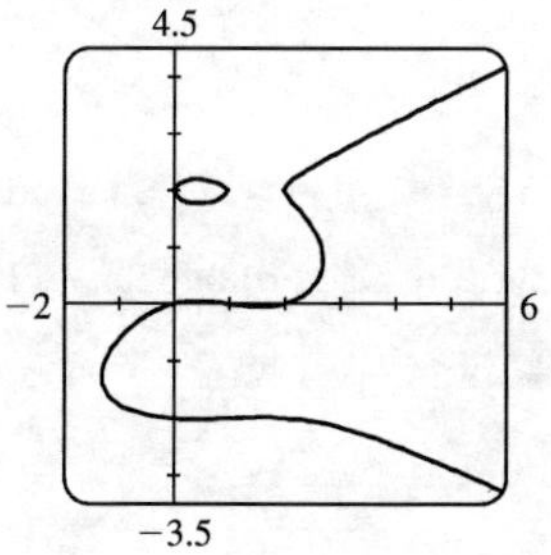

$y(y^2 - 4)(y - 2)$
$= x(x - 1)(x - 2)$

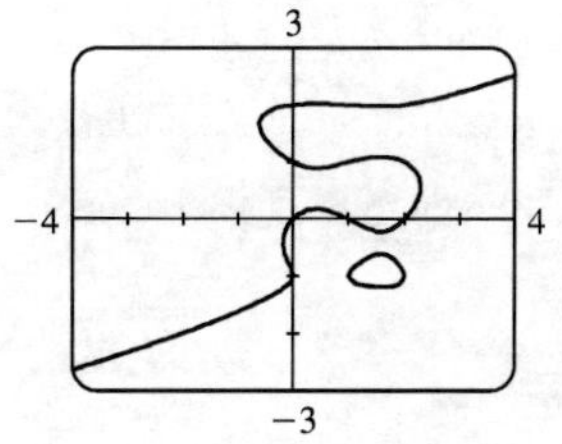

$y(y + 1)(y^2 - 1)(y - 2)$
$= x(x - 1)(x - 2)$

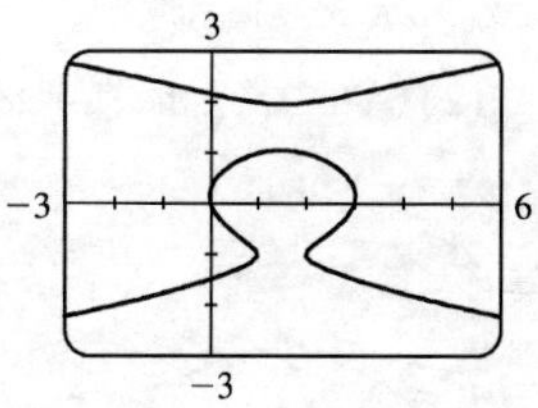

$(y + 1)(y^2 - 1)(y - 2)$
$= (x - 1)(x - 2)$

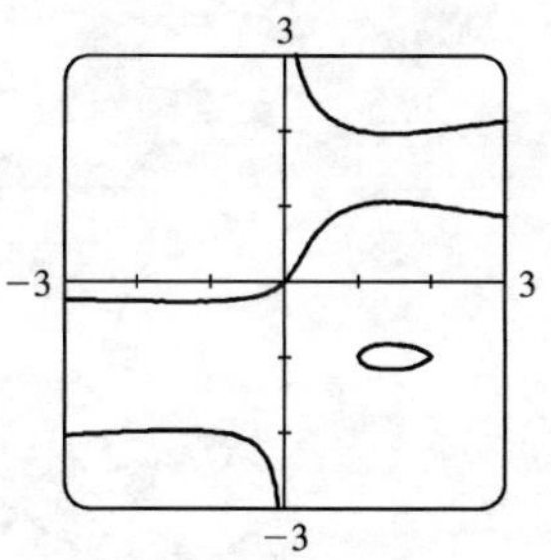

$x(y + 1)(y^2 - 1)(y - 2)$
$= y(x - 1)(x - 2)$

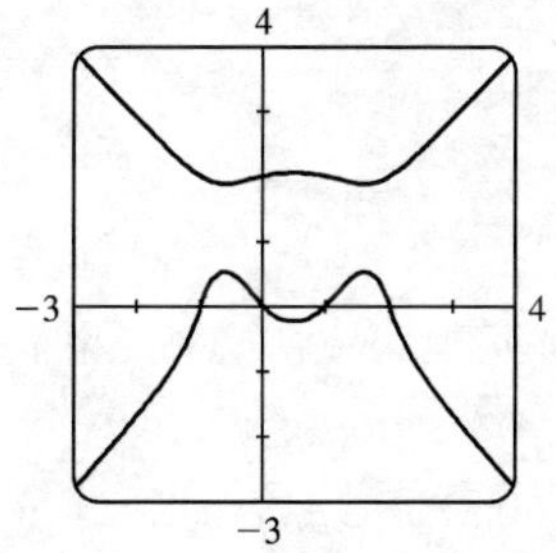

$y(y^2 + 1)(y - 2)$
$= x(x^2 - 1)(x - 2)$

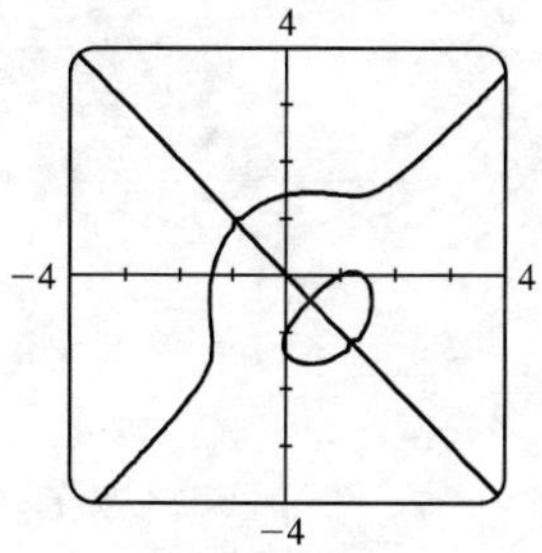

$y(y + 1)(y^2 - 2)$
$= x(x - 1)(x^2 - 2)$

35. From Exercise 29, a tangent to the lemniscate will be horizontal if $y' = 0 \Rightarrow 25x - 4x(x^2 + y^2) = 0 \Rightarrow$ $x[25 - 4(x^2 + y^2)] = 0 \Rightarrow x^2 + y^2 = \frac{25}{4}$ **(1)**. (Note that when x is 0, y is also 0, and there is no horizontal tangent at the origin.) Substituting $\frac{25}{4}$ for $x^2 + y^2$ in the equation of the lemniscate, $2(x^2 + y^2)^2 = 25(x^2 - y^2)$, we get $x^2 - y^2 = \frac{25}{8}$ **(2)**. Solving **(1)** and **(2)**, we have $x^2 = \frac{75}{16}$ and $y^2 = \frac{25}{16}$, so the four points are $\left(\pm\frac{5\sqrt{3}}{4}, \pm\frac{5}{4}\right)$.

37. $\dfrac{x^2}{a^2} - \dfrac{y^2}{b^2} = 1 \Rightarrow \dfrac{2x}{a^2} - \dfrac{2yy'}{b^2} = 0 \Rightarrow y' = \dfrac{b^2 x}{a^2 y} \Rightarrow$ an equation of the tangent line at (x_0, y_0) is $y - y_0 = \dfrac{b^2 x_0}{a^2 y_0}(x - x_0)$. Multiplying both sides by $\dfrac{y_0}{b^2}$ gives $\dfrac{y_0 y}{b^2} - \dfrac{y_0^2}{b^2} = \dfrac{x_0 x}{a^2} - \dfrac{x_0^2}{a^2}$. Since (x_0, y_0) lies on the hyperbola, we have $\dfrac{x_0 x}{a^2} - \dfrac{y_0 y}{b^2} = \dfrac{x_0^2}{a^2} - \dfrac{y_0^2}{b^2} = 1$.

39. If the circle has radius r, its equation is $x^2 + y^2 = r^2 \Rightarrow 2x + 2yy' = 0 \Rightarrow y' = -\dfrac{x}{y}$, so the slope of the tangent line at $P(x_0, y_0)$ is $-\dfrac{x_0}{y_0}$. The negative reciprocal of that slope is $\dfrac{-1}{-x_0/y_0} = \dfrac{y_0}{x_0}$, which is the slope of OP, so the tangent line at P is perpendicular to the radius OP.

41. $2x^2 + y^2 = 3$ and $x = y^2$ intersect when $2x^2 + x - 3 = 0 \Leftrightarrow (2x + 3)(x - 1) = 0 \Leftrightarrow x = -\frac{3}{2}$ or 1, but $-\frac{3}{2}$ is extraneous since $x = y^2$ is nonnegative. When $x = 1$, $1 = y^2 \Rightarrow y = \pm 1$, so there are two points of intersection: $(1, \pm 1)$. $2x^2 + y^2 = 3 \Rightarrow 4x + 2yy' = 0 \Rightarrow y' = -2x/y$, and $x = y^2 \Rightarrow 1 = 2yy' \Rightarrow$ $y' = 1/(2y)$. At $(1, 1)$, the slopes are $m_1 = -2(1)/1 = -2$ and $m_2 = 1/(2 \cdot 1) = \frac{1}{2}$, so the curves are orthogonal (since m_1 and m_2 are negative reciprocals of each other). By symmetry, the curves are also orthogonal at $(1, -1)$.

43.

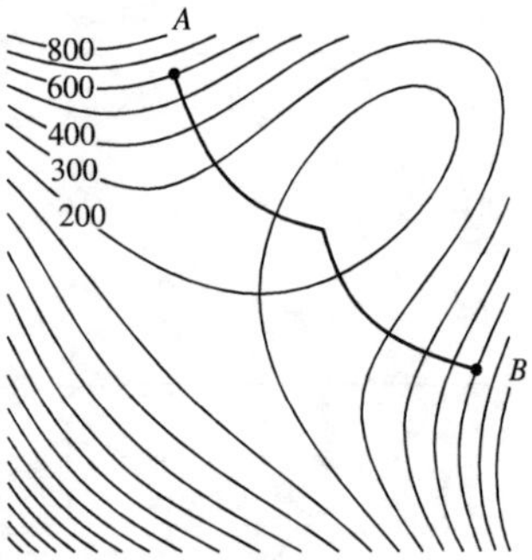

45. $x^2 + y^2 = r^2$ is a circle with center O and $ax + by = 0$ is a line through O. $x^2 + y^2 = r^2 \Rightarrow 2x + 2yy' = 0 \Rightarrow y' = -x/y$, so the slope of the tangent line at $P_0\,(x_0, y_0)$ is $-x_0/y_0$. The slope of the line OP_0 is y_0/x_0, which is the negative reciprocal of $-x_0/y_0$. Hence, the curves are orthogonal, and the families of curves are orthogonal trajectories of each other.

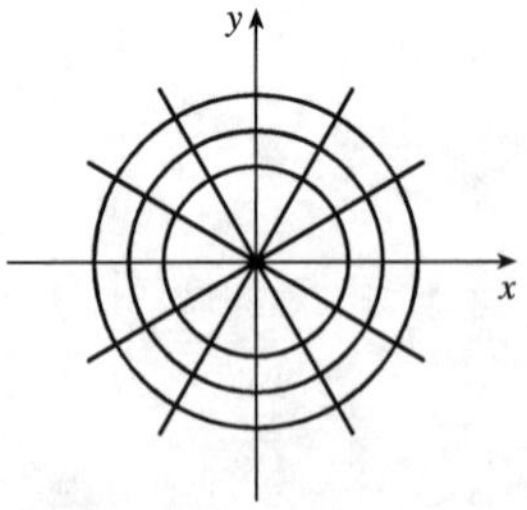

47. $y = cx^2 \Rightarrow y' = 2cx$ and $x^2 + 2y^2 = k \Rightarrow 2x + 4yy' = 0 \Rightarrow$
$2yy' = -x \Rightarrow y' = -\dfrac{x}{2(y)} = -\dfrac{x}{2(cx^2)} = -\dfrac{1}{2cx}$, so the curves are orthogonal.

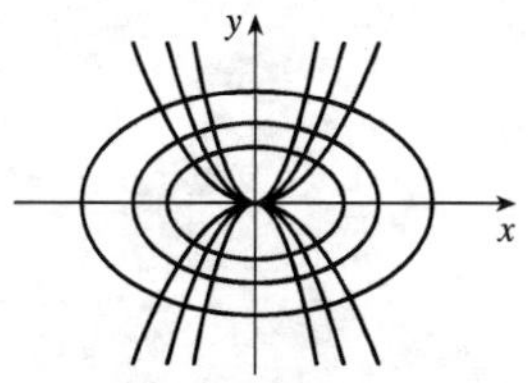

49. To find the points at which the ellipse $x^2 - xy + y^2 = 3$ crosses the x-axis, let $y = 0$ and solve for x.
$y = 0 \Rightarrow x^2 - x(0) + 0^2 = 3 \Leftrightarrow x = \pm\sqrt{3}$. So the graph of the ellipse crosses the x-axis at the points $(\pm\sqrt{3}, 0)$. Using implicit differentiation to find y', we get $2x - xy' - y + 2yy' = 0 \Rightarrow y'(2y - x) = y - 2x$
$\Leftrightarrow y' = \dfrac{y - 2x}{2y - x}$. So y' at $(\sqrt{3}, 0)$ is $\dfrac{0 - 2\sqrt{3}}{2(0) - \sqrt{3}} = 2$ and y' at $(-\sqrt{3}, 0)$ is $\dfrac{0 + 2\sqrt{3}}{2(0) + \sqrt{3}} = 2$. Thus, the tangent lines at these points are parallel.

51. $x^2y^2 + xy = 2 \Rightarrow x^2 \cdot 2yy' + y^2 \cdot 2x + x \cdot y' + y \cdot 1 = 0 \Leftrightarrow y'(2x^2y + x) = -2xy^2 - y \Leftrightarrow$
$y' = -\dfrac{2xy^2 + y}{2x^2y + x}$. So $-\dfrac{2xy^2 + y}{2x^2y + x} = -1 \Leftrightarrow 2xy^2 + y = 2x^2y + x \Leftrightarrow y(2xy + 1) = x(2xy + 1) \Leftrightarrow$
$y(2xy + 1) - x(2xy + 1) = 0 \Leftrightarrow (2xy + 1)(y - x) = 0 \Leftrightarrow xy = -\frac{1}{2}$ or $y = x$. But $xy = -\frac{1}{2} \Rightarrow$
$x^2y^2 + xy = \frac{1}{4} - \frac{1}{2} \neq 2$, so we must have $x = y$. Then $x^2y^2 + xy = 2 \Rightarrow x^4 + x^2 = 2 \Leftrightarrow$
$x^4 + x^2 - 2 = 0 \Leftrightarrow (x^2 + 2)(x^2 - 1) = 0$. So $x^2 = -2$, which is impossible, or $x^2 = 1 \Leftrightarrow x = \pm 1$.
Since $x = y$, the points on the curve where the tangent line has a slope of -1 are $(-1, -1)$ and $(1, 1)$.

53. $x^2 + 4y^2 = 5 \Rightarrow 2x + 4(2yy') = 0 \Rightarrow y' = -\dfrac{x}{4y}$. Now let h be the height of the lamp, and let (a, b) be the point of tangency of the line passing through the points $(3, h)$ and $(-5, 0)$. This line has slope
$(h - 0)/[3 - (-5)] = \frac{1}{8}h$. But the slope of the tangent line through the point (a, b) can be expressed as $y' = -\dfrac{a}{4b}$,
or as $\dfrac{b - 0}{a - (-5)} = \dfrac{b}{a + 5}$ [since the line passes through $(-5, 0)$ and (a, b)], so $-\dfrac{a}{4b} = \dfrac{b}{a + 5} \Leftrightarrow$
$4b^2 = -a^2 - 5a \Leftrightarrow a^2 + 4b^2 = -5a$. But $a^2 + 4b^2 = 5$ [since (a, b) is on the ellipse], so $5 = -5a \Leftrightarrow$
$a = -1$. Then $4b^2 = -a^2 - 5a = -1 - 5(-1) = 4 \Rightarrow b = 1$, since the point is on the top half of the ellipse.
So $\dfrac{h}{8} = \dfrac{b}{a + 5} = \dfrac{1}{-1 + 5} = \dfrac{1}{4} \Rightarrow h = 2$. So the lamp is located 2 units above the x-axis.

Higher Derivatives

1. $a = f, b = f', c = f''$. We can see this because where a has a horizontal tangent, $b = 0$, and where b has a horizontal tangent, $c = 0$. We can immediately see that c can be neither f nor f', since at the points where c has a horizontal tangent, neither a nor b is equal to 0.

3. We can immediately see that a is the graph of the acceleration function, since at the points where a has a horizontal tangent, neither c nor b is equal to 0. Next, we note that $a = 0$ at the point where b has a horizontal tangent, so b must be the graph of the velocity function, and hence, $b' = a$. We conclude that c is the graph of the position function.

5. $f(x) = x^5 + 6x^2 - 7x \quad \Rightarrow \quad f'(x) = 5x^4 + 12x - 7 \quad \Rightarrow \quad f''(x) = 20x^3 + 12$

7. $y = \cos 2\theta \quad \Rightarrow \quad y' = -2\sin 2\theta \quad \Rightarrow \quad y'' = -4\cos 2\theta$

9. $F(t) = (1-7t)^6 \quad \Rightarrow \quad F'(t) = 6(1-7t)^5(-7) = -42(1-7t)^5 \quad \Rightarrow$

$F''(t) = -42 \cdot 5(1-7t)^4(-7) = 1470(1-7t)^4$

11. $h(u) = \dfrac{1-4u}{1+3u} \quad \Rightarrow \quad h'(u) = \dfrac{(1+3u)(-4)-(1-4u)(3)}{(1+3u)^2} = \dfrac{-4-12u-3+12u}{(1+3u)^2} = \dfrac{-7}{(1+3u)^2}$ or

$-7(1+3u)^{-2} \quad \Rightarrow \quad h''(u) = -7(-2)(1+3u)^{-3}(3) = 42(1+3u)^{-3}$ or $\dfrac{42}{(1+3u)^3}$

13. $h(x) = \sqrt{x^2+1} \quad \Rightarrow \quad h'(x) = \dfrac{1}{2}(x^2+1)^{-1/2}(2x) = \dfrac{x}{\sqrt{x^2+1}} \quad \Rightarrow$

$$h''(x) = \frac{\sqrt{x^2+1}\cdot 1 - x\left[\frac{1}{2}(x^2+1)^{-1/2}(2x)\right]}{\left(\sqrt{x^2+1}\right)^2} = \frac{(x^2+1)^{-1/2}\left[(x^2+1)-x^2\right]}{(x^2+1)^1} = \frac{1}{(x^2+1)^{3/2}}$$

15. $y = (x^3+1)^{2/3} \quad \Rightarrow \quad y' = \frac{2}{3}(x^3+1)^{-1/3}(3x^2) = 2x^2(x^3+1)^{-1/3} \quad \Rightarrow$

$y'' = 2x^2\left(-\frac{1}{3}\right)(x^3+1)^{-4/3}(3x^2) + (x^3+1)^{-1/3}(4x) = 4x(x^3+1)^{-1/3} - 2x^4(x^3+1)^{-4/3}$

17. $H(t) = \tan 3t \quad \Rightarrow \quad H'(t) = 3\sec^2 3t \quad \Rightarrow$

$H''(t) = 2 \cdot 3\sec 3t\,\dfrac{d}{dt}(\sec 3t) = 6\sec 3t\,(3\sec 3t\tan 3t) = 18\sec^2 3t\tan 3t$

19. $g(\theta) = \theta\csc\theta \quad \Rightarrow \quad g'(\theta) = -\theta\csc\theta\cot\theta + \csc\theta \quad \Rightarrow$

$$\begin{aligned} g''(\theta) &= (-1)\csc\theta\cot\theta + (-\theta)(-\csc\theta\cot\theta)\cot\theta + (-\theta\csc\theta)(-\csc^2\theta) - \csc\theta\cot\theta \\ &= -\csc\theta\cot\theta + \theta\csc\theta\cot^2\theta + \theta\csc^3\theta - \csc\theta\cot\theta \\ &= \csc\theta(\theta\csc^2\theta + \theta\cot^2\theta - 2\cot\theta) \end{aligned}$$

21. (a) $f(x) = 2\cos x + \sin^2 x \quad \Rightarrow \quad f'(x) = 2(-\sin x) + 2\sin x\,(\cos x) = \sin 2x - 2\sin x \quad \Rightarrow$

$f''(x) = 2\cos 2x - 2\cos x = 2(\cos 2x - \cos x)$

(b)

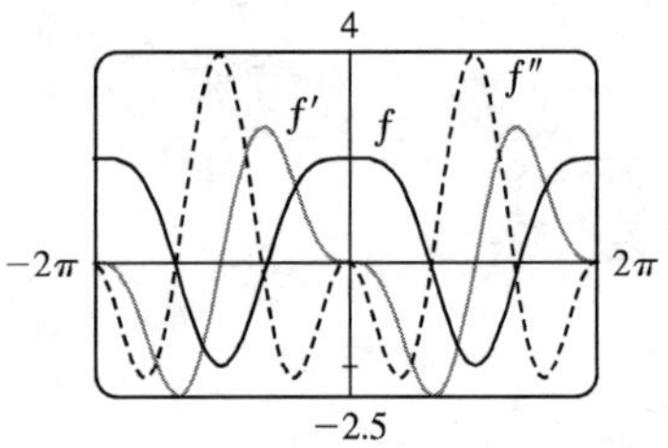

We can see that our answers are plausible, since f has horizontal tangents where $f'(x) = 0$, and f' has horizontal tangents where $f''(x) = 0$.

23. $y = \sqrt{2x+3} = (2x+3)^{1/2} \quad \Rightarrow \quad y' = \frac{1}{2}(2x+3)^{-1/2} \cdot 2 = (2x+3)^{-1/2} \quad \Rightarrow$

$y'' = -\frac{1}{2}(2x+3)^{-3/2} \cdot 2 = -(2x+3)^{-3/2} \quad \Rightarrow \quad y''' = \frac{3}{2}(2x+3)^{-5/2} \cdot 2 = 3(2x+3)^{-5/2}$

25. $f(t) = t\cos t \quad \Rightarrow \quad f'(t) = t(-\sin t) + \cos t \cdot 1 \quad \Rightarrow \quad f''(t) = t(-\cos t) - \sin t \cdot 1 - \sin t \quad \Rightarrow$

$f'''(t) = t\sin t - \cos t \cdot 1 - \cos t - \cos t = t\sin t - 3\cos t$, so $f'''(0) = 0 - 3 = -3$.

27. $f(\theta) = \cot\theta \quad \Rightarrow \quad f'(\theta) = -\csc^2\theta \quad \Rightarrow \quad f''(\theta) = -2\csc\theta\,(-\csc\theta\cot\theta) = 2\csc^2\theta\cot\theta \quad \Rightarrow$

$f'''(\theta) = 2\left(-2\csc^2\theta\cot\theta\right)\cot\theta + 2\csc^2\theta\left(-\csc^2\theta\right) = -2\csc^2\theta\left(2\cot^2\theta + \csc^2\theta\right) \quad \Rightarrow$

$f'''\left(\frac{\pi}{6}\right) = -2(2)^2\left[2\left(\sqrt{3}\right)^2 + (2)^2\right] = -80$

29. $9x^2 + y^2 = 9 \;\Rightarrow\; 18x + 2yy' = 0 \;\Rightarrow\; 2yy' = -18x \;\Rightarrow\; y' = -9x/y \;\Rightarrow$

$y'' = -9\left(\dfrac{y \cdot 1 - x \cdot y'}{y^2}\right) = -9\left(\dfrac{y - x(-9x/y)}{y^2}\right) = -9 \cdot \dfrac{y^2 + 9x^2}{y^3} = -9 \cdot \dfrac{9}{y^3}$ [since x and y must satisfy the original equation, $9x^2 + y^2 = 9$]. Thus, $y'' = -81/y^3$.

31. $x^3 + y^3 = 1 \;\Rightarrow\; 3x^2 + 3y^2y' = 0 \;\Rightarrow\; y' = -\dfrac{x^2}{y^2} \;\Rightarrow$

$$y'' = -\frac{y^2(2x) - x^2 \cdot 2yy'}{(y^2)^2} = -\frac{2xy^2 - 2x^2y(-x^2/y^2)}{y^4} = -\frac{2xy^4 + 2x^4y}{y^6} = -\frac{2xy(y^3 + x^3)}{y^6} = -\frac{2x}{y^5},$$

since x and y must satisfy the original equation, $x^3 + y^3 = 1$.

33. $f(x) = x^n \;\Rightarrow\; f'(x) = nx^{n-1} \;\Rightarrow\; f''(x) = n(n-1)x^{n-2} \;\Rightarrow\; \cdots \;\Rightarrow$

$f^{(n)}(x) = n(n-1)(n-2)\cdots 2 \cdot 1x^{n-n} = n!$

35. $f(x) = (1+x)^{-1} \;\Rightarrow\; f'(x) = -1(1+x)^{-2}$, $f''(x) = 1 \cdot 2(1+x)^{-3}$, $f^{(3)}(x) = -1 \cdot 2 \cdot 3(1+x)^{-4}$,

$f^{(4)}(x) = 1 \cdot 2 \cdot 3 \cdot 4(1+x)^{-5}, \ldots, f^{(n)}(x) = (-1)^n n!(1+x)^{-(n+1)}$

37. $f(x) = 1/(3x^3) = \frac{1}{3}x^{-3} \;\Rightarrow\; f'(x) = \frac{1}{3}(-3)x^{-4} \;\Rightarrow\; f''(x) = \frac{1}{3}(-3)(-4)x^{-5} \;\Rightarrow$

$f'''(x) = \frac{1}{3}(-3)(-4)(-5)x^{-6} \;\Rightarrow\; \cdots \;\Rightarrow$

$$f^{(n)}(x) = \tfrac{1}{3}(-3)(-4)\cdots[-(n+2)]\,x^{-(n+3)} = \frac{(-1)^n \cdot 3 \cdot 4 \cdot 5 \cdot \cdots \cdot (n+2)}{3x^{n+3}} \cdot \frac{2}{2} = \frac{(-1)^n\,(n+2)!}{6x^{n+3}}$$

39. Let $f(x) = \cos x$. Then $Df(2x) = 2f'(2x)$, $D^2f(2x) = 2^2f''(2x)$, $D^3f(2x) = 2^3f'''(2x), \ldots,$ $D^{(n)}f(2x) = 2^nf^{(n)}(2x)$. Since the derivatives of $\cos x$ occur in a cycle of four, and since $103 = 4(25) + 3$, we have $f^{(103)}(x) = f^{(3)}(x) = \sin x$ and $D^{103}\cos 2x = 2^{103}f^{(103)}(2x) = 2^{103}\sin 2x$.

41. By measuring the slope of the graph of $s = f(t)$ at $t = 0$, 1, 2, 3, 4, and 5, and using the method of Example 1 in Section 3.2, we plot the graph of the velocity function $v = f'(t)$ in the first figure. The acceleration when $t = 2$ s is $a = f''(2)$, the slope of the tangent line to the graph of f' when $t = 2$. We estimate the slope of this tangent line to be $a(2) = f''(2) = v'(2) \approx \frac{27}{3} = 9$ ft/s^2. Similar measurements enable us to graph the acceleration function in the second figure.

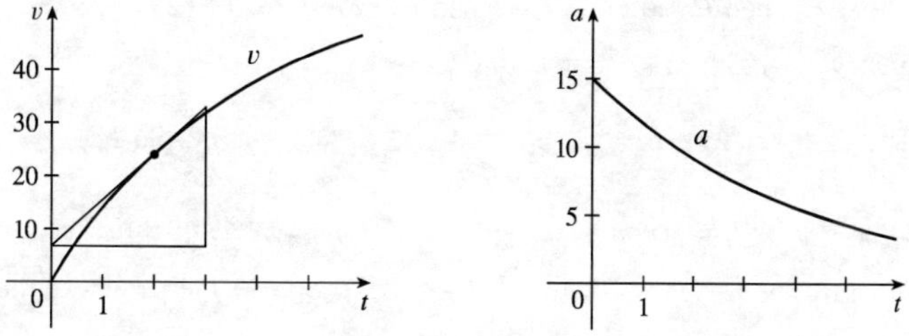

43. (a) $s = 2t^3 - 15t^2 + 36t + 2 \;\Rightarrow\; v(t) = s'(t) = 6t^2 - 30t + 36 \;\Rightarrow\; a(t) = v'(t) = 12t - 30$

(b) $a(1) = 12 \cdot 1 - 30 = -18$ m/s^2

(c) $v(t) = 6(t^2 - 5t + 6) = 6(t-2)(t-3) = 0$ when $t = 2$ or 3 and $a(2) = 24 - 30 = -6$ m/s^2, $a(3) = 36 - 30 = 6$ m/s^2.

45. (a) $s = \sin\left(\frac{\pi}{6}t\right) + \cos\left(\frac{\pi}{6}t\right), 0 \le t \le 2$. $v(t) = s'(t) = \cos\left(\frac{\pi}{6}t\right) \cdot \frac{\pi}{6} - \sin\left(\frac{\pi}{6}t\right) \cdot \frac{\pi}{6} = \frac{\pi}{6}\left[\cos\left(\frac{\pi}{6}t\right) - \sin\left(\frac{\pi}{6}t\right)\right]$

$\Rightarrow \quad a(t) = v'(t) = \frac{\pi}{6}\left[-\sin\left(\frac{\pi}{6}t\right) \cdot \frac{\pi}{6} - \cos\left(\frac{\pi}{6}t\right) \cdot \frac{\pi}{6}\right] = -\dfrac{\pi^2}{36}\left[\sin\left(\frac{\pi}{6}t\right) + \cos\left(\frac{\pi}{6}t\right)\right]$

(b) $a(1) = -\dfrac{\pi^2}{36}\left[\sin\left(\frac{\pi}{6} \cdot 1\right) + \cos\left(\frac{\pi}{6} \cdot 1\right)\right] = -\dfrac{\pi^2}{36}\left[\dfrac{1}{2} + \dfrac{\sqrt{3}}{2}\right] = -\dfrac{\pi^2}{72}\left(1 + \sqrt{3}\right) \approx -0.3745 \text{ m/s}^2$

(c) $v(t) = 0$ for $0 \le t \le 2 \quad \Rightarrow \quad \cos\left(\frac{\pi}{6}t\right) = \sin\left(\frac{\pi}{6}t\right) \quad \Rightarrow \quad 1 = \dfrac{\sin\left(\frac{\pi}{6}t\right)}{\cos\left(\frac{\pi}{6}t\right)} \quad \Rightarrow$

$\tan\left(\frac{\pi}{6}t\right) = 1 \quad \Rightarrow \quad \frac{\pi}{6}t = \tan^{-1} 1 \quad \Rightarrow \quad t = \frac{6}{\pi} \cdot \frac{\pi}{4} = \frac{3}{2} = 1.5$ s. Thus,

$a\left(\frac{3}{2}\right) = -\dfrac{\pi^2}{36}\left[\sin\left(\frac{\pi}{6} \cdot \frac{3}{2}\right) + \cos\left(\frac{\pi}{6} \cdot \frac{3}{2}\right)\right] = -\dfrac{\pi^2}{36}\left[\dfrac{\sqrt{2}}{2} + \dfrac{\sqrt{2}}{2}\right] = -\dfrac{\pi^2}{36}\sqrt{2} \approx -0.3877 \text{ m/s}^2$.

47. (a) $s(t) = t^4 - 4t^3 + 2 \quad \Rightarrow \quad v(t) = s'(t) = 4t^3 - 12t^2 \quad \Rightarrow \quad a(t) = v'(t) = 12t^2 - 24t = 12t(t-2) = 0$ when $t = 0$ or 2.

(b) $s(0) = 2$ m, $v(0) = 0$ m/s, $s(2) = -14$ m, $v(2) = -16$ m/s

49. (a) $s = f(t) = t^3 - 12t^2 + 36t, t \ge 0 \quad \Rightarrow \quad v(t) = f'(t) = 3t^2 - 24t + 36$.
$a(t) = v'(t) = 6t - 24$. $a(3) = 6(3) - 24 = -6$ (m/s)/s or m/s^2.

(b)

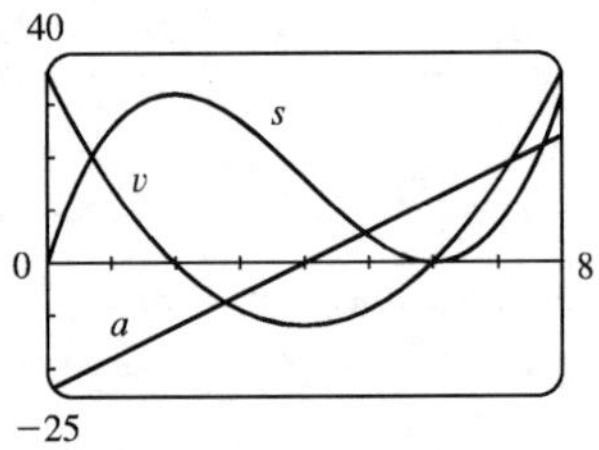

(c) The particle is speeding up when v and a have the same sign. This occurs when $2 < t < 4$ and when $t > 6$. It is slowing down when v and a have opposite signs; that is, when $0 \le t < 2$ and when $4 < t < 6$.

51. (a) $y(t) = A \sin \omega t \quad \Rightarrow \quad v(t) = y'(t) = A\omega \cos \omega t \quad \Rightarrow \quad a(t) = v'(t) = -A\omega^2 \sin \omega t$

(b) $a(t) = -A\omega^2 \sin \omega t = -\omega^2(A \sin \omega t) = -\omega^2 y(t)$, so $a(t)$ is proportional to $y(t)$.

(c) speed $= |v(t)| = A\omega\,|\cos \omega t|$ is a maximum when $\cos \omega t = \pm 1$. But when $\cos \omega t = \pm 1$, we have $\sin \omega t = 0$, and $a(t) = -A\omega^2 \sin \omega t = -A\omega^2(0) = 0$.

53. Let $P(x) = ax^2 + bx + c$. Then $P'(x) = 2ax + b$ and $P''(x) = 2a$. $P''(2) = 2 \quad \Rightarrow \quad 2a = 2 \quad \Rightarrow \quad a = 1$.

$P'(2) = 3 \quad \Rightarrow \quad 2(1)(2) + b = 3 \quad \Rightarrow \quad 4 + b = 3 \quad \Rightarrow \quad b = -1$.

$P(2) = 5 \quad \Rightarrow \quad 1(2)^2 + (-1)(2) + c = 5 \quad \Rightarrow \quad 2 + c = 5 \quad \Rightarrow \quad c = 3$. So $P(x) = x^2 - x + 3$.

55. $y = A \sin x + B \cos x \quad \Rightarrow \quad y' = A \cos x - B \sin x \quad \Rightarrow \quad y'' = -A \sin x - B \cos x$. Substituting into $y'' + y' - 2y = \sin x$ gives us $(-3A - B) \sin x + (A - 3B) \cos x = 1 \sin x$, so we must have $-3A - B = 1$ and $A - 3B = 0$. Solving for A and B, we add the first equation to three times the second to get $B = -\frac{1}{10}$ and $A = -\frac{3}{10}$.

57. $f(x) = xg(x^2) \quad \Rightarrow \quad f'(x) = x \cdot g'(x^2) \cdot 2x + g(x^2) \cdot 1 = g(x^2) + 2x^2 g'(x^2) \quad \Rightarrow$

$f''(x) = g'(x^2) \cdot 2x + 2x^2 \cdot g''(x^2) \cdot 2x + g'(x^2) \cdot 4x = 6xg'(x^2) + 4x^3 g''(x^2)$

59. $f(x) = g(\sqrt{x}) \Rightarrow f'(x) = g'(\sqrt{x}) \cdot \frac{1}{2}x^{-1/2} = \dfrac{g'(\sqrt{x})}{2\sqrt{x}} \Rightarrow$

$$f''(x) = \frac{2\sqrt{x} \cdot g''(\sqrt{x}) \cdot \frac{1}{2}x^{-1/2} - g'(\sqrt{x}) \cdot 2 \cdot \frac{1}{2}x^{-1/2}}{(2\sqrt{x})^2} = \frac{x^{-1/2}\left[\sqrt{x}\, g''(\sqrt{x}) - g'(\sqrt{x})\right]}{4x}$$

$$= \frac{\sqrt{x}\, g''(\sqrt{x}) - g'(\sqrt{x})}{4x\sqrt{x}}$$

61. (a) $f(x) = \dfrac{1}{x^2+x} \Rightarrow f'(x) = \dfrac{-(2x+1)}{(x^2+x)^2} \Rightarrow$

$$f''(x) = \frac{(x^2+x)^2(-2) + (2x+1)(2)(x^2+x)(2x+1)}{(x^2+x)^4} = \frac{2(3x^2+3x+1)}{(x^2+x)^3} \Rightarrow$$

$$f'''(x) = \frac{(x^2+x)^3(2)(6x+3) - 2(3x^2+3x+1)(3)(x^2+x)^2(2x+1)}{(x^2+x)^6}$$

$$= \frac{-6(4x^3+6x^2+4x+1)}{(x^2+x)^4} \Rightarrow$$

$$f^{(4)}(x) = \frac{(x^2+x)^4(-6)(12x^2+12x+4) + 6(4x^3+6x^2+4x+1)(4)(x^2+x)^3(2x+1)}{(x^2+x)^8}$$

$$= \frac{24(5x^4+10x^3+10x^2+5x+1)}{(x^2+x)^5}$$

$f^{(5)}(x) = ?$

(b) $f(x) = \dfrac{1}{x(x+1)} = \dfrac{1}{x} - \dfrac{1}{x+1} \Rightarrow f'(x) = -x^{-2} + (x+1)^{-2} \Rightarrow f''(x) = 2x^{-3} - 2(x+1)^{-3} \Rightarrow$

$f'''(x) = (-3)(2)x^{-4} + (3)(2)(x+1)^{-4} \Rightarrow \cdots \Rightarrow f^{(n)}(x) = (-1)^n n!\left[x^{-(n+1)} - (x+1)^{-(n+1)}\right]$

63. We will show that for each positive integer n, the nth derivative $f^{(n)}$ exists and equals one of $f, f', f'', f''', \ldots,$ $f^{(p-1)}$. Since $f^{(p)} = f$, the first p derivatives of f are $f', f'', f''', \ldots, f^{(p-1)}$, and f. In particular, our statement is true for $n = 1$. Suppose that k is an integer, $k \geq 1$, for which f is k-times differentiable with $f^{(k)}$ in the set $S = \{f, f', f'', \ldots, f^{(p-1)}\}$. Since f is p-times differentiable, every member of S [including $f^{(k)}$] is differentiable, so $f^{(k+1)}$ exists and equals the derivative of some member of S. Thus, $f^{(k+1)}$ is in the set $\{f', f'', f''', \ldots, f^{(p)}\}$, which equals S since $f^{(p)} = f$. We have shown that the statement is true for $n = 1$ and that its truth for $n = k$ implies its truth for $n = k + 1$. By mathematical induction, the statement is true for all positive integers n.

65. The Chain Rule says that $\dfrac{dy}{dx} = \dfrac{dy}{du}\dfrac{du}{dx}$, so

$$\frac{d^2y}{dx^2} = \frac{d}{dx}\left(\frac{dy}{dx}\right) = \frac{d}{dx}\left(\frac{dy}{du}\frac{du}{dx}\right) = \left[\frac{d}{dx}\left(\frac{dy}{du}\right)\right]\frac{du}{dx} + \frac{dy}{du}\frac{d}{dx}\left(\frac{du}{dx}\right) \quad \text{[Product Rule]}$$

$$= \left[\frac{d}{du}\left(\frac{dy}{du}\right)\frac{du}{dx}\right]\frac{du}{dx} + \frac{dy}{du}\frac{d^2u}{dx^2} = \frac{d^2y}{du^2}\left(\frac{du}{dx}\right)^2 + \frac{dy}{du}\frac{d^2u}{dx^2}$$

Related Rates

1. $V = x^3 \;\Rightarrow\; \dfrac{dV}{dt} = \dfrac{dV}{dx}\dfrac{dx}{dt} = 3x^2\,\dfrac{dx}{dt}$

3. $y = x^3 + 2x \;\Rightarrow\; \dfrac{dy}{dt} = \dfrac{dy}{dx}\dfrac{dx}{dt} = (3x^2+2)(5) = 5(3x^2+2)$. When $x = 2$, $\dfrac{dy}{dt} = 5(14) = 70$.

5. $z^2 = x^2 + y^2 \;\Rightarrow\; 2z\,\dfrac{dz}{dt} = 2x\,\dfrac{dx}{dt} + 2y\,\dfrac{dy}{dt} \;\Rightarrow\; \dfrac{dz}{dt} = \dfrac{1}{z}\left(x\,\dfrac{dx}{dt} + y\,\dfrac{dy}{dt}\right)$. When $x = 5$ and $y = 12$,

$z^2 = 5^2 + 12^2 \;\Rightarrow\; z^2 = 169 \;\Rightarrow\; z = \pm 13$. For $\dfrac{dx}{dt} = 2$ and $\dfrac{dy}{dt} = 3$, $\dfrac{dz}{dt} = \dfrac{1}{\pm 13}(5\cdot 2 + 12\cdot 3) = \pm\dfrac{46}{13}$.

7. (a) Given: a plane flying horizontally at an altitude of 1 mi and a speed of 500 mi/h passes directly over a radar station. If we let t be time (in hours) and x be the horizontal distance traveled by the plane (in mi), then we are given that $dx/dt = 500$ mi/h.

(b) Unknown: the rate at which the distance from the plane to the station is increasing when it is 2 mi from the station. If we let y be the distance from the plane to the station, then we want to find dy/dt when $y = 2$ mi.

(c) 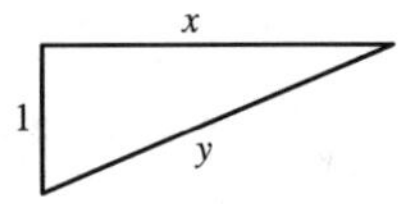

(d) By the Pythagorean Theorem, $y^2 = x^2 + 1 \;\Rightarrow\; 2y(dy/dt) = 2x(dx/dt)$.

(e) $\dfrac{dy}{dt} = \dfrac{x}{y}\dfrac{dx}{dt} = \dfrac{x}{y}(500)$. Since $y^2 = x^2 + 1$, when $y = 2$, $x = \sqrt{3}$, so $\dfrac{dy}{dt} = \frac{\sqrt{3}}{2}(500) = 250\sqrt{3} \approx 433$ mi/h.

9. (a) Given: a man 6 ft tall walks away from a street light mounted on a 15-ft-tall pole at a rate of 5 ft/s. If we let t be time (in s) and x be the distance from the pole to the man (in ft), then we are given that $dx/dt = 5$ ft/s.

(b) Unknown: the rate at which the tip of his shadow is moving when he is 40 ft from the pole. If we let y be the distance from the man to the tip of his shadow (in ft), then we want to find $\dfrac{d}{dt}(x+y)$ when $x = 40$ ft.

(c) 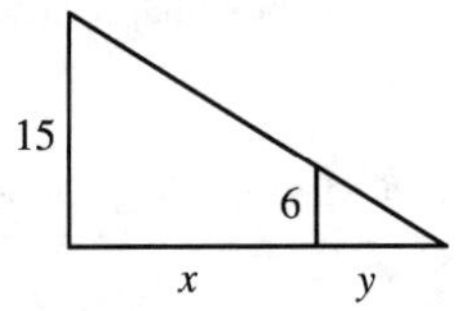

(d) By similar triangles, $\dfrac{15}{6} = \dfrac{x+y}{y} \;\Rightarrow\; 15y = 6x + 6y \;\Rightarrow\; 9y = 6x \;\Rightarrow\; y = \frac{2}{3}x$.

(e) The tip of the shadow moves at a rate of $\dfrac{d}{dt}(x+y) = \dfrac{d}{dt}\left(x + \frac{2}{3}x\right) = \dfrac{5}{3}\dfrac{dx}{dt} = \frac{5}{3}(5) = \frac{25}{3}$ ft/s.

11. 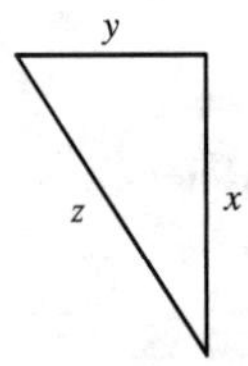

We are given that $\dfrac{dx}{dt} = 60$ mi/h and $\dfrac{dy}{dt} = 25$ mi/h. $z^2 = x^2 + y^2 \;\Rightarrow$

$2z\,\dfrac{dz}{dt} = 2x\,\dfrac{dx}{dt} + 2y\,\dfrac{dy}{dt} \;\Rightarrow\; z\,\dfrac{dz}{dt} = x\,\dfrac{dx}{dt} + y\,\dfrac{dy}{dt} \;\Rightarrow\; \dfrac{dz}{dt} = \dfrac{1}{z}\left(x\,\dfrac{dx}{dt} + y\,\dfrac{dy}{dt}\right)$.

After 2 hours, $x = 2\,(60) = 120$ and $y = 2\,(25) = 50 \;\Rightarrow\; z = \sqrt{120^2 + 50^2} = 130$,

so $\dfrac{dz}{dt} = \dfrac{1}{z}\left(x\,\dfrac{dx}{dt} + y\,\dfrac{dy}{dt}\right) = \dfrac{120(60) + 50(25)}{130} = 65$ mi/h.

13. 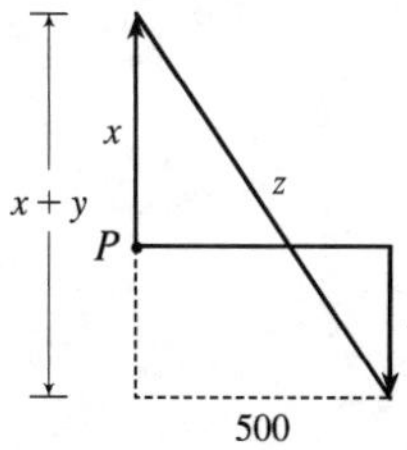

We are given that $\frac{dx}{dt} = 4$ ft/s and $\frac{dy}{dt} = 5$ ft/s. $z^2 = (x+y)^2 + 500^2 \;\Rightarrow$ $2z\,\frac{dz}{dt} = 2(x+y)\left(\frac{dx}{dt} + \frac{dy}{dt}\right)$. 15 minutes after the woman starts, we have $x = (4\text{ ft/s})(20\text{ min})(60\text{ s/min}) = 4800$ ft and $y = 5 \cdot 15 \cdot 60 = 4500 \;\Rightarrow$ $z = \sqrt{(4800+4500)^2 + 500^2} = \sqrt{86{,}740{,}000}$, so

$$\frac{dz}{dt} = \frac{x+y}{z}\left(\frac{dx}{dt} + \frac{dy}{dt}\right) = \frac{4800+4500}{\sqrt{86{,}740{,}000}}(4+5) = \frac{837}{\sqrt{8674}} \approx 8.99 \text{ ft/s}.$$

15. $A = \frac{1}{2}bh$, where b is the base and h is the altitude. We are given that $\frac{dh}{dt} = 1$ cm/min and $\frac{dA}{dt} = 2\text{ cm}^2/\text{min}$. Using the Product Rule, we have $\frac{dA}{dt} = \frac{1}{2}\left(b\,\frac{dh}{dt} + h\,\frac{db}{dt}\right)$. When $h = 10$ and $A = 100$, we have $100 = \frac{1}{2}b(10) \;\Rightarrow\; \frac{1}{2}b = 10 \;\Rightarrow\; b = 20$, so $2 = \frac{1}{2}\left(20 \cdot 1 + 10\,\frac{db}{dt}\right) \;\Rightarrow\; 4 = 20 + 10\,\frac{db}{dt} \;\Rightarrow$ $\frac{db}{dt} = \frac{4-20}{10} = -1.6$ cm/min.

17. 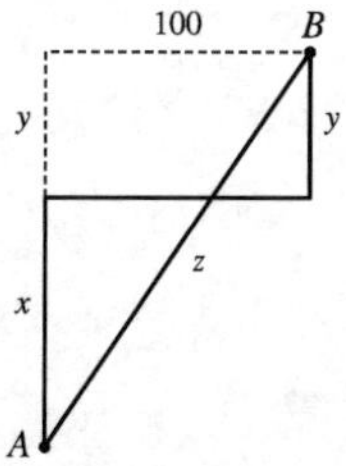

We are given that $\frac{dx}{dt} = 35$ km/h and $\frac{dy}{dt} = 25$ km/h. $z^2 = (x+y)^2 + 100^2$ $\Rightarrow\; 2z\,\frac{dz}{dt} = 2(x+y)\left(\frac{dx}{dt} + \frac{dy}{dt}\right)$. At 4:00 P.M., $x = 4(35) = 140$ and $y = 4(25) = 100 \;\Rightarrow\; z = \sqrt{(140+100)^2 + 100^2} = \sqrt{67{,}600} = 260$, so

$$\frac{dz}{dt} = \frac{x+y}{z}\left(\frac{dx}{dt} + \frac{dy}{dt}\right) = \frac{140+100}{260}(35+25) = \frac{720}{13} \approx 55.4 \text{ km/h}.$$

19. 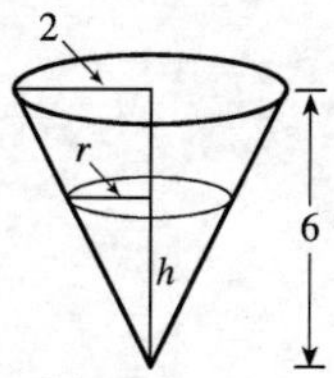

If $C =$ the rate at which water is pumped in, then $\frac{dV}{dt} = C - 10{,}000$, where $V = \frac{1}{3}\pi r^2 h$ is the volume at time t. By similar triangles, $\frac{r}{2} = \frac{h}{6} \;\Rightarrow$ $r = \frac{1}{3}h \;\Rightarrow\; V = \frac{1}{3}\pi\left(\frac{1}{3}h\right)^2 h = \frac{\pi}{27}h^3 \;\Rightarrow\; \frac{dV}{dt} = \frac{\pi}{9}h^2\,\frac{dh}{dt}$. When $h = 200$ cm, $\frac{dh}{dt} = 20$ cm/min, so $C - 10{,}000 = \frac{\pi}{9}(200)^2(20) \;\Rightarrow$ $C = 10{,}000 + \frac{800{,}000}{9}\pi \approx 289{,}253\text{ cm}^3/\text{min}$.

21. 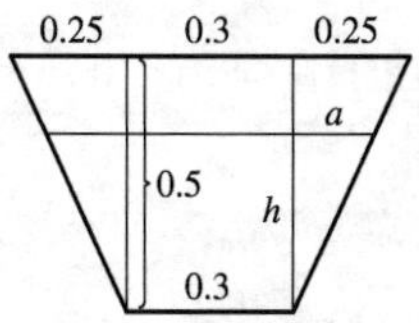

The figure is labeled in meters. The area A of a trapezoid is $\frac{1}{2}(\text{base}_1 + \text{base}_2)(\text{height})$, and the volume V of the 10-meter-long trough is $10A$. Thus, the volume of the trapezoid with height h is $V = (10)\frac{1}{2}\left[0.3 + (0.3 + 2a)\right]h$. By similar triangles, $\frac{a}{h} = \frac{0.25}{0.5} = \frac{1}{2}$, so $2a = h \;\Rightarrow\; V = 5(0.6+h)h = 3h + 5h^2$. Now $\frac{dV}{dt} = \frac{dV}{dh}\frac{dh}{dt} \;\Rightarrow\; 0.2 = (3 + 10h)\frac{dh}{dt} \;\Rightarrow$ $\frac{dh}{dt} = \frac{0.2}{3+10h}$. When $h = 0.3$, $\frac{dh}{dt} = \frac{0.2}{3 + 10(0.3)} = \frac{0.2}{6}$ m/min $= \frac{1}{30}$ m/min or $\frac{10}{3}$ cm/min.

23. 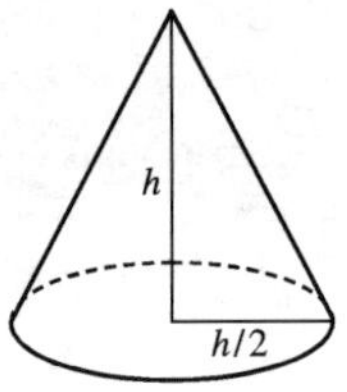

We are given that $\frac{dV}{dt} = 30\ \text{ft}^3/\text{min}$. $V = \frac{1}{3}\pi r^2 h = \frac{1}{3}\pi\left(\frac{h}{2}\right)^2 h = \frac{\pi h^3}{12}$ $\Rightarrow$ $\frac{dV}{dt} = \frac{dV}{dh}\frac{dh}{dt}$ $\Rightarrow$ $30 = \frac{\pi h^2}{4}\frac{dh}{dt}$ $\Rightarrow$ $\frac{dh}{dt} = \frac{120}{\pi h^2}$. When $h = 10$ ft, $\frac{dh}{dt} = \frac{120}{10^2\pi} = \frac{6}{5\pi} \approx 0.38\ \text{ft/min}$.

25. 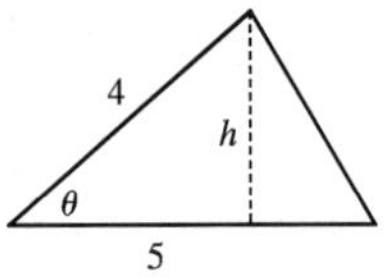

$A = \frac{1}{2}bh$, but $b = 5$ m and $\sin\theta = \frac{h}{4}$ $\Rightarrow$ $h = 4\sin\theta$, so $A = \frac{1}{2}(5)(4\sin\theta) = 10\sin\theta$. We are given $\frac{d\theta}{dt} = 0.06$ rad/s, so $\frac{dA}{dt} = \frac{dA}{d\theta}\frac{d\theta}{dt} = (10\cos\theta)(0.06) = 0.6\cos\theta$. When $\theta = \frac{\pi}{3}$, $\frac{dA}{dt} = 0.6\left(\cos\frac{\pi}{3}\right) = (0.6)\left(\frac{1}{2}\right) = 0.3\ \text{m}^2/\text{s}$.

27. Differentiating both sides of $PV = C$ with respect to t and using the Product Rule gives us $P\frac{dV}{dt} + V\frac{dP}{dt} = 0$ $\Rightarrow$ $\frac{dV}{dt} = -\frac{V}{P}\frac{dP}{dt}$. When $V = 600$, $P = 150$ and $\frac{dP}{dt} = 20$, so we have $\frac{dV}{dt} = -\frac{600}{150}(20) = -80$. Thus, the volume is decreasing at a rate of $80\ \text{cm}^3/\text{min}$.

29. With $R_1 = 80$ and $R_2 = 100$, $\frac{1}{R} = \frac{1}{R_1} + \frac{1}{R_2} = \frac{1}{80} + \frac{1}{100} = \frac{180}{8000} = \frac{9}{400}$, so $R = \frac{400}{9}$. Differentiating $\frac{1}{R} = \frac{1}{R_1} + \frac{1}{R_2}$ with respect to t, we have $-\frac{1}{R^2}\frac{dR}{dt} = -\frac{1}{R_1^2}\frac{dR_1}{dt} - \frac{1}{R_2^2}\frac{dR_2}{dt}$ $\Rightarrow$ $\frac{dR}{dt} = R^2\left(\frac{1}{R_1^2}\frac{dR_1}{dt} + \frac{1}{R_2^2}\frac{dR_2}{dt}\right)$. When $R_1 = 80$ and $R_2 = 100$,

$$\frac{dR}{dt} = \frac{400^2}{9^2}\left[\frac{1}{80^2}(0.3) + \frac{1}{100^2}(0.2)\right] = \frac{107}{810} \approx 0.132\ \Omega/\text{s}.$$

31. 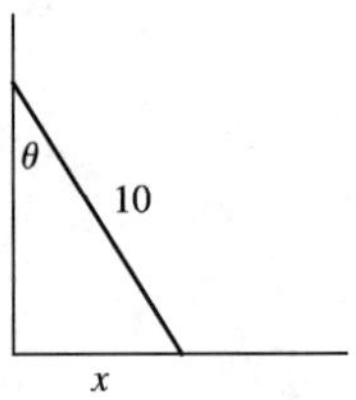

We are given that $\frac{dx}{dt} = 2$ ft/s. $\sin\theta = \frac{x}{10}$ $\Rightarrow$ $x = 10\sin\theta$ $\Rightarrow$ $\frac{dx}{dt} = 10\cos\theta\frac{d\theta}{dt}$. When $\theta = \frac{\pi}{4}$, $2 = 10\cos\frac{\pi}{4}\frac{d\theta}{dt}$ $\Rightarrow$ $\frac{d\theta}{dt} = \frac{2}{10(1/\sqrt{2})} = \frac{\sqrt{2}}{5}$ rad/s.

33. (a) 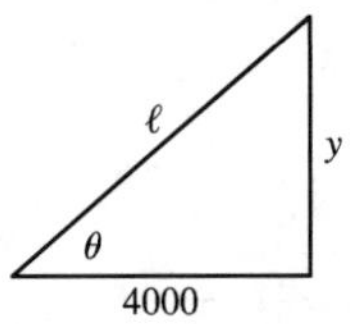

By the Pythagorean Theorem, $4000^2 + y^2 = \ell^2$. Differentiating with respect to t, we obtain $2y\frac{dy}{dt} = 2\ell\frac{d\ell}{dt}$. We know that $\frac{dy}{dt} = 600$ ft/s, so when $y = 3000$ ft, $\ell = \sqrt{4000^2 + 3000^2} = \sqrt{25{,}000{,}000} = 5000$ ft and $\frac{d\ell}{dt} = \frac{y}{\ell}\frac{dy}{dt} = \frac{3000}{5000}(600) = \frac{1800}{5} = 360$ ft/s.

(b) Here $\tan\theta = \frac{y}{4000}$ $\Rightarrow$ $\frac{d}{dt}(\tan\theta) = \frac{d}{dt}\left(\frac{y}{4000}\right)$ $\Rightarrow$ $\sec^2\theta\frac{d\theta}{dt} = \frac{1}{4000}\frac{dy}{dt}$ $\Rightarrow$ $\frac{d\theta}{dt} = \frac{\cos^2\theta}{4000}\frac{dy}{dt}$.

When $y = 3000$ ft, $\frac{dy}{dt} = 600$ ft/s, $\ell = 5000$ and $\cos\theta = \frac{4000}{\ell} = \frac{4000}{5000} = \frac{4}{5}$, so

$\frac{d\theta}{dt} = \frac{(4/5)^2}{4000}(600) = 0.096$ rad/s.

35. 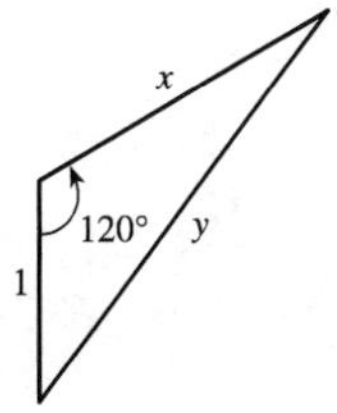

We are given that $\dfrac{dx}{dt} = 300$ km/h. By the Law of Cosines,

$y^2 = x^2 + 1^2 - 2(1)(x)\cos 120° = x^2 + 1 - 2x\left(-\frac{1}{2}\right) = x^2 + x + 1$, so

$2y\,\dfrac{dy}{dt} = 2x\,\dfrac{dx}{dt} + \dfrac{dx}{dt} \;\Rightarrow\; \dfrac{dy}{dt} = \dfrac{2x+1}{2y}\dfrac{dx}{dt}$. After 1 minute,

$x = \frac{300}{60} = 5$ km $\;\Rightarrow\; y = \sqrt{5^2 + 5 + 1} = \sqrt{31}$ km $\;\Rightarrow$

$\dfrac{dy}{dt} = \dfrac{2(5)+1}{2\sqrt{31}}(300) = \dfrac{1650}{\sqrt{31}} \approx 296$ km/h.

37. 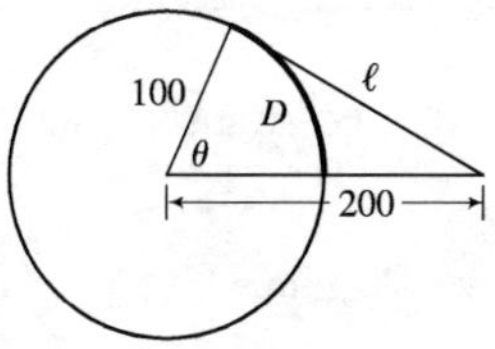

Let the distance between the runner and the friend be ℓ. Then by the Law of Cosines,

$\ell^2 = 200^2 + 100^2 - 2 \cdot 200 \cdot 100 \cdot \cos\theta = 50{,}000 - 40{,}000\cos\theta$ $(\star)$.

Differentiating implicitly with respect to t, we obtain

$2\ell\,\dfrac{d\ell}{dt} = -40{,}000(-\sin\theta)\,\dfrac{d\theta}{dt}$. Now if D is the distance run when the angle is θ radians, then by the formula for the length of an arc on a circle, $s = r\theta$, we have $D = 100\theta$, so

$\theta = \dfrac{1}{100}D \;\Rightarrow\; \dfrac{d\theta}{dt} = \dfrac{1}{100}\dfrac{dD}{dt} = \dfrac{7}{100}$. To substitute into the expression for $\dfrac{d\ell}{dt}$, we must know $\sin\theta$ at the time when $\ell = 200$, which we find from $(\star)$: $200^2 = 50{,}000 - 40{,}000\cos\theta \;\Leftrightarrow\; \cos\theta = \dfrac{1}{4} \;\Rightarrow$

$\sin\theta = \sqrt{1 - \left(\frac{1}{4}\right)^2} = \dfrac{\sqrt{15}}{4}$. Substituting, we get $2(200)\,\dfrac{d\ell}{dt} = 40{,}000\dfrac{\sqrt{15}}{4}\left(\dfrac{7}{100}\right) \;\Rightarrow$

$d\ell/dt = \frac{7\sqrt{15}}{4} \approx 6.78$ m/s. Whether the distance between them is increasing or decreasing depends on the direction in which the runner is running.

Review

CONCEPT CHECK

1. See Definition 3.1.2 and the subsequent discussions on the interpretations of the derivative as the slope of a tangent and as a rate of change.

2. (a) A function f is differentiable at a number a if its derivative f' exists at $x = a$; that is, if $f'(a)$ exists.

(b) See Theorem 3.2.4. This theorem also tells us that if f is *not* continuous at a, then f is *not* differentiable at a.

3. (a) The Power Rule: If n is any real number, then $\dfrac{d}{dx}(x^n) = nx^{n-1}$. The derivative of a variable base raised to a constant power is the power times the base raised to the power minus one.

(b) The Constant Multiple Rule: If c is a constant and f is a differentiable function, then $\dfrac{d}{dx}\,[cf(x)] = c\,\dfrac{d}{dx}\,f(x)$. The derivative of a constant times a function is the constant times the derivative of the function.

(c) The Sum Rule: If f and g are both differentiable, then $\dfrac{d}{dx}\,[f(x) + g(x)] = \dfrac{d}{dx}\,f(x) + \dfrac{d}{dx}\,g(x)$. The derivative of a sum of functions is the sum of the derivatives.

(d) The Difference Rule: If f and g are both differentiable, then $\dfrac{d}{dx}\,[f(x) - g(x)] = \dfrac{d}{dx}\,f(x) - \dfrac{d}{dx}\,g(x)$. The derivative of a difference of functions is the difference of the derivatives.

(e) The Product Rule: If f and g are both differentiable, then $\dfrac{d}{dx}\,[f(x)g(x)] = f(x)\,\dfrac{d}{dx}\,g(x) + g(x)\,\dfrac{d}{dx}\,f(x)$. The derivative of a product of two functions is the first function times the derivative of the second function plus the second function times the derivative of the first function.

(f) The Quotient Rule: If f and g are both differentiable, then $\dfrac{d}{dx}\left[\dfrac{f(x)}{g(x)}\right] = \dfrac{g(x)\,\dfrac{d}{dx}\,f(x) - f(x)\,\dfrac{d}{dx}\,g(x)}{[g(x)]^2}$. The derivative of a quotient of functions is the denominator times the derivative of the numerator minus the numerator times the derivative of the denominator, all divided by the square of the denominator.

(g) The Chain Rule: If f and g are both differentiable and $F = f \circ g$ is the composite function defined by $F(x) = f(g(x))$, then F is differentiable and F' is given by the product $F'(x) = f'(g(x))g'(x)$. The derivative of a composite function is the derivative of the outer function evaluated at the inner function times the derivative of the inner function.

4. (a) $y = x^n \;\Rightarrow\; y' = nx^{n-1}$

(b) $y = \sin x \;\Rightarrow\; y' = \cos x$

(c) $y = \cos x \;\Rightarrow\; y' = -\sin x$

(d) $y = \tan x \;\Rightarrow\; y' = \sec^2 x$

(e) $y = \csc x \;\Rightarrow\; y' = -\csc x\,\cot x$

(f) $y = \sec x \;\Rightarrow\; y' = \sec x\,\tan x$

(g) $y = \cot x \;\Rightarrow\; y' = -\csc^2 x$

5. Implicit differentiation consists of differentiating both sides of an equation involving x and y with respect to x, and then solving the resulting equation for y'.

6. The second derivative of a function f is the rate of change of the first derivative f'. The third derivative is the derivative (rate of change) of the second derivative. If f is the position function of an object, f' is its velocity function, f'' is its acceleration function, and f''' is its jerk function.

7. (a) The linearization L of f at $x = a$ is $L(x) = f(a) + f'(a)(x - a)$.

(b) If $y = f(x)$, then the differential dy is given by $dy = f'(x)\,dx$.

(c) See Figure 6 in Section 3.10.

TRUE-FALSE QUIZ

1. False. See the warning after Theorem 3.2.4.

3. False. See the warning before the Product Rule.

5. True by the Chain Rule.

7. False. $f(x) = |x^2 + x| = x^2 + x$ for $x \geq 0$ or $x \leq -1$ and $|x^2 + x| = -(x^2 + x)$ for $-1 < x < 0$. So $f'(x) = 2x + 1$ for $x > 0$ or $x < -1$ and $f'(x) = -(2x + 1)$ for $-1 < x < 0$. But $|2x + 1| = 2x + 1$ for $x \geq -\frac{1}{2}$ and $|2x + 1| = -2x - 1$ for $x < -\frac{1}{2}$.

9. True. $g(x) = x^5 \quad \Rightarrow \quad g'(x) = 5x^4 \quad \Rightarrow \quad g'(2) = 5(2)^4 = 80$, and by the definition of the derivative, $\lim\limits_{x \to 2} \dfrac{g(x) - g(2)}{x - 2} = g'(2) = 80$.

11. False. A tangent line to the parabola $y = x^2$ has slope $dy/dx = 2x$, so at $(-2, 4)$ the slope of the tangent is $2(-2) = -4$ and an equation of the tangent line is $y - 4 = -4(x + 2)$. [The given equation, $y - 4 = 2x(x + 2)$, is not even linear!]

EXERCISES

1. Estimating the slopes of the tangent lines at $x = 2$, 3, and 5, we obtain approximate values 0.4, 2, and 0.1. The slope of the tangent line at $x = 7$ is negative, so $f'(7) < 0$. Arranging the numbers in increasing order, we have: $f'(7) < 0 < f'(5) < f'(2) < 1 < f'(3)$.

3. (a) $f'(r)$ is the rate at which the total cost changes with respect to the interest rate. Its units are dollars/(percent per year).

(b) The total cost of paying off the loan is increasing by \$1200/(percent per year) as the interest rate reaches 10%. So if the interest rate goes up from 10% to 11%, the cost goes up approximately \$1200.

(c) As r increases, C increases. So $f'(r)$ will always be positive.

For Exercises 4–6, see the hints before Exercise 5 in Section 3.2.

5.

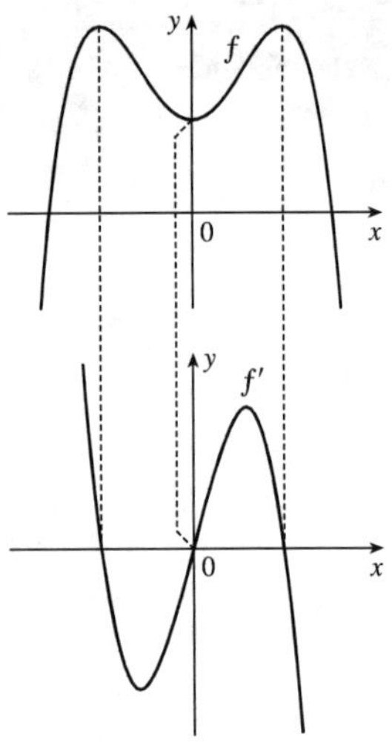

7. The graph of a has tangent lines with positive slope for $x < 0$ and negative slope for $x > 0$, and the values of c fit this pattern, so c must be the graph of the derivative of the function for a. The graph of c has horizontal tangent lines to the left and right of the x-axis and b has zeros at these points. Hence, b is the graph of the derivative of the function for c. Therefore, a is the graph of f, c is the graph of f', and b is the graph of f''.

9. $B'(1990)$ is the rate at which the total value of U.S. banknotes in circulation is changing in billions of dollars per year. To estimate the value of $B'(1990)$, we will average the difference quotients obtained using the times $t = 1985$ and $t = 1995$. Let $A = \dfrac{B(1985) - B(1990)}{1985 - 1990} = \dfrac{182.0 - 268.2}{-5} = 17.24$ and

$$C = \frac{B(1995) - B(1990)}{1995 - 1990} = \frac{401.5 - 268.2}{5} = 26.66. \text{ Then}$$

$$B'(1990) = \lim_{t \to 1990} \frac{B(t) - B(1990)}{t - 1990} \approx \frac{A + C}{2} = \frac{17.24 + 26.66}{2} = 21.95 \text{ billions of dollars/year.}$$

11. $f(x) = x^3 + 5x + 4 \;\Rightarrow$

$$f'(x) = \lim_{h \to 0} \frac{f(x + h) - f(x)}{h} = \lim_{h \to 0} \frac{(x + h)^3 + 5(x + h) + 4 - (x^3 + 5x + 4)}{h}$$

$$= \lim_{h \to 0} \frac{3x^2 h + 3xh^2 + h^3 + 5h}{h} = \lim_{h \to 0} (3x^2 + 3xh + h^2 + 5) = 3x^2 + 5$$

13. $y = \left(x^4 - 3x^2 + 5\right)^3 \;\Rightarrow$

$$y' = 3\left(x^4 - 3x^2 + 5\right)^2 \frac{d}{dx}\left(x^4 - 3x^2 + 5\right) = 3\left(x^4 - 3x^2 + 5\right)^2\left(4x^3 - 6x\right) = 6x\left(x^4 - 3x^2 + 5\right)^2\left(2x^2 - 3\right)$$

15. $y=\sqrt{x}+\dfrac{1}{\sqrt[3]{x^4}}=x^{1/2}+x^{-4/3} \quad\Rightarrow\quad y'=\frac{1}{2}x^{-1/2}-\frac{4}{3}x^{-7/3}=\dfrac{1}{2\sqrt{x}}-\dfrac{4}{3\sqrt[3]{x^7}}$

17. $y=2x\sqrt{x^2+1} \quad\Rightarrow$

$$y'=2x\cdot\tfrac{1}{2}\left(x^2+1\right)^{-1/2}(2x)+\sqrt{x^2+1}\,(2)=\frac{2x^2}{\sqrt{x^2+1}}+2\sqrt{x^2+1}=\frac{2x^2+2\left(x^2+1\right)}{\sqrt{x^2+1}}=\frac{2\left(2x^2+1\right)}{\sqrt{x^2+1}}$$

19. $y=\dfrac{t}{1-t^2} \quad\Rightarrow\quad y'=\dfrac{\left(1-t^2\right)(1)-t(-2t)}{\left(1-t^2\right)^2}=\dfrac{1-t^2+2t^2}{\left(1-t^2\right)^2}=\dfrac{t^2+1}{\left(1-t^2\right)^2}$

21. $y=\tan\sqrt{1-x} \quad\Rightarrow\quad y'=\left(\sec^2\sqrt{1-x}\right)\left(\dfrac{1}{2\sqrt{1-x}}\right)(-1)=-\dfrac{\sec^2\sqrt{1-x}}{2\sqrt{1-x}}$

23. $\dfrac{d}{dx}\left(xy^4+x^2y\right)=\dfrac{d}{dx}\left(x+3y\right) \quad\Rightarrow\quad x\cdot 4y^3y'+y^4\cdot 1+x^2\cdot y'+y\cdot 2x=1+3y' \quad\Rightarrow$

$$y'\left(4xy^3+x^2-3\right)=1-y^4-2xy \quad\Rightarrow\quad y'=\frac{1-y^4-2xy}{4xy^3+x^2-3}$$

25. $y=\dfrac{\sec 2\theta}{1+\tan 2\theta} \quad\Rightarrow$

$$y'=\frac{(1+\tan 2\theta)(\sec 2\theta\,\tan 2\theta\cdot 2)-(\sec 2\theta)\left(\sec^2 2\theta\cdot 2\right)}{(1+\tan 2\theta)^2}=\frac{2\sec 2\theta\left[(1+\tan 2\theta)\tan 2\theta-\sec^2 2\theta\right]}{(1+\tan 2\theta)^2}$$

$$=\frac{2\sec 2\theta\left(\tan 2\theta+\tan^2 2\theta-\sec^2 2\theta\right)}{(1+\tan 2\theta)^2}=\frac{2\sec 2\theta\,(\tan 2\theta-1)}{(1+\tan 2\theta)^2} \qquad \left[1+\tan^2 x=\sec^2 x\right]$$

27. $y=\left(1-x^{-1}\right)^{-1} \quad\Rightarrow$

$$y'=-1(1-x^{-1})^{-2}[-(-1x^{-2})]=-(1-1/x)^{-2}x^{-2}=-((x-1)/x)^{-2}x^{-2}=-(x-1)^{-2}$$

29. $\sin(xy)=x^2-y \quad\Rightarrow\quad \cos(xy)(xy'+y\cdot 1)=2x-y' \quad\Rightarrow\quad x\cos(xy)y'+y'=2x-y\cos(xy) \quad\Rightarrow$

$$y'[x\cos(xy)+1]=2x-y\cos(xy) \quad\Rightarrow\quad y'=\frac{2x-y\cos(xy)}{x\cos(xy)+1}$$

31. $y=\cot\left(3x^2+5\right) \quad\Rightarrow\quad y'=-\csc^2\left(3x^2+5\right)(6x)=-6x\csc^2\left(3x^2+5\right)$

33. $y=\sin\left(\tan\sqrt{1+x^3}\right) \quad\Rightarrow\quad y'=\cos\left(\tan\sqrt{1+x^3}\right)\left(\sec^2\sqrt{1+x^3}\right)\left[3x^2/\left(2\sqrt{1+x^3}\right)\right]$

35. $y=\tan^2(\sin\theta)=[\tan(\sin\theta)]^2 \quad\Rightarrow\quad y'=2[\tan(\sin\theta)]\cdot\sec^2(\sin\theta)\cdot\cos\theta$

37. $y=(x\tan x)^{1/5} \quad\Rightarrow\quad y'=\frac{1}{5}(x\tan x)^{-4/5}(\tan x+x\sec^2 x)$

39. $f(t)=\sqrt{4t+1} \quad\Rightarrow\quad f'(t)=\frac{1}{2}(4t+1)^{-1/2}\cdot 4=2(4t+1)^{-1/2} \quad\Rightarrow$

$f''(t)=2(-\frac{1}{2})(4t+1)^{-3/2}\cdot 4=-4/(4t+1)^{3/2}$, so $f''(2)=-4/9^{3/2}=-\frac{4}{27}$.

41. $x^6+y^6=1 \quad\Rightarrow\quad 6x^5+6y^5y'=0 \quad\Rightarrow\quad y'=-x^5/y^5 \quad\Rightarrow$

$$y''=-\frac{y^5\left(5x^4\right)-x^5\left(5y^4y'\right)}{\left(y^5\right)^2}=-\frac{5x^4y^4\left[y-x\left(-x^5/y^5\right)\right]}{y^{10}}=-\frac{5x^4\left[\left(y^6+x^6\right)/y^5\right]}{y^6}=-\frac{5x^4}{y^{11}}$$

43. $\displaystyle\lim_{x\to 0}\frac{\sec x}{1-\sin x}=\frac{\sec 0}{1-\sin 0}=\frac{1}{1-0}=1$

45. $y = 4\sin^2 x \Rightarrow y' = 4 \cdot 2\sin x \cos x$. At $\left(\frac{\pi}{6}, 1\right)$, $y' = 8 \cdot \frac{1}{2} \cdot \frac{\sqrt{3}}{2} = 2\sqrt{3}$, so an equation of the tangent line is $y - 1 = 2\sqrt{3}\left(x - \frac{\pi}{6}\right)$, or $y = 2\sqrt{3}\,x + 1 - \pi\sqrt{3}/3$.

47. $y = \sqrt{1 + 4\sin x} \Rightarrow y' = \frac{1}{2}(1 + 4\sin x)^{-1/2} \cdot 4\cos x = \dfrac{2\cos x}{\sqrt{1 + 4\sin x}}$. At $(0, 1)$, $y' = \dfrac{2}{\sqrt{1}} = 2$, so an equation of the tangent line is $y - 1 = 2(x - 0)$, or $y = 2x + 1$.

49. (a) $f(x) = x\sqrt{5 - x} \Rightarrow$

$$f'(x) = x\left[\tfrac{1}{2}(5 - x)^{-1/2}(-1)\right] + \sqrt{5 - x} = \frac{-x}{2\sqrt{5 - x}} + \sqrt{5 - x} \cdot \frac{2\sqrt{5 - x}}{2\sqrt{5 - x}}$$

$$= \frac{-x}{2\sqrt{5 - x}} + \frac{2(5 - x)}{2\sqrt{5 - x}} = \frac{-x + 10 - 2x}{2\sqrt{5 - x}} = \frac{10 - 3x}{2\sqrt{5 - x}}$$

(b) At $(1, 2)$: $f'(1) = \frac{7}{4}$. So an equation of the tangent line is $y - 2 = \frac{7}{4}(x - 1)$ or $y = \frac{7}{4}x + \frac{1}{4}$.
At $(4, 4)$: $f'(4) = -\frac{2}{2} = -1$. So an equation of the tangent line is $y - 4 = -1(x - 4)$ or $y = -x + 8$.

(c)

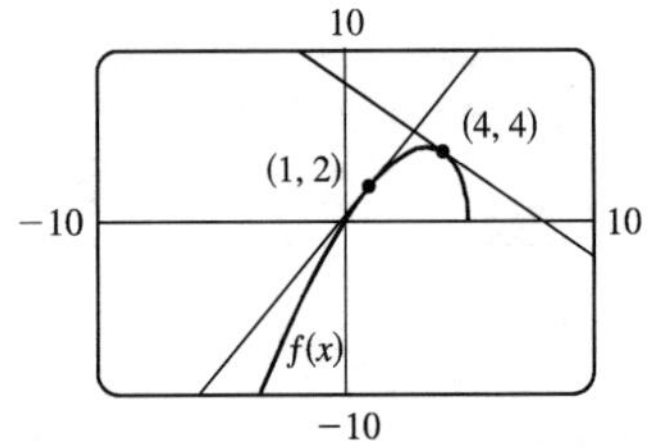

(d)

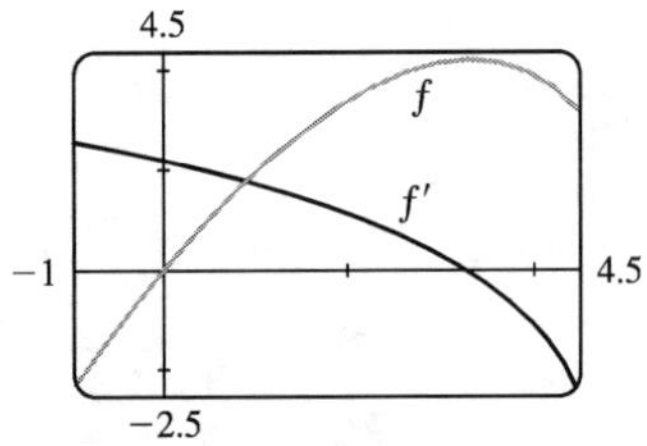

The graphs look reasonable, since f' is positive where f has tangents with positive slope, and f' is negative where f has tangents with negative slope.

51. $y = \sin x + \cos x \Rightarrow y' = \cos x - \sin x = 0 \Leftrightarrow \cos x = \sin x$ and $0 \le x \le 2\pi \Leftrightarrow x = \frac{\pi}{4}$ or $\frac{5\pi}{4}$, so the points are $\left(\frac{\pi}{4}, \sqrt{2}\right)$ and $\left(\frac{5\pi}{4}, -\sqrt{2}\right)$.

53. $f(x) = (x - a)(x - b)(x - c) \Rightarrow f'(x) = (x - b)(x - c) + (x - a)(x - c) + (x - a)(x - b)$. So

$$\frac{f'(x)}{f(x)} = \frac{(x - b)(x - c) + (x - a)(x - c) + (x - a)(x - b)}{(x - a)(x - b)(x - c)} = \frac{1}{x - a} + \frac{1}{x - b} + \frac{1}{x - c}.$$

55. (a) $h(x) = f(x)g(x) \Rightarrow h'(x) = f(x)g'(x) + g(x)f'(x) \Rightarrow$
$h'(2) = f(2)g'(2) + g(2)f'(2) = (3)(4) + (5)(-2) = 12 - 10 = 2$

(b) $F(x) = f(g(x)) \Rightarrow F'(x) = f'(g(x))g'(x) \Rightarrow F'(2) = f'(g(2))g'(2) = f'(5)(4) = 11 \cdot 4 = 44$

57. $f(x) = x^2 g(x) \Rightarrow f'(x) = x^2 g'(x) + g(x)(2x) = x\left[xg'(x) + 2g(x)\right]$

59. $f(x) = [g(x)]^2 \Rightarrow f'(x) = 2[g(x)]^1 \cdot g'(x) = 2g(x)g'(x)$

61. $f(x) = g(g(x)) \Rightarrow f'(x) = g'(g(x))g'(x)$

63. $f(x) = g(\sin x) \Rightarrow f'(x) = g'(\sin x) \cdot \cos x$

65. $h(x) = \dfrac{f(x)g(x)}{f(x)+g(x)} \quad \Rightarrow$

$$h'(x) = \frac{[f(x)+g(x)]\,[f(x)g'(x)+g(x)f'(x)] - f(x)g(x)\,[f'(x)+g'(x)]}{[f(x)+g(x)]^2}$$

$$= \frac{[f(x)]^2\,g'(x) + f(x)g(x)f'(x) + f(x)g(x)g'(x) + [g(x)]^2\,f'(x) - f(x)g(x)f'(x) - f(x)g(x)g'(x)}{[f(x)+g(x)]^2}$$

$$= \frac{f'(x)[g(x)]^2 + g'(x)[f(x)]^2}{[f(x)+g(x)]^2}$$

67. Using the Chain Rule repeatedly, $h(x) = f(g(\sin 4x)) \quad \Rightarrow$

$$h'(x) = f'(g(\sin 4x)) \cdot \frac{d}{dx}(g(\sin 4x)) = f'(g(\sin 4x)) \cdot g'(\sin 4x) \cdot \frac{d}{dx}(\sin 4x)$$
$$= f'(g(\sin 4x))g'(\sin 4x)(\cos 4x)(4)$$

69. f is not differentiable: at $x = -4$ because f is not continuous, at $x = -1$ because f has a corner, at $x = 2$ because f is not continuous, and at $x = 5$ because f has a vertical tangent.

71. (a) $y = t^3 - 12t + 3 \quad \Rightarrow \quad v(t) = y' = 3t^2 - 12 \quad \Rightarrow \quad a(t) = v'(t) = 6t$

(b) $v(t) = 3(t^2 - 4) > 0$ when $t > 2$, so it moves upward when $t > 2$ and downward when $0 \le t < 2$.

(c) Distance upward $= y(3) - y(2) = -6 - (-13) = 7$,
Distance downward $= y(0) - y(2) = 3 - (-13) = 16$. Total distance $= 7 + 16 = 23$.

73. The linear density ρ is the rate of change of mass m with respect to length x. $m = x(1 + \sqrt{x}\,) = x + x^{3/2} \quad \Rightarrow$ $\rho = dm/dx = 1 + \frac{3}{2}\sqrt{x}$, so the linear density when $x = 4$ is $1 + \frac{3}{2}\sqrt{4} = 4$ kg/m.

75. If $x =$ edge length, then $V = x^3 \quad \Rightarrow \quad dV/dt = 3x^2\,dx/dt = 10 \quad \Rightarrow \quad dx/dt = 10/(3x^2)$ and $S = 6x^2 \quad \Rightarrow$ $dS/dt = (12x)\,dx/dt = 12x[10/(3x^2)] = 40/x$. When $x = 30$, $dS/dt = \frac{40}{30} = \frac{4}{3}\ \text{cm}^2/\text{min}$.

77. Given $dh/dt = 5$ and $dx/dt = 15$, find dz/dt. $z^2 = x^2 + h^2 \quad \Rightarrow$

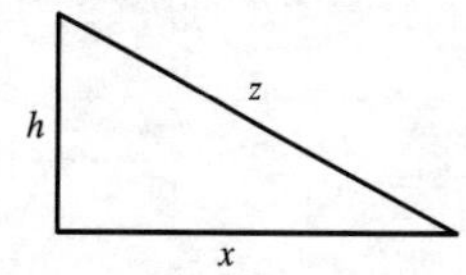

$2z\,\dfrac{dz}{dt} = 2x\,\dfrac{dx}{dt} + 2h\,\dfrac{dh}{dt} \quad \Rightarrow \quad \dfrac{dz}{dt} = \dfrac{1}{z}(15x + 5h)$. When $t = 3$,

$h = 45 + 3(5) = 60$ and $x = 15(3) = 45 \quad \Rightarrow \quad z = \sqrt{45^2 + 60^2} = 75$, so

$\dfrac{dz}{dt} = \frac{1}{75}\,[15(45) + 5(60)] = 13$ ft/s.

79. We are given $d\theta/dt = -0.25$ rad/h. $\tan\theta = 400/x \quad \Rightarrow$

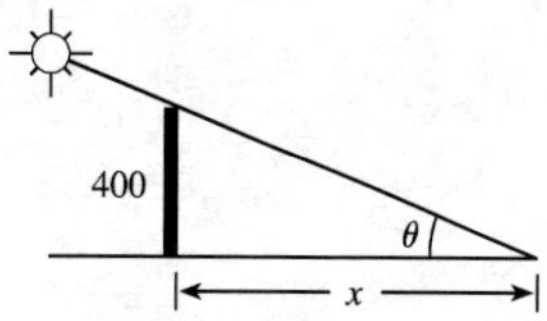

$x = 400\cot\theta \quad \Rightarrow \quad \dfrac{dx}{dt} = -400\csc^2\theta\,\dfrac{d\theta}{dt}$. When $\theta = \frac{\pi}{6}$,

$\dfrac{dx}{dt} = -400(2)^2(-0.25) = 400$ ft/h.

81. (a) $f(x) = \sqrt[3]{1+3x} = (1+3x)^{1/3} \;\Rightarrow\; f'(x) = (1+3x)^{-2/3}$, so the linearization of f at $a = 0$ is

$L(x) = f(0) + f'(0)(x-0) = 1^{1/3} + 1^{-2/3}x = 1 + x$. Thus, $\sqrt[3]{1+3x} \approx 1 + x \;\Rightarrow$

$\sqrt[3]{1.03} = \sqrt[3]{1+3(0.01)} \approx 1 + (0.01) = 1.01$.

(b) The linear approximation is $\sqrt[3]{1+3x} \approx 1 + x$, so for the required accuracy we want $\sqrt[3]{1+3x} - 0.1 < 1 + x < \sqrt[3]{1+3x} + 0.1$. From the graph, it appears that this is true when $-0.23 < x < 0.40$.

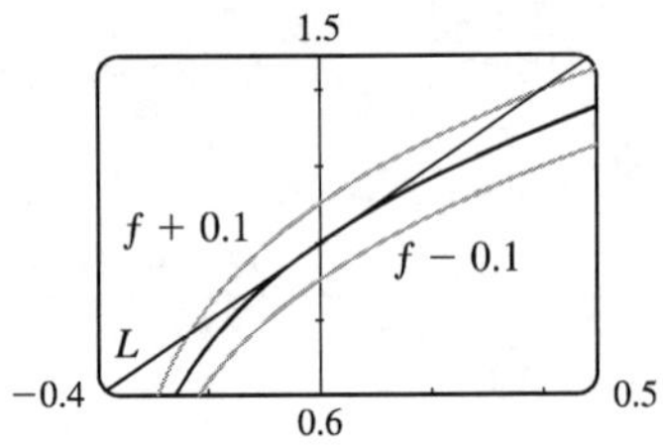

83. $A = x^2 + \frac{1}{2}\pi\left(\frac{1}{2}x\right)^2 = \left(1 + \frac{\pi}{8}\right)x^2 \;\Rightarrow\; dA = \left(2 + \frac{\pi}{4}\right)x\,dx$.

When $x = 60$ and $dx = 0.1$, $dA = \left(2 + \frac{\pi}{4}\right)60(0.1) = 12 + \frac{3\pi}{2}$, so

the maximum error is approximately $12 + \frac{3\pi}{2} \approx 16.7\text{ cm}^2$.

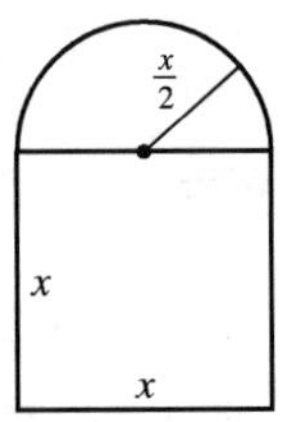

85. $$\lim_{h\to 0} \frac{\sqrt[4]{16+h} - 2}{h} = \left[\frac{d}{dx}\sqrt[4]{x}\right]_{x=16} = \tfrac{1}{4}x^{-3/4}\Big|_{x=16} = \frac{1}{4\left(\sqrt[4]{16}\right)^3} = \frac{1}{32}$$

87. $$\lim_{x\to 0} \frac{\sqrt{1+\tan x} - \sqrt{1+\sin x}}{x^3} = \lim_{x\to 0} \frac{\left(\sqrt{1+\tan x} - \sqrt{1+\sin x}\right)\left(\sqrt{1+\tan x} + \sqrt{1+\sin x}\right)}{x^3\left(\sqrt{1+\tan x} + \sqrt{1+\sin x}\right)}$$

$$= \lim_{x\to 0} \frac{(1+\tan x) - (1+\sin x)}{x^3\left(\sqrt{1+\tan x} + \sqrt{1+\sin x}\right)} = \lim_{x\to 0} \frac{\sin x\,(1/\cos x - 1)}{x^3\left(\sqrt{1+\tan x} + \sqrt{1+\sin x}\right)} \cdot \frac{\cos x}{\cos x}$$

$$= \lim_{x\to 0} \frac{\sin x\,(1 - \cos x)}{x^3\left(\sqrt{1+\tan x} + \sqrt{1+\sin x}\right)\cos x} \cdot \frac{1+\cos x}{1+\cos x}$$

$$= \lim_{x\to 0} \frac{\sin x \cdot \sin^2 x}{x^3\left(\sqrt{1+\tan x} + \sqrt{1+\sin x}\right)\cos x\,(1+\cos x)}$$

$$= \left(\lim_{x\to 0} \frac{\sin x}{x}\right)^3 \lim_{x\to 0} \frac{1}{\left(\sqrt{1+\tan x} + \sqrt{1+\sin x}\right)\cos x\,(1+\cos x)}$$

$$= 1^3 \cdot \frac{1}{\left(\sqrt{1} + \sqrt{1}\right)\cdot 1 \cdot (1+1)} = \frac{1}{4}$$

89. $\frac{d}{dx}\left[f(2x)\right] = x^2 \;\Rightarrow\; f'(2x)\cdot 2 = x^2 \;\Rightarrow\; f'(2x) = \frac{1}{2}x^2$. Let $t = 2x$. Then $f'(t) = \frac{1}{2}\left(\frac{1}{2}t\right)^2 = \frac{1}{8}t^2$, so $f'(x) = \frac{1}{8}x^2$.

How Derivatives Affect the Shape of a Graph

1. (a) f is increasing on $(0, 6)$ and $(8, 9)$.

(b) f is decreasing on $(6, 8)$.

(c) f is concave upward on $(2, 4)$ and $(7, 9)$.

(d) f is concave downward on $(0, 2)$ and $(4, 7)$.

(e) The points of inflection are $(2, 3)$, $(4, 4.5)$ and $(7, 4)$ (where the concavity changes).

3. (a) Use the Increasing/Decreasing (I/D) Test.

(b) Use the Concavity Test.

(c) At any value of x where the concavity changes, we have an inflection point at $(x, f(x))$.

5. (a) Since $f'(x) > 0$ on $(1, 5)$, f is increasing on this interval. Since $f'(x) < 0$ on $(0, 1)$ and $(5, 6)$, f is decreasing on these intervals.

(b) Since $f'(x) = 0$ at $x = 1$ and f' changes from negative to positive there, f changes from decreasing to increasing and has a local minimum at $x = 1$. Since $f'(x) = 0$ at $x = 5$ and f' changes from positive to negative there, f changes from increasing to decreasing and has a local maximum at $x = 5$.

7. There is an inflection point at $x = 1$ because $f''(x)$ changes from negative to positive there, and so the graph of f changes from concave downward to concave upward. There is an inflection point at $x = 7$ because $f''(x)$ changes from positive to negative there, and so the graph of f changes from concave upward to concave downward.

9. The function must be always decreasing and concave downward.

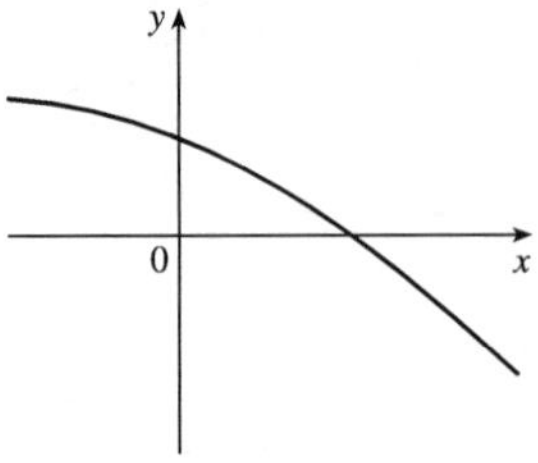

11. (a) $f(x) = x^3 - 12x + 1 \;\Rightarrow\; f'(x) = 3x^2 - 12 = 3(x+2)(x-2)$.

We don't need to include "3" in the chart to determine the sign of $f'(x)$.

Interval	$x+2$	$x-2$	$f'(x)$	f
$x < -2$	$-$	$-$	$+$	increasing on $(-\infty, -2)$
$-2 < x < 2$	$+$	$-$	$-$	decreasing on $(-2, 2)$
$x > 2$	$+$	$+$	$+$	increasing on $(2, \infty)$

So f is increasing on $(-\infty, -2)$ and $(2, \infty)$ and f is decreasing on $(-2, 2)$.

(b) f changes from increasing to decreasing at $x = -2$ and from decreasing to increasing at $x = 2$. Thus, $f(-2) = 17$ is a local maximum value and $f(2) = -15$ is a local minimum value.

(c) $f''(x) = 6x$. $f''(x) > 0 \;\Leftrightarrow\; x > 0$ and $f''(x) < 0 \;\Leftrightarrow\; x < 0$. Thus, f is concave upward on $(0, \infty)$ and concave downward on $(-\infty, 0)$. There is an inflection point where the concavity changes, at $(0, f(0)) = (0, 1)$.

13. (a) $f(x) = x^4 - 2x^2 + 3 \;\Rightarrow\; f'(x) = 4x^3 - 4x = 4x(x^2 - 1) = 4x(x+1)(x-1)$.

Interval	$x+1$	x	$x-1$	$f'(x)$	f
$x < -1$	$-$	$-$	$-$	$-$	decreasing on $(-\infty, -1)$
$-1 < x < 0$	$+$	$-$	$-$	$+$	increasing on $(-1, 0)$
$0 < x < 1$	$+$	$+$	$-$	$-$	decreasing on $(0, 1)$
$x > 1$	$+$	$+$	$+$	$+$	increasing on $(1, \infty)$

So f is increasing on $(-1, 0)$ and $(1, \infty)$ and f is decreasing on $(-\infty, -1)$ and $(0, 1)$.

(b) f changes from increasing to decreasing at $x = 0$ and from decreasing to increasing at $x = -1$ and $x = 1$. Thus, $f(0) = 3$ is a local maximum value and $f(\pm 1) = 2$ are local minimum values.

(c) $f''(x) = 12x^2 - 4 = 12\left(x^2 - \frac{1}{3}\right) = 12\left(x + 1/\sqrt{3}\right)\left(x - 1/\sqrt{3}\right)$. $f''(x) > 0 \;\Leftrightarrow\; x < -1/\sqrt{3}$ or $x > 1/\sqrt{3}$ and $f''(x) < 0 \;\Leftrightarrow\; -1/\sqrt{3} < x < 1/\sqrt{3}$. Thus, f is concave upward on $\left(-\infty, -\sqrt{3}/3\right)$ and $\left(\sqrt{3}/3, \infty\right)$ and concave downward on $\left(-\sqrt{3}/3, \sqrt{3}/3\right)$. There are inflection points at $\left(\pm\sqrt{3}/3, \frac{22}{9}\right)$.

15. (a) $f(x) = x - 2\sin x$ on $(0, 3\pi)$ $\Rightarrow$ $f'(x) = 1 - 2\cos x$. $f'(x) > 0$ $\Leftrightarrow$ $1 - 2\cos x > 0$ $\Leftrightarrow$ $\cos x < \frac{1}{2}$ $\Leftrightarrow$ $\frac{\pi}{3} < x < \frac{5\pi}{3}$ or $\frac{7\pi}{3} < x < 3\pi$. $f'(x) < 0$ $\Leftrightarrow$ $\cos x > \frac{1}{2}$ $\Leftrightarrow$ $0 < x < \frac{\pi}{3}$ or $\frac{5\pi}{3} < x < \frac{7\pi}{3}$. So f is increasing on $\left(\frac{\pi}{3}, \frac{5\pi}{3}\right)$ and $\left(\frac{7\pi}{3}, 3\pi\right)$, and f is decreasing on $\left(0, \frac{\pi}{3}\right)$ and $\left(\frac{5\pi}{3}, \frac{7\pi}{3}\right)$.

(b) f changes from increasing to decreasing at $x = \frac{5\pi}{3}$, and from decreasing to increasing at $x = \frac{\pi}{3}$ and at $x = \frac{7\pi}{3}$. Thus, $f\left(\frac{5\pi}{3}\right) = \frac{5\pi}{3} + \sqrt{3} \approx 6.97$ is a local maximum value and $f\left(\frac{\pi}{3}\right) = \frac{\pi}{3} - \sqrt{3} \approx -0.68$ and $f\left(\frac{7\pi}{3}\right) = \frac{7\pi}{3} - \sqrt{3} \approx 5.60$ are local minimum values.

(c) $f''(x) = 2\sin x > 0$ $\Leftrightarrow$ $0 < x < \pi$ and $2\pi < x < 3\pi$, $f''(x) < 0$ $\Leftrightarrow$ $\pi < x < 2\pi$. Thus, f is concave upward on $(0, \pi)$ and $(2\pi, 3\pi)$, and f is concave downward on $(\pi, 2\pi)$. There are inflection points at (π, π) and $(2\pi, 2\pi)$.

17. $f(x) = x^5 - 5x + 3$ $\Rightarrow$ $f'(x) = 5x^4 - 5 = 5(x^2 + 1)(x + 1)(x - 1)$.

First Derivative Test: $f'(x) < 0$ $\Rightarrow$ $-1 < x < 1$ and $f'(x) > 0$ $\Rightarrow$ $x > 1$ or $x < -1$. Since f' changes from positive to negative at $x = -1$, $f(-1) = 7$ is a local maximum value; and since f' changes from negative to positive at $x = 1$, $f(1) = -1$ is a local minimum value.

Second Derivative Test: $f''(x) = 20x^3$. $f'(x) = 0$ $\Leftrightarrow$ $x = \pm 1$. $f''(-1) = -20 < 0$ $\Rightarrow$ $f(-1) = 7$ is a local maximum value. $f''(1) = 20 > 0$ $\Rightarrow$ $f(1) = -1$ is a local minimum value.

Preference: For this function, the two tests are equally easy.

19. $f(x) = x + \sqrt{1 - x}$ $\Rightarrow$ $f'(x) = 1 + \frac{1}{2}(1 - x)^{-1/2}(-1) = 1 - \dfrac{1}{2\sqrt{1 - x}}$. Note that f is defined for $1 - x \geq 0$; that is, for $x \leq 1$. $f'(x) = 0$ $\Rightarrow$ $2\sqrt{1 - x} = 1$ $\Rightarrow$ $\sqrt{1 - x} = \frac{1}{2}$ $\Rightarrow$ $1 - x = \frac{1}{4}$ $\Rightarrow$ $x = \frac{3}{4}$. f' does not exist at $x = 1$, but we can't have a local maximum or minimum at an endpoint.

First Derivative Test: $f'(x) > 0$ $\Rightarrow$ $x < \frac{3}{4}$ and $f'(x) < 0$ $\Rightarrow$ $\frac{3}{4} < x < 1$. Since f' changes from positive to negative at $x = \frac{3}{4}$, $f\left(\frac{3}{4}\right) = \frac{5}{4}$ is a local maximum value.

Second Derivative Test: $f''(x) = -\frac{1}{2}\left(-\frac{1}{2}\right)(1 - x)^{-3/2}(-1) = -\dfrac{1}{4\left(\sqrt{1 - x}\right)^3}$. $f''\left(\frac{3}{4}\right) = -2 < 0$ $\Rightarrow$ $f\left(\frac{3}{4}\right) = \frac{5}{4}$ is a local maximum value.

Preference: The First Derivative Test may be slightly easier to apply in this case.

21. (a) By the Second Derivative Test, if $f'(2) = 0$ and $f''(2) = -5 < 0$, f has a local maximum at $x = 2$.

(b) If $f'(6) = 0$, we know that f has a horizontal tangent at $x = 6$. Knowing that $f''(6) = 0$ does not provide any additional information since the Second Derivative Test fails. For example, the first and second derivatives of $y = (x - 6)^4$, $y = -(x - 6)^4$, and $y = (x - 6)^3$ all equal zero for $x = 6$, but the first has a local minimum at $x = 6$, the second has a local maximum at $x = 6$, and the third has an inflection point at $x = 6$.

23. $f'(0) = f'(2) = f'(4) = 0$ $\Leftrightarrow$ horizontal tangents at $x = 0, 2, 4$. $f'(x) > 0$ if $x < 0$ or $2 < x < 4$ $\Rightarrow$ f is increasing on $(-\infty, 0)$ and $(2, 4)$. $f'(x) < 0$ if $0 < x < 2$ or $x > 4$ $\Rightarrow$ f is decreasing on $(0, 2)$ and $(4, \infty)$. $f''(x) > 0$ if $1 < x < 3$ $\Rightarrow$ f is concave upward on $(1, 3)$. $f''(x) < 0$ if $x < 1$ or $x > 3$ $\Rightarrow$ f is concave downward on $(-\infty, 1)$ and $(3, \infty)$. There are inflection points when $x = 1$ and 3.

25. $f'(x) > 0$ if $|x| < 2 \Rightarrow f$ is increasing on $(-2, 2)$. $f'(x) < 0$ if $|x| > 2 \Rightarrow f$ is decreasing on $(-\infty, -2)$ and $(2, \infty)$. $f'(-2) = 0 \Rightarrow$ horizontal tangent at $x = -2$. $\lim_{x \to 2} |f'(x)| = \infty \Rightarrow$ there is a vertical asymptote or vertical tangent (cusp) at $x = 2$. $f''(x) > 0$ if $x \neq 2 \Rightarrow f$ is concave upward on $(-\infty, 2)$ and $(2, \infty)$.

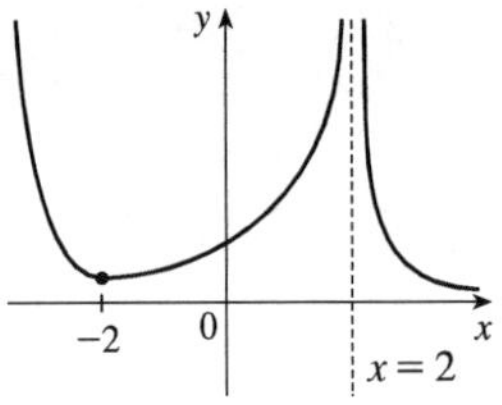

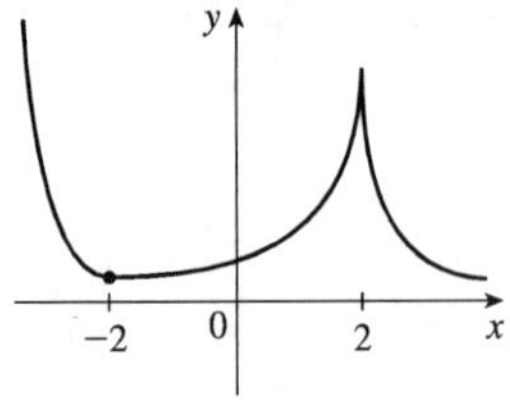

27. (a) f is increasing where f' is positive, that is, on $(0, 2)$, $(4, 6)$, and $(8, \infty)$; and decreasing where f' is negative, that is, on $(2, 4)$ and $(6, 8)$.

(b) f has local maxima where f' changes from positive to negative, at $x = 2$ and at $x = 6$, and local minima where f' changes from negative to positive, at $x = 4$ and at $x = 8$.

(c) f is concave upward (CU) where f' is increasing, that is, on $(3, 6)$ and $(6, \infty)$, and concave downward (CD) where f' is decreasing, that is, on $(0, 3)$.

(d) There is a point of inflection where f changes from being CD to being CU, that is, at $x = 3$.

(e)

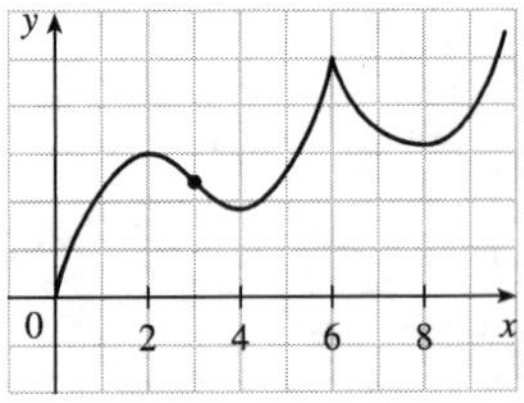

29. (a) $f(x) = 2x^3 - 3x^2 - 12x \Rightarrow f'(x) = 6x^2 - 6x - 12 = 6(x^2 - x - 2) = 6(x - 2)(x + 1)$. $f'(x) > 0 \Leftrightarrow x < -1$ or $x > 2$ and $f'(x) < 0 \Leftrightarrow -1 < x < 2$. So f is increasing on $(-\infty, -1)$ and $(2, \infty)$, and f is decreasing on $(-1, 2)$.

(b) Since f changes from increasing to decreasing at $x = -1$, $f(-1) = 7$ is a local maximum value. Since f changes from decreasing to increasing at $x = 2$, $f(2) = -20$ is a local minimum value.

(c) $f''(x) = 6(2x - 1) \Rightarrow f''(x) > 0$ on $\left(\frac{1}{2}, \infty\right)$ and $f''(x) < 0$ on $\left(-\infty, \frac{1}{2}\right)$. So f is concave upward on $\left(\frac{1}{2}, \infty\right)$ and concave downward on $\left(-\infty, \frac{1}{2}\right)$. There is a change in concavity at $x = \frac{1}{2}$, and we have an inflection point at $\left(\frac{1}{2}, -\frac{13}{2}\right)$.

(d)

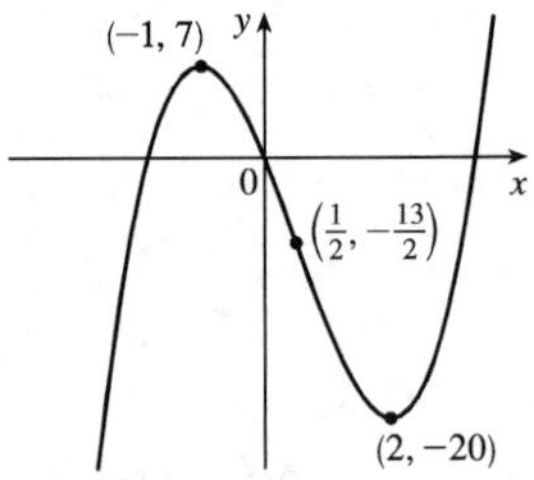

31. (a) $f(x) = x^4 - 6x^2 \Rightarrow f'(x) = 4x^3 - 12x = 4x(x^2 - 3) = 0$ when $x = 0, \pm\sqrt{3}$.

Interval	$4x$	$x^2 - 3$	$f'(x)$	f
$x < -\sqrt{3}$	−	+	−	decreasing on $(-\infty, -\sqrt{3})$
$-\sqrt{3} < x < 0$	−	−	+	increasing on $(-\sqrt{3}, 0)$
$0 < x < \sqrt{3}$	+	−	−	decreasing on $(0, \sqrt{3})$
$x > \sqrt{3}$	+	+	+	increasing on $(\sqrt{3}, \infty)$

(b) Local minimum values $f(\pm\sqrt{3}) = -9$, local maximum value $f(0) = 0$

(c) $f''(x) = 12x^2 - 12 = 12(x^2 - 1) > 0 \;\Leftrightarrow\; x^2 > 1 \;\Leftrightarrow\; |x| > 1 \;\Leftrightarrow\; x > 1$ or $x < -1$, so f is CU on $(-\infty, -1)$, $(1, \infty)$ and CD on $(-1, 1)$. Inflection points at $(\pm 1, -5)$

(d)

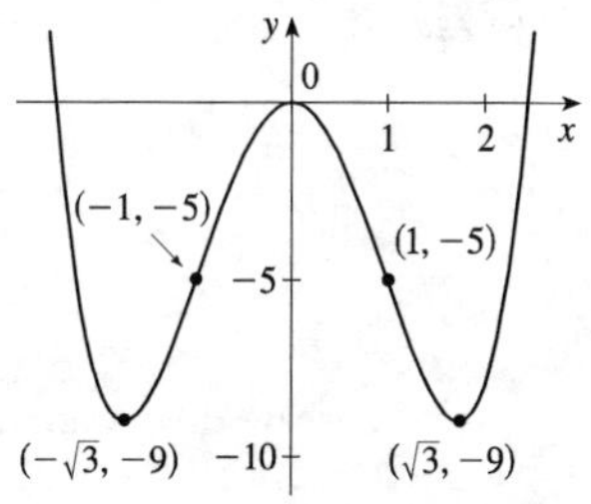

33. (a) $h(x) = 3x^5 - 5x^3 + 3 \;\Rightarrow\; h'(x) = 15x^4 - 15x^2 = 15x^2(x^2 - 1) = 0$ when $x = 0, \pm 1$. Since $15x^2$ is nonnegative, $h'(x) > 0 \;\Leftrightarrow\; x^2 > 1 \;\Leftrightarrow\; |x| > 1 \;\Leftrightarrow\; x > 1$ or $x < -1$, so h is increasing on $(-\infty, -1)$ and $(1, \infty)$ and decreasing on $(-1, 1)$, with a horizontal tangent at $x = 0$.

(b) Local maximum value $h(-1) = 5$, local minimum value $h(1) = 1$

(c) $h''(x) = 60x^3 - 30x = 30x(2x^2 - 1)$

$$= 60x\left(x + \tfrac{1}{\sqrt{2}}\right)\left(x - \tfrac{1}{\sqrt{2}}\right) \;\Rightarrow$$

$h''(x) > 0$ when $x > \frac{1}{\sqrt{2}}$ or $-\frac{1}{\sqrt{2}} < x < 0$, so h is CU on $\left(-\frac{1}{\sqrt{2}}, 0\right)$ and $\left(\frac{1}{\sqrt{2}}, \infty\right)$ and CD on $\left(-\infty, -\frac{1}{\sqrt{2}}\right)$ and $\left(0, \frac{1}{\sqrt{2}}\right)$.

Inflection points at $(0, 3)$ and $\left(\pm\frac{1}{\sqrt{2}}, 3 \mp \frac{7}{8}\sqrt{2}\right)$ [about $(-0.71, 4.24)$ and $(0.71, 1.76)$].

(d)

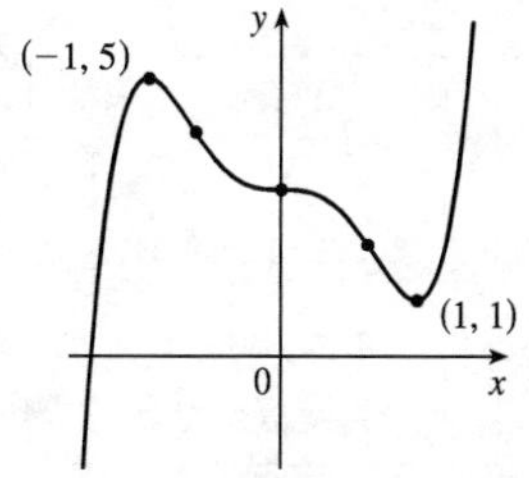

35. (a) $A(x) = x\sqrt{x+3} \;\Rightarrow$

$$A'(x) = x \cdot \tfrac{1}{2}(x+3)^{-1/2} + \sqrt{x+3} \cdot 1 = \frac{x}{2\sqrt{x+3}} + \sqrt{x+3} = \frac{x + 2(x+3)}{2\sqrt{x+3}} = \frac{3x+6}{2\sqrt{x+3}}.$$

The domain of A is $[-3, \infty)$. $A'(x) > 0$ for $x > -2$ and $A'(x) < 0$ for $-3 < x < -2$, so A is increasing on $(-2, \infty)$ and decreasing on $(-3, -2)$.

(b) $A(-2) = -2$ is a local minimum value.

(c)
$$A''(x) = \frac{2\sqrt{x+3} \cdot 3 - (3x+6) \cdot \dfrac{1}{\sqrt{x+3}}}{\left(2\sqrt{x+3}\right)^2}$$

$$= \frac{6(x+3) - (3x+6)}{4(x+3)^{3/2}} = \frac{3x+12}{4(x+3)^{3/2}} = \frac{3(x+4)}{4(x+3)^{3/2}}$$

$A''(x) > 0$ for all $x > -3$, so A is concave upward on $(-3, \infty)$. There is no inflection point.

(d)

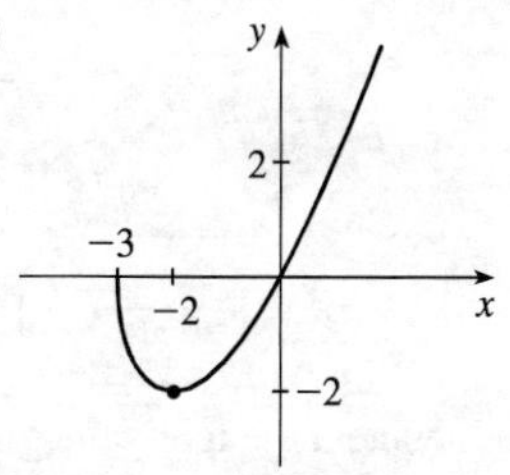

37. (a) $C(x) = x^{1/3}(x+4) = x^{4/3} + 4x^{1/3} \;\Rightarrow\; C'(x) = \frac{4}{3}x^{1/3} + \frac{4}{3}x^{-2/3} = \frac{4}{3}x^{-2/3}(x+1) = \dfrac{4(x+1)}{3\sqrt[3]{x^2}}$.

$C'(x) > 0$ if $-1 < x < 0$ or $x > 0$ and $C'(x) < 0$ for $x < -1$, so C is increasing on $(-1, \infty)$ and C is decreasing on $(-\infty, -1)$.

(b) $C(-1) = -3$ is a local minimum value.

(c) $C''(x) = \frac{4}{9}x^{-2/3} - \frac{8}{9}x^{-5/3} = \frac{4}{9}x^{-5/3}(x-2) = \dfrac{4(x-2)}{9\sqrt[3]{x^5}}$.

$C''(x) < 0$ for $0 < x < 2$ and $C''(x) > 0$ for $x < 0$ and $x > 2$, so C is concave downward on $(0, 2)$ and concave upward on $(-\infty, 0)$ and $(2, \infty)$. There are inflection points at $(0, 0)$ and $(2, 6\sqrt[3]{2}) \approx (2, 7.56)$.

(d)

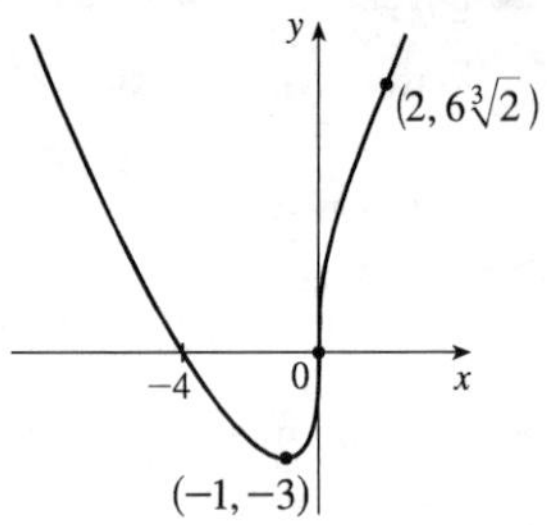

39. (a) $f(\theta) = 2\cos\theta - \cos 2\theta,\ 0 \le \theta \le 2\pi$.

$f'(\theta) = -2\sin\theta + 2\sin 2\theta = -2\sin\theta + 2(2\sin\theta\cos\theta) = 2\sin\theta\,(2\cos\theta - 1)$.

Interval	$\sin\theta$	$2\cos\theta - 1$	$f'(\theta)$	f
$0 < \theta < \frac{\pi}{3}$	+	+	+	increasing on $(0, \frac{\pi}{3})$
$\frac{\pi}{3} < \theta < \pi$	+	−	−	decreasing on $(\frac{\pi}{3}, \pi)$
$\pi < \theta < \frac{5\pi}{3}$	−	−	+	increasing on $(\pi, \frac{5\pi}{3})$
$\frac{5\pi}{3} < \theta < 2\pi$	−	+	−	decreasing on $(\frac{5\pi}{3}, 2\pi)$

(b) $f(\frac{\pi}{3}) = \frac{3}{2}$ and $f(\frac{5\pi}{3}) = \frac{3}{2}$ are local maximum values and $f(\pi) = -3$ is a local minimum value.

(c) $f'(\theta) = -2\sin\theta + 2\sin 2\theta \Rightarrow$

$$f''(\theta) = -2\cos\theta + 4\cos 2\theta = -2\cos\theta + 4(2\cos^2\theta - 1) = 2(4\cos^2\theta - \cos\theta - 2)$$

$$f''(\theta) = 0 \Leftrightarrow \cos\theta = \frac{1 \pm \sqrt{33}}{8} \Leftrightarrow \theta = \cos^{-1}\left(\frac{1 \pm \sqrt{33}}{8}\right)$$

$$\Leftrightarrow \theta = \cos^{-1}\left(\frac{1 + \sqrt{33}}{8}\right) \approx 0.5678,$$

(d)

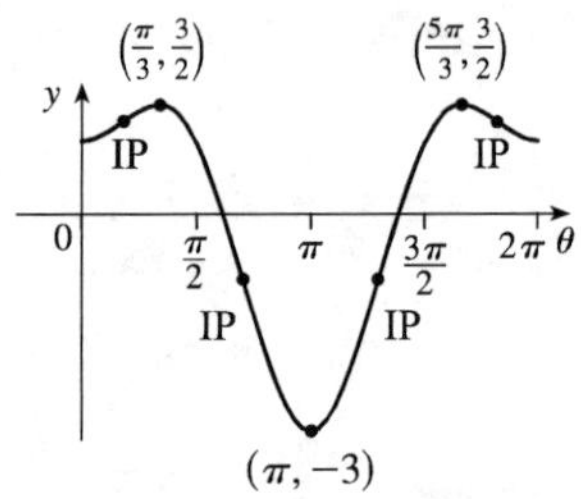

$$2\pi - \cos^{-1}\left(\frac{1 + \sqrt{33}}{8}\right) \approx 5.7154,\ \cos^{-1}\left(\frac{1 - \sqrt{33}}{8}\right) \approx 2.2057, \text{ or } 2\pi - \cos^{-1}\left(\frac{1 - \sqrt{33}}{8}\right) \approx 4.0775.$$

Denote these four values of θ by θ_1, θ_4, θ_2, and θ_3, respectively. Then f is CU on $(0, \theta_1)$, CD on (θ_1, θ_2), CU on (θ_2, θ_3), CD on (θ_3, θ_4), and CU on $(\theta_4, 2\pi)$. To find the *exact* y-coordinate for $\theta = \theta_1$, we have

$$f(\theta_1) = 2\cos\theta_1 - \cos 2\theta_1 = 2\cos\theta_1 - (2\cos^2\theta_1 - 1) = 2\left(\frac{1 + \sqrt{33}}{8}\right) - 2\left(\frac{1 + \sqrt{33}}{8}\right)^2 + 1$$

$$= \tfrac{1}{4} + \tfrac{1}{4}\sqrt{33} - \tfrac{1}{32} - \tfrac{1}{16}\sqrt{33} - \tfrac{33}{32} + 1 = \tfrac{3}{16} + \tfrac{3}{16}\sqrt{33} = \tfrac{3}{16}(1 + \sqrt{33}) = y_1 \approx 1.26.$$

Similarly, $f(\theta_2) = \frac{3}{16}(1 - \sqrt{33}) = y_2 \approx -0.89$. So f has inflection points at (θ_1, y_1), (θ_2, y_2), (θ_3, y_2), and (θ_4, y_1).

41. (a)

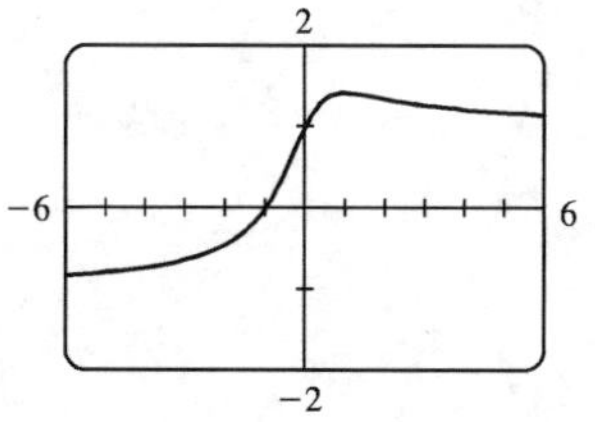

From the graph, we get an estimate of $f(1) \approx 1.41$ as a local maximum value, and no local minimum value.

$f(x) = \dfrac{x+1}{\sqrt{x^2+1}} \quad \Rightarrow \quad f'(x) = \dfrac{1-x}{(x^2+1)^{3/2}}$.

$f'(x) = 0 \quad \Leftrightarrow \quad x = 1$. $f(1) = \frac{2}{\sqrt{2}} = \sqrt{2}$ is the exact value.

(b) From the graph in part (a), f increases most rapidly somewhere between $x = -\frac{1}{2}$ and $x = -\frac{1}{4}$. To find the exact value, we need to find the maximum value of f', which we can do by finding the critical numbers of f'.

$f''(x) = \dfrac{2x^2 - 3x - 1}{(x^2+1)^{5/2}} = 0 \quad \Leftrightarrow \quad x = \dfrac{3 \pm \sqrt{17}}{4}$. $x = \dfrac{3+\sqrt{17}}{4}$ corresponds to the *minimum* value of f'.

The maximum value of f' is at $\left(\frac{3-\sqrt{17}}{4}, \sqrt{\frac{7}{6} - \frac{\sqrt{17}}{6}}\right) \approx (-0.28, 0.69)$.

43. $f(x) = \cos x + \frac{1}{2}\cos 2x \quad \Rightarrow \quad f'(x) = -\sin x - \sin 2x \quad \Rightarrow \quad f''(x) = -\cos x - 2\cos 2x$

(a)

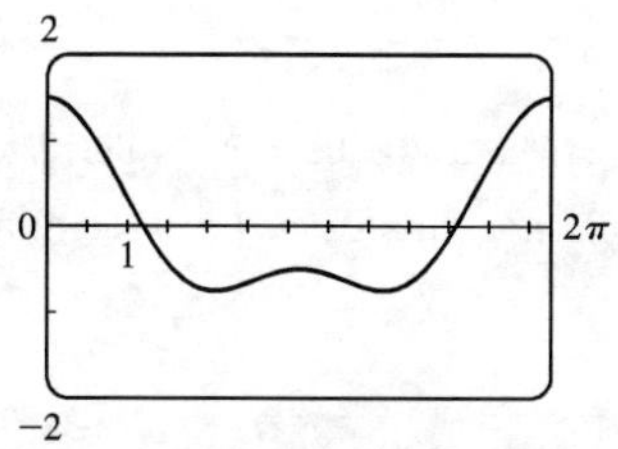

From the graph of f, it seems that f is CD on $(0, 1)$, CU on $(1, 2.5)$, CD on $(2.5, 3.7)$, CU on $(3.7, 5.3)$, and CD on $(5.3, 2\pi)$. The points of inflection appear to be at $(1, 0.4)$, $(2.5, -0.6)$, $(3.7, -0.6)$, and $(5.3, 0.4)$.

(b)

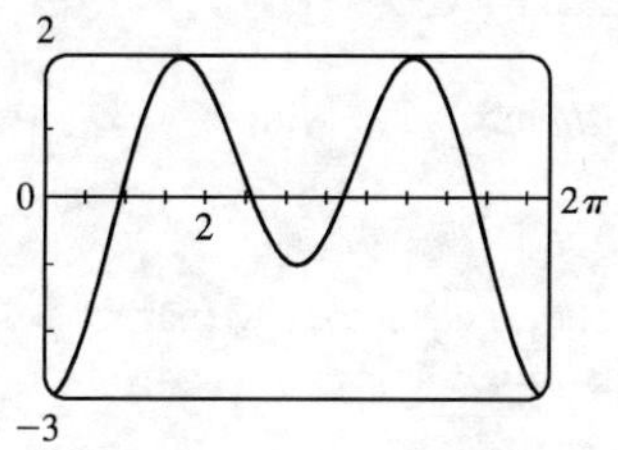

From the graph of f'' (and zooming in near the zeros), it seems that f is CD on $(0, 0.94)$, CU on $(0.94, 2.57)$, CD on $(2.57, 3.71)$, CU on $(3.71, 5.35)$, and CD on $(5.35, 2\pi)$. Refined estimates of the inflection points are $(0.94, 0.44)$, $(2.57, -0.63)$, $(3.71, -0.63)$, and $(5.35, 0.44)$.

45. In Maple, we define f and then use the command `plot(diff(diff(f,x),x),x=-3..3);`. In Mathematica, we define f and then use `Plot[Dt[Dt[f,x],x],{x,-3,3}]`.We see that $f'' > 0$ for $x > 0.1$ and $f'' < 0$ for $x < 0.1$. So f is concave up on $(0.1, \infty)$ and concave down on $(-\infty, 0.1)$.

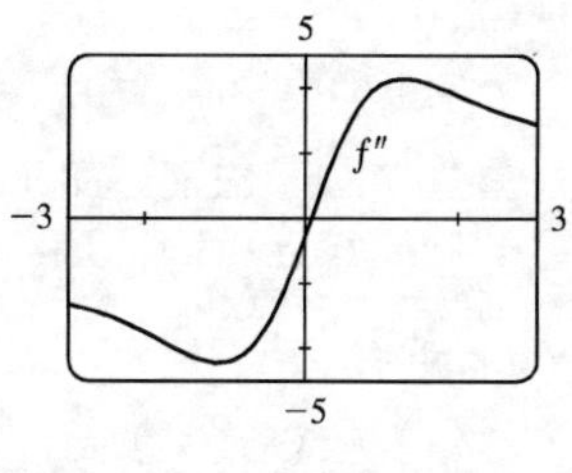

47. Most students learn more in the third hour of studying than in the eighth hour, so $K(3) - K(2)$ is larger than $K(8) - K(7)$. In other words, as you begin studying for a test, the rate of knowledge gain is large and then starts to taper off, so $K'(t)$ decreases and the graph of K is concave downward.

49. $f(x) = \tan x - x \Rightarrow f'(x) = \sec^2 x - 1 > 0$ for $0 < x < \frac{\pi}{2}$ since $\sec^2 x > 1$ for $0 < x < \frac{\pi}{2}$. So f is increasing on $\left(0, \frac{\pi}{2}\right)$. Thus, $f(x) > f(0) = 0$ for $0 < x < \frac{\pi}{2} \Rightarrow \tan x - x > 0 \Rightarrow \tan x > x$ for $0 < x < \frac{\pi}{2}$.

51. $f(x) = ax^3 + bx^2 + cx + d \Rightarrow f'(x) = 3ax^2 + 2bx + c$. We are given that $f(1) = 0$ and $f(-2) = 3$, so $f(1) = a + b + c + d = 0$ and $f(-2) = -8a + 4b - 2c + d = 3$. Also $f'(1) = 3a + 2b + c = 0$ and $f'(-2) = 12a - 4b + c = 0$ by Fermat's Theorem. Solving these four equations, we get $a = \frac{2}{9}$, $b = \frac{1}{3}$, $c = -\frac{4}{3}$, $d = \frac{7}{9}$, so the function is $f(x) = \frac{1}{9}\left(2x^3 + 3x^2 - 12x + 7\right)$.

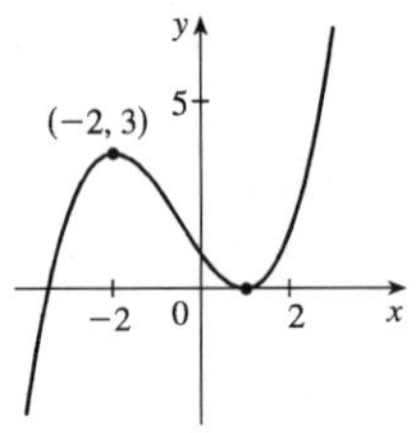

53. Suppose that f is differentiable on an interval I and $f'(x) > 0$ for all x in I except $x = c$. To show that f is increasing on I, let x_1, x_2 be two numbers in I with $x_1 < x_2$.

Case 1 $x_1 < x_2 < c$. Let J be the interval $\{x \in I \mid x < c\}$. By applying the Increasing/Decreasing Test to f on J, we see that f is increasing on J, so $f(x_1) < f(x_2)$.

Case 2 $c < x_1 < x_2$. Apply the Increasing/Decreasing Test to f on $K = \{x \in I \mid x > c\}$.

Case 3 $x_1 < x_2 = c$. Apply the proof of the Increasing/Decreasing Test, using the Mean Value Theorem (MVT) on the interval $[x_1, x_2]$ and noting that the MVT does not require f to be differentiable at the endpoints of $[x_1, x_2]$.

Case 4 $c = x_1 < x_2$. Same proof as in Case 3.

Case 5 $x_1 < c < x_2$. By Cases 3 and 4, f is increasing on $[x_1, c]$ and on $[c, x_2]$, so $f(x_1) < f(c) < f(x_2)$.

In all cases, we have shown that $f(x_1) < f(x_2)$. Since x_1, x_2 were any numbers in I with $x_1 < x_2$, we have shown that f is increasing on I.

55. (a) Since f and g are positive, increasing, and CU on I with f'' and g'' never equal to 0, we have $f > 0$, $f' \geq 0$, $f'' > 0$, $g > 0$, $g' \geq 0$, $g'' > 0$ on I. Then $(fg)' = f'g + fg' \Rightarrow$ $(fg)'' = f''g + 2f'g' + fg'' \geq f''g + fg'' > 0$ on $I \Rightarrow fg$ is CU on I.

(b) In part (a), if f and g are both decreasing instead of increasing, then $f' \leq 0$ and $g' \leq 0$ on I, so we still have $2f'g' \geq 0$ on I. Thus, $(fg)'' = f''g + 2f'g' + fg'' \geq f''g + fg'' > 0$ on $I \Rightarrow fg$ is CU on I as in part (a).

(c) Suppose f is increasing and g is decreasing [with f and g positive and CU]. Then $f' \geq 0$ and $g' \leq 0$ on I, so $2f'g' \leq 0$ on I and the argument in parts (a) and (b) fails.

Example 1. $I = (0, \infty)$, $f(x) = x^3$, $g(x) = 1/x$. Then $(fg)(x) = x^2$, so $(fg)'(x) = 2x$ and $(fg)''(x) = 2 > 0$ on I. Thus, fg is CU on I.

Example 2. $I = (0, \infty)$, $f(x) = 4x\sqrt{x}$, $g(x) = 1/x$. Then $(fg)(x) = 4\sqrt{x}$, so $(fg)'(x) = 2/\sqrt{x}$ and $(fg)''(x) = -1/\sqrt{x^3} < 0$ on I. Thus, fg is CD on I.

Example 3. $I = (0, \infty)$, $f(x) = x^2$, $g(x) = 1/x$. Thus, $(fg)(x) = x$, so fg is linear on I.

57. Let the cubic function be $f(x) = ax^3 + bx^2 + cx + d \;\Rightarrow\; f'(x) = 3ax^2 + 2bx + c \;\Rightarrow\; f''(x) = 6ax + 2b$. So f is CU when $6ax + 2b > 0 \;\Leftrightarrow\; x > -b/(3a)$, CD when $x < -b/(3a)$, and so the only point of inflection occurs when $x = -b/(3a)$. If the graph has three x-intercepts x_1, x_2 and x_3, then the expression for $f(x)$ must factor as $f(x) = a(x - x_1)(x - x_2)(x - x_3)$. Multiplying these factors together gives us $f(x) = a\left[x^3 - (x_1 + x_2 + x_3)x^2 + (x_1x_2 + x_1x_3 + x_2x_3)x - x_1x_2x_3\right]$. Equating the coefficients of the x^2-terms for the two forms of f gives us $b = -a(x_1 + x_2 + x_3)$. Hence, the x-coordinate of the point of inflection is $-\dfrac{b}{3a} = -\dfrac{-a(x_1 + x_2 + x_3)}{3a} = \dfrac{x_1 + x_2 + x_3}{3}$.

59. By hypothesis $g = f'$ is differentiable on an open interval containing c. Since $(c, f(c))$ is a point of inflection, the concavity changes at $x = c$, so $f''(x)$ changes signs at $x = c$. Hence, by the First Derivative Test, f' has a local extremum at $x = c$. Thus, by Fermat's Theorem $f''(c) = 0$.

61. Using the fact that $|x| = \sqrt{x^2}$, we have that $g(x) = x\sqrt{x^2} \;\Rightarrow\; g'(x) = \sqrt{x^2} + \sqrt{x^2} = 2\sqrt{x^2} = 2\,|x| \;\Rightarrow\; g''(x) = 2x\left(x^2\right)^{-1/2} = \dfrac{2x}{|x|} < 0$ for $x < 0$ and $g''(x) > 0$ for $x > 0$, so $(0, 0)$ is an inflection point. But $g''(0)$ does not exist.

☐ APPLICATIONS OF DIFFERENTIATION

Maximum and Minimum Values

1. A function f has an **absolute minimum** at $x = c$ if $f(c)$ is the smallest function value on the entire domain of f, whereas f has a **local minimum** at c if $f(c)$ is the smallest function value when x is near c.

3. Absolute maximum at b; absolute minimum at d; local maxima at b and e; local minima at d and s; neither a maximum nor a minimum at a, c, r, and t.

5. Absolute maximum value is $f(4) = 4$; absolute minimum value is $f(7) = 0$; local maximum values are $f(4) = 4$ and $f(6) = 3$; local minimum values are $f(2) = 1$ and $f(5) = 2$.

7. Absolute minimum at 2, absolute maximum at 3, local minimum at 4

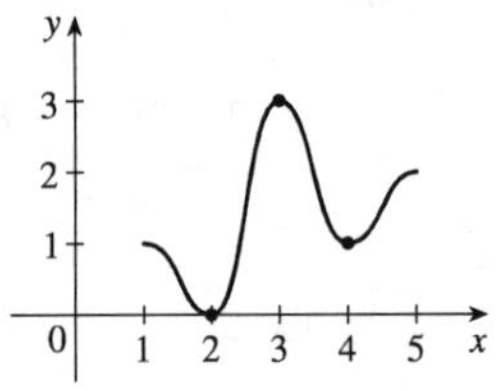

9. Absolute maximum at 5, absolute minimum at 2, local maximum at 3, local minima at 2 and 4

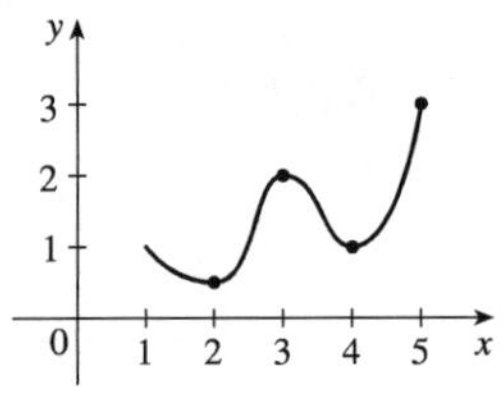

11. (a)

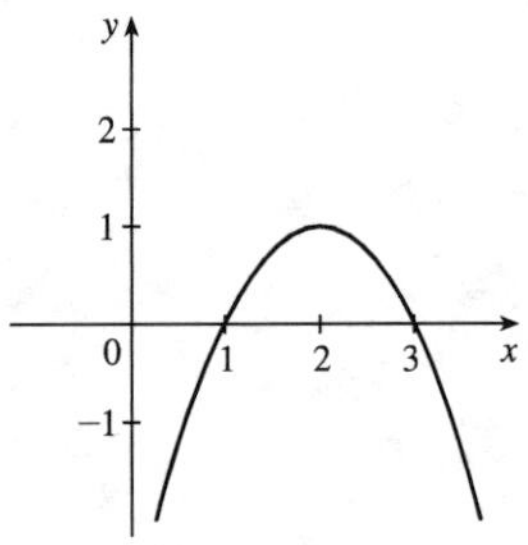

(b)

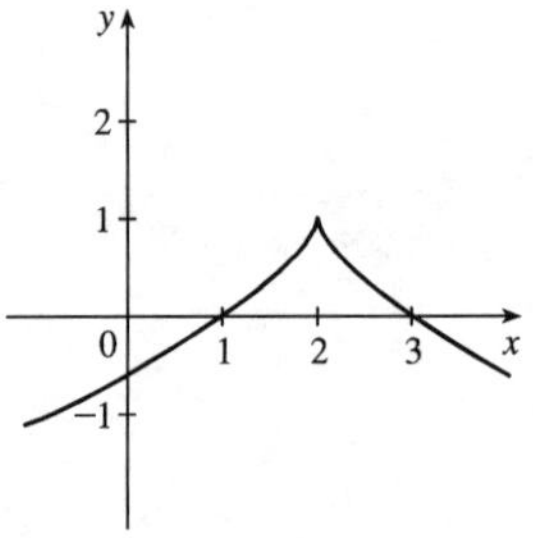

(c)

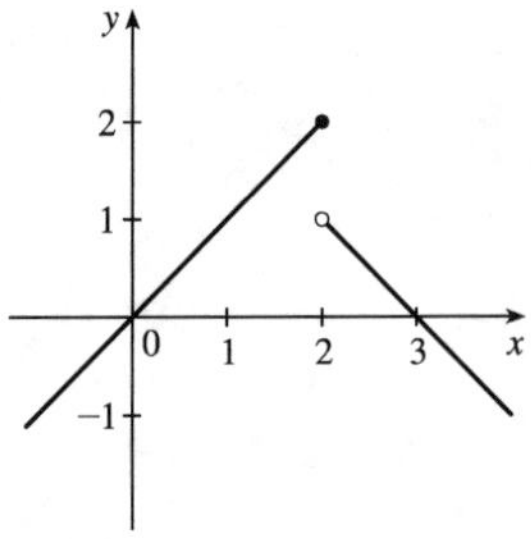

13. (a) *Note:* By the Extreme Value Theorem, f must *not* be continuous; because if it were, it would attain an absolute minimum.

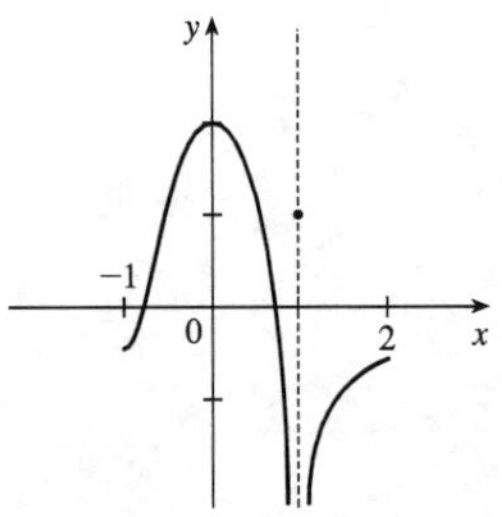

(b)

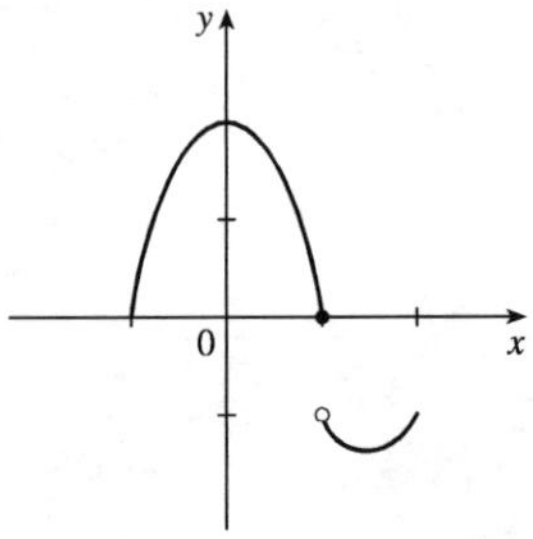

15. $f(x) = 8 - 3x$, $x \geq 1$. Absolute maximum $f(1) = 5$; no local maximum. No absolute or local minimum.

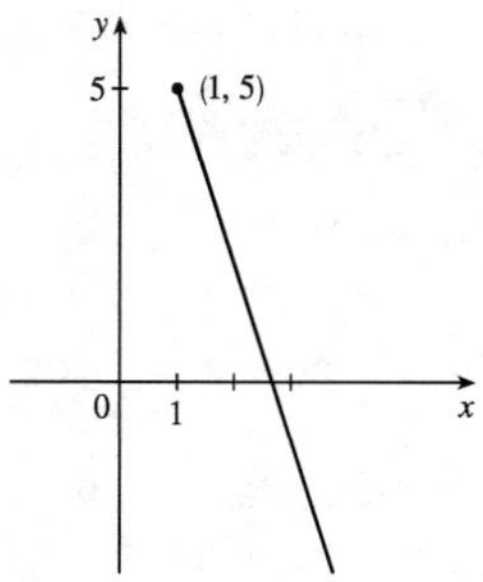

17. $f(x) = x^2$, $0 < x < 2$. No absolute or local maximum or minimum value.

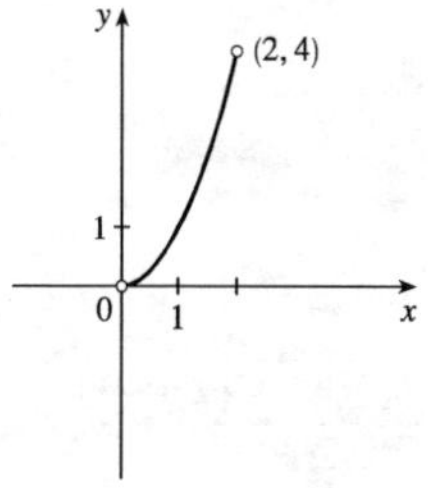

19. $f(x) = x^2$, $0 \leq x < 2$. Absolute minimum $f(0) = 0$; no local minimum. No absolute or local maximum.

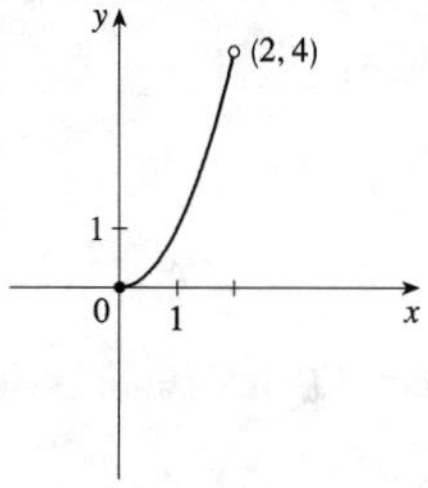

21. $f(x) = x^2$, $-3 \leq x \leq 2$. Absolute maximum $f(-3) = 9$. No local maximum. Absolute and local minimum $f(0) = 0$.

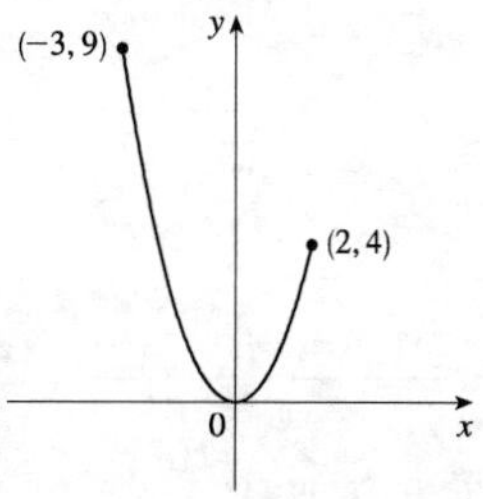

23. $f(t) = 1/t$, $0 < t < 1$. No maximum or minimum.

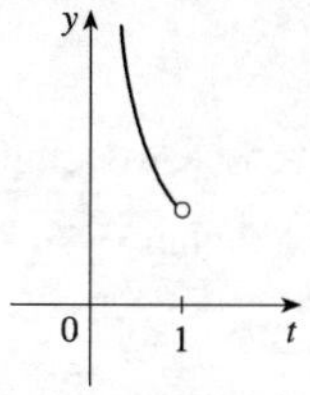

25. $f(\theta) = \sin\theta$, $-2\pi \leq \theta \leq 2\pi$. Absolute and local maxima $f\left(-\frac{3\pi}{2}\right) = f\left(\frac{\pi}{2}\right) = 1$. Absolute and local minima $f\left(-\frac{\pi}{2}\right) = f\left(\frac{3\pi}{2}\right) = -1$.

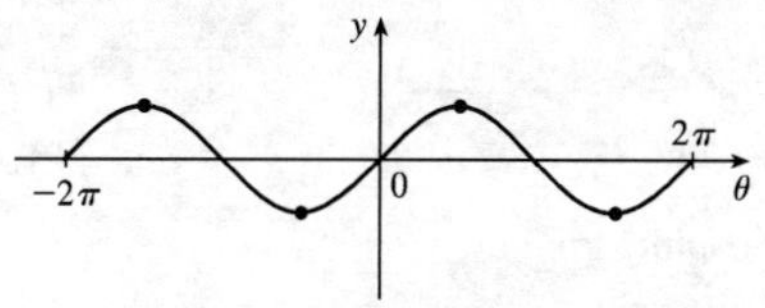

27. $f(x) = 1 - \sqrt{x}$. Absolute maximum $f(0) = 1$; no local maximum. No absolute or local minimum.

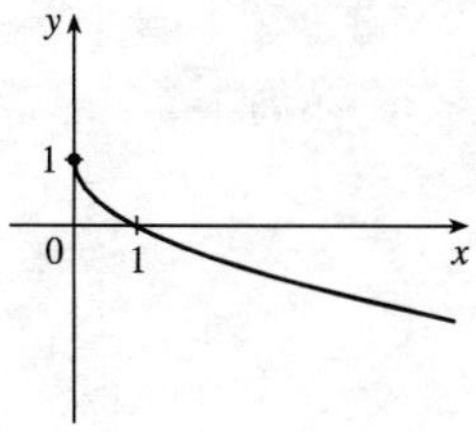

29. $f(x) = \begin{cases} 1 - x & \text{if } 0 \leq x < 2 \\ 2x - 4 & \text{if } 2 \leq x \leq 3 \end{cases}$

Absolute maximum $f(3) = 2$; no local maximum. No absolute or local minimum.

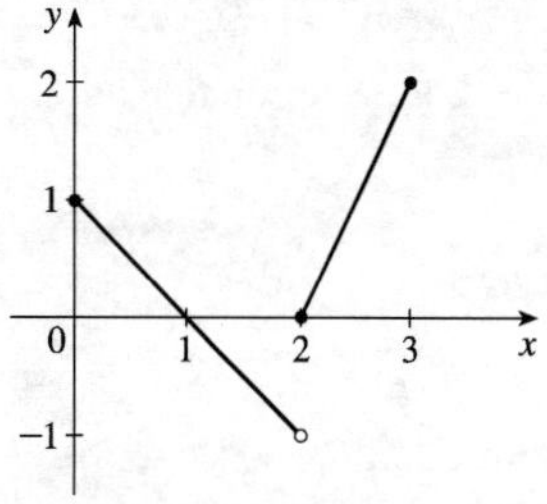

31. $f(x) = 5x^2 + 4x \Rightarrow f'(x) = 10x + 4$. $f'(x) = 0 \Rightarrow x = -\frac{2}{5}$, so $-\frac{2}{5}$ is the only critical number.

33. $f(x) = x^3 + 3x^2 - 24x \Rightarrow f'(x) = 3x^2 + 6x - 24 = 3(x^2 + 2x - 8)$.

$f'(x) = 0 \Rightarrow 3(x+4)(x-2) = 0 \Rightarrow x = -4, 2$. These are the only critical numbers.

35. $s(t) = 3t^4 + 4t^3 - 6t^2 \Rightarrow s'(t) = 12t^3 + 12t^2 - 12t$. $s'(t) = 0 \Rightarrow 12t(t^2 + t - 1) \Rightarrow t = 0$ or $t^2 + t - 1 = 0$. Using the quadratic formula to solve the latter equation gives us

$$t = \frac{-1 \pm \sqrt{1^2 - 4(1)(-1)}}{2(1)} = \frac{-1 \pm \sqrt{5}}{2} \approx 0.618, -1.618.$$ The three critical numbers are $0, \frac{-1 \pm \sqrt{5}}{2}$.

37. $$g(x) = |2x+3| = \begin{cases} 2x+3 & \text{if } 2x+3 \ge 0 \\ -(2x+3) & \text{if } 2x+3 < 0 \end{cases} \Rightarrow g'(x) = \begin{cases} 2 & \text{if } x > -\frac{3}{2} \\ -2 & \text{if } x < -\frac{3}{2} \end{cases}$$

$g'(x)$ is never 0, but $g'(x)$ does not exist for $x = -\frac{3}{2}$, so $-\frac{3}{2}$ is the only critical number.

39. $g(t) = 5t^{2/3} + t^{5/3} \Rightarrow g'(t) = \frac{10}{3}t^{-1/3} + \frac{5}{3}t^{2/3}$. $g'(0)$ does not exist, so $t = 0$ is a critical number.

$g'(t) = \frac{5}{3}t^{-1/3}(2 + t) = 0 \Leftrightarrow t = -2$, so $t = -2$ is also a critical number.

41. $F(x) = x^{4/5}(x-4)^2 \Rightarrow$

$$F'(x) = x^{4/5} \cdot 2(x-4) + (x-4)^2 \cdot \tfrac{4}{5}x^{-1/5} = \tfrac{1}{5}x^{-1/5}(x-4)\,[5 \cdot x \cdot 2 + (x-4) \cdot 4]$$
$$= \frac{(x-4)(14x-16)}{5x^{1/5}} = \frac{2(x-4)(7x-8)}{5x^{1/5}} = 0 \text{ when } x = 4, \tfrac{8}{7}; \text{ and } F'(0) \text{ does not exist.}$$

Critical numbers are $0, \frac{8}{7}, 4$.

43. $f(\theta) = 2\cos\theta + \sin^2\theta \Rightarrow f'(\theta) = -2\sin\theta + 2\sin\theta\cos\theta$. $f'(\theta) = 0 \Rightarrow 2\sin\theta(\cos\theta - 1) = 0 \Rightarrow \sin\theta = 0$ or $\cos\theta = 1 \Rightarrow \theta = n\pi$ (n an integer) or $\theta = 2n\pi$. The solutions $\theta = n\pi$ include the solutions $\theta = 2n\pi$, so the critical numbers are $\theta = n\pi$.

45. $f(x) = 3x^2 - 12x + 5$, $[0, 3]$. $f'(x) = 6x - 12 = 0 \Leftrightarrow x = 2$. Applying the Closed Interval Method, we find that $f(0) = 5$, $f(2) = -7$, and $f(3) = -4$. So $f(0) = 5$ is the absolute maximum value and $f(2) = -7$ is the absolute minimum value.

47. $f(x) = 2x^3 - 3x^2 - 12x + 1$, $[-2, 3]$. $f'(x) = 6x^2 - 6x - 12 = 6(x^2 - x - 2) = 6(x-2)(x+1) = 0 \Leftrightarrow x = 2, -1$. $f(-2) = -3$, $f(-1) = 8$, $f(2) = -19$, and $f(3) = -8$. So $f(-1) = 8$ is the absolute maximum value and $f(2) = -19$ is the absolute minimum value.

49. $f(x) = x^4 - 2x^2 + 3$, $[-2, 3]$. $f'(x) = 4x^3 - 4x = 4x(x^2 - 1) = 4x(x+1)(x-1) = 0 \Leftrightarrow x = -1, 0, 1$. $f(-2) = 11$, $f(-1) = 2$, $f(0) = 3$, $f(1) = 2$, $f(3) = 66$. So $f(3) = 66$ is the absolute maximum value and $f(\pm 1) = 2$ is the absolute minimum value.

51. $f(x) = \dfrac{x}{x^2+1}$, $[0, 2]$. $f'(x) = \dfrac{(x^2+1) - x(2x)}{(x^2+1)^2} = \dfrac{1 - x^2}{(x^2+1)^2} = 0 \Leftrightarrow x = \pm 1$, but -1 is not in $[0, 2]$.

$f(0) = 0$, $f(1) = \frac{1}{2}$, $f(2) = \frac{2}{5}$. So $f(1) = \frac{1}{2}$ is the absolute maximum value and $f(0) = 0$ is the absolute minimum value.

53. $f(t) = t\sqrt{4-t^2}$, $[-1,2]$.

$$f'(t) = t\cdot\tfrac{1}{2}(4-t^2)^{-1/2}(-2t) + (4-t^2)^{1/2}\cdot 1 = \frac{-t^2}{\sqrt{4-t^2}} + \sqrt{4-t^2} = \frac{-t^2+(4-t^2)}{\sqrt{4-t^2}} = \frac{4-2t^2}{\sqrt{4-t^2}}.$$

$f'(t) = 0 \;\Rightarrow\; 4-2t^2 = 0 \;\Rightarrow\; t^2 = 2 \;\Rightarrow\; t = \pm\sqrt{2}$, but $t = -\sqrt{2}$ is not in the given interval, $[-1,2]$.

$f'(t)$ does not exist if $4 - t^2 = 0 \;\Rightarrow\; t = \pm 2$, but -2 is not in the given interval. $f(-1) = -\sqrt{3}$, $f(\sqrt{2}) = 2$, and $f(2) = 0$. So $f(\sqrt{2}) = 2$ is the absolute maximum value and $f(-1) = -\sqrt{3}$ is the absolute minimum value.

55. $f(x) = \sin x + \cos x$, $\left[0, \frac{\pi}{3}\right]$. $f'(x) = \cos x - \sin x = 0 \;\Leftrightarrow\; \sin x = \cos x \;\Rightarrow\; \dfrac{\sin x}{\cos x} = 1 \;\Rightarrow$

$\tan x = 1 \;\Rightarrow\; x = \frac{\pi}{4}$. $f(0) = 1$, $f\left(\frac{\pi}{4}\right) = \sqrt{2} \approx 1.41$, $f\left(\frac{\pi}{3}\right) = \frac{\sqrt{3}+1}{2} \approx 1.37$. So $f\left(\frac{\pi}{4}\right) = \sqrt{2}$ is the absolute maximum value and $f(0) = 1$ is the absolute minimum value.

57. $f(x) = x^a(1-x)^b$, $0 \le x \le 1$, $a > 0$, $b > 0$.

$$\begin{aligned} f'(x) &= x^a\cdot b(1-x)^{b-1}(-1) + (1-x)^b\cdot ax^{a-1} = x^{a-1}(1-x)^{b-1}\left[x\cdot b(-1) + (1-x)\cdot a\right] \\ &= x^{a-1}(1-x)^{b-1}(a - ax - bx) \end{aligned}$$

At the endpoints, we have $f(0) = f(1) = 0$ [the minimum value of f]. In the interval $(0,1)$, $f'(x) = 0 \;\Leftrightarrow$ $x = \dfrac{a}{a+b}$.

$$f\left(\frac{a}{a+b}\right) = \left(\frac{a}{a+b}\right)^a\left(1 - \frac{a}{a+b}\right)^b = \frac{a^a}{(a+b)^a}\left(\frac{a+b-a}{a+b}\right)^b = \frac{a^a}{(a+b)^a}\cdot\frac{b^b}{(a+b)^b} = \frac{a^ab^b}{(a+b)^{a+b}}.$$

So $f\left(\dfrac{a}{a+b}\right) = \dfrac{a^ab^b}{(a+b)^{a+b}}$ is the absolute maximum value.

59. (a)

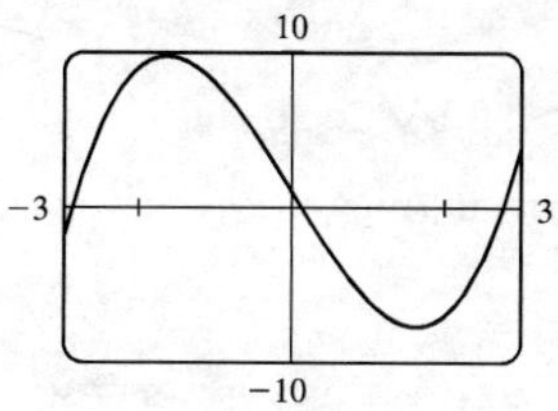

From the graph, it appears that the absolute maximum value is about $f(-1.63) = 9.71$, and the absolute minimum value is about $f(1.63) = -7.71$. These values make sense because the graph is symmetric about the point $(0,1)$. ($y = x^3 - 8x$ is symmetric about the origin.)

(b) $f(x) = x^3 - 8x + 1 \;\Rightarrow\; f'(x) = 3x^2 - 8$. So $f'(x) = 0 \;\Rightarrow\; x = \pm\sqrt{\frac{8}{3}}$.

$$\begin{aligned} f\left(\pm\sqrt{\tfrac{8}{3}}\right) &= \left(\pm\sqrt{\tfrac{8}{3}}\right)^3 - 8\left(\pm\sqrt{\tfrac{8}{3}}\right) + 1 = \pm\tfrac{8}{3}\sqrt{\tfrac{8}{3}} \mp 8\sqrt{\tfrac{8}{3}} + 1 \\ &= -\tfrac{16}{3}\sqrt{\tfrac{8}{3}} + 1 = 1 - \tfrac{32\sqrt{6}}{9} \text{ [minimum] or } \tfrac{16}{3}\sqrt{\tfrac{8}{3}} + 1 = 1 + \tfrac{32\sqrt{6}}{9} \text{ [maximum]} \end{aligned}$$

(From the graph, we see that the extreme values do not occur at the endpoints.)

61. (a)

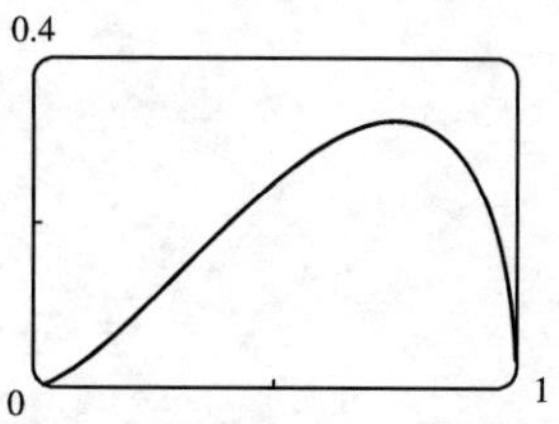

From the graph, it appears that the absolute maximum value is about $f(0.75) = 0.32$, and the absolute minimum value is $f(0) = f(1) = 0$; that is, at both endpoints.

(b) $f(x) = x\sqrt{x - x^2} \Rightarrow f'(x) = x \cdot \dfrac{1 - 2x}{2\sqrt{x - x^2}} + \sqrt{x - x^2} = \dfrac{(x - 2x^2) + (2x - 2x^2)}{2\sqrt{x - x^2}} = \dfrac{3x - 4x^2}{2\sqrt{x - x^2}}$.

So $f'(x) = 0 \Rightarrow 3x - 4x^2 = 0 \Rightarrow x(3 - 4x) = 0 \Rightarrow x = 0$ or $\frac{3}{4}$. $f(0) = f(1) = 0$ [minimum],

and $f\left(\frac{3}{4}\right) = \frac{3}{4}\sqrt{\frac{3}{4} - \left(\frac{3}{4}\right)^2} = \frac{3\sqrt{3}}{16}$ [maximum].

63. The density is defined as $\rho = \dfrac{\text{mass}}{\text{volume}} = \dfrac{1000}{V(T)}$ (in g/cm^3). But a critical point of ρ will also be a critical point of V $\left[\text{since } \dfrac{d\rho}{dT} = -1000V^{-2}\dfrac{dV}{dT} \text{ and } V \text{ is never } 0\right]$, and V is easier to differentiate than ρ.

$V(T) = 999.87 - 0.06426T + 0.0085043T^2 - 0.0000679T^3 \Rightarrow$

$V'(T) = -0.06426 + 0.0170086T - 0.0002037T^2$. Setting this equal to 0 and using the quadratic formula to find T, we get $T = \dfrac{-0.0170086 \pm \sqrt{0.0170086^2 - 4 \cdot 0.0002037 \cdot 0.06426}}{2(-0.0002037)} \approx 3.9665°\text{C}$ or $79.5318°\text{C}$. Since we are only interested in the region $0°\text{C} \leq T \leq 30°\text{C}$, we check the density ρ at the endpoints and at $3.9665°\text{C}$:

$\rho(0) \approx \dfrac{1000}{999.87} \approx 1.00013$; $\rho(30) \approx \dfrac{1000}{1003.7628} \approx 0.99625$; $\rho(3.9665) \approx \dfrac{1000}{999.7447} \approx 1.000255$. So water has its maximum density at about $3.9665°\text{C}$.

65. We apply the Closed Interval Method to the continuous function

$I(t) = 0.00009045t^5 + 0.001438t^4 - 0.06561t^3 + 0.4598t^2 - 0.6270t + 99.33$ on $[0, 10]$. Its derivative is $I'(t) = 0.00045225t^4 + 0.005752t^3 - 0.19683t^2 + 0.9196t - 0.6270$. Since I' exists for all t, the only critical numbers of I occur when $I'(t) = 0$. We use a root-finder on a computer algebra system (or a graphing device) to find that $I'(t) = 0$ when $t \approx -29.7186, 0.8231, 5.1309$, or 11.0459, but only the second and third roots lie in the interval $[0, 10]$. The values of I at these critical numbers are $I(0.8231) \approx 99.09$ and $I(5.1309) \approx 100.67$. The values of I at the endpoints of the interval are $I(0) = 99.33$ and $I(10) \approx 96.86$. Comparing these four numbers, we see that food was most expensive at $t \approx 5.1309$ (corresponding roughly to August, 1989) and cheapest at $t = 10$ (midyear 1994).

67. (a) $v(r) = k(r_0 - r)r^2 = kr_0r^2 - kr^3 \Rightarrow v'(r) = 2kr_0r - 3kr^2$. $v'(r) = 0 \Rightarrow kr(2r_0 - 3r) = 0$

$\Rightarrow r = 0$ or $\frac{2}{3}r_0$ (but 0 is not in the interval). Evaluating v at $\frac{1}{2}r_0$, $\frac{2}{3}r_0$, and r_0, we get $v\left(\frac{1}{2}r_0\right) = \frac{1}{8}kr_0^3$, $v\left(\frac{2}{3}r_0\right) = \frac{4}{27}kr_0^3$, and $v(r_0) = 0$. Since $\frac{4}{27} > \frac{1}{8}$, v attains its maximum value at $r = \frac{2}{3}r_0$. This supports the statement in the text.

(b) From part (a), the maximum value of v is $\frac{4}{27}kr_0^3$.

(c)

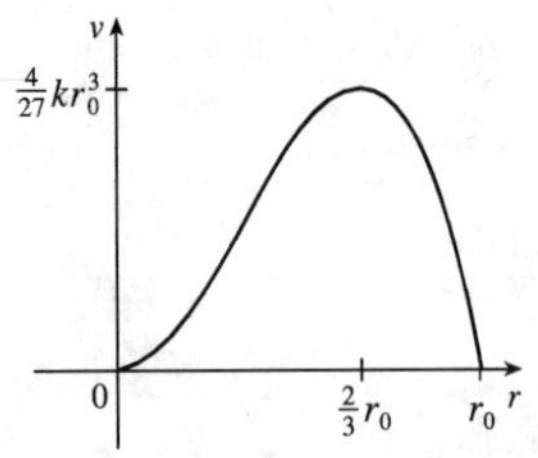

69. $f(x) = x^{101} + x^{51} + x + 1 \quad \Rightarrow \quad f'(x) = 101x^{100} + 51x^{50} + 1 \geq 1$ for all x, so $f'(x) = 0$ has no solution. Thus, $f(x)$ has no critical number, so $f(x)$ can have no local maximum or minimum.

71. If f has a local minimum at c, then $g(x) = -f(x)$ has a local maximum at c, so $g'(c) = 0$ by the case of Fermat's Theorem proved in the text. Thus, $f'(c) = -g'(c) = 0$.

Summary of Curve Sketching

1. $y = f(x) = x^3 + x = x(x^2 + 1)$ **A.** f is a polynomial, so $D = \mathbb{R}$. **B.** x-intercept $= 0$, y-intercept $= f(0) = 0$ **C.** $f(-x) = -f(x)$, so f is odd; the curve is symmetric about the origin. **D.** f is a polynomial, so there is no asymptote. **E.** $f'(x) = 3x^2 + 1 > 0$, so f is increasing on $(-\infty, \infty)$. **F.** There is no critical number and hence, no local maximum or minimum value. **G.** $f''(x) = 6x > 0$ on $(0, \infty)$ and $f''(x) < 0$ on $(-\infty, 0)$, so f is CU on $(0, \infty)$ and CD on $(-\infty, 0)$. Since the concavity changes at $x = 0$, there is an inflection point at $(0, 0)$.

H.

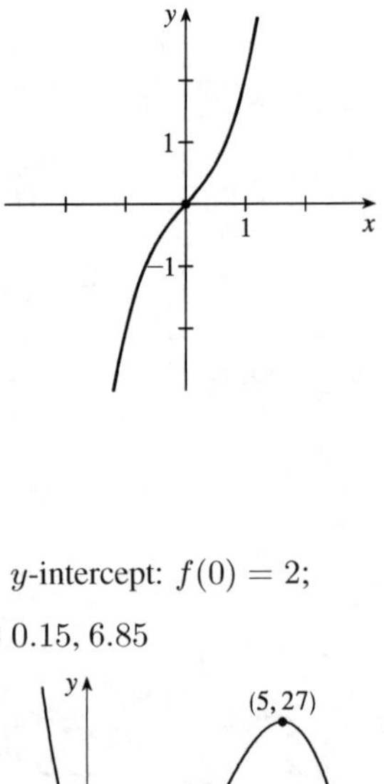

3. $y = f(x) = 2 - 15x + 9x^2 - x^3 = -(x - 2)(x^2 - 7x + 1)$ **A.** $D = \mathbb{R}$ **B.** y-intercept: $f(0) = 2$; x-intercepts: $f(x) = 0 \Rightarrow x = 2$ or (by the quadratic formula) $x = \frac{7 \pm \sqrt{45}}{2} \approx 0.15, 6.85$ **C.** No symmetry **D.** No asymptote

E. $f'(x) = -15 + 18x - 3x^2 = -3(x^2 - 6x + 5)$

$= -3(x - 1)(x - 5) > 0 \Leftrightarrow 1 < x < 5$

so f is increasing on $(1, 5)$ and decreasing on $(-\infty, 1)$ and $(5, \infty)$.

F. Local maximum value $f(5) = 27$, local minimum value $f(1) = -5$

G. $f''(x) = 18 - 6x = -6(x - 3) > 0 \Leftrightarrow x < 3$, so f is CU on $(-\infty, 3)$ and CD on $(3, \infty)$. IP at $(3, 11)$

H.

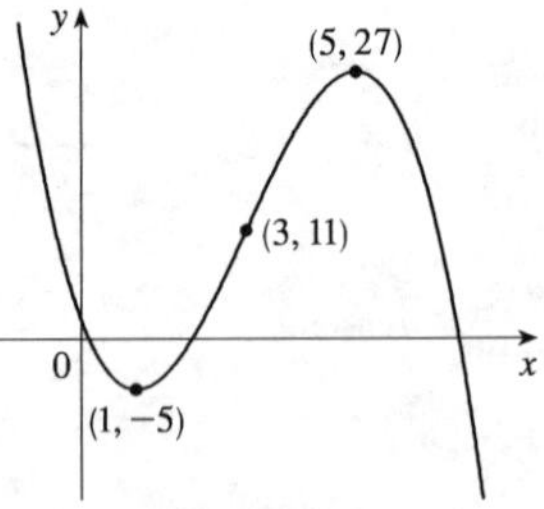

5. $y = f(x) = x^4 + 4x^3 = x^3(x+4)$ **A.** $D = \mathbb{R}$ **B.** y-intercept: $f(0) = 0$; x-intercepts: $f(x) = 0 \Leftrightarrow x = -4, 0$ **C.** No symmetry **D.** No asymptote **E.** $f'(x) = 4x^3 + 12x^2 = 4x^2(x+3) > 0 \Leftrightarrow x > -3$, so f is increasing on $(-3, \infty)$ and decreasing on $(-\infty, -3)$.

F. Local minimum value $f(-3) = -27$, no local maximum

G. $f''(x) = 12x^2 + 24x = 12x(x+2) < 0 \Leftrightarrow -2 < x < 0$, so f is CD on $(-2, 0)$ and CU on $(-\infty, -2)$ and $(0, \infty)$.
IP at $(0, 0)$ and $(-2, -16)$

H.

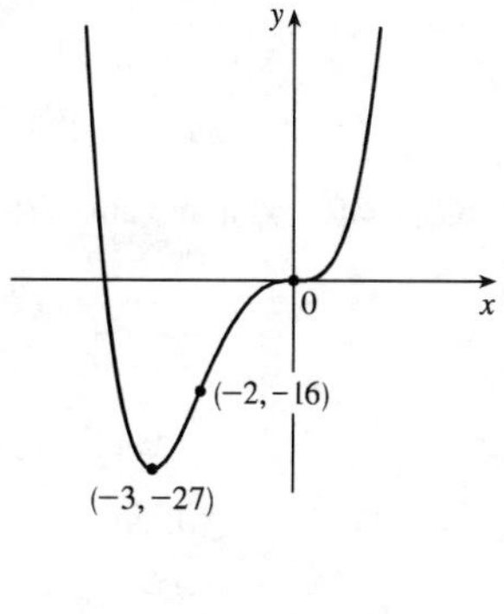

7. $y = f(x) = 2x^5 - 5x^2 + 1$ **A.** $D = \mathbb{R}$ **B.** y-intercept: $f(0) = 1$ **C.** No symmetry **D.** No asymptote
E. $f'(x) = 10x^4 - 10x = 10x(x^3 - 1) = 10x(x-1)(x^2 + x + 1)$, so $f'(x) < 0 \Leftrightarrow 0 < x < 1$ and $f'(x) > 0 \Leftrightarrow x < 0$ or $x > 1$. Thus, f is increasing on $(-\infty, 0)$ and $(1, \infty)$ and decreasing on $(0, 1)$.

F. Local maximum value $f(0) = 1$, local minimum value $f(1) = -2$

G. $f''(x) = 40x^3 - 10 = 10(4x^3 - 1)$ so $f''(x) = 0 \Leftrightarrow x = 1/\sqrt[3]{4}$.
$f''(x) > 0 \Leftrightarrow x > 1/\sqrt[3]{4}$ and $f''(x) < 0 \Leftrightarrow x < 1/\sqrt[3]{4}$, so f is CD on $\left(-\infty, 1/\sqrt[3]{4}\right)$ and CU on $\left(1/\sqrt[3]{4}, \infty\right)$.

IP at $\left(\dfrac{1}{\sqrt[3]{4}}, 1 - \dfrac{9}{2\left(\sqrt[3]{4}\right)^2}\right) \approx (0.630, -0.786)$

H.

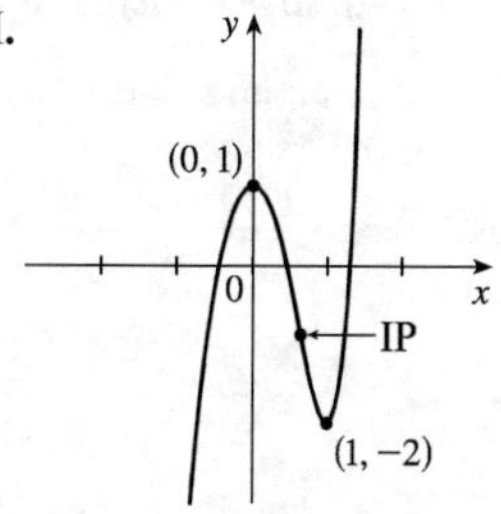

9. $y = f(x) = x/(x-1)$ **A.** $D = \{x \mid x \neq 1\} = (-\infty, 1) \cup (1, \infty)$ **B.** x-intercept $= 0$, y-intercept $= f(0) = 0$ **C.** No symmetry **D.** $\lim\limits_{x \to \pm\infty} \dfrac{x}{x-1} = 1$, so $y = 1$ is a HA.

$\lim\limits_{x \to 1^-} \dfrac{x}{x-1} = -\infty$, $\lim\limits_{x \to 1^+} \dfrac{x}{x-1} = \infty$, so $x = 1$ is a VA.

E. $f'(x) = \dfrac{(x-1) - x}{(x-1)^2} = \dfrac{-1}{(x-1)^2} < 0$ for $x \neq 1$, so f is decreasing on $(-\infty, 1)$ and $(1, \infty)$. **F.** No extreme values

G. $f''(x) = \dfrac{2}{(x-1)^3} > 0 \Leftrightarrow x > 1$, so f is CU on $(1, \infty)$ and CD on $(-\infty, 1)$. No IP

H.

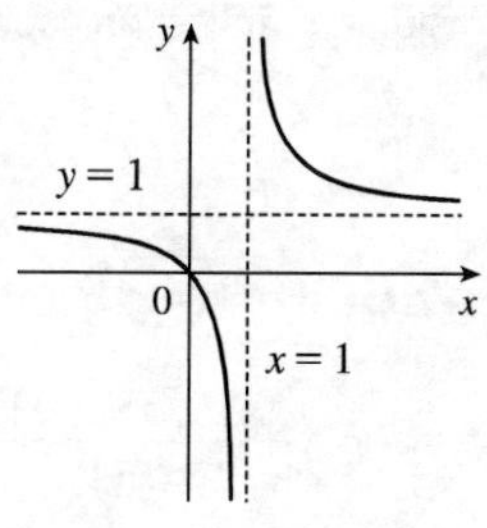

11. $y = f(x) = 1/(x^2 - 9)$ **A.** $D = \{x \mid x \neq \pm 3\} = (-\infty, -3) \cup (-3, 3) \cup (3, \infty)$
B. y-intercept $= f(0) = -\frac{1}{9}$, no x-intercept **C.** $f(-x) = f(x) \Rightarrow f$ is even; the curve is symmetric about the y-axis. **D.** $\lim\limits_{x \to \pm\infty} \dfrac{1}{x^2 - 9} = 0$, so $y = 0$ is a HA. $\lim\limits_{x \to 3^-} \dfrac{1}{x^2 - 9} = -\infty$, $\lim\limits_{x \to 3^+} \dfrac{1}{x^2 - 9} = \infty$,

$\lim\limits_{x \to -3^-} \dfrac{1}{x^2 - 9} = \infty$, $\lim\limits_{x \to -3^+} \dfrac{1}{x^2 - 9} = -\infty$, so $x = 3$ and $x = -3$ are VA.

E. $f'(x) = -\dfrac{2x}{(x^2-9)^2} > 0 \;\Leftrightarrow\; x < 0 \; (x \neq -3)$ so f is increasing on $(-\infty, -3)$ and $(-3, 0)$ and decreasing on $(0, 3)$ and $(3, \infty)$.

F. Local maximum value $f(0) = -\frac{1}{9}$.

G. $y'' = \dfrac{-2(x^2-9)^2 + (2x)2(x^2-9)(2x)}{(x^2-9)^4} = \dfrac{6(x^2+3)}{(x^2-9)^3} > 0 \;\Leftrightarrow$

$x^2 > 9 \;\Leftrightarrow\; x > 3$ or $x < -3$, so f is CU on $(-\infty, -3)$ and $(3, \infty)$ and CD on $(-3, 3)$. No IP

H.

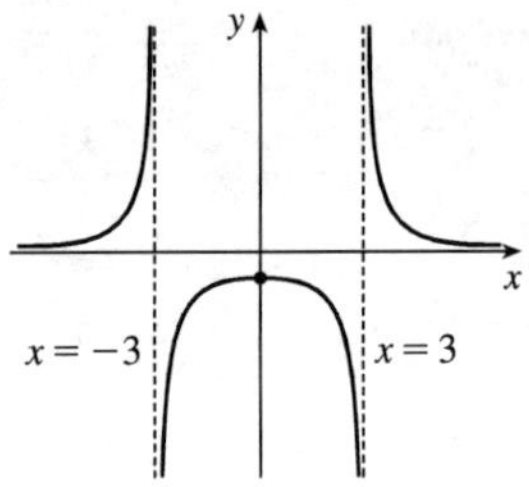

13. $y = f(x) = x/(x^2+9)$ **A.** $D = \mathbb{R}$ **B.** y-intercept: $f(0) = 0$; x-intercept: $f(x) = 0 \;\Leftrightarrow\; x = 0$

C. $f(-x) = -f(x)$, so f is odd and the curve is symmetric about the origin. **D.** $\lim\limits_{x\to\pm\infty} \left[x/(x^2+9)\right] = 0$, so $y = 0$ is a HA; no VA **E.** $f'(x) = \dfrac{(x^2+9)(1) - x(2x)}{(x^2+9)^2} = \dfrac{9-x^2}{(x^2+9)^2} = \dfrac{(3+x)(3-x)}{(x^2+9)^2} > 0 \;\Leftrightarrow$

$-3 < x < 3$, so f is increasing on $(-3, 3)$ and decreasing on $(-\infty, -3)$ and $(3, \infty)$.

F. Local minimum value $f(-3) = -\frac{1}{6}$, local maximum value $f(3) = \frac{1}{6}$

G.
$$f''(x) = \frac{(x^2+9)^2(-2x) - (9-x^2)\cdot 2(x^2+9)(2x)}{\left[(x^2+9)^2\right]^2}$$
$$= \frac{(2x)(x^2+9)\left[-(x^2+9) - 2(9-x^2)\right]}{(x^2+9)^4}$$
$$= \frac{2x(x^2-27)}{(x^2+9)^3} = 0 \;\Leftrightarrow\; x = 0, \pm\sqrt{27} = \pm 3\sqrt{3}$$

$f''(x) > 0 \;\Leftrightarrow\; -3\sqrt{3} < x < 0$ or $x > 3\sqrt{3}$, so f is CU on $(-3\sqrt{3}, 0)$ and $(3\sqrt{3}, \infty)$, and CD on $(-\infty, -3\sqrt{3})$ and $(0, 3\sqrt{3})$.

There are three inflection points: $(0, 0)$ and $\left(\pm 3\sqrt{3}, \pm\frac{1}{12}\sqrt{3}\right)$.

H.

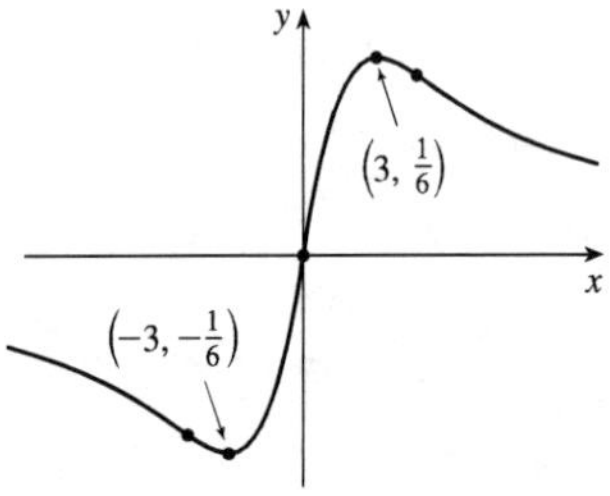

15. $y = f(x) = \dfrac{x-1}{x^2}$ **A.** $D = \{x \mid x \neq 0\} = (-\infty, 0) \cup (0, \infty)$ **B.** No y-intercept; x-intercept: $f(x) = 0 \;\Leftrightarrow\; x = 1$ **C.** No symmetry **D.** $\lim\limits_{x\to\pm\infty} \dfrac{x-1}{x^2} = 0$, so $y = 0$ is a HA. $\lim\limits_{x\to 0} \dfrac{x-1}{x^2} = -\infty$, so $x = 0$ is a VA.

E. $f'(x) = \dfrac{x^2 \cdot 1 - (x-1)\cdot 2x}{(x^2)^2} = \dfrac{-x^2+2x}{x^4} = \dfrac{-(x-2)}{x^3}$, so $f'(x) > 0 \;\Leftrightarrow\; 0 < x < 2$ and $f'(x) < 0 \;\Leftrightarrow\; x < 0$ or $x > 2$. Thus, f is increasing on $(0, 2)$ and decreasing on $(-\infty, 0)$ and $(2, \infty)$.

F. No local minimum, local maximum value $f(2) = \frac{1}{4}$.

G. $f''(x) = \dfrac{x^3\cdot(-1) - [-(x-2)]\cdot 3x^2}{(x^3)^2} = \dfrac{2x^3 - 6x^2}{x^6} = \dfrac{2(x-3)}{x^4}$.

$f''(x)$ is negative on $(-\infty, 0)$ and $(0, 3)$ and positive on $(3, \infty)$, so f is CD on $(-\infty, 0)$ and $(0, 3)$ and CU on $(3, \infty)$. IP at $\left(3, \frac{2}{9}\right)$

H.

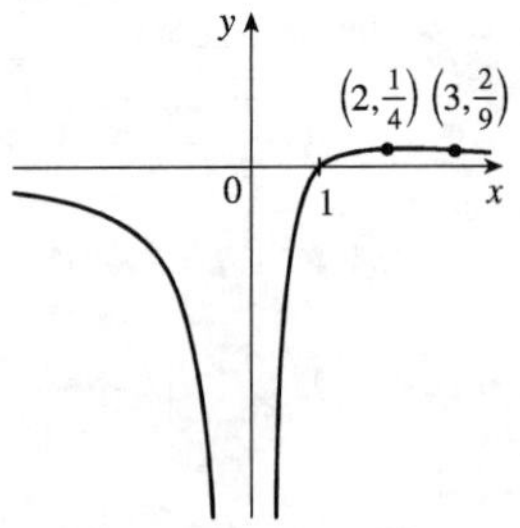

17. $y = f(x) = \dfrac{x^2}{x^2+3} = \dfrac{(x^2+3)-3}{x^2+3} = 1 - \dfrac{3}{x^2+3}$ **A.** $D = \mathbb{R}$ **B.** y-intercept: $f(0) = 0$; x-intercepts: $f(x) = 0 \Leftrightarrow x = 0$ **C.** $f(-x) = f(x)$, so f is even; the graph is symmetric about the y-axis.

D. $\displaystyle\lim_{x\to\pm\infty} \frac{x^2}{x^2+3} = 1$, so $y = 1$ is a HA. No VA. **E.** Using the Reciprocal Rule,

$f'(x) = -3 \cdot \dfrac{-2x}{(x^2+3)^2} = \dfrac{6x}{(x^2+3)^2}$. $f'(x) > 0 \Leftrightarrow x > 0$ and $f'(x) < 0 \Leftrightarrow x < 0$, so f is decreasing on $(-\infty, 0)$ and increasing on $(0, \infty)$. **F.** Local minimum value $f(0) = 0$, no local maximum.

G.
$$\begin{aligned} f''(x) &= \frac{(x^2+3)^2 \cdot 6 - 6x \cdot 2(x^2+3) \cdot 2x}{[(x^2+3)^2]^2} \\ &= \frac{6(x^2+3)\left[(x^2+3) - 4x^2\right]}{(x^2+3)^4} \\ &= \frac{6(3-3x^2)}{(x^2+3)^3} = \frac{-18(x+1)(x-1)}{(x^2+3)^3} \end{aligned}$$

H.

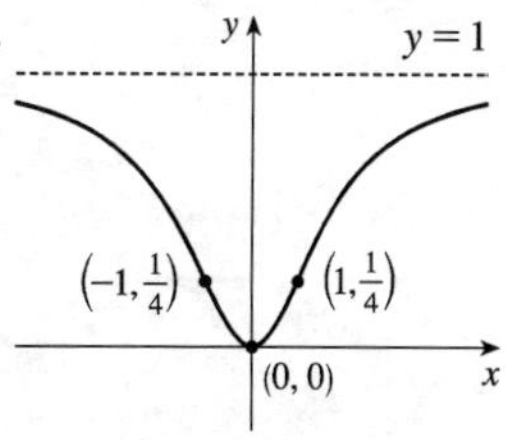

$f''(x)$ is negative on $(-\infty, -1)$ and $(1, \infty)$ and positive on $(-1, 1)$, so f is CD on $(-\infty, -1)$ and $(1, \infty)$ and CU on $(-1, 1)$. IP at $\left(\pm 1, \frac{1}{4}\right)$

19. $y = f(x) = x\sqrt{5-x}$ **A.** The domain is $\{x \mid 5 - x \geq 0\} = (-\infty, 5]$ **B.** y-intercept: $f(0) = 0$; x-intercepts: $f(x) = 0 \Leftrightarrow x = 0, 5$ **C.** No symmetry **D.** No asymptote

E. $f'(x) = x \cdot \frac{1}{2}(5-x)^{-1/2}(-1) + (5-x)^{1/2} \cdot 1 = \frac{1}{2}(5-x)^{-1/2}\left[-x + 2(5-x)\right] = \dfrac{10-3x}{2\sqrt{5-x}} > 0 \Leftrightarrow$

$x < \frac{10}{3}$, so f is increasing on $\left(-\infty, \frac{10}{3}\right)$ and decreasing on $\left(\frac{10}{3}, 5\right)$.

F. Local maximum value $f\left(\frac{10}{3}\right) = \frac{10}{9}\sqrt{15} \approx 4.3$; no local minimum

G.
$$\begin{aligned} f''(x) &= \frac{2(5-x)^{1/2}(-3) - (10-3x) \cdot 2\left(\frac{1}{2}\right)(5-x)^{-1/2}(-1)}{\left(2\sqrt{5-x}\right)^2} \\ &= \frac{(5-x)^{-1/2}\left[-6(5-x) + (10-3x)\right]}{4(5-x)} = \frac{3x-20}{4(5-x)^{3/2}} \end{aligned}$$

$f''(x) < 0$ for $x < 5$, so f is CD on $(-\infty, 5)$. No IP

H.

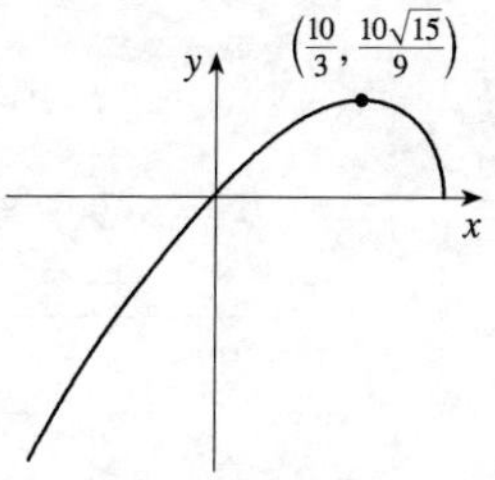

21. $y = f(x) = \sqrt{x^2+1} - x$ **A.** $D = \mathbb{R}$ **B.** No x-intercept, y-intercept $= 1$ **C.** No symmetry

D. $\displaystyle\lim_{x\to-\infty}\left(\sqrt{x^2+1} - x\right) = \infty$ and

$$\lim_{x\to\infty}\left(\sqrt{x^2+1} - x\right) = \lim_{x\to\infty}\left(\sqrt{x^2+1} - x\right)\frac{\sqrt{x^2+1}+x}{\sqrt{x^2+1}+x} = \lim_{x\to\infty}\frac{1}{\sqrt{x^2+1}+x} = 0,$$

so $y = 0$ is a HA. **E.** $f'(x) = \dfrac{x}{\sqrt{x^2+1}} - 1 = \dfrac{x - \sqrt{x^2+1}}{\sqrt{x^2+1}} \Rightarrow$

$f'(x) < 0$, so f is decreasing on $\mathbb{R}$. **F.** No extreme values

G. $f''(x) = \dfrac{1}{(x^2+1)^{3/2}} > 0$, so f is CU on $\mathbb{R}$. No IP

H.

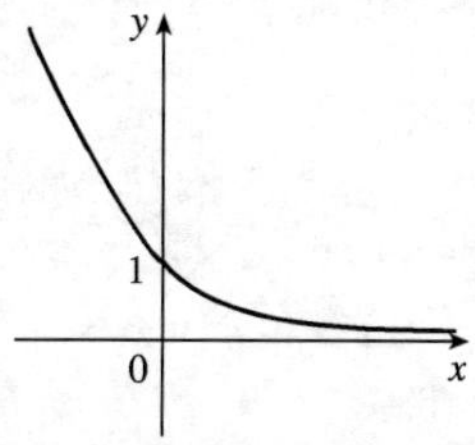

23. $y = f(x) = x/\sqrt{x^2+1}$ **A.** $D = \mathbb{R}$ **B.** y-intercept: $f(0) = 0$; x-intercepts: $f(x) = 0 \Rightarrow x = 0$
C. $f(-x) = -f(x)$, so f is odd; the graph is symmetric about the origin.

D. $$\lim_{x\to\infty} f(x) = \lim_{x\to\infty} \frac{x}{\sqrt{x^2+1}} = \lim_{x\to\infty} \frac{x/x}{\sqrt{x^2+1}/x} = \lim_{x\to\infty} \frac{x/x}{\sqrt{x^2+1}/\sqrt{x^2}} = \lim_{x\to\infty} \frac{1}{\sqrt{1+1/x^2}} = \frac{1}{\sqrt{1+0}} = 1$$

and

$$\lim_{x\to-\infty} f(x) = \lim_{x\to-\infty} \frac{x}{\sqrt{x^2+1}} = \lim_{x\to-\infty} \frac{x/x}{\sqrt{x^2+1}/x} = \lim_{x\to-\infty} \frac{x/x}{\sqrt{x^2+1}\Big/\left(-\sqrt{x^2}\right)} = \lim_{x\to-\infty} \frac{1}{-\sqrt{1+1/x^2}} = \frac{1}{-\sqrt{1+0}} = -1$$

so $y = \pm 1$ are HA. No VA.

E. $$f'(x) = \frac{\sqrt{x^2+1} - x \cdot \dfrac{2x}{2\sqrt{x^2+1}}}{[(x^2+1)^{1/2}]^2} = \frac{x^2+1-x^2}{(x^2+1)^{3/2}} = \frac{1}{(x^2+1)^{3/2}} > 0$$ for all x, so f is increasing on $\mathbb{R}$.

F. No extreme values

G. $f''(x) = -\frac{3}{2}(x^2+1)^{-5/2} \cdot 2x = \dfrac{-3x}{(x^2+1)^{5/2}}$, so $f''(x) > 0$ for $x < 0$ and $f''(x) < 0$ for $x > 0$. Thus, f is CU on $(-\infty, 0)$ and CD on $(0, \infty)$. IP at $(0, 0)$

H.

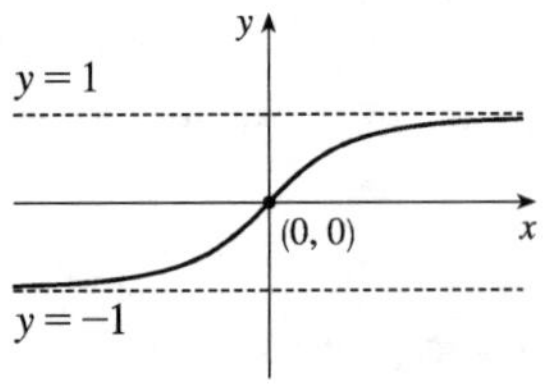

25. $y = f(x) = \sqrt{1-x^2}/x$ **A.** $D = \left\{x \,\middle|\, |x| \le 1, x \ne 0\right\} = [-1, 0) \cup (0, 1]$ **B.** x-intercepts ± 1, no y-intercept

C. $f(-x) = -f(x)$, so the curve is symmetric about $(0, 0)$. **D.** $\displaystyle\lim_{x\to 0^+} \frac{\sqrt{1-x^2}}{x} = \infty$, $\displaystyle\lim_{x\to 0^-} \frac{\sqrt{1-x^2}}{x} = -\infty$,

so $x = 0$ is a VA. **E.** $f'(x) = \dfrac{\left(-x^2/\sqrt{1-x^2}\right) - \sqrt{1-x^2}}{x^2} = -\dfrac{1}{x^2\sqrt{1-x^2}} < 0$, so f is decreasing on $(-1, 0)$ and $(0, 1)$. **F.** No extreme values

G. $f''(x) = \dfrac{2-3x^2}{x^3(1-x^2)^{3/2}} > 0 \Leftrightarrow -1 < x < -\sqrt{\frac{2}{3}}$ or $0 < x < \sqrt{\frac{2}{3}}$, so f is CU on $\left(-1, -\sqrt{\frac{2}{3}}\right)$ and $\left(0, \sqrt{\frac{2}{3}}\right)$ and CD on $\left(-\sqrt{\frac{2}{3}}, 0\right)$ and $\left(\sqrt{\frac{2}{3}}, 1\right)$. IP at $\left(\pm\sqrt{\frac{2}{3}}, \pm\frac{1}{\sqrt{2}}\right)$

H.

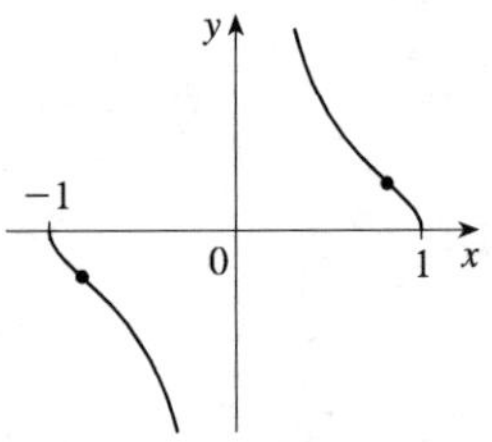

27. $y = f(x) = x - 3x^{1/3}$ **A.** $D = \mathbb{R}$ **B.** y-intercept: $f(0) = 0$; x-intercepts: $f(x) = 0 \Rightarrow x = 3x^{1/3} \Rightarrow x^3 = 27x \Rightarrow x^3 - 27x = 0 \Rightarrow x(x^2 - 27) = 0 \Rightarrow x = 0, \pm 3\sqrt{3}$ **C.** $f(-x) = -f(x)$, so f is odd; the graph is symmetric about the origin. **D.** No asymptote **E.** $f'(x) = 1 - x^{-2/3} = 1 - \dfrac{1}{x^{2/3}} = \dfrac{x^{2/3}-1}{x^{2/3}}$.
$f'(x) > 0$ when $|x| > 1$ and $f'(x) < 0$ when $0 < |x| < 1$, so f is increasing on $(-\infty, -1)$ and $(1, \infty)$, and decreasing on $(-1, 0)$ and $(0, 1)$ [hence decreasing on $(-1, 1)$ since f is continuous on $(-1, 1)$]. **F.** Local maximum value $f(-1) = 2$, local minimum value $f(1) = -2$ **G.** $f''(x) = \frac{2}{3}x^{-5/3} < 0$ when $x < 0$ and $f''(x) > 0$ when $x > 0$, so f is CD on $(-\infty, 0)$ and CU on $(0, \infty)$. IP at $(0, 0)$

H.

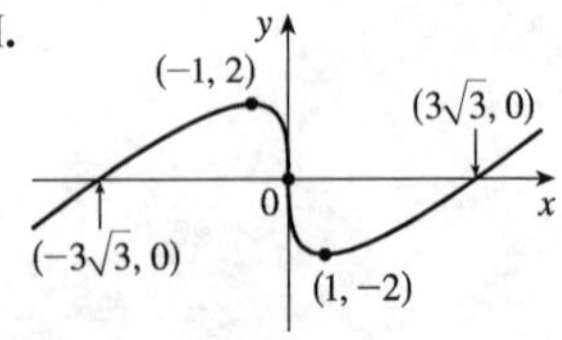

29. $y = f(x) = x + \sqrt{|x|}$ **A.** $D = \mathbb{R}$ **B.** x-intercepts 0, -1; y-intercept 0 **C.** No symmetry

D. $\lim_{x\to\infty}\left(x+\sqrt{|x|}\right) = \infty$, $\lim_{x\to-\infty}\left(x+\sqrt{|x|}\right) = -\infty$. No asymptote **E.** For $x > 0$, $f(x) = x + \sqrt{x} \;\Rightarrow$

$f'(x) = 1 + \dfrac{1}{2\sqrt{x}} > 0$, so f increases on $(0, \infty)$. For $x < 0$, $f(x) = x + \sqrt{-x} \;\Rightarrow\; f'(x) = 1 - \dfrac{1}{2\sqrt{-x}} > 0$

$\Leftrightarrow\; 2\sqrt{-x} > 1 \;\Leftrightarrow\; -x > \frac{1}{4} \;\Leftrightarrow\; x < -\frac{1}{4}$, so f increases on $\left(-\infty, -\frac{1}{4}\right)$ and decreases on $\left(-\frac{1}{4}, 0\right)$.

F. Local maximum value $f\left(-\frac{1}{4}\right) = \frac{1}{4}$, local minimum value $f(0) = 0$

G. For $x > 0$, $f''(x) = -\frac{1}{4}x^{-3/2} \;\Rightarrow\; f''(x) < 0$, so f is CD on $(0, \infty)$. For $x < 0$, $f''(x) = -\frac{1}{4}(-x)^{-3/2} \;\Rightarrow\; f''(x) < 0$, so f is CD on $(-\infty, 0)$. No IP

H.

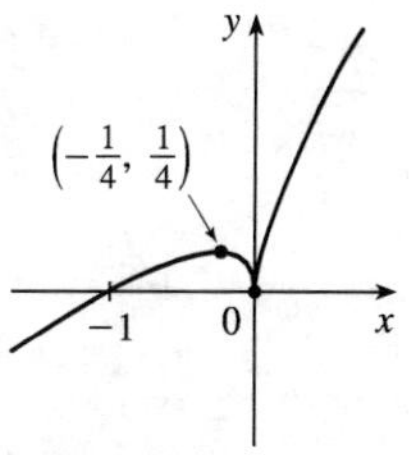

31. $y = f(x) = 3\sin x - \sin^3 x$ **A.** $D = \mathbb{R}$ **B.** y-intercept: $f(0) = 0$; x-intercepts: $f(x) = 0 \;\Rightarrow$

$\sin x\,(3 - \sin^2 x) = 0 \;\Rightarrow\; \sin x = 0$ [since $\sin^2 x \le 1 < 3$] $\Rightarrow\; x = n\pi$, n an integer.

C. $f(-x) = -f(x)$, so f is odd; the graph (shown for $-2\pi \le x \le 2\pi$) is symmetric about the origin and periodic with period 2π. **D.** No asymptote **E.** $f'(x) = 3\cos x - 3\sin^2 x\cos x = 3\cos x\,(1 - \sin^2 x) = 3\cos^3 x$.

$f'(x) > 0 \;\Leftrightarrow\; \cos x > 0 \;\Leftrightarrow\; x \in \left(2n\pi - \frac{\pi}{2}, 2n\pi + \frac{\pi}{2}\right)$ for each integer n, and $f'(x) < 0 \;\Leftrightarrow$ $\cos x < 0 \;\Leftrightarrow\; x \in \left(2n\pi + \frac{\pi}{2}, 2n\pi + \frac{3\pi}{2}\right)$ for each integer n. Thus, f is increasing on $\left(2n\pi - \frac{\pi}{2}, 2n\pi + \frac{\pi}{2}\right)$ for each integer n, and f is decreasing on $\left(2n\pi + \frac{\pi}{2}, 2n\pi + \frac{3\pi}{2}\right)$ for each integer n.

F. f has local maximum values $f(2n\pi + \frac{\pi}{2}) = 2$ and local minimum values $f(2n\pi + \frac{3\pi}{2}) = -2$.

G. $f''(x) = -9\sin x\cos^2 x = -9\sin x\,(1 - \sin^2 x) = -9\sin x\,(1 - \sin x)(1 + \sin x)$. $f''(x) < 0 \;\Leftrightarrow$ $\sin x > 0$ and $\sin x \ne \pm 1 \;\Leftrightarrow\; x \in \left(2n\pi, 2n\pi + \frac{\pi}{2}\right) \cup \left(2n\pi + \frac{\pi}{2}, 2n\pi + \pi\right)$ for some integer n.

$f''(x) > 0 \;\Leftrightarrow\; \sin x < 0$ and $\sin x \ne \pm 1 \;\Leftrightarrow\; x \in \left((2n-1)\pi, (2n-1)\pi + \frac{\pi}{2}\right) \cup \left((2n-1)\pi + \frac{\pi}{2}, 2n\pi\right)$

for some integer n. Thus, f is CD on the intervals $\left(2n\pi, \left(2n + \frac{1}{2}\right)\pi\right)$ and $\left(\left(2n + \frac{1}{2}\right)\pi, (2n+1)\pi\right)$ [hence CD on the intervals $(2n\pi, (2n+1)\pi)$] for each integer n, and f is CU on the intervals $\left((2n-1)\pi, \left(2n - \frac{1}{2}\right)\pi\right)$ and $\left(\left(2n - \frac{1}{2}\right)\pi, 2n\pi\right)$ [hence CU on the intervals $((2n-1)\pi, 2n\pi)$] for each integer n. f has inflection points at $(n\pi, 0)$ for each integer n.

H.

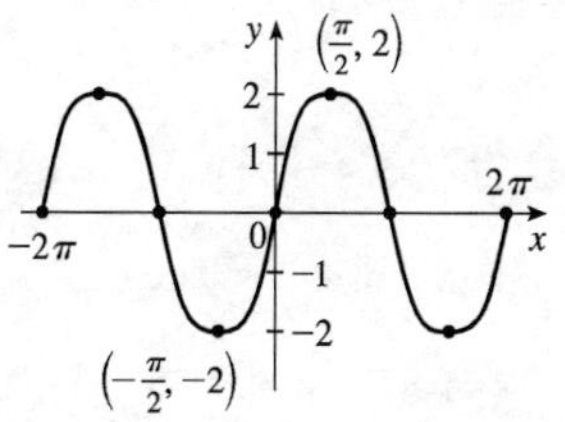

33. $y = f(x) = x \tan x$, $-\frac{\pi}{2} < x < \frac{\pi}{2}$ **A.** $D = \left(-\frac{\pi}{2}, \frac{\pi}{2}\right)$ **B.** Intercepts are 0 **C.** $f(-x) = f(x)$, so the curve is symmetric about the y-axis. **D.** $\lim\limits_{x\to(\pi/2)^-} x \tan x = \infty$ and $\lim\limits_{x\to-(\pi/2)^+} x \tan x = \infty$, so $x = \frac{\pi}{2}$ and $x = -\frac{\pi}{2}$ are VA. **E.** $f'(x) = \tan x + x \sec^2 x > 0 \Leftrightarrow 0 < x < \frac{\pi}{2}$, so f increases on $\left(0, \frac{\pi}{2}\right)$ and decreases on $\left(-\frac{\pi}{2}, 0\right)$.

F. Absolute and local minimum value $f(0) = 0$.

G. $y'' = 2\sec^2 x + 2x \tan x \sec^2 x > 0$ for $-\frac{\pi}{2} < x < \frac{\pi}{2}$, so f is CU on $\left(-\frac{\pi}{2}, \frac{\pi}{2}\right)$. No IP

H.

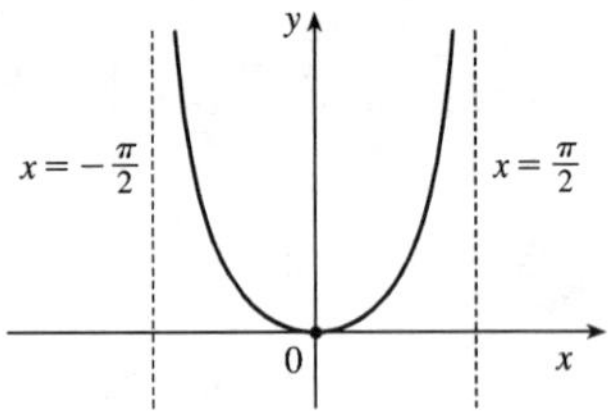

35. $y = f(x) = \frac{1}{2}x - \sin x$, $0 < x < 3\pi$ **A.** $D = (0, 3\pi)$ **B.** No y-intercept. The x-intercept, approximately 1.9, can be found using Newton's Method. **C.** No symmetry **D.** No asymptote **E.** $f'(x) = \frac{1}{2} - \cos x > 0 \Leftrightarrow \cos x < \frac{1}{2} \Leftrightarrow \frac{\pi}{3} < x < \frac{5\pi}{3}$ or $\frac{7\pi}{3} < x < 3\pi$, so f is increasing on $\left(\frac{\pi}{3}, \frac{5\pi}{3}\right)$ and $\left(\frac{7\pi}{3}, 3\pi\right)$ and decreasing on $\left(0, \frac{\pi}{3}\right)$ and $\left(\frac{5\pi}{3}, \frac{7\pi}{3}\right)$. **F.** Local minimum value $f\left(\frac{\pi}{3}\right) = \frac{\pi}{6} - \frac{\sqrt{3}}{2}$, local maximum value $f\left(\frac{5\pi}{3}\right) = \frac{5\pi}{6} + \frac{\sqrt{3}}{2}$, local minimum value $f\left(\frac{7\pi}{3}\right) = \frac{7\pi}{6} - \frac{\sqrt{3}}{2}$ **G.** $f''(x) = \sin x > 0 \Leftrightarrow 0 < x < \pi$ or $2\pi < x < 3\pi$, so f is CU on $(0, \pi)$ and $(2\pi, 3\pi)$ and CD on $(\pi, 2\pi)$. IPs at $\left(\pi, \frac{\pi}{2}\right)$ and $(2\pi, \pi)$.

H.

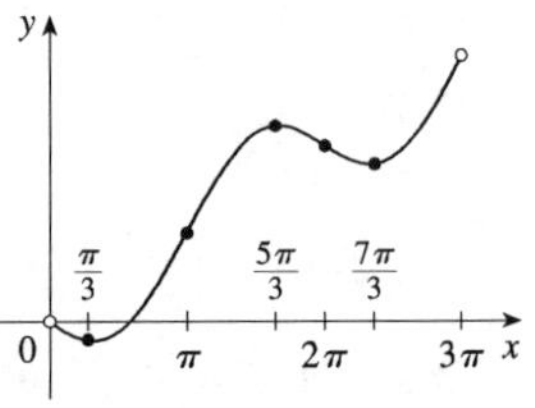

37. $y = f(x) = \sin 2x - 2\sin x$ **A.** $D = \mathbb{R}$ **B.** y-intercept $= f(0) = 0$. $y = 0 \Leftrightarrow$ $2\sin x = \sin 2x = 2\sin x \cos x \Leftrightarrow \sin x = 0$ or $\cos x = 1 \Leftrightarrow x = n\pi$ (x-intercepts)

C. $f(-x) = -f(x)$, so the curve is symmetric about $(0, 0)$.

Note: f is periodic with period 2π, so we determine E–G for $-\pi \le x \le \pi$. **D.** No asymptotes

E. $f'(x) = 2\cos 2x - 2\cos x = 2\left(2\cos^2 x - 1 - \cos x\right) = 2(2\cos x + 1)(\cos x - 1) > 0 \Leftrightarrow \cos x < -\frac{1}{2}$ $\Leftrightarrow -\pi < x < -\frac{2\pi}{3}$ or $\frac{2\pi}{3} < x < \pi$, so f is increasing on $\left(-\pi, -\frac{2\pi}{3}\right)$, $\left(\frac{2\pi}{3}, \pi\right)$ and decreasing on $\left(-\frac{2\pi}{3}, \frac{2\pi}{3}\right)$.

F. Local maximum value $f\left(-\frac{2\pi}{3}\right) = \frac{3\sqrt{3}}{2}$, local minimum value $f\left(\frac{2\pi}{3}\right) = -\frac{3\sqrt{3}}{2}$

G. $f''(x) = -4\sin 2x + 2\sin x = 2\sin x\,(1 - 4\cos x) = 0$ when $x = 0, \pm\pi$ or $\cos x = \frac{1}{4}$. If $\alpha = \cos^{-1}\frac{1}{4}$, then f is CU on $(-\alpha, 0)$ and (α, π) and CD on $(-\pi, -\alpha)$ and $(0, \alpha)$.

IPs at $(0, 0)$, $(\pm\pi, 0)$, $\left(\alpha, -\frac{3\sqrt{15}}{8}\right)$, $\left(-\alpha, \frac{3\sqrt{15}}{8}\right)$.

H.

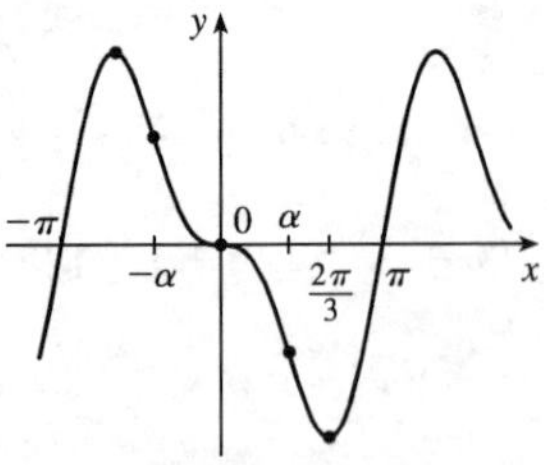

39. $y = f(x) = \dfrac{\sin x}{1 + \cos x}$ $\left[\overset{\text{when}}{\underset{\cos x \neq 1}{=}} \dfrac{\sin x}{1 + \cos x} \cdot \dfrac{1 - \cos x}{1 - \cos x} = \dfrac{\sin x\,(1 - \cos x)}{\sin^2 x} = \dfrac{1 - \cos x}{\sin x} = \csc x - \cot x \right]$

A. The domain of f is the set of all real numbers except odd integer multiples of π. **B.** y-intercept: $f(0) = 0$; x-intercepts: $x = n\pi$, n an even integer. **C.** $f(-x) = -f(x)$, so f is an odd function; the graph is symmetric about the origin and has period 2π. **D.** When n is an odd integer, $\lim\limits_{x \to (n\pi)^-} f(x) = \infty$ and $\lim\limits_{x \to (n\pi)^+} f(x) = -\infty$, so $x = n\pi$ is a VA for each odd integer n. No HA.

E. $f'(x) = \dfrac{(1 + \cos x) \cdot \cos x - \sin x(-\sin x)}{(1 + \cos x)^2} = \dfrac{1 + \cos x}{(1 + \cos x)^2} = \dfrac{1}{1 + \cos x}$. $f'(x) > 0$ for all x except odd multiples of π, so f is increasing on $((2k - 1)\pi, (2k + 1)\pi)$ for each integer k. **F.** No extreme values

G. $f''(x) = \dfrac{\sin x}{(1 + \cos x)^2} > 0 \;\Rightarrow\; \sin x > 0 \;\Rightarrow$ $x \in (2k\pi, (2k + 1)\pi)$ and $f''(x) < 0$ on $((2k - 1)\pi, 2k\pi)$ for each integer k. f is CU on $(2k\pi, (2k + 1)\pi)$ and CD on $((2k - 1)\pi, 2k\pi)$ for each integer k. f has IPs at $(2k\pi, 0)$ for each integer k.

H.

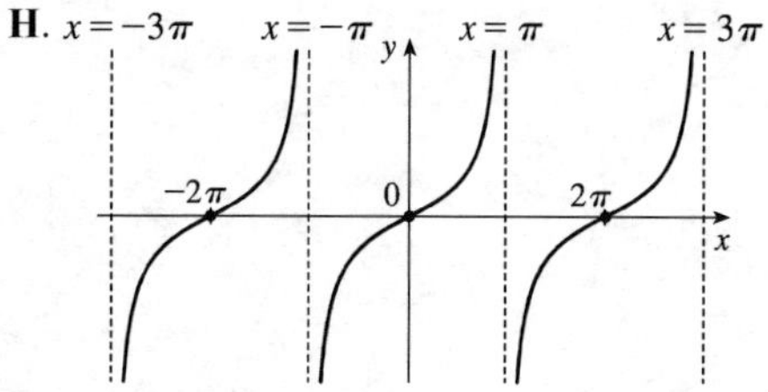

41. $y = -\dfrac{W}{24EI}x^4 + \dfrac{WL}{12EI}x^3 - \dfrac{WL^2}{24EI}x^2 = -\dfrac{W}{24EI}x^2(x^2 - 2Lx + L^2)$
$= \dfrac{-W}{24EI}x^2(x - L)^2 = cx^2(x - L)^2$

where $c = -\dfrac{W}{24EI}$ is a negative constant and $0 \le x \le L$. We sketch

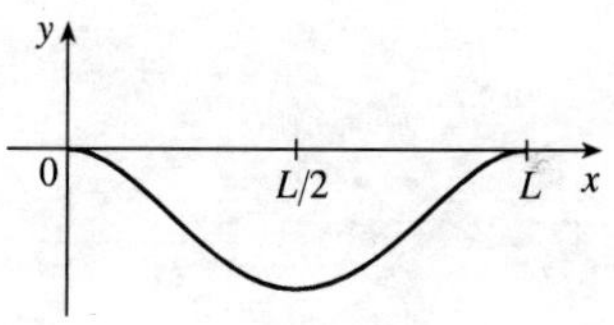

$f(x) = cx^2(x - L)^2$ for $c = -1$. $f(0) = f(L) = 0$.

$f'(x) = cx^2[2(x - L)] + (x - L)^2(2cx) = 2cx(x - L)\,[x + (x - L)] = 2cx(x - L)(2x - L)$. So for $0 < x < L$, $f'(x) > 0 \;\Leftrightarrow\; x(x - L)(2x - L) < 0$ (since $c < 0$) $\Leftrightarrow\; L/2 < x < L$ and $f'(x) < 0 \;\Leftrightarrow$ $0 < x < L/2$. So f is increasing on $(L/2, L)$ and decreasing on $(0, L/2)$, and there is a local and absolute minimum at $(L/2, f(L/2)) = \left(L/2, cL^4/16\right)$.

$f'(x) = 2c\,[x(x - L)(2x - L)] \;\Rightarrow$

$f''(x) = 2c\,[1(x - L)(2x - L) + x(1)(2x - L) + x(x - L)(2)] = 2c\left(6x^2 - 6Lx + L^2\right) = 0 \;\Leftrightarrow$

$x = \dfrac{6L \pm \sqrt{12L^2}}{12} = \frac{1}{2}L \pm \frac{\sqrt{3}}{6}L$, and these are the x-coordinates of the two inflection points.

43. $y = \dfrac{x^2+1}{x+1}$. Long division gives us:

$$\begin{array}{r|l} & \quad x - 1 \\ \hline x+1 & x^2 \qquad + 1 \\ & \underline{x^2 + x} \\ & \quad -x + 1 \\ & \quad \underline{-x - 1} \\ & \qquad\quad 2 \end{array}$$

Thus, $y = f(x) = \dfrac{x^2+1}{x+1} = x - 1 + \dfrac{2}{x+1}$ and $f(x) - (x-1) = \dfrac{2}{x+1} = \dfrac{\frac{2}{x}}{1 + \frac{1}{x}}$ [for $x \neq 0$] $\to 0$ as $x \to \pm\infty$. So the line $y = x - 1$ is a slant asymptote (SA).

45. $y = \dfrac{4x^3 - 2x^2 + 5}{2x^2 + x - 3}$. Long division gives us:

$$\begin{array}{r|l} & \qquad\qquad 2x - 2 \\ \hline 2x^2 + x - 3 & 4x^3 - 2x^2 \qquad + 5 \\ & \underline{4x^3 + 2x^2 - 6x} \\ & \quad -4x^2 + 6x + 5 \\ & \quad \underline{-4x^2 - 2x + 6} \\ & \qquad\qquad 8x - 1 \end{array}$$

Thus, $y = f(x) = \dfrac{4x^3 - 2x^2 + 5}{2x^2 + x - 3} = 2x - 2 + \dfrac{8x - 1}{2x^2 + x - 3}$ and

$f(x) - (2x - 2) = \dfrac{8x - 1}{2x^2 + x - 3} = \dfrac{\frac{8}{x} - \frac{1}{x^2}}{2 + \frac{1}{x} - \frac{3}{x^2}}$ [for $x \neq 0$] $\to 0$ as $x \to \pm\infty$. So the line $y = 2x - 2$ is a SA.

47. $y = f(x) = \dfrac{-2x^2 + 5x - 1}{2x - 1} = -x + 2 + \dfrac{1}{2x - 1}$ **A.** $D = \left\{x \in \mathbb{R} \mid x \neq \frac{1}{2}\right\} = \left(-\infty, \frac{1}{2}\right) \cup \left(\frac{1}{2}, \infty\right)$

B. y-intercept: $f(0) = 1$; x-intercepts: $f(x) = 0 \Rightarrow -2x^2 + 5x - 1 = 0 \Rightarrow x = \dfrac{-5 \pm \sqrt{17}}{-4} \Rightarrow$ $x \approx 0.22, 2.28$. **C.** No symmetry

D. $\lim\limits_{x \to (1/2)^-} f(x) = -\infty$ and $\lim\limits_{x \to (1/2)^+} f(x) = \infty$, so $x = \frac{1}{2}$ is a VA.

$\lim\limits_{x \to \pm\infty} [f(x) - (-x + 2)] = \lim\limits_{x \to \pm\infty} \dfrac{1}{2x - 1} = 0$, so the line $y = -x + 2$ is a SA.

E. $f'(x) = -1 - \dfrac{2}{(2x-1)^2} < 0$ for $x \neq \frac{1}{2}$, so f is decreasing on $\left(-\infty, \frac{1}{2}\right)$ and $\left(\frac{1}{2}, \infty\right)$. **F.** No extreme values

G. $f'(x) = -1 - 2(2x - 1)^{-2} \Rightarrow$

$f''(x) = -2(-2)(2x - 1)^{-3}(2) = \dfrac{8}{(2x-1)^3}$, so $f''(x) > 0$ when $x > \frac{1}{2}$ and $f''(x) < 0$ when $x < \frac{1}{2}$. Thus, f is CU on $\left(\frac{1}{2}, \infty\right)$ and CD on $\left(-\infty, \frac{1}{2}\right)$. No IP

H.

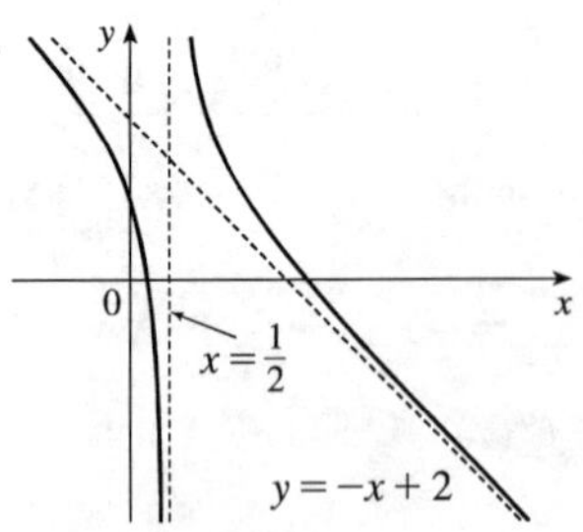

49. $y = f(x) = (x^2+4)/x = x + 4/x$ **A.** $D = \{x \mid x \neq 0\} = (-\infty, 0) \cup (0, \infty)$ **B.** No intercept

C. $f(-x) = -f(x) \Rightarrow$ symmetry about the origin **D.** $\lim_{x \to \infty} (x + 4/x) = \infty$ but $f(x) - x = 4/x \to 0$ as $x \to \pm\infty$, so $y = x$ is a slant asymptote. $\lim_{x \to 0^+} (x + 4/x) = \infty$ and $\lim_{x \to 0^-} (x + 4/x) = -\infty$, so $x = 0$ is a VA. **E.** $f'(x) = 1 - 4/x^2 > 0$ $\Leftrightarrow$ $x^2 > 4$ $\Leftrightarrow$ $x > 2$ or $x < -2$, so f is increasing on $(-\infty, -2)$ and $(2, \infty)$ and decreasing on $(-2, 0)$ and $(0, 2)$.

F. Local maximum value $f(-2) = -4$, local minimum value $f(2) = 4$

G. $f''(x) = 8/x^3 > 0 \Leftrightarrow x > 0$ so f is CU on $(0, \infty)$ and CD on $(-\infty, 0)$. No IP

H.

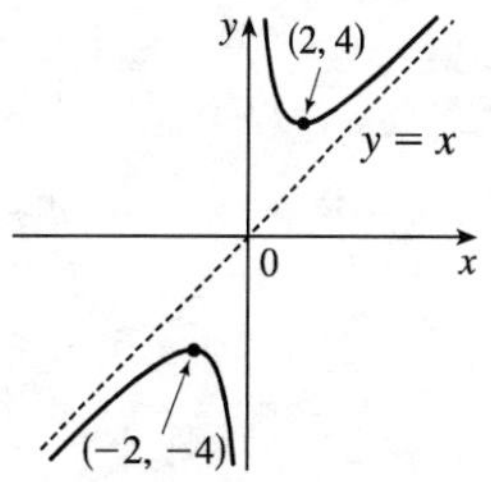

51. $y = f(x) = \dfrac{2x^3 + x^2 + 1}{x^2 + 1} = 2x + 1 + \dfrac{-2x}{x^2+1}$ **A.** $D = \mathbb{R}$ **B.** y-intercept: $f(0) = 1$; x-intercept: $f(x) = 0$ $\Rightarrow 0 = 2x^3 + x^2 + 1 = (x+1)(2x^2 - x + 1) \Rightarrow x = -1$ **C.** No symmetry **D.** No VA

$\lim_{x \to \pm\infty} [f(x) - (2x+1)] = \lim_{x \to \pm\infty} \dfrac{-2x}{x^2+1} = \lim_{x \to \pm\infty} \dfrac{-2/x}{1 + 1/x^2} = 0$, so the line $y = 2x + 1$ is a slant asymptote.

E.
$$f'(x) = 2 + \frac{(x^2+1)(-2) - (-2x)(2x)}{(x^2+1)^2} = \frac{2(x^4 + 2x^2 + 1) - 2x^2 - 2 + 4x^2}{(x^2+1)^2}$$
$$= \frac{2x^4 + 6x^2}{(x^2+1)^2} = \frac{2x^2(x^2+3)}{(x^2+1)^2}$$

so $f'(x) > 0$ if $x \neq 0$. Thus, f is increasing on $(-\infty, 0)$ and $(0, \infty)$. Since f is continuous at 0, f is increasing on $\mathbb{R}$. **F.** No extreme values

G.
$$f''(x) = \frac{(x^2+1)^2 \cdot (8x^3 + 12x) - (2x^4 + 6x^2) \cdot 2(x^2+1)(2x)}{[(x^2+1)^2]^2}$$
$$= \frac{4x(x^2+1)[(x^2+1)(2x^2+3) - 2x^4 - 6x^2]}{(x^2+1)^4}$$
$$= \frac{4x(-x^2+3)}{(x^2+1)^3}$$

so $f''(x) > 0$ for $x < -\sqrt{3}$ and $0 < x < \sqrt{3}$, and $f''(x) < 0$ for $-\sqrt{3} < x < 0$ and $x > \sqrt{3}$. f is CU on $(-\infty, -\sqrt{3})$ and $(0, \sqrt{3})$, and CD on $(-\sqrt{3}, 0)$ and $(\sqrt{3}, \infty)$. There are three IPs: $(0, 1)$, $\left(-\sqrt{3}, -\frac{3}{2}\sqrt{3} + 1\right) \approx (-1.73, -1.60)$, and $\left(\sqrt{3}, \frac{3}{2}\sqrt{3} + 1\right) \approx (1.73, 3.60)$.

H.

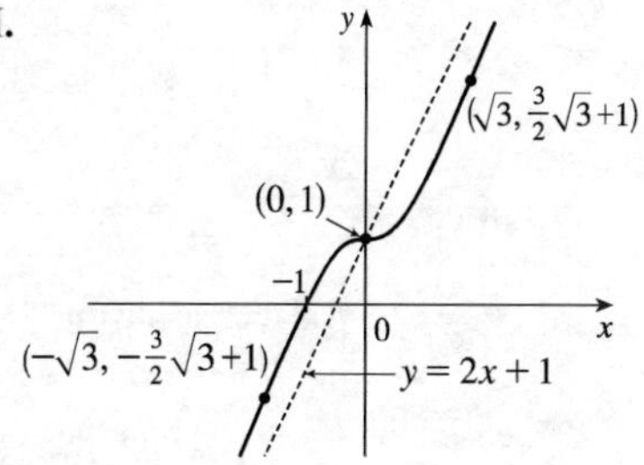

53. $y = f(x) = \sqrt{4x^2+9} \;\Rightarrow\; f'(x) = \dfrac{4x}{\sqrt{4x^2+9}} \;\Rightarrow$

$f''(x) = \dfrac{\sqrt{4x^2+9}\cdot 4 - 4x\cdot 4x/\sqrt{4x^2+9}}{4x^2+9} = \dfrac{4(4x^2+9)-16x^2}{(4x^2+9)^{3/2}} = \dfrac{36}{(4x^2+9)^{3/2}}$. f is defined on $(-\infty,\infty)$. $f(-x)=f(x)$, so f is even, which means its graph is symmetric about the y-axis. The y-intercept is $f(0)=3$. There are no x-intercepts since $f(x)>0$ for all x.

$$\lim_{x\to\infty}\left(\sqrt{4x^2+9}-2x\right) = \lim_{x\to\infty}\frac{\left(\sqrt{4x^2+9}-2x\right)\left(\sqrt{4x^2+9}+2x\right)}{\sqrt{4x^2+9}+2x} = \lim_{x\to\infty}\frac{(4x^2+9)-4x^2}{\sqrt{4x^2+9}+2x} = \lim_{x\to\infty}\frac{9}{\sqrt{4x^2+9}+2x} = 0$$

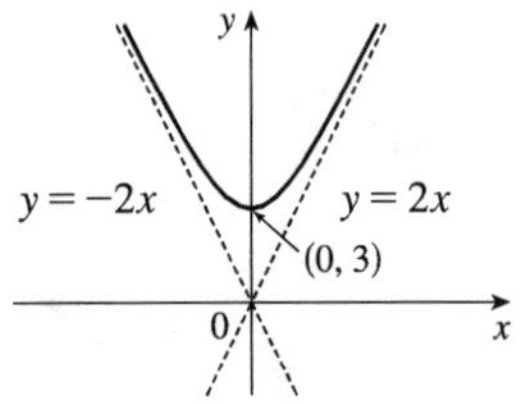

and, similarly, $\displaystyle\lim_{x\to-\infty}\left(\sqrt{4x^2+9}+2x\right) = \lim_{x\to-\infty}\frac{9}{\sqrt{4x^2+9}-2x} = 0$,

so $y=\pm 2x$ are slant asymptotes. f is decreasing on $(-\infty,0)$ and increasing on $(0,\infty)$ with local minimum $f(0)=3$. $f''(x)>0$ for all x, so f is CU on $\mathbb{R}$.

55. $\dfrac{x^2}{a^2}-\dfrac{y^2}{b^2}=1 \;\Rightarrow\; y=\pm\dfrac{b}{a}\sqrt{x^2-a^2}$. Now

$$\lim_{x\to\infty}\left[\frac{b}{a}\sqrt{x^2-a^2}-\frac{b}{a}x\right] = \frac{b}{a}\cdot\lim_{x\to\infty}\left(\sqrt{x^2-a^2}-x\right)\frac{\sqrt{x^2-a^2}+x}{\sqrt{x^2-a^2}+x} = \frac{b}{a}\cdot\lim_{x\to\infty}\frac{-a^2}{\sqrt{x^2-a^2}+x} = 0,$$

which shows that $y=\dfrac{b}{a}x$ is a slant asymptote. Similarly,

$$\lim_{x\to\infty}\left[-\frac{b}{a}\sqrt{x^2-a^2}-\left(-\frac{b}{a}x\right)\right] = -\frac{b}{a}\cdot\lim_{x\to\infty}\frac{-a^2}{\sqrt{x^2-a^2}+x} = 0, \text{ so } y=-\frac{b}{a}x \text{ is a slant asymptote.}$$

57. $\displaystyle\lim_{x\to\pm\infty}\left[f(x)-x^3\right] = \lim_{x\to\pm\infty}\frac{x^4+1}{x}-\frac{x^4}{x} = \lim_{x\to\pm\infty}\frac{1}{x}=0$, so the graph of f is asymptotic to that of $y=x^3$.

A. $D=\{x \mid x\neq 0\}$ **B.** No intercept **C.** f is symmetric about the origin. **D.** $\displaystyle\lim_{x\to 0^-}\left(x^3+\frac{1}{x}\right)=-\infty$ and $\displaystyle\lim_{x\to 0^+}\left(x^3+\frac{1}{x}\right)=\infty$, so $x=0$ is a vertical asymptote, and as shown above, the graph of f is asymptotic to that of $y=x^3$. **E.** $f'(x)=3x^2-1/x^2>0 \;\Leftrightarrow\; x^4>\frac{1}{3} \;\Leftrightarrow\; |x|>\frac{1}{\sqrt[4]{3}}$, so f is increasing on $\left(-\infty,-\frac{1}{\sqrt[4]{3}}\right)$ and $\left(\frac{1}{\sqrt[4]{3}},\infty\right)$ and decreasing on $\left(-\frac{1}{\sqrt[4]{3}},0\right)$ and $\left(0,\frac{1}{\sqrt[4]{3}}\right)$. **F.** Local maximum value $f\left(-\frac{1}{\sqrt[4]{3}}\right)=-4\cdot 3^{-5/4}$, local minimum value $f\left(\frac{1}{\sqrt[4]{3}}\right)=4\cdot 3^{-5/4}$

G. $f''(x)=6x+2/x^3>0 \;\Leftrightarrow\; x>0$, so f is CU on $(0,\infty)$ and CD on $(-\infty,0)$. No IP

H.

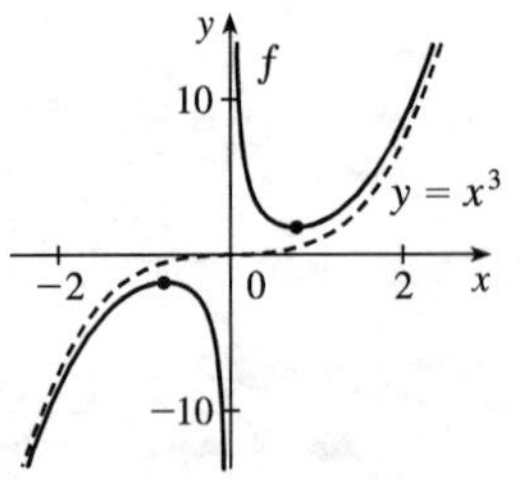

Optimization Problems

1. (a)

First Number	Second Number	Product
1	22	22
2	21	42
3	20	60
4	19	76
5	18	90
6	17	102
7	16	112
8	15	120
9	14	126
10	13	130
11	12	132

We needn't consider pairs where the first number is larger than the second, since we can just interchange the numbers in such cases. The answer appears to be 11 and 12, but we have considered only integers in the table.

(b) Call the two numbers x and y. Then $x + y = 23$, so $y = 23 - x$. Call the product P. Then $P = xy = x(23 - x) = 23x - x^2$, so we wish to maximize the function $P(x) = 23x - x^2$. Since $P'(x) = 23 - 2x$, we see that $P'(x) = 0 \Leftrightarrow x = \frac{23}{2} = 11.5$. Thus, the maximum value of P is $P(11.5) = (11.5)^2 = 132.25$ and it occurs when $x = y = 11.5$.

Or: Note that $P''(x) = -2 < 0$ for all x, so P is everywhere concave downward and the local maximum at $x = 11.5$ must be an absolute maximum.

3. The two numbers are x and $\dfrac{100}{x}$, where $x > 0$. Minimize $f(x) = x + \dfrac{100}{x}$. $f'(x) = 1 - \dfrac{100}{x^2} = \dfrac{x^2 - 100}{x^2}$. The critical number is $x = 10$. Since $f'(x) < 0$ for $0 < x < 10$ and $f'(x) > 0$ for $x > 10$, there is an absolute minimum at $x = 10$. The numbers are 10 and 10.

5. If the rectangle has dimensions x and y, then its perimeter is $2x + 2y = 100$ m, so $y = 50 - x$. Thus, the area is $A = xy = x(50 - x)$. We wish to maximize the function $A(x) = x(50 - x) = 50x - x^2$, where $0 < x < 50$. Since $A'(x) = 50 - 2x = -2(x - 25)$, $A'(x) > 0$ for $0 < x < 25$ and $A'(x) < 0$ for $25 < x < 50$. Thus, A has an absolute maximum at $x = 25$, and $A(25) = 25^2 = 625\text{ m}^2$. The dimensions of the rectangle that maximize its area are $x = y = 25$ m. (The rectangle is a square.)

7. (a)

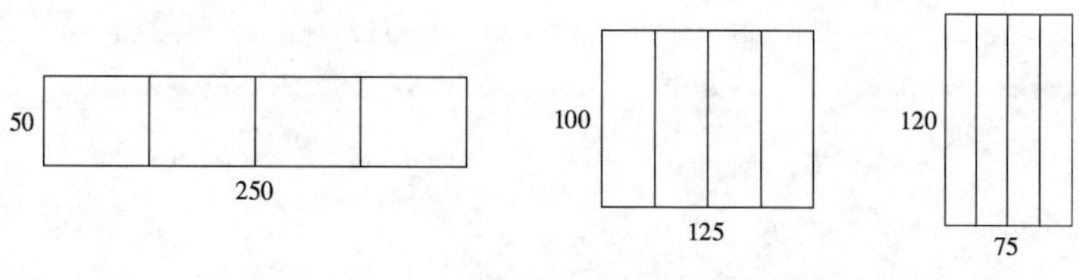

The areas of the three figures are 12,500, 12,500, and 9000 ft^2. There appears to be a maximum area of at least 12,500 ft^2.

(b) Let x denote the length of each of two sides and three dividers.

Let y denote the length of the other two sides.

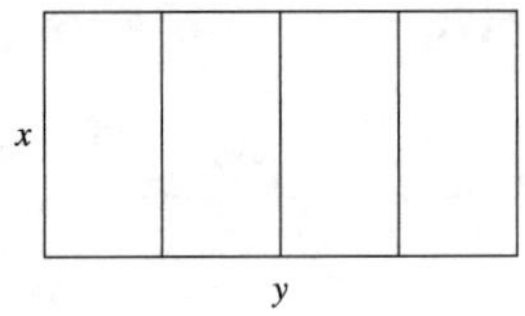

(c) Area $A = \text{length} \times \text{width} = y \cdot x$

(d) Length of fencing $= 750 \quad \Rightarrow \quad 5x + 2y = 750$

(e) $5x + 2y = 750 \quad \Rightarrow \quad y = 375 - \frac{5}{2}x \quad \Rightarrow \quad A(x) = \left(375 - \frac{5}{2}x\right)x = 375x - \frac{5}{2}x^2$

(f) $A'(x) = 375 - 5x = 0 \quad \Rightarrow \quad x = 75$. Since $A''(x) = -5 < 0$ there is an absolute maximum when $x = 75$. Then $y = \frac{375}{2} = 187.5$. The largest area is $75\left(\frac{375}{2}\right) = 14{,}062.5\ \text{ft}^2$. These values of x and y are between the values in the first and second figures in part (a). Our original estimate was low.

9. 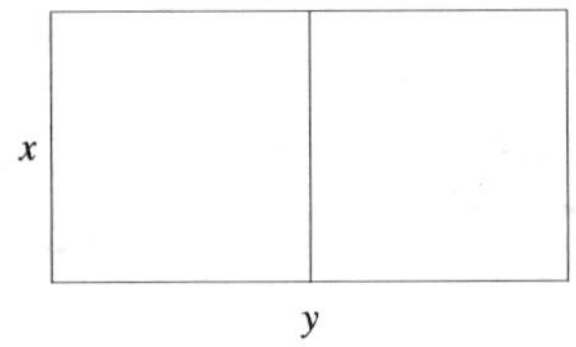

$xy = 1.5 \times 10^6$, so $y = 1.5 \times 10^6/x$. Minimize the amount of fencing, which is $3x + 2y = 3x + 2(1.5 \times 10^6/x) = 3x + 3 \times 10^6/x = F(x)$. $F'(x) = 3 - 3 \times 10^6/x^2 = 3\left(x^2 - 10^6\right)/x^2$. The critical number is $x = 10^3$ and $F'(x) < 0$ for $0 < x < 10^3$ and $F'(x) > 0$ if $x > 10^3$, so the absolute minimum occurs when $x = 10^3$ and $y = 1.5 \times 10^3$.

The field should be 1000 feet by 1500 feet with the middle fence parallel to the short side of the field.

11. Let b be the length of the base of the box and h the height. The surface area is $1200 = b^2 + 4hb \quad \Rightarrow$ $h = \left(1200 - b^2\right)/(4b)$. The volume is $V = b^2h = b^2\left(1200 - b^2\right)/4b = 300b - b^3/4 \quad \Rightarrow \quad V'(b) = 300 - \frac{3}{4}b^2$. $V'(b) = 0 \quad \Rightarrow \quad 300 = \frac{3}{4}b^2 \quad \Rightarrow \quad b^2 = 400 \quad \Rightarrow \quad b = \sqrt{400} = 20$. Since $V'(b) > 0$ for $0 < b < 20$ and $V'(b) < 0$ for $b > 20$, there is an absolute maximum when $b = 20$ by the First Derivative Test for Absolute Extreme Values (see page 280). If $b = 20$, then $h = \left(1200 - 20^2\right)/(4 \cdot 20) = 10$, so the largest possible volume is $b^2h = (20)^2(10) = 4000\ \text{cm}^3$.

13. 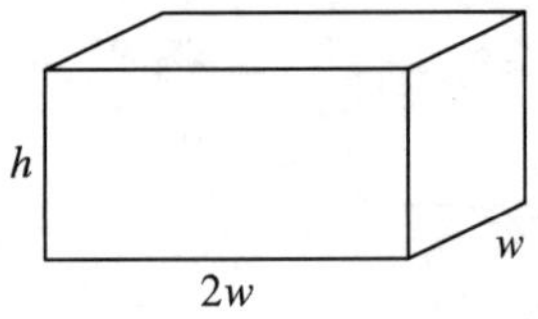

$10 = (2w)(w)h = 2w^2h$, so $h = 5/w^2$. The cost is

$$C(w) = 10\left(2w^2\right) + 6[2(2wh) + 2hw] + 6\left(2w^2\right)$$
$$= 32w^2 + 36wh = 32w^2 + 180/w$$

$C'(w) = 64w - 180/w^2 = 4\left(16w^3 - 45\right)/w^2 \quad \Rightarrow \quad w = \sqrt[3]{\frac{45}{16}}$ is the critical number. $C'(w) < 0$ for $0 < w < \sqrt[3]{\frac{45}{16}}$ and $C'(w) > 0$ for $w > \sqrt[3]{\frac{45}{16}}$. The minimum cost is $C\left(\sqrt[3]{\frac{45}{16}}\right) = 32(2.8125)^{2/3} + 180/\sqrt[3]{2.8125} \approx \$191.28$.

15. The distance from a point (x, y) on the line $y = 4x + 7$ to the origin is $\sqrt{(x-0)^2 + (y-0)^2} = \sqrt{x^2 + y^2}$. However, it is easier to work with the *square* of the distance; that is,

$D(x) = \left(\sqrt{x^2 + y^2}\right)^2 = x^2 + y^2 = x^2 + (4x + 7)^2$. Because the distance is positive, its minimum value will occur at the same point as the minimum value of D.

$D'(x) = 2x + 2(4x + 7)(4) = 34x + 56$, so $D'(x) = 0 \quad \Leftrightarrow \quad x = -\frac{28}{17}$.

$D''(x) = 34 > 0$, so D is concave upward for all x. Thus, D has an absolute minimum at $x = -\frac{28}{17}$. The point closest to the origin is $(x, y) = \left(-\frac{28}{17}, 4\left(-\frac{28}{17}\right) + 7\right) = \left(-\frac{28}{17}, \frac{7}{17}\right)$.

17.

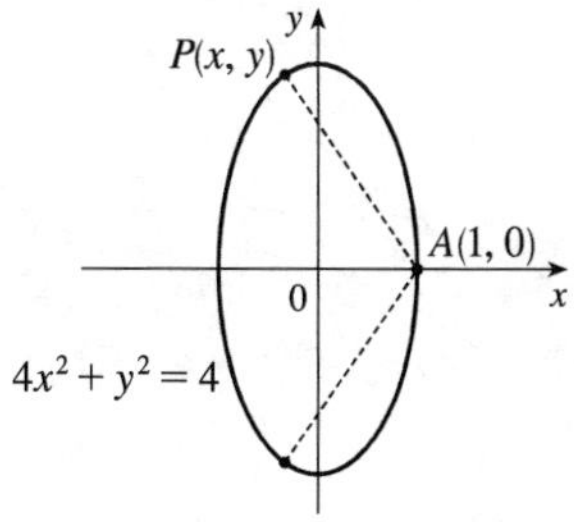

From the figure, we see that there are two points that are farthest away from $A(1,0)$. The distance d from A to an arbitrary point $P(x,y)$ on the ellipse is $d = \sqrt{(x-1)^2 + (y-0)^2}$ and the square of the distance is $S = d^2 = x^2 - 2x + 1 + y^2 = x^2 - 2x + 1 + (4 - 4x^2) = -3x^2 - 2x + 5$. $S' = -6x - 2$ and $S' = 0 \;\Rightarrow\; x = -\frac{1}{3}$. Now $S'' = -6 < 0$, so we know that S has a maximum at $x = -\frac{1}{3}$. Since $-1 \le x \le 1$, $S(-1) = 4$, $S\left(-\frac{1}{3}\right) = \frac{16}{3}$, and $S(1) = 0$, we see that the maximum distance is $\sqrt{\frac{16}{3}}$. The corresponding y-values are $y = \pm\sqrt{4 - 4\left(-\frac{1}{3}\right)^2} = \pm\sqrt{\frac{32}{9}} = \pm\frac{4}{3}\sqrt{2} \approx \pm 1.89$. The points are $\left(-\frac{1}{3}, \pm\frac{4}{3}\sqrt{2}\right)$.

19.

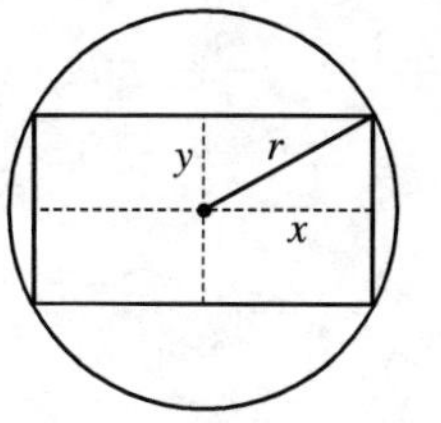

The area of the rectangle is $(2x)(2y) = 4xy$. Also $r^2 = x^2 + y^2$ so $y = \sqrt{r^2 - x^2}$, so the area is $A(x) = 4x\sqrt{r^2 - x^2}$. Now $A'(x) = 4\left(\sqrt{r^2 - x^2} - \dfrac{x^2}{\sqrt{r^2 - x^2}}\right) = 4\dfrac{r^2 - 2x^2}{\sqrt{r^2 - x^2}}$. The critical number is $x = \frac{1}{\sqrt{2}}r$. Clearly this gives a maximum.

$y = \sqrt{r^2 - \left(\frac{1}{\sqrt{2}}r\right)^2} = \sqrt{\frac{1}{2}r^2} = \frac{1}{\sqrt{2}}r = x$, which tells us that the rectangle is a square. The dimensions are $2x = \sqrt{2}\,r$ and $2y = \sqrt{2}\,r$.

21.

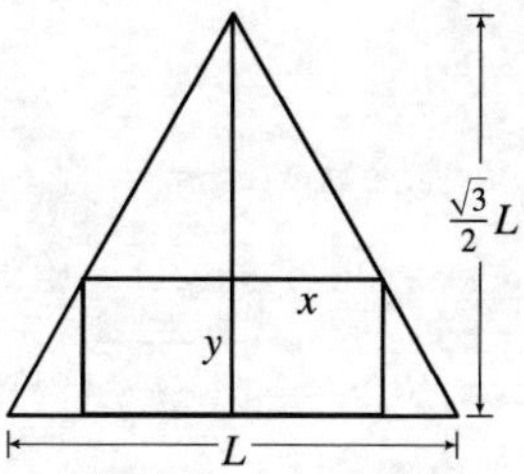

The height h of the equilateral triangle with sides of length L is $\frac{\sqrt{3}}{2}L$, since $h^2 + (L/2)^2 = L^2 \;\Rightarrow\; h^2 = L^2 - \frac{1}{4}L^2 = \frac{3}{4}L^2 \;\Rightarrow\; h = \frac{\sqrt{3}}{2}L$. Using similar triangles, $\dfrac{\frac{\sqrt{3}}{2}L - y}{x} = \dfrac{\frac{\sqrt{3}}{2}L}{L/2} = \sqrt{3} \;\Rightarrow$

$$\sqrt{3}\,x = \frac{\sqrt{3}}{2}L - y \;\Rightarrow\; y = \frac{\sqrt{3}}{2}L - \sqrt{3}\,x \;\Rightarrow\; y = \frac{\sqrt{3}}{2}(L - 2x).$$

The area of the inscribed rectangle is $A(x) = (2x)y = \sqrt{3}\,x(L - 2x) = \sqrt{3}\,Lx - 2\sqrt{3}\,x^2$, where $0 \le x \le L/2$. Now $0 = A'(x) = \sqrt{3}\,L - 4\sqrt{3}\,x \;\Rightarrow\; x = \sqrt{3}\,L/(4\sqrt{3}) = L/4$. Since $A(0) = A(L/2) = 0$, the maximum occurs when $x = L/4$, and $y = \frac{\sqrt{3}}{2}L - \frac{\sqrt{3}}{4}L = \frac{\sqrt{3}}{4}L$, so the dimensions are $L/2$ and $\frac{\sqrt{3}}{4}L$.

23.

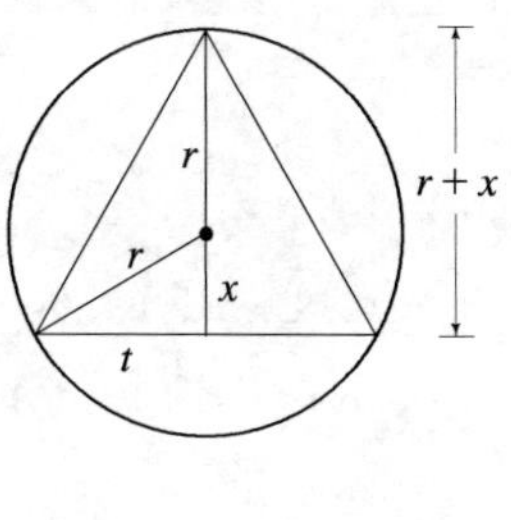

The area of the triangle is

$A(x) = \frac{1}{2}(2t)(r + x) = t(r + x) = \sqrt{r^2 - x^2}(r + x)$. Then

$$0 = A'(x) = r\frac{-2x}{2\sqrt{r^2 - x^2}} + \sqrt{r^2 - x^2} + x\frac{-2x}{2\sqrt{r^2 - x^2}}$$

$$= -\frac{x^2 + rx}{\sqrt{r^2 - x^2}} + \sqrt{r^2 - x^2} \;\Rightarrow$$

$$\frac{x^2 + rx}{\sqrt{r^2 - x^2}} = \sqrt{r^2 - x^2} \;\Rightarrow\; x^2 + rx = r^2 - x^2 \;\Rightarrow\; 0 = 2x^2 + rx - r^2 = (2x - r)(x + r) \;\Rightarrow$$

$x = \frac{1}{2}r$ or $x = -r$. Now $A(r) = 0 = A(-r) \;\Rightarrow\;$ the maximum occurs where $x = \frac{1}{2}r$, so the triangle has height $r + \frac{1}{2}r = \frac{3}{2}r$ and base $2\sqrt{r^2 - \left(\frac{1}{2}r\right)^2} = 2\sqrt{\frac{3}{4}r^2} = \sqrt{3}\,r$.

25.

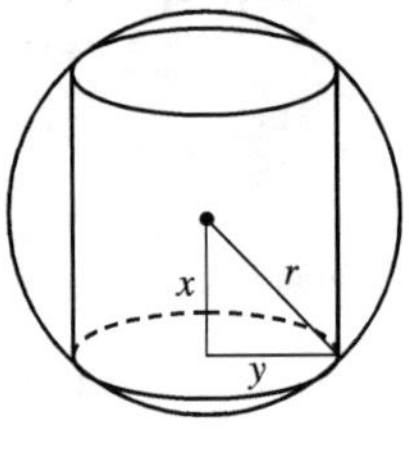

The cylinder has volume $V = \pi y^2(2x)$. Also $x^2 + y^2 = r^2 \;\Rightarrow\; y^2 = r^2 - x^2$, so $V(x) = \pi\left(r^2 - x^2\right)(2x) = 2\pi\left(r^2x - x^3\right)$, where $0 \le x \le r$. $V'(x) = 2\pi\left(r^2 - 3x^2\right) = 0 \;\Rightarrow\; x = r/\sqrt{3}$. Now $V(0) = V(r) = 0$, so there is a maximum when $x = r/\sqrt{3}$ and $V\left(r/\sqrt{3}\right) = \pi\left(r^2 - r^2/3\right)\left(2r/\sqrt{3}\right) = 4\pi r^3/\left(3\sqrt{3}\right)$.

27.

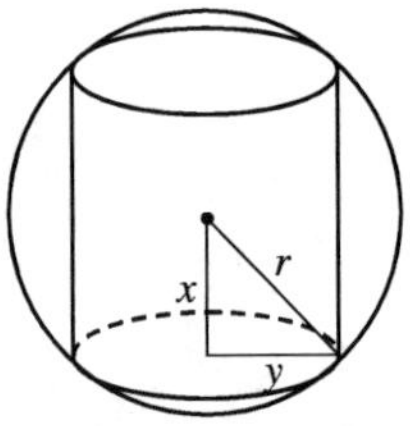

The cylinder has surface area

2(area of the base) + (lateral surface area)

$= 2\pi(\text{radius})^2 + 2\pi(\text{radius})(\text{height}) = 2\pi y^2 + 2\pi y(2x)$.

Now $x^2 + y^2 = r^2 \;\Rightarrow\; y^2 = r^2 - x^2 \;\Rightarrow\; y = \sqrt{r^2 - x^2}$, so the surface area is

$$S(x) = 2\pi\left(r^2 - x^2\right) + 4\pi x\sqrt{r^2 - x^2}, \quad 0 \le x \le r$$
$$= 2\pi r^2 - 2\pi x^2 + 4\pi\left(x\sqrt{r^2 - x^2}\right)$$

Thus, $S'(x) = 0 - 4\pi x + 4\pi\left[x \cdot \frac{1}{2}\left(r^2 - x^2\right)^{-1/2}(-2x) + \left(r^2 - x^2\right)^{1/2} \cdot 1\right]$

$$= 4\pi\left[-x - \frac{x^2}{\sqrt{r^2 - x^2}} + \sqrt{r^2 - x^2}\right] = 4\pi \cdot \frac{-x\sqrt{r^2 - x^2} - x^2 + r^2 - x^2}{\sqrt{r^2 - x^2}}$$

$S'(x) = 0 \;\Rightarrow\; x\sqrt{r^2 - x^2} = r^2 - 2x^2 \quad (\star) \;\Rightarrow\; \left(x\sqrt{r^2 - x^2}\right)^2 = \left(r^2 - 2x^2\right)^2 \;\Rightarrow\;$ $x^2\left(r^2 - x^2\right) = r^4 - 4r^2x^2 + 4x^4 \;\Rightarrow\; r^2x^2 - x^4 = r^4 - 4r^2x^2 + 4x^4 \;\Rightarrow\; 5x^4 - 5r^2x^2 + r^4 = 0$.

This is a quadratic equation in x^2. By the quadratic formula, $x^2 = \frac{5 \pm \sqrt{5}}{10}r^2$, but we reject the root with the + sign since it doesn't satisfy $(\star)$. [The right side is negative and the left side is positive.] So $x = \sqrt{\frac{5 - \sqrt{5}}{10}}\,r$. Since $S(0) = S(r) = 0$, the maximum surface area occurs at the critical number and $x^2 = \frac{5 - \sqrt{5}}{10}r^2 \;\Rightarrow\;$ $y^2 = r^2 - \frac{5 - \sqrt{5}}{10}r^2 = \frac{5 + \sqrt{5}}{10}r^2 \;\Rightarrow\;$ the surface area is

$$2\pi\left(\frac{5+\sqrt{5}}{10}\right)r^2 + 4\pi\sqrt{\frac{5 - \sqrt{5}}{10}}\sqrt{\frac{5 + \sqrt{5}}{10}}\,r^2 = \pi r^2\left[2 \cdot \frac{5+\sqrt{5}}{10} + 4\frac{\sqrt{\left(5-\sqrt{5}\right)\left(5+\sqrt{5}\right)}}{10}\right] = \pi r^2\left[\frac{5+\sqrt{5}}{5} + \frac{2\sqrt{20}}{5}\right] =$$
$$\pi r^2\left[\frac{5+\sqrt{5}+2\cdot 2\sqrt{5}}{5}\right] = \pi r^2\left[\frac{5+5\sqrt{5}}{5}\right] = \pi r^2\left(1 + \sqrt{5}\right).$$

29.

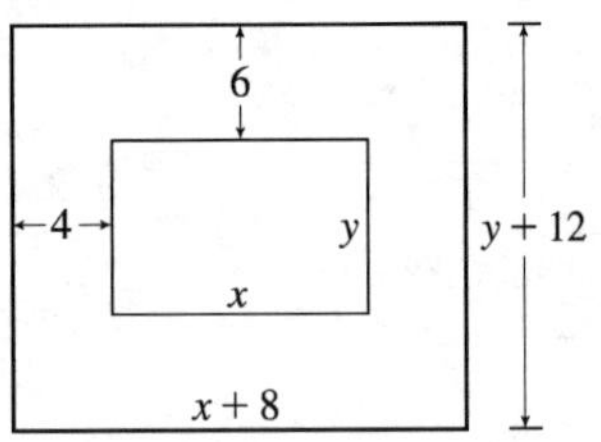

$xy = 384 \;\Rightarrow\; y = 384/x$. Total area is $A(x) = (8 + x)(12 + 384/x) = 12(40 + x + 256/x)$, so $A'(x) = 12\left(1 - 256/x^2\right) = 0 \;\Rightarrow\; x = 16$. There is an absolute minimum when $x = 16$ since $A'(x) < 0$ for $0 < x < 16$ and $A'(x) > 0$ for $x > 16$. When $x = 16$, $y = 384/16 = 24$, so the dimensions are 24 cm and 36 cm.

31. 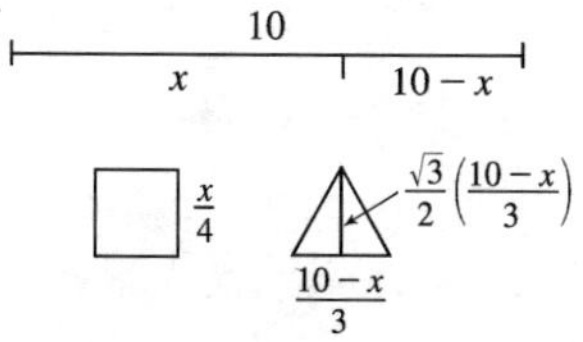

Let x be the length of the wire used for the square. The total area is

$$A(x) = \left(\frac{x}{4}\right)^2 + \frac{1}{2}\left(\frac{10-x}{3}\right)\frac{\sqrt{3}}{2}\left(\frac{10-x}{3}\right)$$

$$= \tfrac{1}{16}x^2 + \tfrac{\sqrt{3}}{36}(10-x)^2, \quad 0 \le x \le 10$$

$A'(x) = \frac{1}{8}x - \frac{\sqrt{3}}{18}(10-x) = 0 \quad \Leftrightarrow \quad \frac{9}{72}x + \frac{4\sqrt{3}}{72}x - \frac{40\sqrt{3}}{72} = 0 \quad \Leftrightarrow \quad x = \frac{40\sqrt{3}}{9+4\sqrt{3}}$. Now

$A(0) = \left(\frac{\sqrt{3}}{36}\right)100 \approx 4.81$, $A(10) = \frac{100}{16} = 6.25$ and $A\left(\frac{40\sqrt{3}}{9+4\sqrt{3}}\right) \approx 2.72$, so

(a) The maximum area occurs when $x = 10$ m, and all the wire is used for the square.

(b) The minimum area occurs when $x = \frac{40\sqrt{3}}{9+4\sqrt{3}} \approx 4.35$ m.

33. 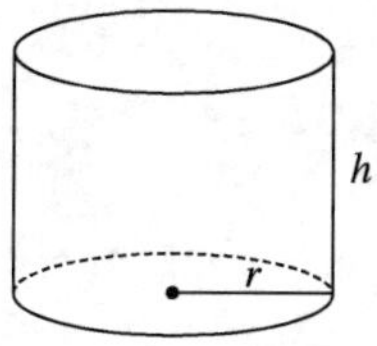

The volume is $V = \pi r^2 h$ and the surface area is

$$S(r) = \pi r^2 + 2\pi r h = \pi r^2 + 2\pi r\left(\frac{V}{\pi r^2}\right) = \pi r^2 + \frac{2V}{r}.$$

$$S'(r) = 2\pi r - \frac{2V}{r^2} = 0 \;\Rightarrow\; 2\pi r^3 = 2V \;\Rightarrow\; r = \sqrt[3]{\frac{V}{\pi}} \text{ cm.}$$

This gives an absolute minimum since $S'(r) < 0$ for $0 < r < \sqrt[3]{\frac{V}{\pi}}$ and $S'(r) > 0$ for $r > \sqrt[3]{\frac{V}{\pi}}$. When

$r = \sqrt[3]{\frac{V}{\pi}}$, $h = \frac{V}{\pi r^2} = \frac{V}{\pi (V/\pi)^{2/3}} = \sqrt[3]{\frac{V}{\pi}}$ cm.

35. 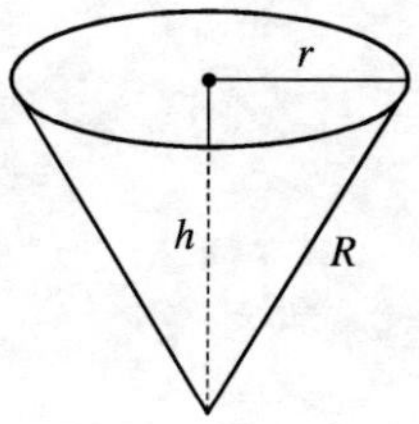

$h^2 + r^2 = R^2 \;\Rightarrow\; V = \frac{\pi}{3}r^2 h = \frac{\pi}{3}(R^2 - h^2)h = \frac{\pi}{3}(R^2 h - h^3)$.

$V'(h) = \frac{\pi}{3}(R^2 - 3h^2) = 0$ when $h = \frac{1}{\sqrt{3}}R$. This gives an absolute maximum, since $V'(h) > 0$ for $0 < h < \frac{1}{\sqrt{3}}R$ and $V'(h) < 0$ for $h > \frac{1}{\sqrt{3}}R$. The maximum volume is

$$V\left(\tfrac{1}{\sqrt{3}}R\right) = \tfrac{\pi}{3}\left(\tfrac{1}{\sqrt{3}}R^3 - \tfrac{1}{3\sqrt{3}}R^3\right) = \tfrac{2}{9\sqrt{3}}\pi R^3.$$

37. 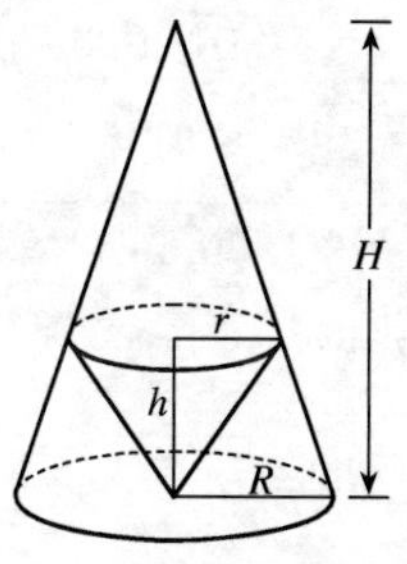

By similar triangles, $\frac{H}{R} = \frac{H-h}{r}$ **(1)**. The volume of the inner cone is $V = \frac{1}{3}\pi r^2 h$, so we'll solve **(1)** for h. $\frac{Hr}{R} = H - h \;\Rightarrow$

$$h = H - \frac{Hr}{R} = \frac{HR - Hr}{R} = \frac{H}{R}(R - r) \quad \textbf{(2)}.$$

Thus, $V(r) = \frac{\pi}{3}r^2 \cdot \frac{H}{R}(R - r) = \frac{\pi H}{3R}(Rr^2 - r^3) \;\Rightarrow$

$$V'(r) = \frac{\pi H}{3R}(2Rr - 3r^2) = \frac{\pi H}{3R}r(2R - 3r).$$

$V'(r) = 0 \;\Rightarrow\; r = 0$ or $2R = 3r \;\Rightarrow\; r = \frac{2}{3}R$ and from **(2)**, $h = \frac{H}{R}\left(R - \frac{2}{3}R\right) = \frac{H}{R}\left(\frac{1}{3}R\right) = \frac{1}{3}H$.

$V'(r)$ changes from positive to negative at $r = \frac{2}{3}R$, so the inner cone has a maximum volume of

$V = \frac{1}{3}\pi r^2 h = \frac{1}{3}\pi\left(\frac{2}{3}R\right)^2\left(\frac{1}{3}H\right) = \frac{4}{27}\cdot\frac{1}{3}\pi R^2 H$, which is approximately 15% of the volume of the larger cone.

39. $S = 6sh - \frac{3}{2}s^2 \cot\theta + 3s^2 \frac{\sqrt{3}}{2} \csc\theta$

(a) $\dfrac{dS}{d\theta} = \frac{3}{2}s^2 \csc^2\theta - 3s^2 \frac{\sqrt{3}}{2} \csc\theta \cot\theta$ or $\frac{3}{2}s^2 \csc\theta \left(\csc\theta - \sqrt{3}\cot\theta\right)$.

(b) $\dfrac{dS}{d\theta} = 0$ when $\csc\theta - \sqrt{3}\cot\theta = 0 \;\Rightarrow\; \dfrac{1}{\sin\theta} - \sqrt{3}\,\dfrac{\cos\theta}{\sin\theta} = 0 \;\Rightarrow\; \cos\theta = \frac{1}{\sqrt{3}}$. The First Derivative Test shows that the minimum surface area occurs when $\theta = \cos^{-1}\left(\frac{1}{\sqrt{3}}\right) \approx 55^\circ$.

(c)

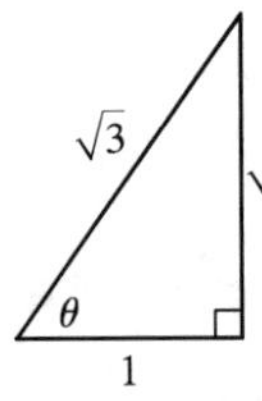

If $\cos\theta = \frac{1}{\sqrt{3}}$, then $\cot\theta = \frac{1}{\sqrt{2}}$ and $\csc\theta = \frac{\sqrt{3}}{\sqrt{2}}$, so the surface area is

$$S = 6sh - \tfrac{3}{2}s^2 \tfrac{1}{\sqrt{2}} + 3s^2 \tfrac{\sqrt{3}}{2} \tfrac{\sqrt{3}}{\sqrt{2}} = 6sh - \tfrac{3}{2\sqrt{2}}s^2 + \tfrac{9}{2\sqrt{2}}s^2$$

$$= 6sh + \tfrac{6}{2\sqrt{2}}s^2 = 6s\left(h + \tfrac{1}{2\sqrt{2}}s\right)$$

41. Here $T(x) = \dfrac{\sqrt{x^2+25}}{6} + \dfrac{5-x}{8}, 0 \le x \le 5 \;\Rightarrow\; T'(x) = \dfrac{x}{6\sqrt{x^2+25}} - \dfrac{1}{8} = 0 \;\Leftrightarrow\; 8x = 6\sqrt{x^2+25}$ $\Leftrightarrow\; 16x^2 = 9(x^2+25) \;\Leftrightarrow\; x = \frac{15}{\sqrt{7}}$. But $\frac{15}{\sqrt{7}} > 5$, so T has no critical number. Since $T(0) \approx 1.46$ and $T(5) \approx 1.18$, he should row directly to B.

43.

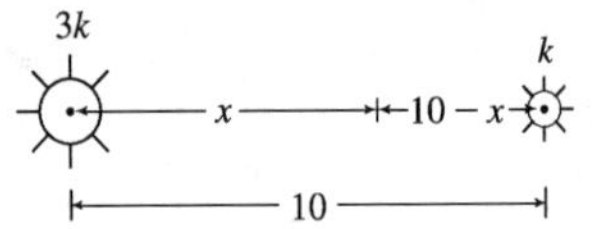

The total illumination is $I(x) = \dfrac{3k}{x^2} + \dfrac{k}{(10-x)^2}, 0 < x < 10$. Then

$I'(x) = \dfrac{-6k}{x^3} + \dfrac{2k}{(10-x)^3} = 0 \;\Rightarrow\; 6k(10-x)^3 = 2kx^3 \;\Rightarrow$

$3(10-x)^3 = x^3 \;\Rightarrow\; \sqrt[3]{3}\,(10-x) = x \;\Rightarrow\; 10\sqrt[3]{3} - \sqrt[3]{3}\,x = x$

$\Rightarrow\; 10\sqrt[3]{3} = x + \sqrt[3]{3}\,x \;\Rightarrow\; 10\sqrt[3]{3} = \left(1 + \sqrt[3]{3}\right)x \;\Rightarrow$

$x = \dfrac{10\sqrt[3]{3}}{1+\sqrt[3]{3}} \approx 5.9$ ft. This gives a minimum since $I''(x) > 0$ for $0 < x < 10$.

45.

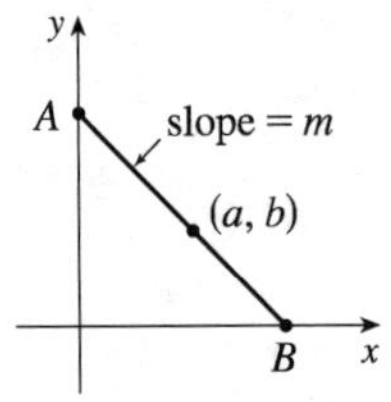

Every line segment in the first quadrant passing through (a,b) with endpoints on the x- and y-axes satisfies an equation of the form $y - b = m(x-a)$, where $m < 0$. By setting $x = 0$ and then $y = 0$, we find its endpoints, $A(0, b - am)$ and $B\left(a - \frac{b}{m}, 0\right)$. The distance d from A to B is given by $d = \sqrt{\left[\left(a - \frac{b}{m}\right) - 0\right]^2 + [0 - (b - am)]^2}$.

It follows that the square of the length of the line segment, as a function of m, is given by

$$S(m) = \left(a - \frac{b}{m}\right)^2 + (am-b)^2 = a^2 - \frac{2ab}{m} + \frac{b^2}{m^2} + a^2m^2 - 2abm + b^2.$$ Thus,

$$S'(m) = \frac{2ab}{m^2} - \frac{2b^2}{m^3} + 2a^2m - 2ab = \frac{2}{m^3}(abm - b^2 + a^2m^4 - abm^3)$$
$$= \frac{2}{m^3}[b(am-b) + am^3(am-b)] = \frac{2}{m^3}(am-b)(b+am^3)$$

Thus, $S'(m) = 0 \;\Leftrightarrow\; m = b/a$ or $m = -\sqrt[3]{\frac{b}{a}}$. Since $b/a > 0$ and $m < 0$, m must equal $-\sqrt[3]{\frac{b}{a}}$. Since

$\dfrac{2}{m^3} < 0$, we see that $S'(m) < 0$ for $m < -\sqrt[3]{\frac{b}{a}}$ and $S'(m) > 0$ for $m > -\sqrt[3]{\frac{b}{a}}$. Thus, S has its absolute minimum value when $m = -\sqrt[3]{\frac{b}{a}}$. That value is

$$S\left(-\sqrt[3]{\tfrac{b}{a}}\right) = \left(a + b\sqrt[3]{\tfrac{a}{b}}\right)^2 + \left(-a\sqrt[3]{\tfrac{b}{a}} - b\right)^2 = \left(a + \sqrt[3]{ab^2}\right)^2 + \left(\sqrt[3]{a^2b} + b\right)^2$$

$$= a^2 + 2a^{4/3}b^{2/3} + a^{2/3}b^{4/3} + a^{4/3}b^{2/3} + 2a^{2/3}b^{4/3} + b^2 = a^2 + 3a^{4/3}b^{2/3} + 3a^{2/3}b^{4/3} + b^2$$

The last expression is of the form $x^3 + 3x^2y + 3xy^2 + y^3 \quad [= (x+y)^3]$ with $x = a^{2/3}$ and $y = b^{2/3}$, so we can write it as $(a^{2/3} + b^{2/3})^3$ and the shortest such line segment has length $\sqrt{S} = (a^{2/3} + b^{2/3})^{3/2}$.

47.

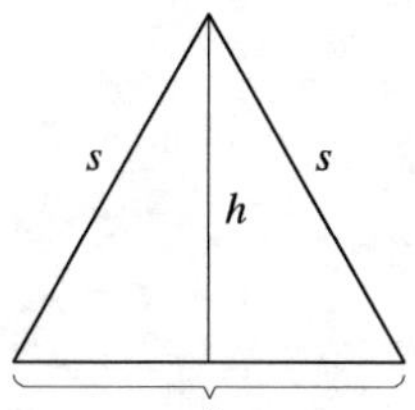

Here $s^2 = h^2 + b^2/4$, so $h^2 = s^2 - b^2/4$. The area is $A = \frac{1}{2}b\sqrt{s^2 - b^2/4}$. Let the perimeter be p, so $2s + b = p$ or $s = (p - b)/2 \Rightarrow$ $A(b) = \frac{1}{2}b\sqrt{(p-b)^2/4 - b^2/4} = b\sqrt{p^2 - 2pb}/4$. Now

$$A'(b) = \frac{\sqrt{p^2 - 2pb}}{4} - \frac{bp/4}{\sqrt{p^2 - 2pb}} = \frac{-3pb + p^2}{4\sqrt{p^2 - 2pb}}.$$ Therefore, $A'(b) = 0 \Rightarrow$ $-3pb + p^2 = 0 \Rightarrow b = p/3$. Since $A'(b) > 0$ for $b < p/3$ and $A'(b) < 0$ for $b > p/3$, there is an absolute maximum when $b = p/3$. But then $2s + p/3 = p$, so $s = p/3 \Rightarrow s = b \Rightarrow$ the triangle is equilateral.

49. Note that $|AD| = |AP| + |PD| \Rightarrow 5 = x + |PD| \Rightarrow |PD| = 5 - x$. Using the Pythagorean Theorem for ΔPDB and ΔPDC gives us

$$L(x) = |AP| + |BP| + |CP| = x + \sqrt{(5-x)^2 + 2^2} + \sqrt{(5-x)^2 + 3^2}$$

$$= x + \sqrt{x^2 - 10x + 29} + \sqrt{x^2 - 10x + 34}$$

$$\Rightarrow \quad L'(x) = 1 + \frac{x-5}{\sqrt{x^2 - 10x + 29}} + \frac{x-5}{\sqrt{x^2 - 10x + 34}}.$$

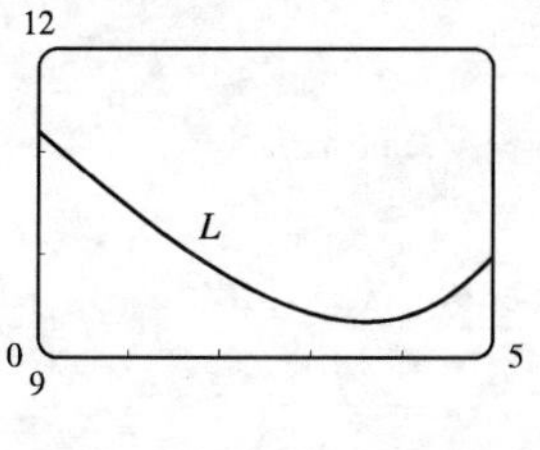

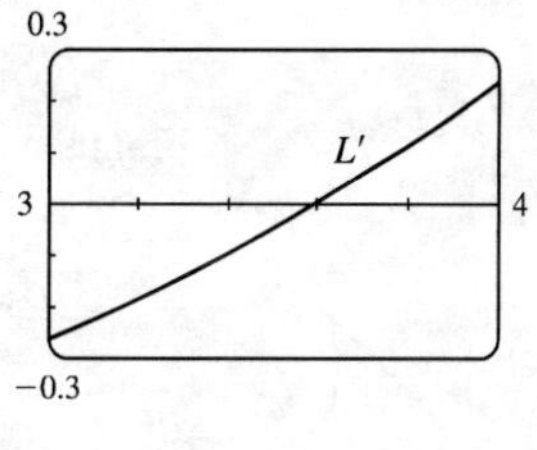

From the graphs of L and L', it seems that the minimum value of L is about $L(3.59) = 9.35$ m.

51.

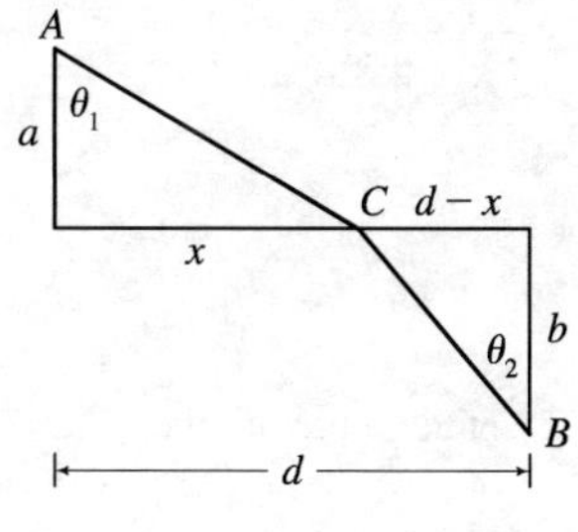

The total time is

$$T(x) = (\text{time from } A \text{ to } C) + (\text{time from } C \text{ to } B)$$

$$= \frac{\sqrt{a^2 + x^2}}{v_1} + \frac{\sqrt{b^2 + (d-x)^2}}{v_2}, \quad 0 < x < d$$

$$T'(x) = \frac{x}{v_1\sqrt{a^2 + x^2}} - \frac{d - x}{v_2\sqrt{b^2 + (d-x)^2}} = \frac{\sin\theta_1}{v_1} - \frac{\sin\theta_2}{v_2}$$

The minimum occurs when $T'(x) = 0 \Rightarrow \dfrac{\sin\theta_1}{v_1} = \dfrac{\sin\theta_2}{v_2}$.

[*Note:* $T''(x) > 0$]

53.

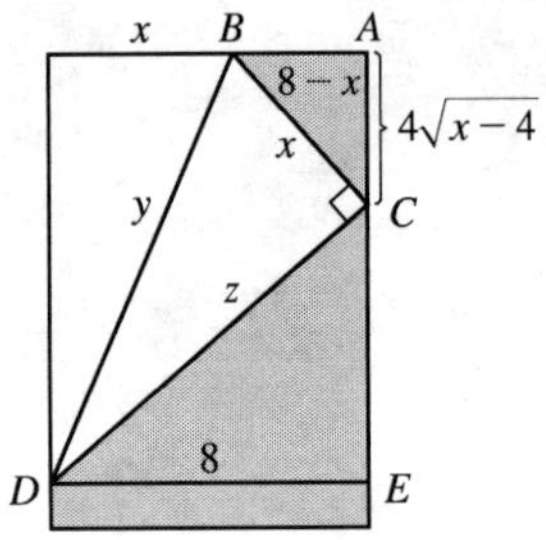

$y^2 = x^2 + z^2$, but triangles CDE and BCA are similar, so $z/8 = x/(4\sqrt{x-4}) \;\Rightarrow\; z = 2x/\sqrt{x-4}$. Thus, we minimize $f(x) = y^2 = x^2 + 4x^2/(x-4) = x^3/(x-4)$, $4 < x \leq 8$.

$$f'(x) = \frac{(x-4)(3x^2) - x^3}{(x-4)^2} = \frac{x^2[3(x-4) - x]}{(x-4)^2} = \frac{2x^2(x-6)}{(x-4)^2} = 0$$

when $x = 6$. $f'(x) < 0$ when $x < 6$, $f'(x) > 0$ when $x > 6$, so the minimum occurs when $x = 6$ in.

55.

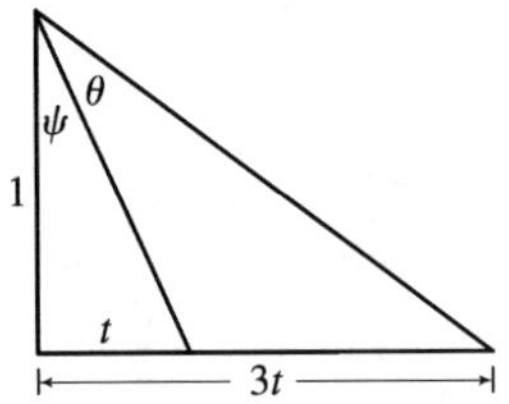

It suffices to maximize $\tan\theta$. Now

$$\frac{3t}{1} = \tan(\psi + \theta) = \frac{\tan\psi + \tan\theta}{1 - \tan\psi\,\tan\theta} = \frac{t + \tan\theta}{1 - t\tan\theta}.$$

So $3t(1 - t\tan\theta) = t + \tan\theta \;\Rightarrow\; 2t = (1 + 3t^2)\tan\theta \;\Rightarrow$

$\tan\theta = \dfrac{2t}{1+3t^2}$. Let $f(t) = \tan\theta = \dfrac{2t}{1+3t^2} \;\Rightarrow$

$$f'(t) = \frac{2(1+3t^2) - 2t(6t)}{(1+3t^2)^2} = \frac{2(1-3t^2)}{(1+3t^2)^2} = 0 \;\Leftrightarrow$$

$1 - 3t^2 = 0 \;\Leftrightarrow\; t = \frac{1}{\sqrt{3}}$ since $t \geq 0$.

Now $f'(t) > 0$ for $0 \leq t < \frac{1}{\sqrt{3}}$ and $f'(t) < 0$ for $t > \frac{1}{\sqrt{3}}$, so f has an absolute maximum when $t = \frac{1}{\sqrt{3}}$ and $\tan\theta = \dfrac{2(1/\sqrt{3})}{1 + 3(1/\sqrt{3})^2} = \frac{1}{\sqrt{3}} \;\Rightarrow\; \theta = \frac{\pi}{6}$. Substituting for t and θ in $3t = \tan(\psi + \theta)$ gives us $\sqrt{3} = \tan(\psi + \frac{\pi}{6}) \;\Rightarrow\; \psi = \frac{\pi}{6}$.

57.

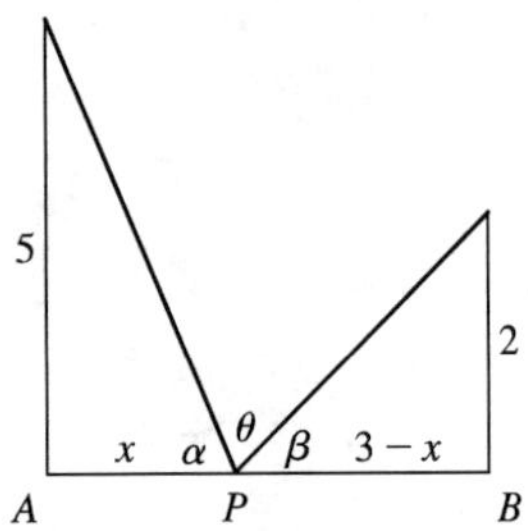

From the figure, $\tan\alpha = \dfrac{5}{x}$ and $\tan\beta = \dfrac{2}{3-x}$. Since

$$\alpha + \beta + \theta = 180^\circ = \pi,\; \theta = \pi - \tan^{-1}\left(\frac{5}{x}\right) - \tan^{-1}\left(\frac{2}{3-x}\right) \;\Rightarrow$$

$$\frac{d\theta}{dx} = -\frac{1}{1 + \left(\frac{5}{x}\right)^2}\left(-\frac{5}{x^2}\right) - \frac{1}{1 + \left(\frac{2}{3-x}\right)^2}\left[\frac{2}{(3-x)^2}\right]$$

$$= \frac{x^2}{x^2+25}\cdot\frac{5}{x^2} - \frac{(3-x)^2}{(3-x)^2+4}\cdot\frac{2}{(3-x)^2}.$$

Now $\dfrac{d\theta}{dx} = 0 \;\Rightarrow\; \dfrac{5}{x^2+25} = \dfrac{2}{x^2 - 6x + 13} \;\Rightarrow\; 2x^2 + 50 = 5x^2 - 30x + 65 \;\Rightarrow$ $3x^2 - 30x + 15 = 0 \;\Rightarrow\; x^2 - 10x + 5 = 0 \;\Rightarrow\; x = 5 \pm 2\sqrt{5}$. We reject the root with the + sign, since it is larger than 3. $d\theta/dx > 0$ for $x < 5 - 2\sqrt{5}$ and $d\theta/dx < 0$ for $x > 5 - 2\sqrt{5}$, so θ is maximized when $|AP| = x = 5 - 2\sqrt{5} \approx 0.53$.

59. 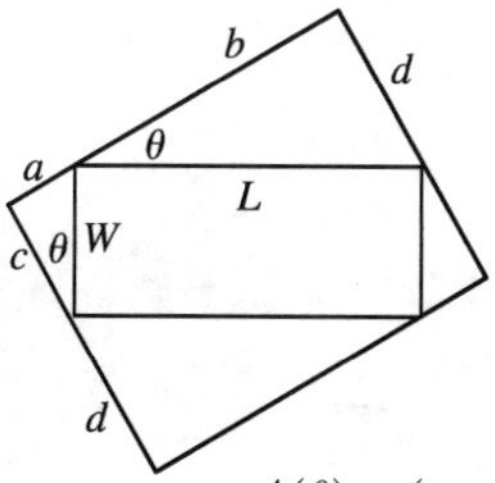

In the small triangle with sides a and c and hypotenuse W, $\sin\theta = \dfrac{a}{W}$ and $\cos\theta = \dfrac{c}{W}$. In the triangle with sides b and d and hypotenuse L, $\sin\theta = \dfrac{d}{L}$ and $\cos\theta = \dfrac{b}{L}$. Thus, $a = W\sin\theta$, $c = W\cos\theta$, $d = L\sin\theta$, and $b = L\cos\theta$, so the area of the circumscribed rectangle is

$$\begin{aligned}
A(\theta) &= (a+b)(c+d) = (W\sin\theta + L\cos\theta)(W\cos\theta + L\sin\theta)\\
&= W^2\sin\theta\,\cos\theta + WL\sin^2\theta + LW\cos^2\theta + L^2\sin\theta\,\cos\theta\\
&= LW\sin^2\theta + LW\cos^2\theta + (L^2+W^2)\sin\theta\,\cos\theta\\
&= LW(\sin^2\theta + \cos^2\theta) + (L^2+W^2)\cdot\tfrac{1}{2}\cdot 2\sin\theta\,\cos\theta\\
&= LW + \tfrac{1}{2}(L^2+W^2)\sin 2\theta, \quad 0 \le \theta \le \tfrac{\pi}{2}
\end{aligned}$$

This expression shows, without calculus, that the maximum value of $A(\theta)$ occurs when $\sin 2\theta = 1 \Leftrightarrow 2\theta = \frac{\pi}{2}$ $\Rightarrow \theta = \frac{\pi}{4}$. So the maximum area is $A\left(\frac{\pi}{4}\right) = LW + \frac{1}{2}(L^2+W^2) = \frac{1}{2}(L^2 + 2LW + W^2) = \frac{1}{2}(L+W)^2$.

61. (a) 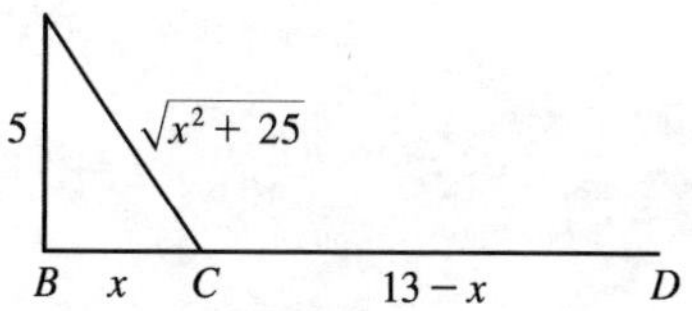

If k = energy/km over land, then energy/km over water $= 1.4k$. So the total energy is $E = 1.4k\sqrt{25+x^2} + k(13-x)$, $0 \le x \le 13$, and so $\dfrac{dE}{dx} = \dfrac{1.4kx}{(25+x^2)^{1/2}} - k$.

Set $\dfrac{dE}{dx} = 0$: $1.4kx = k(25+x^2)^{1/2} \;\Rightarrow\; 1.96x^2 = x^2 + 25 \;\Rightarrow\; 0.96x^2 = 25 \;\Rightarrow\; x = \frac{5}{\sqrt{0.96}} \approx 5.1$.
Testing against the value of E at the endpoints: $E(0) = 1.4k(5) + 13k = 20k$, $E(5.1) \approx 17.9k$, $E(13) \approx 19.5k$. Thus, to minimize energy, the bird should fly to a point about 5.1 km from B.

(b) If W/L is large, the bird would fly to a point C that is closer to B than to D to minimize the energy used flying over water. If W/L is small, the bird would fly to a point C that is closer to D than to B to minimize the distance of the flight. $E = W\sqrt{25+x^2} + L(13-x) \;\Rightarrow\; \dfrac{dE}{dx} = \dfrac{Wx}{\sqrt{25+x^2}} - L = 0$ when $\dfrac{W}{L} = \dfrac{\sqrt{25+x^2}}{x}$. By the same sort of argument as in part (a), this ratio will give the minimal expenditure of energy if the bird heads for the point x km from B.

(c) For flight direct to D, $x = 13$, so from part (b), $W/L = \frac{\sqrt{25+13^2}}{13} \approx 1.07$. There is no value of W/L for which the bird should fly directly to B. But note that $\lim\limits_{x\to 0^+}(W/L) = \infty$, so if the point at which E is a minimum is close to B, then W/L is large.

(d) Assuming that the birds instinctively choose the path that minimizes the energy expenditure, we can use the equation for $dE/dx = 0$ from part (a) with $1.4k = c$, $x = 4$, and $k = 1$: $(c)(4) = 1\cdot(25+4^2)^{1/2} \;\Rightarrow$ $c = \sqrt{41}/4 \approx 1.6$.

Newton's Method

1. (a)

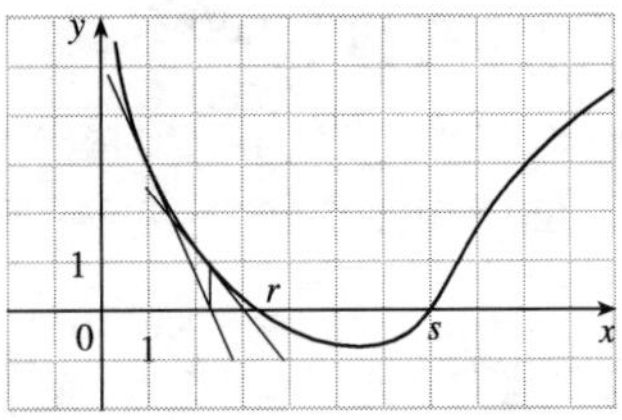

The tangent line at $x = 1$ intersects the x-axis at $x \approx 2.3$, so $x_2 \approx 2.3$. The tangent line at $x = 2.3$ intersects the x-axis at $x \approx 3$, so $x_3 \approx 3.0$.

(b) $x_1 = 5$ would *not* be a better first approximation than $x_1 = 1$ since the tangent line is nearly horizontal. In fact, the second approximation for $x_1 = 5$ appears to be to the left of $x = 1$.

3. Since $x_1 = 3$ and $y = 5x - 4$ is tangent to $y = f(x)$ at $x = 3$, we simply need to find where the tangent line intersects the x-axis. $y = 0 \Rightarrow 5x_2 - 4 = 0 \Rightarrow x_2 = \frac{4}{5}$.

5. $f(x) = x^3 + 2x - 4 \Rightarrow f'(x) = 3x^2 + 2$, so $x_{n+1} = x_n - \dfrac{x_n^3 + 2x_n - 4}{3x_n^2 + 2}$. Now $x_1 = 1 \Rightarrow$

$$x_2 = 1 - \frac{1 + 2 - 4}{3 \cdot 1^2 + 2} = 1 - \frac{-1}{5} = 1.2 \Rightarrow x_3 = 1.2 - \frac{(1.2)^3 + 2(1.2) - 4}{3(1.2)^2 + 2} \approx 1.1797.$$

7. $f(x) = x^4 - 20 \Rightarrow f'(x) = 4x^3$, so $x_{n+1} = x_n - \dfrac{f(x_n)}{f'(x_n)} = x_n - \dfrac{x_n^4 - 20}{4x_n^3}$.

Now $x_1 = 2 \Rightarrow x_2 = 2 - \dfrac{2^4 - 20}{4(2)^3} = 2.125 \Rightarrow x_3 = 2.125 - \dfrac{(2.125)^4 - 20}{4(2.125)^3} \approx 2.1148.$

9. $f(x) = x^3 + x + 3 \quad \Rightarrow \quad f'(x) = 3x^2 + 1$, so

$x_{n+1} = x_n - \dfrac{x_n^3 + x_n + 3}{3x_n^2 + 1}$. Now $x_1 = -1 \quad \Rightarrow$

$x_2 = -1 - \dfrac{(-1)^3 + (-1) + 3}{3(-1)^2 + 1} = -1 - \dfrac{-1 - 1 + 3}{3 + 1} = -1 - \dfrac{1}{4} = -1.25.$

Newton's method follows the tangent line at $(-1, 1)$ down to its intersection with the x-axis at $(-1.25, 0)$, giving the second approximation $x_2 = -1.25$.

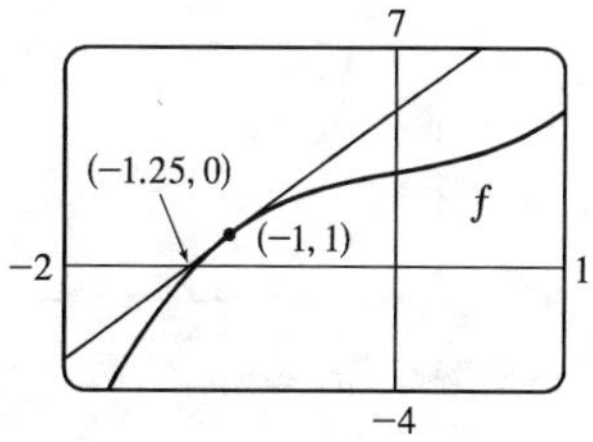

11. To approximate $x = \sqrt[3]{30}$ (so that $x^3 = 30$), we can take $f(x) = x^3 - 30$. So $f'(x) = 3x^2$, and thus, $x_{n+1} = x_n - \dfrac{x_n^3 - 30}{3x_n^2}$. Since $\sqrt[3]{27} = 3$ and 27 is close to 30, we'll use $x_1 = 3$. We need to find approximations until they agree to eight decimal places. $x_1 = 3 \quad \Rightarrow \quad x_2 \approx 3.11111111$, $x_3 \approx 3.10723734$, $x_4 \approx 3.10723251 \approx x_5$. So $\sqrt[3]{30} \approx 3.10723251$, to eight decimal places.

Here is a quick and easy method for finding the iterations for Newton's method on a programmable calculator. (The screens shown are from the TI-83 Plus, but the method is similar on other calculators.) Assign $f(x) = x^3 - 30$ to Y_1, and $f'(x) = 3x^2$ to Y_2. Now store $x_1 = 3$ in X and then enter $\mathrm{X} - \mathrm{Y}_1/\mathrm{Y}_2 \to \mathrm{X}$ to get $x_2 = 3.\overline{1}$. By successively pressing the ENTER key, you get the approximations $x_3, x_4, \ldots$.

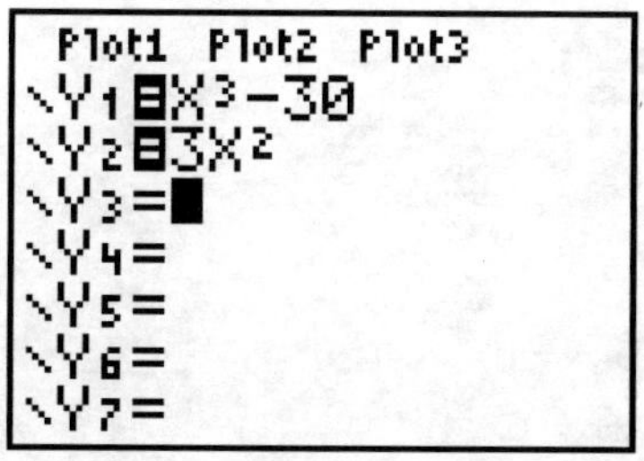

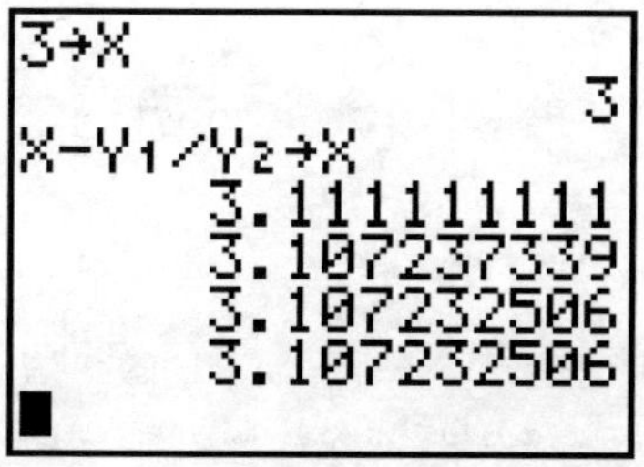

In Derive, load the utility file `SOLVE`. Enter `NEWTON(x^3-30,x,3)` and then `APPROXIMATE` to get [3, 3.11111111, 3.10723733, 3.10723250, 3.10723250]. You can request a specific iteration by adding a fourth argument. For example, `NEWTON(x^3-30,x,3,2)` gives [3, 3.11111111, 3.10723733].

In Maple, make the assignments $f := x \to x\hat{}3 - 30$;, $g := x \to x - f(x)/D(f)(x)$;, and $x := 3.$;. Repeatedly execute the command $x := g(x)$; to generate successive approximations.

In Mathematica, make the assignments $f[x\_] := x\hat{}3 - 30$, $g[x\_] := x - f[x]/f'[x]$, and $x = 3$. Repeatedly execute the command $x = g[x]$ to generate successive approximations.

13. $f(x) = 2x^3 - 6x^2 + 3x + 1 \quad \Rightarrow \quad f'(x) = 6x^2 - 12x + 3 \quad \Rightarrow \quad x_{n+1} = x_n - \dfrac{2x_n^3 - 6x_n^2 + 3x_n + 1}{6x_n^2 - 12x_n + 3}$. We need to find approximations until they agree to six decimal places. $x_1 = 2.5 \quad \Rightarrow \quad x_2 \approx 2.285714$, $x_3 \approx 2.228824$, $x_4 \approx 2.224765$, $x_5 \approx 2.224745 \approx x_6$. So the root is 2.224745, to six decimal places.

15. $\sin x = x^2$, so $f(x) = \sin x - x^2 \quad \Rightarrow \quad f'(x) = \cos x - 2x \quad \Rightarrow$

$x_{n+1} = x_n - \dfrac{\sin x_n - x_n^2}{\cos x_n - 2x_n}$. From the figure, the positive root of $\sin x = x^2$ is near 1. $x_1 = 1 \quad \Rightarrow \quad x_2 \approx 0.891396$, $x_3 \approx 0.876985$, $x_4 \approx 0.876726 \approx x_5$. So the positive root is 0.876726, to six decimal places.

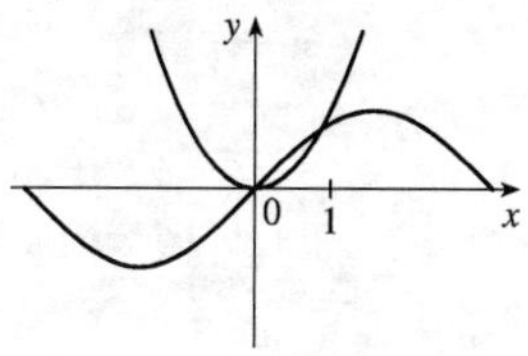

17. 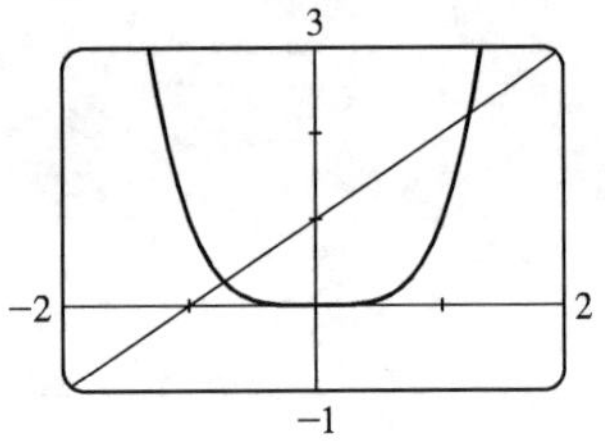

From the graph, we see that there appear to be points of intersection near $x = -0.7$ and $x = 1.2$. Solving $x^4 = 1 + x$ is the same as solving $f(x) = x^4 - x - 1 = 0$. $f(x) = x^4 - x - 1 \;\Rightarrow\; f'(x) = 4x^3 - 1$, so $x_{n+1} = x_n - \dfrac{x_n^4 - x_n - 1}{4x_n^3 - 1}$.

$x_1 = -0.7$	$x_1 = 1.2$
$x_2 \approx -0.725253$	$x_2 \approx 1.221380$
$x_3 \approx -0.724493$	$x_3 \approx 1.220745$
$x_4 \approx -0.724492 \approx x_5$	$x_4 \approx 1.220744 \approx x_5$

To six decimal places, the roots of the equation are -0.724492 and 1.220744.

19. 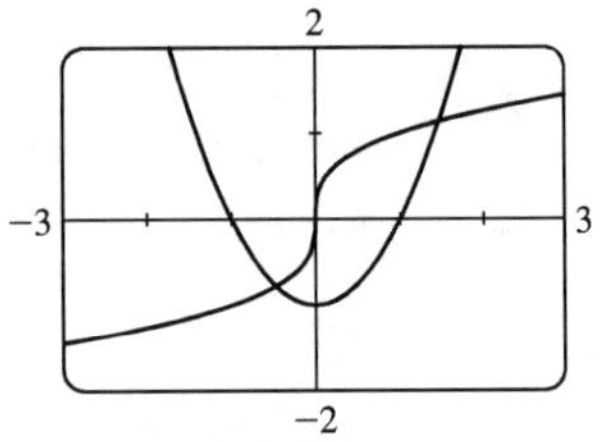

From the graph, we see that there appear to be points of intersection near $x = -0.5$ and $x = 1.5$. Solving $\sqrt[3]{x} = x^2 - 1$ is the same as solving $f(x) = \sqrt[3]{x} - x^2 + 1 = 0$. $f(x) = \sqrt[3]{x} - x^2 + 1 \;\Rightarrow$ $f'(x) = \frac{1}{3}x^{-2/3} - 2x$, so $x_{n+1} = x_n - \dfrac{\sqrt[3]{x_n} - x_n^2 + 1}{\frac{1}{3}x_n^{-2/3} - 2x_n}$.

$x_1 = -0.5$	$x_1 = 1.5$
$x_2 \approx -0.471421$	$x_2 \approx 1.461653$
$x_3 \approx -0.471074 \approx x_4$	$x_3 \approx 1.461070 \approx x_4$

To six decimal places, the roots are -0.471074 and 1.461070.

21. From the graph, there appears to be a point of intersection near $x = 0.6$. Solving $\cos x = \sqrt{x}$ is the same as solving $f(x) = \cos x - \sqrt{x} = 0$. $f(x) = \cos x - \sqrt{x} \;\Rightarrow\; f'(x) = -\sin x - 1/(2\sqrt{x})$, so $x_{n+1} = x_n - \dfrac{\cos x_n - \sqrt{x_n}}{-\sin x_n - 1/(2\sqrt{x})}$. Now $x_1 = 0.6 \;\Rightarrow$ $x_2 \approx 0.641928$, $x_3 \approx 0.641714 \approx x_4$. To six decimal places, the root of the equation is 0.641714.

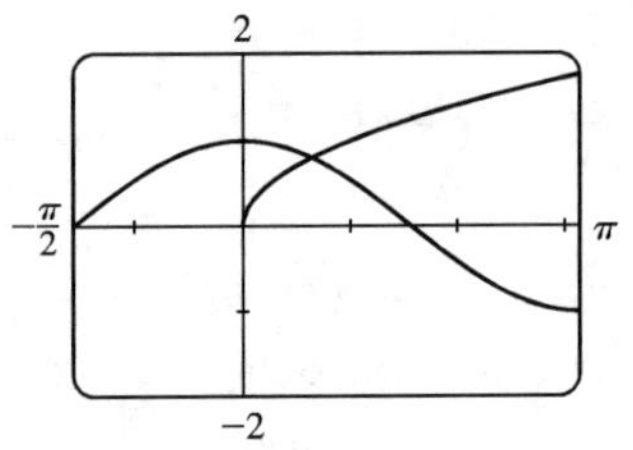

23. 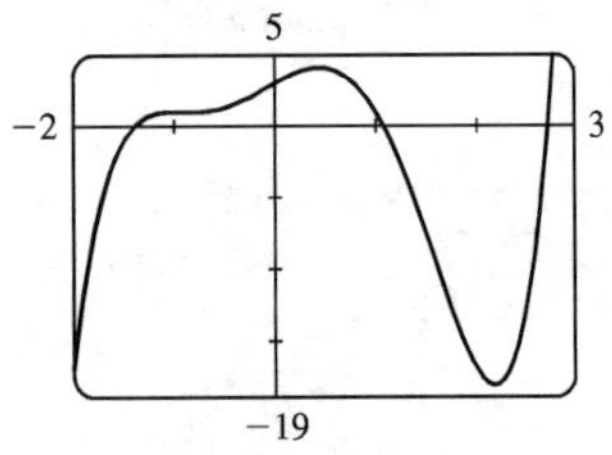

$f(x) = x^5 - x^4 - 5x^3 - x^2 + 4x + 3 \;\Rightarrow$
$f'(x) = 5x^4 - 4x^3 - 15x^2 - 2x + 4 \;\Rightarrow$
$x_{n+1} = x_n - \dfrac{x_n^5 - x_n^4 - 5x_n^3 - x_n^2 + 4x_n + 3}{5x_n^4 - 4x_n^3 - 15x_n^2 - 2x_n + 4}$. From the graph of f, there appear to be roots near -1.4, 1.1, and 2.7.

$x_1 = -1.4$	$x_1 = 1.1$	$x_1 = 2.7$
$x_2 \approx -1.39210970$	$x_2 \approx 1.07780402$	$x_2 \approx 2.72046250$
$x_3 \approx -1.39194698$	$x_3 \approx 1.07739442$	$x_3 \approx 2.71987870$
$x_4 \approx -1.39194691 \approx x_5$	$x_4 \approx 1.07739428 \approx x_5$	$x_4 \approx 2.71987822 \approx x_5$

To eight decimal places, the roots of the equation are -1.39194691, 1.07739428, and 2.71987822.

25. 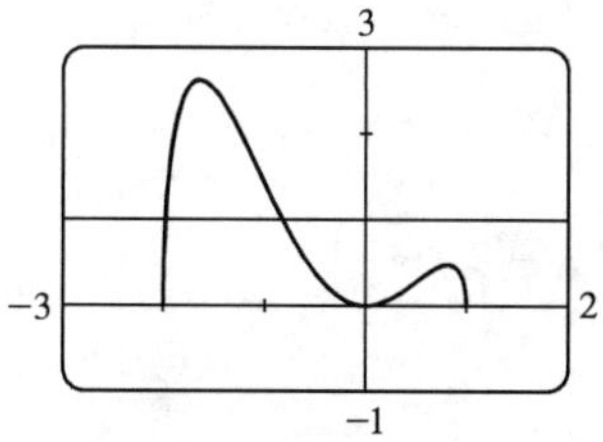

From the graph, $y = x^2\sqrt{2 - x - x^2}$ and $y = 1$ intersect twice, at $x \approx -2$ and at $x \approx -1$. $f(x) = x^2\sqrt{2 - x - x^2} - 1 \Rightarrow$

$$f'(x) = x^2 \cdot \tfrac{1}{2}(2 - x - x^2)^{-1/2}(-1 - 2x) + (2 - x - x^2)^{1/2} \cdot 2x$$
$$= \tfrac{1}{2}x(2 - x - x^2)^{-1/2}\left[x(-1 - 2x) + 4(2 - x - x^2)\right]$$
$$= \frac{x(8 - 5x - 6x^2)}{2\sqrt{(2 + x)(1 - x)}},$$

so $x_{n+1} = x_n - \dfrac{x_n^2\sqrt{2 - x_n - x_n^2} - 1}{\dfrac{x_n(8 - 5x_n - 6x_n^2)}{2\sqrt{(2 + x_n)(1 - x_n)}}}$. Trying $x_1 = -2$ won't work because $f'(-2)$ is undefined, so we'll try $x_1 = -1.95$.

$x_1 = -1.95$ | $x_1 = -0.8$
$x_2 \approx -1.98580357$ | $x_2 \approx -0.82674444$
$x_3 \approx -1.97899778$ | $x_3 \approx -0.82646236$
$x_4 \approx -1.97807848$ | $x_4 \approx -0.82646233 \approx x_5$
$x_5 \approx -1.97806682$
$x_6 \approx -1.97806681 \approx x_7$

To eight decimal places, the roots of the equation are -1.97806681 and -0.82646233.

27. (a) $f(x) = x^2 - a \Rightarrow f'(x) = 2x$, so Newton's method gives

$$x_{n+1} = x_n - \frac{x_n^2 - a}{2x_n} = x_n - \frac{1}{2}x_n + \frac{a}{2x_n} = \frac{1}{2}x_n + \frac{a}{2x_n} = \frac{1}{2}\left(x_n + \frac{a}{x_n}\right).$$

(b) Using (a) with $a = 1000$ and $x_1 = \sqrt{900} = 30$, we get $x_2 \approx 31.666667$, $x_3 \approx 31.622807$, and $x_4 \approx 31.622777 \approx x_5$. So $\sqrt{1000} \approx 31.622777$.

29. $f(x) = x^3 - 3x + 6 \Rightarrow f'(x) = 3x^2 - 3$. If $x_1 = 1$, then $f'(x_1) = 0$ and the tangent line used for approximating x_2 is horizontal. Attempting to find x_2 results in trying to divide by zero.

31. For $f(x) = x^{1/3}$, $f'(x) = \frac{1}{3}x^{-2/3}$ and

$$x_{n+1} = x_n - \frac{f(x_n)}{f'(x_n)} = x_n - \frac{x_n^{1/3}}{\frac{1}{3}x_n^{-2/3}} = x_n - 3x_n = -2x_n.$$

Therefore, each successive approximation becomes twice as large as the previous one in absolute value, so the sequence of approximations fails to converge to the root, which is 0. In the figure, we have $x_1 = 0.5$, $x_2 = -2(0.5) = -1$, and $x_3 = -2(-1) = 2$.

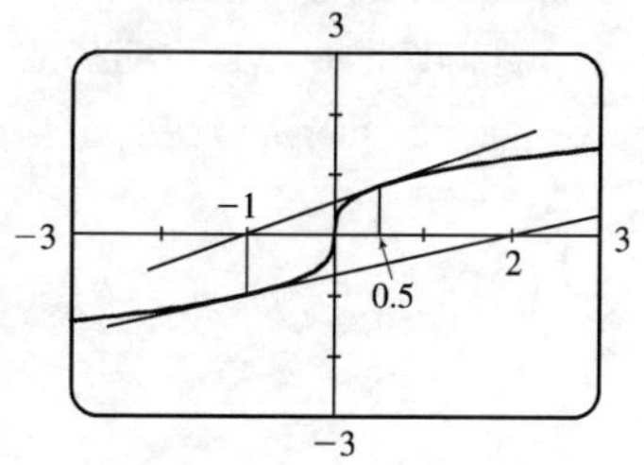

33. (a) $f(x) = 3x^4 - 28x^3 + 6x^2 + 24x \Rightarrow f'(x) = 12x^3 - 84x^2 + 12x + 24 \Rightarrow$

$f''(x) = 36x^2 - 168x + 12$. Now to solve $f'(x) = 0$, try $x_1 = \frac{1}{2} \Rightarrow x_2 = x_1 - \dfrac{f'(x_1)}{f''(x_1)} = \dfrac{2}{3} \Rightarrow$ $x_3 \approx 0.6455 \Rightarrow x_4 \approx 0.6452 \Rightarrow x_5 \approx 0.6452$. Now try $x_1 = 6 \Rightarrow x_2 = 7.12 \Rightarrow$ $x_3 \approx 6.8353 \Rightarrow x_4 \approx 6.8102 \Rightarrow x_5 \approx 6.8100$. Finally try $x_1 = -0.5 \Rightarrow x_2 \approx -0.4571 \Rightarrow$ $x_3 \approx -0.4552 \Rightarrow x_4 \approx -0.4552$. Therefore, $x = -0.455$, 6.810 and 0.645 are all critical numbers correct to three decimal places.

(b) $f(-1) = 13$, $f(7) = -1939$, $f(6.810) \approx -1949.07$, $f(-0.455) \approx -6.912$, $f(0.645) \approx 10.982$. Therefore, $f(6.810) \approx -1949.07$ is the absolute minimum correct to two decimal places.

35. $y = x^3 + \cos x \quad \Rightarrow \quad y' = 3x^2 - \sin x \quad \Rightarrow \quad y'' = 6x - \cos x \quad \Rightarrow \quad y''' = 6 + \sin x$. Now to solve $y'' = 0$, try $x_1 = 0$, and then $x_2 = x_1 - \dfrac{y''(x_1)}{y'''(x_1)} = \dfrac{1}{6} \quad \Rightarrow \quad x_3 \approx 0.164419 \approx x_4$. For $x < 0.164419$, $y'' < 0$, and for $x > 0.164419$, $y'' > 0$. Therefore, the point of inflection, correct to six decimal places, is $(0.164419, y(0.164419)) \approx (0.164419, 0.990958)$.

37. 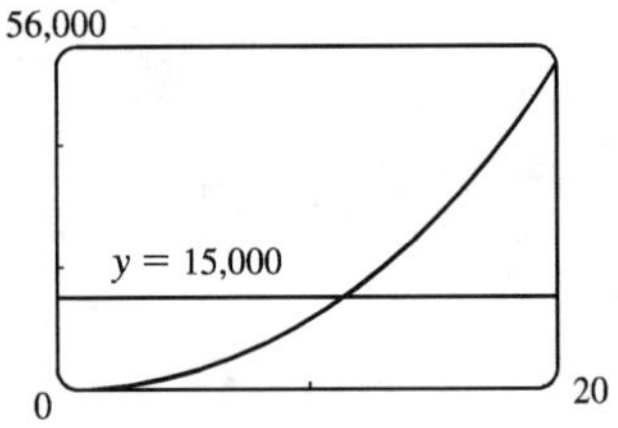

The volume of the silo, in terms of its radius, is $V(r) = \pi r^2(30) + \frac{1}{2}\left(\frac{4}{3}\pi r^3\right) = 30\pi r^2 + \frac{2}{3}\pi r^3$. From a graph of V, we see that $V(r) = 15{,}000$ at $r \approx 11$ ft. Now we use Newton's method to solve the equation $V(r) - 15{,}000 = 0$. $\dfrac{dV}{dr} = 60\pi r + 2\pi r^2$, so $r_{n+1} = r_n - \dfrac{30\pi r_n^2 + \frac{2}{3}\pi r_n^3 - 15{,}000}{60\pi r_n + 2\pi r_n^2}$. Taking $r_1 = 11$, we get $r_2 \approx 11.2853$, $r_3 \approx 11.2807 \approx r_4$. So in order for the silo to hold 15,000 ft^3 of grain, its radius must be about 11.2807 ft.

39. In this case, $A = 18{,}000$, $R = 375$, and $n = 5(12) = 60$. So the formula $A = \dfrac{R}{i}\left[1 - (1+i)^{-n}\right]$ becomes $18{,}000 = \dfrac{375}{x}\left[1 - (1+x)^{-60}\right] \quad \Leftrightarrow \quad 48x = 1 - (1+x)^{-60}$ [multiply each term by $(1+x)^{60}$] $\Leftrightarrow$ $48x(1+x)^{60} - (1+x)^{60} + 1 = 0$. Let the LHS be called $f(x)$, so that

$$\begin{aligned} f'(x) &= 48x(60)(1+x)^{59} + 48(1+x)^{60} - 60(1+x)^{59} \\ &= 12(1+x)^{59}\left[4x(60) + 4(1+x) - 5\right] = 12(1+x)^{59}(244x - 1) \end{aligned}$$

$x_{n+1} = x_n - \dfrac{48x_n(1+x_n)^{60} - (1+x_n)^{60} + 1}{12(1+x_n)^{59}(244x_n - 1)}$. An interest rate of 1% per month seems like a reasonable estimate for $x = i$. So let $x_1 = 1\% = 0.01$, and we get $x_2 \approx 0.0082202$, $x_3 \approx 0.0076802$, $x_4 \approx 0.0076291$, $x_5 \approx 0.0076286 \approx x_6$. Thus, the dealer is charging a monthly interest rate of 0.76286% (or 9.55% per year, compounded monthly).

Review

CONCEPT CHECK

1. A function f has an **absolute maximum** at $x = c$ if $f(c)$ is the largest function value on the entire domain of f, whereas f has a **local maximum** at c if $f(c)$ is the largest function value when x is near c. See Figure 4 in Section 4.1.

2. (a) See Theorem 4.1.3.

(b) See the Closed Interval Method before Example 8 in Section 4.1.

3. (a) See Theorem 4.1.4.

(b) See Definition 4.1.6.

4. (a) See Rolle's Theorem at the beginning of Section 4.2.

(b) See the Mean Value Theorem in Section 4.2. Geometric interpretation—there is some point P on the graph of a function f [on the interval (a, b)] where the tangent line is parallel to the secant line that connects $(a, f(a))$ and $(b, f(b))$.

5. (a) See the I/D Test before Example 1 in Section 4.3.

(b) See the Concavity Test before Example 4 in Section 4.3.

6. (a) See the First Derivative Test after Example 1 in Section 4.3.

(b) See the Second Derivative Test before Example 6 in Section 4.3.

(c) See the note before Example 7 in Section 4.3.

7. (a) See Definitions 4.4.1 and 4.4.5.

(b) See Definitions 4.4.2 and 4.4.6.

(c) See Definition 4.4.7.

(d) See Definition 4.4.3.

8. Without calculus you could get misleading graphs that fail to show the most interesting features of a function. See Examples 1–3 in Section 4.6.

9. (a) See Figure 3 in Section 4.9.

(b) $x_2 = x_1 - \dfrac{f(x_1)}{f'(x_1)}$

(c) $x_{n+1} = x_n - \dfrac{f(x_n)}{f'(x_n)}$

(d) Newton's method is likely to fail or to work very slowly when $f'(x_1)$ is close to 0.

10. (a) See the definition at the beginning of Section 4.10.

(b) If F_1 and F_2 are both antiderivatives of f on an interval I, then they differ by a constant.

TRUE-FALSE QUIZ

1. False. For example, take $f(x) = x^3$, then $f'(x) = 3x^2$ and $f'(0) = 0$, but $f(0) = 0$ is not a maximum or minimum; $(0, 0)$ is an inflection point.

3. False. For example, $f(x) = x$ is continuous on $(0, 1)$ but attains neither a maximum nor a minimum value on $(0, 1)$. Don't confuse this with f being continuous on the *closed* interval $[a, b]$, which would make the statement true.

5. True. This is an example of part (b) of the I/D Test.

7. False. $f'(x) = g'(x) \quad \Rightarrow \quad f(x) = g(x) + C$. For example, if $f(x) = x + 2$ and $g(x) = x + 1$, then $f'(x) = g'(x) = 1$, but $f(x) \neq g(x)$.

9. True. The graph of one such function is sketched.

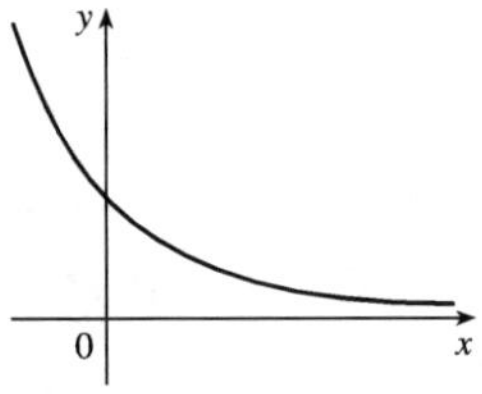

11. True. Let $x_1 < x_2$ where $x_1, x_2 \in I$. Then $f(x_1) < f(x_2)$ and $g(x_1) < g(x_2)$ (since f and g are increasing on I), so $(f + g)(x_1) = f(x_1) + g(x_1) < f(x_2) + g(x_2) = (f + g)(x_2)$.

13. False. Take $f(x) = x$ and $g(x) = x - 1$. Then both f and g are increasing on $(0, 1)$. But $f(x)g(x) = x(x - 1)$ is not increasing on $(0, 1)$.

15. True. Let $x_1, x_2 \in I$ and $x_1 < x_2$. Then $f(x_1) < f(x_2)$ (f is increasing) $\Rightarrow$ $\dfrac{1}{f(x_1)} > \dfrac{1}{f(x_2)}$ (f is positive) $\Rightarrow \quad g(x_1) > g(x_2) \quad \Rightarrow \quad g(x) = 1/f(x)$ is decreasing on I.

17. True. By the Mean Value Theorem, there exists a number c in $(0, 1)$ such that $f(1) - f(0) = f'(c)(1 - 0) = f'(c)$. Since $f'(c)$ is nonzero, $f(1) - f(0) \neq 0$, so $f(1) \neq f(0)$.

EXERCISES

1. $f(x) = 10 + 27x - x^3$, $0 \leq x \leq 4$. $f'(x) = 27 - 3x^2 = -3(x^2 - 9) = -3(x + 3)(x - 3) = 0$ only when $x = 3$ (since -3 is not in the domain). $f'(x) > 0$ for $x < 3$ and $f'(x) < 0$ for $x > 3$, so $f(3) = 64$ is a local maximum value. Checking the endpoints, we find $f(0) = 10$ and $f(4) = 54$. Thus, $f(0) = 10$ is the absolute minimum value and $f(3) = 64$ is the absolute maximum value.

3. $f(x) = \dfrac{x}{x^2 + x + 1}$, $-2 \leq x \leq 0$. $f'(x) = \dfrac{(x^2 + x + 1)(1) - x(2x + 1)}{(x^2 + x + 1)^2} = \dfrac{1 - x^2}{(x^2 + x + 1)^2} = 0 \quad \Leftrightarrow$ $x = -1$ (since 1 is not in the domain). $f'(x) < 0$ for $-2 < x < -1$ and $f'(x) > 0$ for $-1 < x < 0$, so $f(-1) = -1$ is a local and absolute minimum value. $f(-2) = -\frac{2}{3}$ and $f(0) = 0$, so $f(0) = 0$ is an absolute maximum value.

5. $f(x) = x + \sin 2x$, $[0, \pi]$. $f'(x) = 1 + 2\cos 2x = 0 \iff \cos 2x = -\frac{1}{2} \iff 2x = \frac{2\pi}{3}$ or $\frac{4\pi}{3} \iff x = \frac{\pi}{3}$ or $\frac{2\pi}{3}$. $f''(x) = -4\sin 2x$, so $f''\left(\frac{\pi}{3}\right) = -4\sin\frac{2\pi}{3} = -2\sqrt{3} < 0$ and $f''\left(\frac{2\pi}{3}\right) = -4\sin\frac{4\pi}{3} = 2\sqrt{3} > 0$, so $f\left(\frac{\pi}{3}\right) = \frac{\pi}{3} + \frac{\sqrt{3}}{2} \approx 1.91$ is a local maximum value and $f\left(\frac{2\pi}{3}\right) = \frac{2\pi}{3} - \frac{\sqrt{3}}{2} \approx 1.23$ is a local minimum value. Also $f(0) = 0$ and $f(\pi) = \pi$, so $f(0) = 0$ is the absolute minimum value and $f(\pi) = \pi$ is the absolute maximum value.

7. $$\lim_{x\to\infty} \frac{3x^4 + x - 5}{6x^4 - 2x^2 + 1} = \lim_{x\to\infty} \frac{3 + \dfrac{1}{x^3} - \dfrac{5}{x^4}}{6 - \dfrac{2}{x^2} + \dfrac{1}{x^4}} = \frac{3 + 0 + 0}{6 - 0 + 0} = \frac{1}{2}$$

9. $$\lim_{x\to-\infty} \frac{\sqrt{4x^2+1}}{3x-1} = \lim_{x\to-\infty} \frac{\sqrt{4x^2+1}/\sqrt{x^2}}{(3x-1)/\sqrt{x^2}} = \lim_{x\to-\infty} \frac{\sqrt{4+1/x^2}}{-3+1/x} \quad [\text{since } -x = |x| = \sqrt{x^2} \text{ for } x < 0]$$
$$= \frac{2}{-3+0} = -\frac{2}{3}$$

11. $$\lim_{x\to\infty} \left(\sqrt{4x^2+3x} - 2x\right) = \lim_{x\to\infty} \frac{\sqrt{4x^2+3x} - 2x}{1} \cdot \frac{\sqrt{4x^2+3x} + 2x}{\sqrt{4x^2+3x} + 2x} = \lim_{x\to\infty} \frac{(4x^2+3x) - 4x^2}{\sqrt{4x^2+3x} + 2x}$$
$$= \lim_{x\to\infty} \frac{3x}{\sqrt{4x^2+3x} + 2x} = \lim_{x\to\infty} \frac{3x/\sqrt{x^2}}{\left(\sqrt{4x^2+3x} + 2x\right)/\sqrt{x^2}}$$
$$= \lim_{x\to\infty} \frac{3}{\sqrt{4+3/x}+2} \quad [\text{since } x = |x| = \sqrt{x^2} \text{ for } x > 0]$$
$$= \frac{3}{2+2} = \frac{3}{4}$$

13. $f(0) = 0$, $f'(-2) = f'(1) = f'(9) = 0$, $\lim_{x\to\infty} f(x) = 0$, $\lim_{x\to6} f(x) = -\infty$, $f'(x) < 0$ on $(-\infty, -2)$, $(1, 6)$, and $(9, \infty)$, $f'(x) > 0$ on $(-2, 1)$ and $(6, 9)$, $f''(x) > 0$ on $(-\infty, 0)$ and $(12, \infty)$, $f''(x) < 0$ on $(0, 6)$ and $(6, 12)$

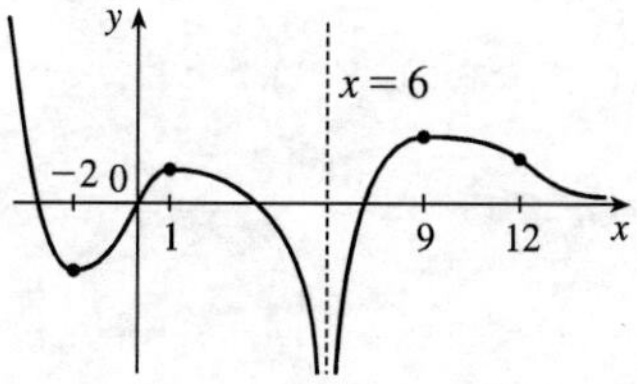

15. f is odd, $f'(x) < 0$ for $0 < x < 2$, $f'(x) > 0$ for $x > 2$, $f''(x) > 0$ for $0 < x < 3$, $f''(x) < 0$ for $x > 3$, $\lim_{x\to\infty} f(x) = -2$

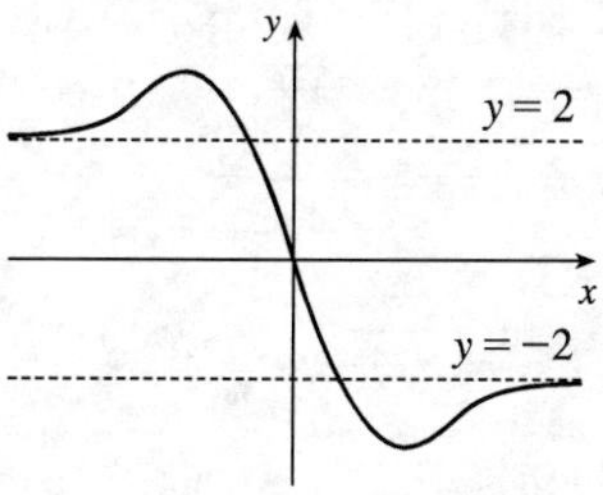

17. $y = f(x) = 2 - 2x - x^3$ **A.** $D = \mathbb{R}$ **B.** y-intercept: $f(0) = 2$. The x-intercept (approximately 0.770917) can be found using Newton's Method. **C.** No symmetry **D.** No asymptote
E. $f'(x) = -2 - 3x^2 = -(3x^2 + 2) < 0$, so f is decreasing on $\mathbb{R}$.
F. No extreme value **G.** $f''(x) = -6x < 0$ on $(0, \infty)$ and $f''(x) > 0$ on $(-\infty, 0)$, so f is CD on $(0, \infty)$ and CU on $(-\infty, 0)$. There is an IP at $(0, 2)$.

H.

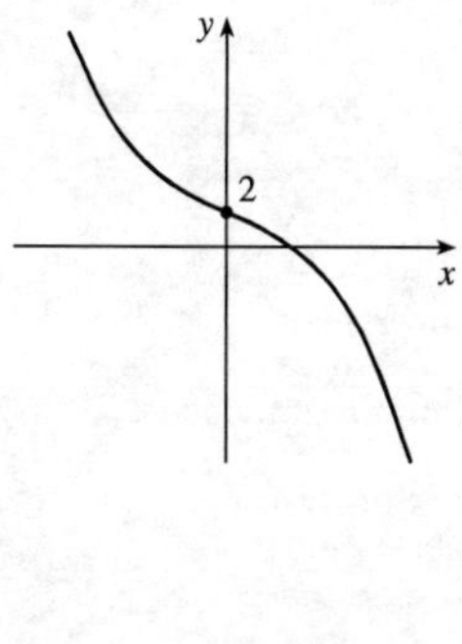

19. $y = f(x) = x^4 - 3x^3 + 3x^2 - x = x(x-1)^3$ **A.** $D = \mathbb{R}$ **B.** y-intercept: $f(0) = 0$; x-intercepts: $f(x) = 0$ $\Leftrightarrow$ $x = 0$ or $x = 1$ **C.** No symmetry **D.** f is a polynomial function and hence, it has no asymptote.
E. $f'(x) = 4x^3 - 9x^2 + 6x - 1$. Since the sum of the coefficients is 0, 1 is a root of f', so $f'(x) = (x-1)(4x^2 - 5x + 1) = (x-1)^2(4x-1)$. $f'(x) < 0 \;\Rightarrow\; x < \frac{1}{4}$, so f is decreasing on $\left(-\infty, \frac{1}{4}\right)$ and f is increasing on $\left(\frac{1}{4}, \infty\right)$. **F.** $f'(x)$ does not change sign at $x = 1$, so there is not a local extremum there. $f\left(\frac{1}{4}\right) = -\frac{27}{256}$ is a local minimum value. **G.** $f''(x) = 12x^2 - 18x + 6 = 6(2x-1)(x-1)$. $f''(x) = 0 \;\Leftrightarrow\; x = \frac{1}{2}$ or 1. $f''(x) < 0 \;\Leftrightarrow\; \frac{1}{2} < x < 1 \;\Rightarrow$ f is CD on $\left(\frac{1}{2}, 1\right)$ and CU on $\left(-\infty, \frac{1}{2}\right)$ and $(1, \infty)$. There are inflection points at $\left(\frac{1}{2}, -\frac{1}{16}\right)$ and $(1, 0)$.

H.

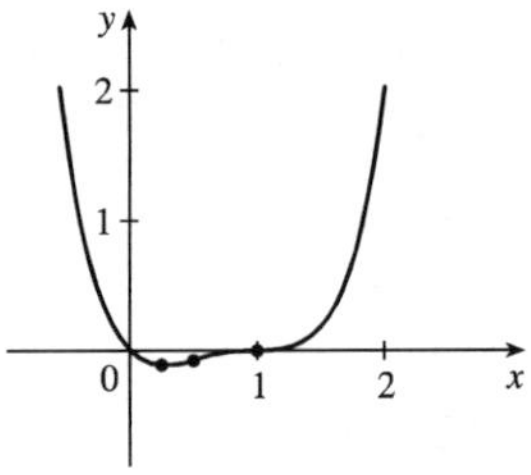

21. $y = f(x) = \dfrac{1}{x(x-3)^2}$ **A.** $D = \{x \mid x \neq 0, 3\} = (-\infty, 0) \cup (0, 3) \cup (3, \infty)$ **B.** No intercepts.
C. No symmetry. **D.** $\displaystyle\lim_{x\to\pm\infty} \frac{1}{x(x-3)^2} = 0$, so $y = 0$ is a HA. $\displaystyle\lim_{x\to 0^+} \frac{1}{x(x-3)^2} = \infty$, $\displaystyle\lim_{x\to 0^-} \frac{1}{x(x-3)^2} = -\infty$, $\displaystyle\lim_{x\to 3} \frac{1}{x(x-3)^2} = \infty$, so $x = 0$ and $x = 3$ are VA.

E. $f'(x) = -\dfrac{(x-3)^2 + 2x(x-3)}{x^2(x-3)^4} = \dfrac{3(1-x)}{x^2(x-3)^3} \;\Rightarrow$

$f'(x) > 0 \;\Leftrightarrow\; 1 < x < 3$, so f is increasing on $(1, 3)$ and decreasing on $(-\infty, 0)$, $(0, 1)$, and $(3, \infty)$. **F.** Local minimum value $f(1) = \frac{1}{4}$

G. $f''(x) = \dfrac{6(2x^2 - 4x + 3)}{x^3(x-3)^4}$. Note that $2x^2 - 4x + 3 > 0$ for all x since it has negative discriminant. So $f''(x) > 0 \;\Leftrightarrow\; x > 0 \;\Rightarrow\; f$ is CU on $(0, 3)$ and $(3, \infty)$ and CD on $(-\infty, 0)$. No IP

H.

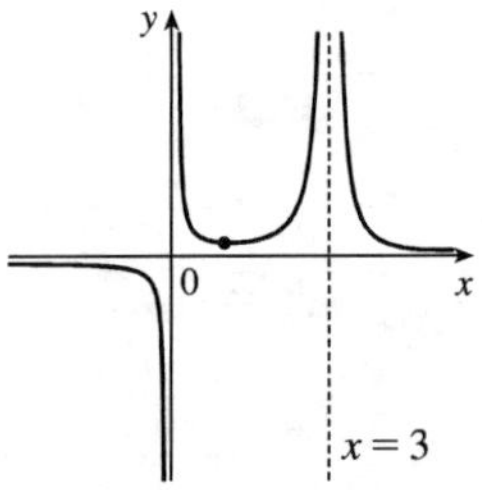

23. $y = f(x) = \dfrac{x^2}{x+8} = x - 8 + \dfrac{64}{x+8}$ **A.** $D = \{x \mid x \neq -8\}$ **B.** Intercepts are 0 **C.** No symmetry

D. $\displaystyle\lim_{x\to\infty} \frac{x^2}{x+8} = \infty$, but $f(x) - (x-8) = \dfrac{64}{x+8} \to 0$ as $x \to \infty$, so $y = x - 8$ is a slant asymptote.

$\displaystyle\lim_{x\to -8^+} \frac{x^2}{x+8} = \infty$ and $\displaystyle\lim_{x\to -8^-} \frac{x^2}{x+8} = -\infty$, so $x = -8$ is a VA.

E. $f'(x) = 1 - \dfrac{64}{(x+8)^2} = \dfrac{x(x+16)}{(x+8)^2} > 0 \;\Leftrightarrow\; x > 0$ or $x < -16$, so f is increasing on $(-\infty, -16)$ and $(0, \infty)$ and decreasing on $(-16, -8)$ and $(-8, 0)$. **F.** Local maximum value $f(-16) = -32$, local minimum value $f(0) = 0$ **G.** $f''(x) = 128/(x+8)^3 > 0 \;\Leftrightarrow$ $x > -8$, so f is CU on $(-8, \infty)$ and CD on $(-\infty, -8)$. No IP

H.

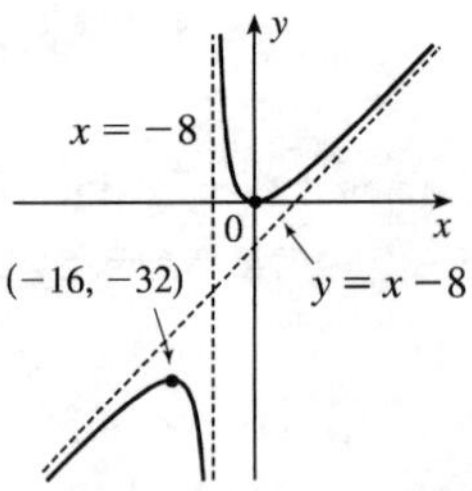

25. $y = f(x) = x\sqrt{2+x}$ **A.** $D = [-2, \infty)$ **B.** y-intercept: $f(0) = 0$; x-intercepts: -2 and 0 **C.** No symmetry **D.** No asymptote **E.** $f'(x) = \dfrac{x}{2\sqrt{2+x}} + \sqrt{2+x} = \dfrac{1}{2\sqrt{2+x}}[x + 2(2+x)] = \dfrac{3x+4}{2\sqrt{2+x}} = 0$ when $x = -\frac{4}{3}$, so f is decreasing on $\left(-2, -\frac{4}{3}\right)$ and increasing on $\left(-\frac{4}{3}, \infty\right)$. **F.** Local minimum value $f\left(-\frac{4}{3}\right) = -\frac{4}{3}\sqrt{\frac{2}{3}} = -\frac{4\sqrt{6}}{9} \approx -1.09$, no local maximum

G.
$$f''(x) = \frac{2\sqrt{2+x} \cdot 3 - (3x+4)\dfrac{1}{\sqrt{2+x}}}{4(2+x)}$$
$$= \frac{6(2+x) - (3x+4)}{4(2+x)^{3/2}} = \frac{3x+8}{4(2+x)^{3/2}}$$

$f''(x) > 0$ for $x > -2$, so f is CU on $(-2, \infty)$. No IP

H.

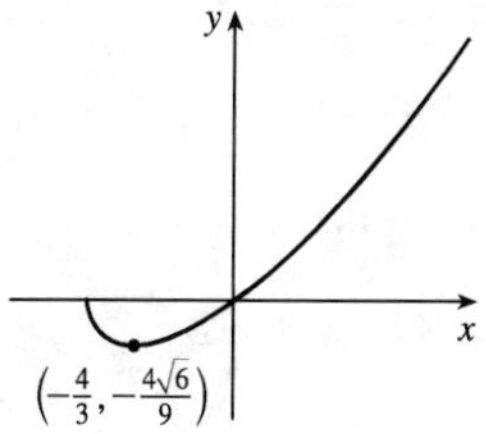

27. $y = f(x) = \sin^2 x - 2\cos x$ **A.** $D = \mathbb{R}$ **B.** y-intercept: $f(0) = -2$ **C.** $f(-x) = f(x)$, so f is symmetric with respect to the y-axis. f has period 2π. **D.** No asymptote

E. $y' = 2\sin x\cos x + 2\sin x = 2\sin x\,(\cos x + 1)$. $y' = 0 \Leftrightarrow \sin x = 0$ or $\cos x = -1 \Leftrightarrow x = n\pi$ or $x = (2n+1)\pi$. $y' > 0$ when $\sin x > 0$, since $\cos x + 1 \geq 0$ for all x. Therefore, $y' > 0$ (and so f is increasing) on $(2n\pi, (2n+1)\pi)$; $y' < 0$ (and so f is decreasing) on $((2n-1)\pi, 2n\pi)$. **F.** Local maximum values are $f((2n+1)\pi) = 2$; local minimum values are $f(2n\pi) = -2$. **G.** $y' = \sin 2x + 2\sin x \Rightarrow$

$$y'' = 2\cos 2x + 2\cos x = 2\left(2\cos^2 x - 1\right) + 2\cos x = 4\cos^2 x + 2\cos x - 2$$
$$= 2\left(2\cos^2 x + \cos x - 1\right) = 2(2\cos x - 1)(\cos x + 1)$$

$y'' = 0 \Leftrightarrow \cos x = \frac{1}{2}$ or $-1 \Leftrightarrow x = 2n\pi \pm \frac{\pi}{3}$ or $x = (2n+1)\pi$. $y'' > 0$ (and so f is CU) on $\left(2n\pi - \frac{\pi}{3}, 2n\pi + \frac{\pi}{3}\right)$; $y'' \leq 0$ (and so f is CD) on $\left(2n\pi + \frac{\pi}{3}, 2n\pi + \frac{5\pi}{3}\right)$. There are inflection points at $\left(2n\pi \pm \frac{\pi}{3}, -\frac{1}{4}\right)$.

H.

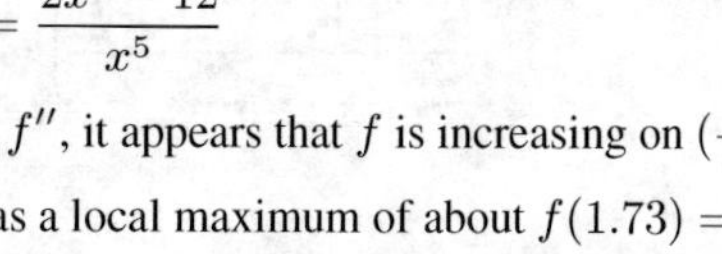

29. $f(x) = \dfrac{x^2 - 1}{x^3} \Rightarrow f'(x) = \dfrac{x^3(2x) - (x^2-1)3x^2}{x^6} = \dfrac{3 - x^2}{x^4} \Rightarrow$

$f''(x) = \dfrac{x^4(-2x) - (3 - x^2)4x^3}{x^8} = \dfrac{2x^2 - 12}{x^5}$

Estimates: From the graphs of f' and f'', it appears that f is increasing on $(-1.73, 0)$ and $(0, 1.73)$ and decreasing on $(-\infty, -1.73)$ and $(1.73, \infty)$; f has a local maximum of about $f(1.73) = 0.38$ and a local minimum of about

$f(-1.7) = -0.38$; f is CU on $(-2.45, 0)$ and $(2.45, \infty)$, and CD on $(-\infty, -2.45)$ and $(0, 2.45)$; and f has inflection points at about $(-2.45, -0.34)$ and $(2.45, 0.34)$.

Exact: Now $f'(x) = \dfrac{3 - x^2}{x^4}$ is positive for $0 < x^2 < 3$, that is, f is increasing on $(-\sqrt{3}, 0)$ and $(0, \sqrt{3})$; and $f'(x)$ is negative (and so f is decreasing) on $(-\infty, -\sqrt{3})$ and $(\sqrt{3}, \infty)$. $f'(x) = 0$ when $x = \pm\sqrt{3}$. f' goes from positive to negative at $x = \sqrt{3}$, so f has a local maximum of $f(\sqrt{3}) = \dfrac{(\sqrt{3})^2 - 1}{(\sqrt{3})^3} = \frac{2\sqrt{3}}{9}$; and since f is odd, we know that maxima on the interval $(0, \infty)$ correspond to minima on $(-\infty, 0)$, so f has a local minimum of $f(-\sqrt{3}) = -\frac{2\sqrt{3}}{9}$. Also, $f''(x) = \dfrac{2x^2 - 12}{x^5}$ is positive (so f is CU) on $(-\sqrt{6}, 0)$ and $(\sqrt{6}, \infty)$, and negative (so f is CD) on $(-\infty, -\sqrt{6})$ and $(0, \sqrt{6})$. There are IP at $\left(\sqrt{6}, \frac{5\sqrt{6}}{36}\right)$ and $\left(-\sqrt{6}, -\frac{5\sqrt{6}}{36}\right)$.

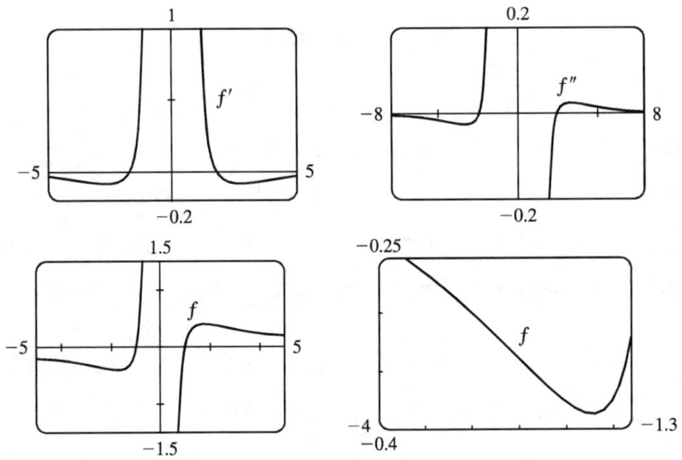

31. $f(x) = 3x^6 - 5x^5 + x^4 - 5x^3 - 2x^2 + 2 \;\Rightarrow\; f'(x) = 18x^5 - 25x^4 + 4x^3 - 15x^2 - 4x \;\Rightarrow$
$f''(x) = 90x^4 - 100x^3 + 12x^2 - 30x - 4$

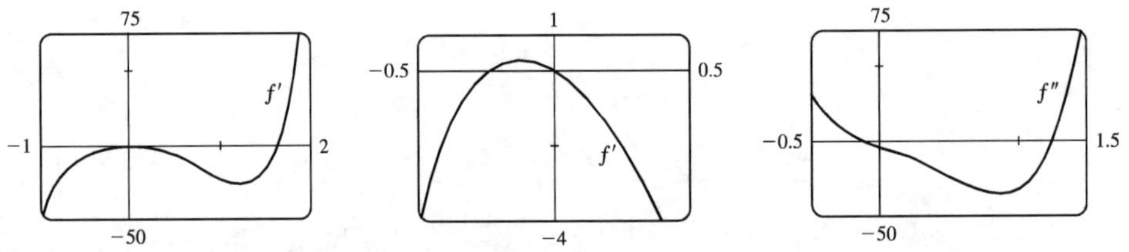

From the graphs of f' and f'', it appears that f is increasing on $(-0.23, 0)$ and $(1.62, \infty)$ and decreasing on $(-\infty, -0.23)$ and $(0, 1.62)$; f has a local maximum of about $f(0) = 2$ and local minima

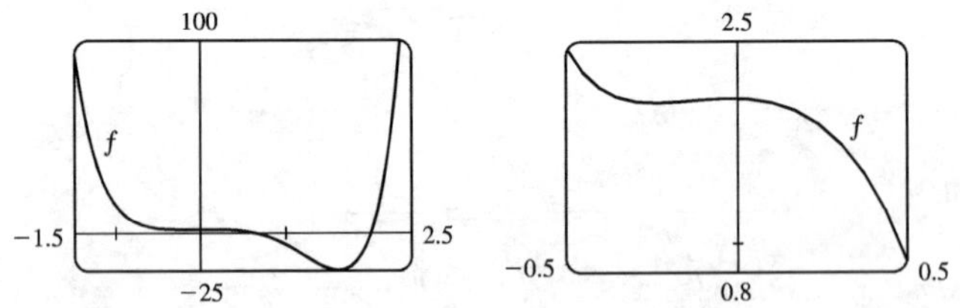

of about $f(-0.23) = 1.96$ and $f(1.62) = -19.2$; f is CU on $(-\infty, -0.12)$ and $(1.24, \infty)$ and CD on $(-0.12, 1.24)$; and f has inflection points at about $(-0.12, 1.98)$ and $(1.24, -12.1)$.

33. $f(x) = x^{101} + x^{51} + x - 1 = 0$. Since f is continuous and $f(0) = -1$ and $f(1) = 2$, the equation has at least one root in $(0, 1)$, by the Intermediate Value Theorem. Suppose the equation has two roots, a and b, with $a < b$. Then $f(a) = 0 = f(b)$, so by the Mean Value Theorem, there is a number x in (a, b) such that

$f'(x) = \dfrac{f(b) - f(a)}{b - a} = \dfrac{0}{b - a} = 0$, so f' has a root in (a, b). But this is impossible since

$f'(x) = 101x^{100} + 51x^{50} + 1 \geq 1$ for all x.

35. Since f is continuous on $[32, 33]$ and differentiable on $(32, 33)$, then by the Mean Value Theorem there exists a number c in $(32, 33)$ such that $f'(c) = \frac{1}{5}c^{-4/5} = \dfrac{\sqrt[5]{33} - \sqrt[5]{32}}{33 - 32} = \sqrt[5]{33} - 2$, but $\frac{1}{5}c^{-4/5} > 0 \;\Rightarrow\; \sqrt[5]{33} - 2 > 0$ $\Rightarrow\; \sqrt[5]{33} > 2$. Also f' is decreasing, so that $f'(c) < f'(32) = \frac{1}{5}(32)^{-4/5} = 0.0125 \;\Rightarrow$ $0.0125 > f'(c) = \sqrt[5]{33} - 2 \;\Rightarrow\; \sqrt[5]{33} < 2.0125$. Therefore, $2 < \sqrt[5]{33} < 2.0125$.

37. (a) $g(x) = f(x^2) \;\Rightarrow\; g'(x) = 2xf'(x^2)$ by the Chain Rule. Since $f'(x) > 0$ for all $x \neq 0$, we must have $f'(x^2) > 0$ for $x \neq 0$, so $g'(x) = 0 \;\Leftrightarrow\; x = 0$. Now $g'(x)$ changes sign (from negative to positive) at $x = 0$, since one of its factors, $f'(x^2)$, is positive for all x, and its other factor, $2x$, changes from negative to positive at this point, so by the First Derivative Test, f has a local and absolute minimum at $x = 0$.

(b) $g'(x) = 2xf'(x^2) \;\Rightarrow\; g''(x) = 2\left[xf''(x^2)(2x) + f'(x^2)\right] = 4x^2f''(x^2) + 2f'(x^2)$ by the Product Rule and the Chain Rule. But $x^2 > 0$ for all $x \neq 0$, $f''(x^2) > 0$ (since f is CU for $x > 0$), and $f'(x^2) > 0$ for all $x \neq 0$, so since all of its factors are positive, $g''(x) > 0$ for $x \neq 0$. Whether $g''(0)$ is positive or 0 doesn't matter (since the sign of g'' does not change there); g is concave upward on $\mathbb{R}$.

39. If $B = 0$, the line is vertical and the distance from $x = -\dfrac{C}{A}$ to (x_1, y_1) is $\left|x_1 + \dfrac{C}{A}\right| = \dfrac{|Ax_1 + By_1 + C|}{\sqrt{A^2 + B^2}}$, so assume $B \neq 0$. The square of the distance from (x_1, y_1) to the line is $f(x) = (x - x_1)^2 + (y - y_1)^2$ where $Ax + By + C = 0$, so we minimize $f(x) = (x - x_1)^2 + \left(-\dfrac{A}{B}x - \dfrac{C}{B} - y_1\right)^2 \;\Rightarrow$

$f'(x) = 2(x - x_1) + 2\left(-\dfrac{A}{B}x - \dfrac{C}{B} - y_1\right)\left(-\dfrac{A}{B}\right)$. $f'(x) = 0 \;\Rightarrow\; x = \dfrac{B^2x_1 - ABy_1 - AC}{A^2 + B^2}$ and this gives a minimum since $f''(x) = 2\left(1 + \dfrac{A^2}{B^2}\right) > 0$. Substituting this value of x into $f(x)$ and simplifying gives

$f(x) = \dfrac{(Ax_1 + By_1 + C)^2}{A^2 + B^2}$, so the minimum distance is $\sqrt{f(x)} = \dfrac{|Ax_1 + By_1 + C|}{\sqrt{A^2 + B^2}}$.

41.

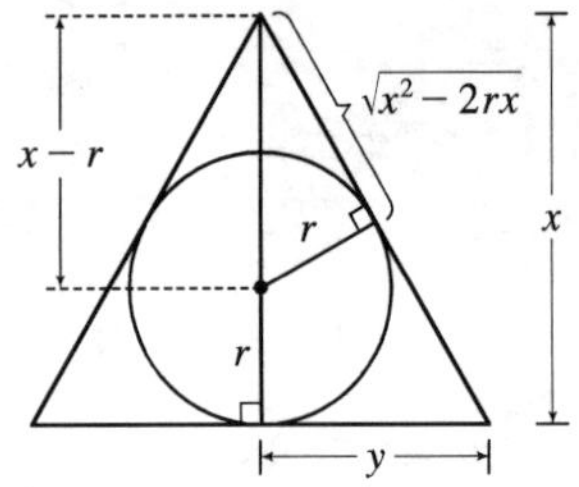

By similar triangles, $\dfrac{y}{x} = \dfrac{r}{\sqrt{x^2-2rx}}$, so the area of the triangle is

$$A(x) = \tfrac{1}{2}(2y)x = xy = \frac{rx^2}{\sqrt{x^2-2rx}} \quad \Rightarrow$$

$$A'(x) = \frac{2rx\sqrt{x^2-2rx} - rx^2(x-r)/\sqrt{x^2-2rx}}{x^2-2rx}$$

$$= \frac{rx^2\,(x-3r)}{(x^2-2rx)^{3/2}} = 0 \text{ when } x = 3r.$$

$A'(x) < 0$ when $2r < x < 3r$, $A'(x) > 0$ when $x > 3r$. So $x = 3r$ gives a minimum and $A(3r) = r(9r^2)/(\sqrt{3}\,r) = 3\sqrt{3}\,r^2$.

43.

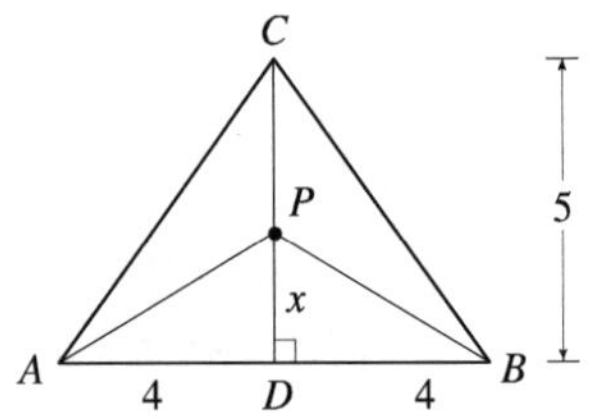

We minimize

$$L(x) = |PA| + |PB| + |PC| = 2\sqrt{x^2+16} + (5-x),$$

$0 \le x \le 5$. $L'(x) = 2x/\sqrt{x^2+16} - 1 = 0 \quad \Leftrightarrow$

$2x = \sqrt{x^2+16} \quad \Leftrightarrow \quad 4x^2 = x^2+16 \quad \Leftrightarrow \quad x = \frac{4}{\sqrt{3}}$.

$L(0) = 13$, $L\left(\frac{4}{\sqrt{3}}\right) \approx 11.9$, $L(5) \approx 12.8$, so the minimum occurs when $x = \frac{4}{\sqrt{3}} \approx 2.3$.

45. $v = K\sqrt{\dfrac{L}{C} + \dfrac{C}{L}} \quad \Rightarrow \quad \dfrac{dv}{dL} = \dfrac{K}{2\sqrt{(L/C)+(C/L)}}\left(\dfrac{1}{C} - \dfrac{C}{L^2}\right) = 0 \quad \Leftrightarrow \quad \dfrac{1}{C} = \dfrac{C}{L^2} \quad \Leftrightarrow \quad L^2 = C^2 \quad \Leftrightarrow$ $L = C$. This gives the minimum velocity since $v' < 0$ for $0 < L < C$ and $v' > 0$ for $L > C$.

47. Let x denote the number of \$1 decreases in ticket price. Then the ticket price is $\$12 - \$1(x)$, and the average attendance is $11{,}000 + 1000(x)$. Now the revenue per game is

$$R(x) = (\text{price per person}) \times (\text{number of people per game})$$

$$= (12-x)(11{,}000+1000x) = -1000x^2 + 1000x + 132{,}000$$

for $0 \le x \le 4$ (since the seating capacity is 15,000) $\Rightarrow \quad R'(x) = -2000x + 1000 = 0 \quad \Leftrightarrow \quad x = 0.5$. This is a maximum since $R''(x) = -2000 < 0$ for all x. Now we must check the value of $R(x) = (12-x)(11{,}000+1000x)$ at $x = 0.5$ and at the endpoints of the domain to see which value of x gives the maximum value of R. $R(0) = (12)(11{,}000) = 132{,}000$, $R(0.5) = (11.5)(11{,}500) = 132{,}250$, and $R(4) = (8)(15{,}000) = 120{,}000$. Thus, the maximum revenue of \$132,250 per game occurs when the average attendance is 11,500 and the ticket price is \$11.50.

49. $f(x) = x^5 - x^4 + 3x^2 - 3x - 2 \Rightarrow f'(x) = 5x^4 - 4x^3 + 6x - 3$, so

$x_{n+1} = x_n - \dfrac{x_n^5 - x_n^4 + 3x_n^2 - 3x_n - 2}{5x_n^4 - 4x_n^3 + 6x_n - 3}$. Now $x_1 = 1 \Rightarrow x_2 = 1.5 \Rightarrow x_3 \approx 1.343860 \Rightarrow$

$x_4 \approx 1.300320 \Rightarrow x_5 \approx 1.297396 \Rightarrow x_6 \approx 1.297383 \approx x_7$, so the root in $[1, 2]$ is 1.297383, to six decimal places.

51. $f(t) = \cos t + t - t^2 \Rightarrow f'(t) = -\sin t + 1 - 2t$. $f'(t)$ exists for all t, so to find the maximum of f, we can examine the zeros of f'. From the graph of f', we see that a good choice for t_1 is $t_1 = 0.3$. Use $g(t) = -\sin t + 1 - 2t$ and $g'(t) = -\cos t - 2$ to obtain $t_2 \approx 0.33535293$, $t_3 \approx 0.33541803 \approx t_4$. Since $f''(t) = -\cos t - 2 < 0$ for all t, $f(0.33541803) \approx 1.16718557$ is the absolute maximum.

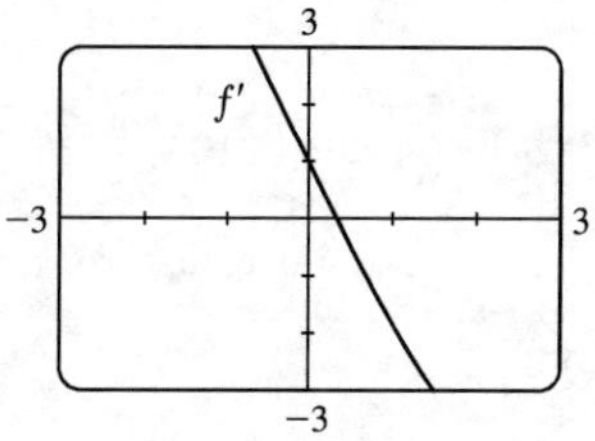

53. $f'(x) = \sqrt{x^5} - 4/\sqrt[5]{x} = x^{5/2} - 4x^{-1/5} \Rightarrow f(x) = \frac{2}{7}x^{7/2} - 4\left(\frac{5}{4}x^{4/5}\right) + C = \frac{2}{7}x^{7/2} - 5x^{4/5} + C$

55. $f'(t) = 2t - 3\sin t \Rightarrow f(t) = t^2 + 3\cos t + C$.

$f(0) = 3 + C$ and $f(0) = 5 \Rightarrow C = 2$, so $f(t) = t^2 + 3\cos t + 2$.

57. $f''(x) = 1 - 6x + 48x^2 \Rightarrow f'(x) = x - 3x^2 + 16x^3 + C$. $f'(0) = C$ and $f'(0) = 2 \Rightarrow C = 2$, so $f'(x) = x - 3x^2 + 16x^3 + 2$ and hence, $f(x) = \frac{1}{2}x^2 - x^3 + 4x^4 + 2x + D$. $f(0) = D$ and $f(0) = 1 \Rightarrow$ $D = 1$, so $f(x) = \frac{1}{2}x^2 - x^3 + 4x^4 + 2x + 1$.

59. $f(x) = x^2 \sin(x^2)$, $0 \le x \le \pi$

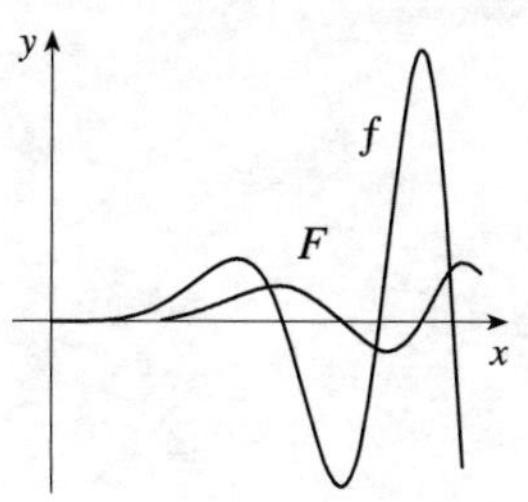

61. Choosing the positive direction to be upward, we have $a(t) = -9.8 \Rightarrow v(t) = -9.8t + v_0$, but $v(0) = 0 = v_0$ $\Rightarrow v(t) = -9.8t = s'(t) \Rightarrow s(t) = -4.9t^2 + s_0$, but $s(0) = s_0 = 500 \Rightarrow s(t) = -4.9t^2 + 500$. When $s = 0$, $-4.9t^2 + 500 = 0 \Rightarrow t_1 = \sqrt{\frac{500}{4.9}} \approx 10.1 \Rightarrow v(t_1) = -9.8\sqrt{\frac{500}{4.9}} \approx -98.995$ m/s. Since the canister has been designed to withstand an impact velocity of 100 m/s, the canister will *not burst*.

63. (a) $I = \dfrac{k\cos\theta}{d^2} = \dfrac{k(h/d)}{d^2} = k\dfrac{h}{d^3} = k\dfrac{h}{\left(\sqrt{40^2+h^2}\,\right)^3} = k\dfrac{h}{(1600+h^2)^{3/2}} \quad\Rightarrow$

$$\frac{dI}{dh} = k\frac{\left(1600+h^2\right)^{3/2} - h\frac{3}{2}\left(1600+h^2\right)^{1/2}\cdot 2h}{\left[(1600+h^2)^{3/2}\right]^2} = \frac{k\left(1600+h^2\right)^{1/2}\left(1600+h^2-3h^2\right)}{\left(1600+h^2\right)^3}$$

$$= \frac{k\left(1600-2h^2\right)}{\left(1600+h^2\right)^{5/2}} \quad [k \text{ is the constant of proportionality}]$$

Set $dI/dh = 0$: $1600 - 2h^2 = 0 \;\Rightarrow\; h^2 = 800 \;\Rightarrow\; h = \sqrt{800} = 20\sqrt{2}$. By the First Derivative Test, I has a local maximum at $h = 20\sqrt{2} \approx 28$ ft.

(b)

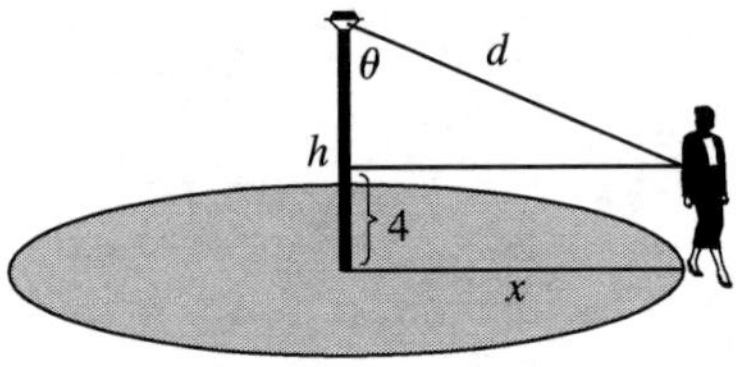

$\dfrac{dx}{dt} = 4$ ft/s

$$I = \frac{k\cos\theta}{d^2} = \frac{k[(h-4)/d]}{d^2} = \frac{k(h-4)}{d^3} = \frac{k(h-4)}{[(h-4)^2+x^2]^{3/2}} = k(h-4)\left[(h-4)^2+x^2\right]^{-3/2}$$

$$\frac{dI}{dt} = \frac{dI}{dx}\cdot\frac{dx}{dt} = k(h-4)\left(-\tfrac{3}{2}\right)\left[(h-4)^2+x^2\right]^{-5/2}\cdot 2x\cdot\frac{dx}{dt}$$

$$= k(h-4)(-3x)\left[(h-4)^2+x^2\right]^{-5/2}\cdot 4 = \frac{-12xk(h-4)}{[(h-4)^2+x^2]^{5/2}}$$

$$\left.\frac{dI}{dt}\right|_{x=40} = -\frac{480k(h-4)}{[(h-4)^2+1600]^{5/2}}$$

Antiderivatives

1. $f(x) = 6x^2 - 8x + 3 \quad\Rightarrow\quad F(x) = 6\dfrac{x^{2+1}}{2+1} - 8\dfrac{x^{1+1}}{1+1} + 3x + C = 2x^3 - 4x^2 + 3x + C$

Check: $F'(x) = 2 \cdot 3x^2 - 4 \cdot 2x + 3 + 0 = 6x^2 - 8x + 3 = f(x)$

3. $f(x) = 1 - x^3 + 5x^5 - 3x^7 \quad\Rightarrow\quad F(x) = x - \dfrac{x^{3+1}}{3+1} + 5\dfrac{x^{5+1}}{5+1} - 3\dfrac{x^{7+1}}{7+1} + C = x - \frac{1}{4}x^4 + \frac{5}{6}x^6 - \frac{3}{8}x^8 + C$

5. $f(x) = 5x^{1/4} - 7x^{3/4} \quad\Rightarrow\quad F(x) = 5\dfrac{x^{1/4+1}}{\frac{1}{4}+1} - 7\dfrac{x^{3/4+1}}{\frac{3}{4}+1} + C = 5\dfrac{x^{5/4}}{5/4} - 7\dfrac{x^{7/4}}{7/4} + C = 4x^{5/4} - 4x^{7/4} + C$

7. $f(x) = 6\sqrt{x} - \sqrt[6]{x} = 6x^{1/2} - x^{1/6} \quad\Rightarrow$

$$F(x) = 6\,\frac{x^{1/2+1}}{\frac{1}{2}+1} - \frac{x^{1/6+1}}{\frac{1}{6}+1} + C = 6\,\frac{x^{3/2}}{3/2} - \frac{x^{7/6}}{7/6} + C = 4x^{3/2} - \tfrac{6}{7}x^{7/6} + C$$

9. $f(x)=\dfrac{10}{x^9}=10x^{-9}$ has domain $(-\infty,0)\cup(0,\infty)$, so $F(x)=\begin{cases}\dfrac{10x^{-8}}{-8}+C_1=-\dfrac{5}{4x^8}+C_1 & \text{if } x<0\\ -\dfrac{5}{4x^8}+C_2 & \text{if } x>0\end{cases}$

See Example 1 for a similar problem.

11. $f(u)=\dfrac{u^4+3\sqrt{u}}{u^2}=\dfrac{u^4}{u^2}+\dfrac{3u^{1/2}}{u^2}=u^2+3u^{-3/2}\quad\Rightarrow$

$F(u)=\dfrac{u^3}{3}+3\dfrac{u^{-3/2+1}}{-3/2+1}+C=\dfrac{1}{3}u^3+3\dfrac{u^{-1/2}}{-1/2}+C=\dfrac{1}{3}u^3-\dfrac{6}{\sqrt{u}}+C$

13. $h(x)=x^3+5\sin x\quad\Rightarrow\quad H(x)=\frac{1}{4}x^4+5(-\cos x)+C=\frac{1}{4}x^4-5\cos x+C$

15. $f(t)=4\sqrt{t}-\sec t\tan t\quad\Rightarrow\quad F(t)=\frac{4}{3/2}t^{3/2}-\sec t+C=\frac{8}{3}t^{3/2}-\sec t+C_n$ on the interval $\left(n\pi-\frac{\pi}{2},n\pi+\frac{\pi}{2}\right)$.

17. $f''(x)=6x+12x^2\quad\Rightarrow\quad f'(x)=6\cdot\dfrac{x^2}{2}+12\cdot\dfrac{x^3}{3}+C=3x^2+4x^3+C\quad\Rightarrow$

$f(x)=3\cdot\dfrac{x^3}{3}+4\cdot\dfrac{x^4}{4}+Cx+D=x^3+x^4+Cx+D$ [C and D are just arbitrary constants]

19. $f''(x)=1+x^{4/5}\quad\Rightarrow\quad f'(x)=x+\frac{5}{9}x^{9/5}+C\quad\Rightarrow$

$f(x)=\frac{1}{2}x^2+\frac{5}{9}\cdot\frac{5}{14}x^{14/5}+Cx+D=\frac{1}{2}x^2+\frac{25}{126}x^{14/5}+Cx+D$

21. $f'''(t)=60t^2\quad\Rightarrow\quad f''(t)=20t^3+C\quad\Rightarrow\quad f'(t)=5t^4+Ct+D\quad\Rightarrow\quad f(t)=t^5+\frac{1}{2}Ct^2+Dt+E$

23. $f'(x)=1-6x\quad\Rightarrow\quad f(x)=x-3x^2+C$. $f(0)=C$ and $f(0)=8\quad\Rightarrow\quad C=8$, so $f(x)=x-3x^2+8$.

25. $f'(x)=\sqrt{x}(6+5x)=6x^{1/2}+5x^{3/2}\quad\Rightarrow\quad f(x)=4x^{3/2}+2x^{5/2}+C$.
$f(1)=6+C$ and $f(1)=10\quad\Rightarrow\quad C=4$, so $f(x)=4x^{3/2}+2x^{5/2}+4$.

27. $f'(t)=2\cos t+\sec^2 t\quad\Rightarrow\quad f(t)=2\sin t+\tan t+C$ because $-\pi/2<t<\pi/2$.
$f\left(\frac{\pi}{3}\right)=2\left(\sqrt{3}/2\right)+\sqrt{3}+C=2\sqrt{3}+C$ and $f\left(\frac{\pi}{3}\right)=4\quad\Rightarrow\quad C=4-2\sqrt{3}$, so
$f(t)=2\sin t+\tan t+4-2\sqrt{3}$.

29. $f''(x)=24x^2+2x+10\quad\Rightarrow\quad f'(x)=8x^3+x^2+10x+C$. $f'(1)=8+1+10+C$ and $f'(1)=-3\quad\Rightarrow$
$19+C=-3\quad\Rightarrow\quad C=-22$, so $f'(x)=8x^3+x^2+10x-22$ and hence,
$f(x)=2x^4+\frac{1}{3}x^3+5x^2-22x+D$. $f(1)=2+\frac{1}{3}+5-22+D$ and $f(1)=5\quad\Rightarrow\quad D=22-\frac{7}{3}=\frac{59}{3}$,
so $f(x)=2x^4+\frac{1}{3}x^3+5x^2-22x+\frac{59}{3}$.

31. $f''(\theta)=\sin\theta+\cos\theta\quad\Rightarrow\quad f'(\theta)=-\cos\theta+\sin\theta+C$. $f'(0)=-1+C$ and $f'(0)=4\quad\Rightarrow\quad C=5$, so
$f'(\theta)=-\cos\theta+\sin\theta+5$ and hence, $f(\theta)=-\sin\theta-\cos\theta+5\theta+D$. $f(0)=-1+D$ and $f(0)=3\quad\Rightarrow$
$D=4$, so $f(\theta)=-\sin\theta-\cos\theta+5\theta+4$.

33. $f''(x)=2-12x\quad\Rightarrow\quad f'(x)=2x-6x^2+C\quad\Rightarrow\quad f(x)=x^2-2x^3+Cx+D$.
$f(0)=D$ and $f(0)=9\quad\Rightarrow\quad D=9$. $f(2)=4-16+2C+9=2C-3$ and $f(2)=15\quad\Rightarrow\quad 2C=18\quad\Rightarrow$
$C=9$, so $f(x)=x^2-2x^3+9x+9$.

35. $f''(x)=2+\cos x\quad\Rightarrow\quad f'(x)=2x+\sin x+C\quad\Rightarrow\quad f(x)=x^2-\cos x+Cx+D$. $f(0)=-1+D$ and
$f(0)=-1\quad\Rightarrow\quad D=0$. $f\left(\frac{\pi}{2}\right)=\pi^2/4+\left(\frac{\pi}{2}\right)C$ and $f\left(\frac{\pi}{2}\right)=0\quad\Rightarrow\quad\left(\frac{\pi}{2}\right)C=-\pi^2/4\quad\Rightarrow\quad C=-\frac{\pi}{2}$, so
$f(x)=x^2-\cos x-\left(\frac{\pi}{2}\right)x$.

37. Given $f'(x)=2x+1$, we have $f(x)=x^2+x+C$. Since f passes through $(1,6)$,
$f(1)=6\quad\Rightarrow\quad 1^2+1+C=6\quad\Rightarrow\quad C=4$. Therefore, $f(x)=x^2+x+4$ and $f(2)=2^2+2+4=10$.

39. b is the antiderivative of f. For small x, f is negative, so the graph of its antiderivative must be decreasing. But both a and c are increasing for small x, so only b can be f's antiderivative. Also, f is positive where b is increasing, which supports our conclusion.

41. The graph of F will have a minimum at 0 and a maximum at 2, since $f = F'$ goes from negative to positive at $x = 0$, and from positive to negative at $x = 2$.

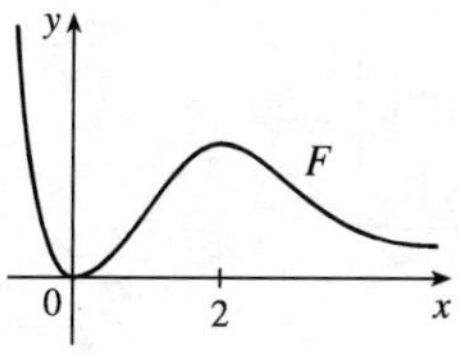

43.

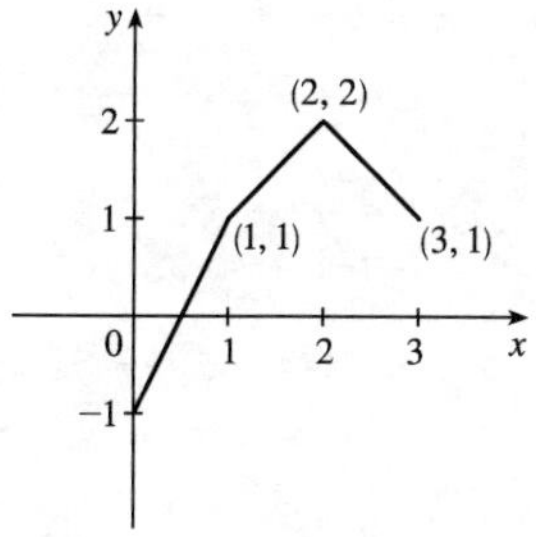

$$f'(x) = \begin{cases} 2 & \text{if } 0 \le x < 1 \\ 1 & \text{if } 1 < x < 2 \\ -1 & \text{if } 2 < x \le 3 \end{cases} \Rightarrow f(x) = \begin{cases} 2x + C & \text{if } 0 \le x < 1 \\ x + D & \text{if } 1 < x < 2 \\ -x + E & \text{if } 2 < x \le 3 \end{cases}$$

$f(0) = -1 \Rightarrow 2(0) + C = -1 \Rightarrow C = -1$. Starting at the point $(0, -1)$ and moving to the right on a line with slope 2 gets us to the point $(1, 1)$. The slope for $1 < x < 2$ is 1, so we get to the point $(2, 2)$. Here we have used the fact that f is continuous. We can include the point $x = 1$ on either the first or the second part of f. The line connecting $(1, 1)$ to $(2, 2)$ is $y = x$, so $D = 0$. The slope for $2 < x \le 3$ is -1, so we get to $(3, 1)$. $f(3) = 1 \Rightarrow -3 + E = 1 \Rightarrow E = 4$. Thus,

$$f(x) = \begin{cases} 2x - 1 & \text{if } 0 \le x \le 1 \\ x & \text{if } 1 < x < 2 \\ -x + 4 & \text{if } 2 \le x \le 3 \end{cases}$$

Note that $f'(x)$ does not exist at $x = 1$ or at $x = 2$.

45. $f(x) = \sin(x^2)$, $0 \le x \le 4$

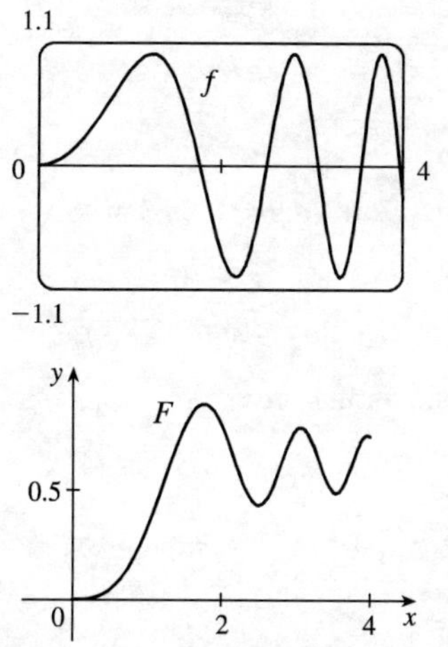

47.

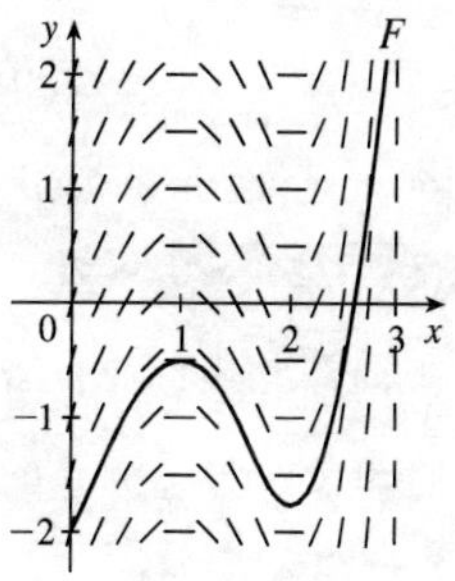

49.

x	$f(x)$
0	1
0.5	0.959
1.0	0.841
1.5	0.665
2.0	0.455
2.5	0.239
3.0	0.047

x	$f(x)$
3.5	−0.100
4.0	−0.189
4.5	−0.217
5.0	−0.192
5.5	−0.128
6.0	−0.047

We compute slopes [values of $f(x) = (\sin x)/x$ for $0 < x < 2\pi$] as in the table [$\lim_{x\to 0^+} f(x) = 1$] and draw a direction field as in Example 6. Then we use the direction field to graph F starting at $(0, 0)$.

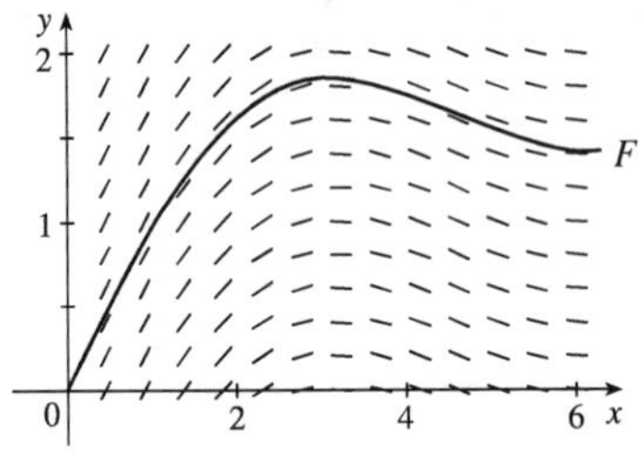

51.

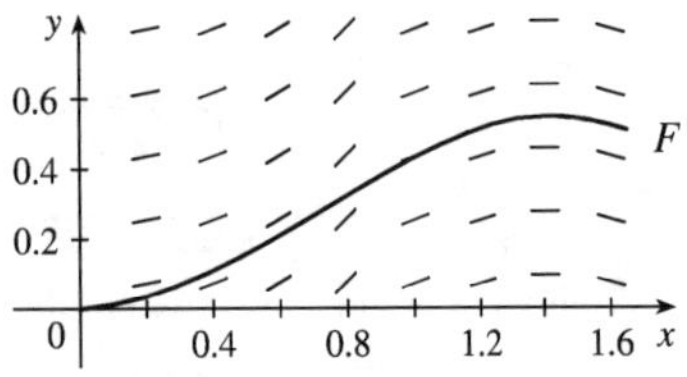

Remember that the given table values of f are the slopes of F at any x. For example, at $x = 1.4$, the slope of F is $f(1.4) = 0$.

53. $v(t) = s'(t) = \sin t - \cos t \;\Rightarrow\; s(t) = -\cos t - \sin t + C$. $s(0) = -1 + C$ and $s(0) = 0 \;\Rightarrow\; C = 1$, so $s(t) = -\cos t - \sin t + 1$.

55. $a(t) = v'(t) = t - 2 \;\Rightarrow\; v(t) = \frac{1}{2}t^2 - 2t + C$. $v(0) = C$ and $v(0) = 3 \;\Rightarrow\; C = 3$, so $v(t) = \frac{1}{2}t^2 - 2t + 3$ and $s(t) = \frac{1}{6}t^3 - t^2 + 3t + D$. $s(0) = D$ and $s(0) = 1 \;\Rightarrow\; D = 1$, and $s(t) = \frac{1}{6}t^3 - t^2 + 3t + 1$.

57. $a(t) = v'(t) = 10\sin t + 3\cos t \;\Rightarrow\; v(t) = -10\cos t + 3\sin t + C \;\Rightarrow$ $s(t) = -10\sin t - 3\cos t + Ct + D$. $s(0) = -3 + D = 0$ and $s(2\pi) = -3 + 2\pi C + D = 12 \;\Rightarrow\; D = 3$ and $C = \frac{6}{\pi}$. Thus, $s(t) = -10\sin t - 3\cos t + \frac{6}{\pi}t + 3$.

59. (a) We first observe that since the stone is dropped 450 m above the ground, $v(0) = 0$ and $s(0) = 450$.
$v'(t) = a(t) = -9.8 \;\Rightarrow\; v(t) = -9.8t + C$. Now $v(0) = 0 \;\Rightarrow\; C = 0$, so $v(t) = -9.8t \;\Rightarrow$ $s(t) = -4.9t^2 + D$. Last, $s(0) = 450 \;\Rightarrow\; D = 450 \;\Rightarrow\; s(t) = 450 - 4.9t^2$.

(b) The stone reaches the ground when $s(t) = 0$. $450 - 4.9t^2 = 0 \;\Rightarrow\; t^2 = 450/4.9 \;\Rightarrow$ $t_1 = \sqrt{450/4.9} \approx 9.58$ s.

(c) The velocity with which the stone strikes the ground is $v(t_1) = -9.8\sqrt{450/4.9} \approx -93.9$ m/s.

(d) This is just reworking parts (a) and (b) with $v(0) = -5$. Using $v(t) = -9.8t + C$, $v(0) = -5 \;\Rightarrow$ $0 + C = -5 \;\Rightarrow\; v(t) = -9.8t - 5$. So $s(t) = -4.9t^2 - 5t + D$ and $s(0) = 450 \;\Rightarrow\; D = 450 \;\Rightarrow$ $s(t) = -4.9t^2 - 5t + 450$. Solving $s(t) = 0$ by using the quadratic formula gives us $t = \left(5 \pm \sqrt{8845}\,\right)/(-9.8) \;\Rightarrow\; t_1 \approx 9.09$ s.

61. By Exercise 60 with $a = -9.8$, $s(t) = -4.9t^2 + v_0 t + s_0$ and $v(t) = s'(t) = -9.8t + v_0$. So $[v(t)]^2 = (-9.8t + v_0)^2 = (9.8)^2\,t^2 - 19.6v_0 t + v_0^2 = v_0^2 + 96.04t^2 - 19.6v_0 t = v_0^2 - 19.6\left(-4.9t^2 + v_0 t\right)$. But $-4.9t^2 + v_0 t$ is just $s(t)$ without the s_0 term; that is, $s(t) - s_0$. Thus, $[v(t)]^2 = v_0^2 - 19.6\,[s(t) - s_0]$.

63. Using Exercise 60 with $a = -32$, $v_0 = 0$, and $s_0 = h$ (the height of the cliff), we know that the height at time t is $s(t) = -16t^2 + h$. $v(t) = s'(t) = -32t$ and $v(t) = -120 \Rightarrow -32t = -120 \Rightarrow t = 3.75$, so $0 = s(3.75) = -16(3.75)^2 + h \Rightarrow h = 16(3.75)^2 = 225$ ft.

65. Marginal cost $= 1.92 - 0.002x = C'(x) \Rightarrow C(x) = 1.92x - 0.001x^2 + K$. But $C(1) = 1.92 - 0.001 + K = 562 \Rightarrow K = 560.081$. Therefore, $C(x) = 1.92x - 0.001x^2 + 560.081 \Rightarrow C(100) = 742.081$, so the cost of producing 100 items is \$742.08.

67. Taking the upward direction to be positive we have that for $0 \le t \le 10$ (using the subscript 1 to refer to $0 \le t \le 10$), $a_1(t) = -(9 - 0.9t) = v_1'(t) \Rightarrow v_1(t) = -9t + 0.45t^2 + v_0$, but $v_1(0) = v_0 = -10 \Rightarrow v_1(t) = -9t + 0.45t^2 - 10 = s_1'(t) \Rightarrow s_1(t) = -\frac{9}{2}t^2 + 0.15t^3 - 10t + s_0$. But $s_1(0) = 500 = s_0 \Rightarrow s_1(t) = -\frac{9}{2}t^2 + 0.15t^3 - 10t + 500$. $s_1(10) = -450 + 150 - 100 + 500 = 100$, so it takes more than 10 seconds for the raindrop to fall. Now for $t > 10$, $a(t) = 0 = v'(t) \Rightarrow v(t) = \text{constant} = v_1(10) = -9(10) + 0.45(10)^2 - 10 = -55 \Rightarrow v(t) = -55$. At 55 ft/s, it will take $100/55 \approx 1.8$ s to fall the last 100 ft. Hence, the total time is $10 + \frac{100}{55} = \frac{130}{11} \approx 11.8$ s.

69. $a(t) = k$, the initial velocity is 30 mi/h $= 30 \cdot \frac{5280}{3600} = 44$ ft/s, and the final velocity (after 5 seconds) is 50 mi/h $= 50 \cdot \frac{5280}{3600} = \frac{220}{3}$ ft/s. So $v(t) = kt + C$ and $v(0) = 44 \Rightarrow C = 44$. Thus, $v(t) = kt + 44 \Rightarrow v(5) = 5k + 44$. But $v(5) = \frac{220}{3}$, so $5k + 44 = \frac{220}{3} \Rightarrow 5k = \frac{88}{3} \Rightarrow k = \frac{88}{15} \approx 5.87 \text{ ft/s}^2$.

71. Let the acceleration be $a(t) = k \text{ km/h}^2$. We have $v(0) = 100$ km/h and we can take the initial position $s(0)$ to be 0. We want the time t_f for which $v(t) = 0$ to satisfy $s(t) < 0.08$ km. In general, $v'(t) = a(t) = k$, so $v(t) = kt + C$, where $C = v(0) = 100$. Now $s'(t) = v(t) = kt + 100$, so $s(t) = \frac{1}{2}kt^2 + 100t + D$, where $D = s(0) = 0$. Thus, $s(t) = \frac{1}{2}kt^2 + 100t$. Since $v(t_f) = 0$, we have $kt_f + 100 = 0$ or $t_f = -100/k$, so

$s(t_f) = \dfrac{1}{2}k\left(-\dfrac{100}{k}\right)^2 + 100\left(-\dfrac{100}{k}\right) = 10{,}000\left(\dfrac{1}{2k} - \dfrac{1}{k}\right) = -\dfrac{5{,}000}{k}$. The condition $s(t_f)$ must satisfy is

$-\dfrac{5{,}000}{k} < 0.08 \Rightarrow -\dfrac{5{,}000}{0.08} > k$ [k is negative] $\Rightarrow k < -62{,}500 \text{ km/h}^2$, or equivalently,

$k < -\frac{3125}{648} \approx -4.82 \text{ m/s}^2$.

73. (a) First note that 90 mi/h $= 90 \times \frac{5280}{3600}$ ft/s $= 132$ ft/s. Then $a(t) = 4 \text{ ft/s}^2 \Rightarrow v(t) = 4t + C$, but $v(0) = 0 \Rightarrow C = 0$. Now $4t = 132$ when $t = \frac{132}{4} = 33$ s, so it takes 33 s to reach 132 ft/s. Therefore, taking $s(0) = 0$, we have $s(t) = 2t^2$, $0 \le t \le 33$. So $s(33) = 2178$ ft. 15 minutes $= 15(60) = 900$ s, so for $33 < t \le 933$ we have $v(t) = 132$ ft/s $\Rightarrow s(933) = 132(900) + 2178 = 120{,}978$ ft $= 22.9125$ mi.

(b) As in part (a), the train accelerates for 33 s and travels 2178 ft while doing so. Similarly, it decelerates for 33 s and travels 2178 ft at the end of its trip. During the remaining $900 - 66 = 834$ s it travels at 132 ft/s, so the distance traveled is $132 \cdot 834 = 110{,}088$ ft. Thus, the total distance is $2178 + 110{,}088 + 2178 = 114{,}444$ ft $= 21.675$ mi.

(c) $45 \text{ mi} = 45(5280) = 237{,}600$ ft. Subtract $2(2178)$ to take care of the speeding up and slowing down, and we have 233,244 ft at 132 ft/s for a trip of $233{,}244/132 = 1767$ s at 90 mi/h. The total time is $1767 + 2(33) = 1833 \text{ s} = 30 \text{ min } 33 \text{ s} = 30.55$ min.

(d) $37.5(60) = 2250$ s. $2250 - 2(33) = 2184$ s at maximum speed. $2184(132) + 2(2178) = 292{,}644$ total feet or $292{,}644/5280 = 55.425$ mi.

Indefinite Integrals and the Net Change Theorem

1. $\frac{d}{dx}\left[\sqrt{x^2+1}+C\right] = \frac{d}{dx}\left[\left(x^2+1\right)^{1/2}+C\right] = \frac{1}{2}\left(x^2+1\right)^{-1/2}\cdot 2x = \frac{x}{\sqrt{x^2+1}}$

3. $\frac{d}{dx}\left[\frac{x}{a^2\sqrt{a^2-x^2}}+C\right] = \frac{1}{a^2}\frac{\sqrt{a^2-x^2}-x(-x)/\sqrt{a^2-x^2}}{a^2-x^2} = \frac{1}{a^2}\frac{\left(a^2-x^2\right)+x^2}{\left(a^2-x^2\right)^{3/2}} = \frac{1}{\sqrt{\left(a^2-x^2\right)^3}}$

5. $\int x^{-3/4}\,dx = \frac{x^{-3/4+1}}{-3/4+1}+C = \frac{x^{1/4}}{1/4}+C = 4x^{1/4}+C$

7. $\int \left(x^3+6x+1\right)dx = \frac{x^4}{4}+6\frac{x^2}{2}+x+C = \frac{1}{4}x^4+3x^2+x+C$

9. $\int (1-t)\left(2+t^2\right)dt = \int \left(2-2t+t^2-t^3\right)dt = 2t-2\frac{t^2}{2}+\frac{t^3}{3}-\frac{t^4}{4}+C = 2t-t^2+\frac{1}{3}t^3-\frac{1}{4}t^4+C$

11. $\int \left(2-\sqrt{x}\right)^2 dx = \int \left(4-4\sqrt{x}+x\right)dx = 4x-4\frac{x^{3/2}}{3/2}+\frac{x^2}{2}+C = 4x-\frac{8}{3}x^{3/2}+\frac{1}{2}x^2+C$

13. $\int \frac{\sin x}{1-\sin^2 x}\,dx = \int \frac{\sin x}{\cos^2 x}\,dx = \int \frac{1}{\cos x}\cdot\frac{\sin x}{\cos x}\,dx = \int \sec x\tan x\,dx = \sec x + C$

15. $\int x\sqrt{x}\,dx = \int x^{3/2}\,dx = \frac{2}{5}x^{5/2} + C.$

The members of the family in the figure correspond to $C = 5, 3, 0, -2$, and -4.

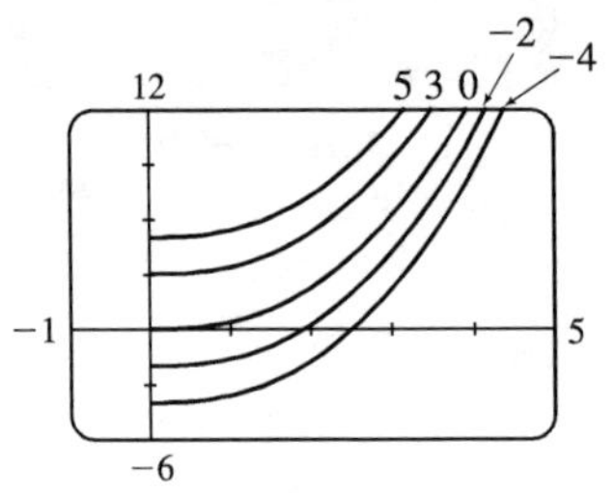

17. $\int_0^2 (6x^2 - 4x + 5)\,dx = \left[6 \cdot \frac{1}{3}x^3 - 4 \cdot \frac{1}{2}x^2 + 5x\right]_0^2 = \left[2x^3 - 2x^2 + 5x\right]_0^2 = (16 - 8 + 10) - 0 = 18$

19. $\int_{-3}^{0} (5y^4 - 6y^2 + 14)\,dy = \left[5\left(\frac{1}{5}y^5\right) - 6\left(\frac{1}{3}y^3\right) + 14y\right]_{-3}^{0} = \left[y^5 - 2y^3 + 14y\right]_{-3}^{0}$
$= 0 - (-243 + 54 - 42) = 231$

21. $\int_{-2}^{2} (3u + 1)^2\,du = \int_{-2}^{2} (9u^2 + 6u + 1)\,du = \left[9 \cdot \frac{1}{3}u^3 + 6 \cdot \frac{1}{2}u^2 + u\right]_{-2}^{2} = \left[3u^3 + 3u^2 + u\right]_{-2}^{2}$
$= (24 + 12 + 2) - (-24 + 12 - 2) = 38 - (-14) = 52$

23. $\int_1^4 \sqrt{t}\,(1 + t)\,dt = \int_1^4 (t^{1/2} + t^{3/2})\,dt = \left[\frac{2}{3}t^{3/2} + \frac{2}{5}t^{5/2}\right]_1^4 = \left(\frac{16}{3} + \frac{64}{5}\right) - \left(\frac{2}{3} + \frac{2}{5}\right) = \frac{14}{3} + \frac{62}{5} = \frac{256}{15}$

25. $\displaystyle\int_{-2}^{-1} \left(4y^3 + \frac{2}{y^3}\right) dy = \left[4 \cdot \frac{1}{4}y^4 + 2 \cdot \frac{1}{-2}y^{-2}\right]_{-2}^{-1} = \left[y^4 - \frac{1}{y^2}\right]_{-2}^{-1} = (1 - 1) - \left(16 - \frac{1}{4}\right) = -\frac{63}{4}$

27. $\int_0^1 x(\sqrt[3]{x} + \sqrt[4]{x}\,)\,dx = \int_0^1 (x^{4/3} + x^{5/4})\,dx = \left[\frac{3}{7}x^{7/3} + \frac{4}{9}x^{9/4}\right]_0^1 = \left(\frac{3}{7} + \frac{4}{9}\right) - 0 = \frac{55}{63}$

29. $\int_1^4 \sqrt{5/x}\,dx = \sqrt{5}\int_1^4 x^{-1/2}\,dx = \sqrt{5}\,[2\sqrt{x}\,]_1^4 = \sqrt{5}\,(2 \cdot 2 - 2 \cdot 1) = 2\sqrt{5}$

31. $\int_0^\pi (4\sin\theta - 3\cos\theta)\,d\theta = [-4\cos\theta - 3\sin\theta]_0^\pi = (4 - 0) - (-4 - 0) = 8$

33. $\displaystyle\int_0^{\pi/4} \frac{1 + \cos^2\theta}{\cos^2\theta}\,d\theta = \int_0^{\pi/4} \left(\frac{1}{\cos^2\theta} + \frac{\cos^2\theta}{\cos^2\theta}\right) d\theta = \int_0^{\pi/4} (\sec^2\theta + 1)\,d\theta$
$= [\tan\theta + \theta]_0^{\pi/4} = \left(\tan\frac{\pi}{4} + \frac{\pi}{4}\right) - (0 + 0) = 1 + \frac{\pi}{4}$

35. $\displaystyle\int_1^{64} \frac{1 + \sqrt[3]{x}}{\sqrt{x}}\,dx = \int_1^{64} \left(\frac{1}{x^{1/2}} + \frac{x^{1/3}}{x^{1/2}}\right) dx = \int_1^{64} (x^{-1/2} + x^{(1/3)-(1/2)})\,dx = \int_1^{64} (x^{-1/2} + x^{-1/6})\,dx$
$= \left[2x^{1/2} + \frac{6}{5}x^{5/6}\right]_1^{64} = \left(16 + \frac{192}{5}\right) - \left(2 + \frac{6}{5}\right) = 14 + \frac{186}{5} = \frac{256}{5}$

37. $\int_0^1 \left(\sqrt[4]{x^5} + \sqrt[5]{x^4}\right) dx = \int_0^1 \left(x^{5/4} + x^{4/5}\right) dx = \left[\dfrac{x^{9/4}}{9/4} + \dfrac{x^{9/5}}{9/5}\right]_0^1 = \left[\frac{4}{9}x^{9/4} + \frac{5}{9}x^{9/5}\right]_0^1 = \frac{4}{9} + \frac{5}{9} - 0 = 1$

39. $\int_{-1}^{2} (x - 2\,|x|)\,dx = \int_{-1}^{0} [x - 2(-x)]\,dx + \int_0^2 [x - 2(x)]\,dx = \int_{-1}^{0} 3x\,dx + \int_0^2 (-x)\,dx = 3\left[\frac{1}{2}x^2\right]_{-1}^{0} - \left[\frac{1}{2}x^2\right]_0^2$
$= 3\left(0 - \frac{1}{2}\right) - (2 - 0) = -\frac{7}{2} = -3.5$

41. The graph shows that $y = x + x^2 - x^4$ has x-intercepts at $x = 0$ and at $x = a \approx 1.32$. So the area of the region that lies under the curve and above the x-axis is

$$\int_0^a (x + x^2 - x^4)\,dx = \left[\tfrac{1}{2}x^2 + \tfrac{1}{3}x^3 - \tfrac{1}{5}x^5\right]_0^a = \left(\tfrac{1}{2}a^2 + \tfrac{1}{3}a^3 - \tfrac{1}{5}a^5\right) - 0 \approx 0.84$$

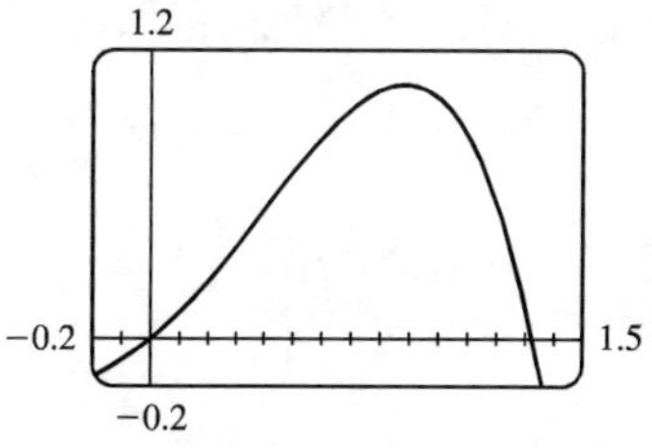

43. $A = \int_0^2 (2y - y^2)\,dy = \left[y^2 - \frac{1}{3}y^3\right]_0^2 = \left(4 - \frac{8}{3}\right) - 0 = \frac{4}{3}$

45. If $w'(t)$ is the rate of change of weight in pounds per year, then $w(t)$ represents the weight in pounds of the child at age t. We know from the Net Change Theorem that $\int_5^{10} w'(t)\,dt = w(10) - w(5)$, so the integral represents the increase in the child's weight (in pounds) between the ages of 5 and 10.

47. Since $r(t)$ is the rate at which oil leaks, we can write $r(t) = -V'(t)$, where $V(t)$ is the volume of oil at time t. [Note that the minus sign is needed because V is decreasing, so $V'(t)$ is negative, but $r(t)$ is positive.] Thus, by the Net Change Theorem, $\int_0^{120} r(t)\,dt = -\int_0^{120} V'(t)\,dt = -\left[V(120) - V(0)\right] = V(0) - V(120)$, which is the number of gallons of oil that leaked from the tank in the first two hours (120 minutes).

49. By the Net Change Theorem, $\int_{1000}^{5000} R'(x)\,dx = R(5000) - R(1000)$, so it represents the increase in revenue when production is increased from 1000 units to 5000 units.

51. In general, the unit of measurement for $\int_a^b f(x)\,dx$ is the product of the unit for $f(x)$ and the unit for x. Since $f(x)$ is measured in newtons and x is measured in meters, the units for $\int_0^{100} f(x)\,dx$ are newton-meters. (A newton-meter is abbreviated N-m and is called a joule.)

53. (a) displacement $= \int_0^3 (3t - 5)\,dt = \left[\frac{3}{2}t^2 - 5t\right]_0^3 = \frac{27}{2} - 15 = -\frac{3}{2}$ m

(b) distance traveled $= \int_0^3 |3t - 5|\,dt = \int_0^{5/3} (5 - 3t)\,dt + \int_{5/3}^3 (3t - 5)\,dt$

$$= \left[5t - \tfrac{3}{2}t^2\right]_0^{5/3} + \left[\tfrac{3}{2}t^2 - 5t\right]_{5/3}^3 = \tfrac{25}{3} - \tfrac{3}{2}\cdot\tfrac{25}{9} + \tfrac{27}{2} - 15 - \left(\tfrac{3}{2}\cdot\tfrac{25}{9} - \tfrac{25}{3}\right) = \tfrac{41}{6}\text{ m}$$

55. (a) $v'(t) = a(t) = t + 4 \;\Rightarrow\; v(t) = \frac{1}{2}t^2 + 4t + C \;\Rightarrow\; v(0) = C = 5 \;\Rightarrow\; v(t) = \frac{1}{2}t^2 + 4t + 5$ m/s

(b) distance traveled $= \int_0^{10} |v(t)|\,dt = \int_0^{10} \left|\frac{1}{2}t^2 + 4t + 5\right|\,dt = \int_0^{10} \left(\frac{1}{2}t^2 + 4t + 5\right)dt$

$$= \left[\tfrac{1}{6}t^3 + 2t^2 + 5t\right]_0^{10} = \tfrac{500}{3} + 200 + 50 = 416\tfrac{2}{3}\text{ m}$$

57. Since $m'(x) = \rho(x)$, $m = \int_0^4 \rho(x)\,dx = \int_0^4 (9 + 2\sqrt{x}\,)\,dx = \left[9x + \frac{4}{3}x^{3/2}\right]_0^4 = 36 + \frac{32}{3} - 0 = \frac{140}{3} = 46\frac{2}{3}$ kg.

59. Let s be the position of the car. We know from Equation 2 that $s(100) - s(0) = \int_0^{100} v(t)\,dt$. We use the Midpoint Rule for $0 \le t \le 100$ with $n = 5$. Note that the length of each of the five time intervals is 20 seconds $= \frac{20}{3600}$ hour $= \frac{1}{180}$ hour. So the distance traveled is

$$\int_0^{100} v(t)\,dt \approx \tfrac{1}{180}\left[v(10) + v(30) + v(50) + v(70) + v(90)\right] = \tfrac{1}{180}(38 + 58 + 51 + 53 + 47) = \tfrac{247}{180} \approx 1.4\text{ miles}$$

61. From the Net Change Theorem, the increase in cost if the production level is raised from 2000 yards to 4000 yards is $C(4000) - C(2000) = \int_{2000}^{4000} C'(x)\,dx$.

$$\int_{2000}^{4000} C'(x)\,dx = \int_{2000}^{4000} (3 - 0.01x + 0.000006x^2)\,dx$$

$$= \left[3x - 0.005x^2 + 0.000002x^3\right]_{2000}^{4000} = 60{,}000 - 2{,}000 = \$58{,}000$$

63. (a) We can find the area between the Lorenz curve and the line $y = x$ by subtracting the area under $y = L(x)$ from the area under $y = x$. Thus,

$$\text{coefficient of inequality} = \frac{\text{area between Lorenz curve and line } y = x}{\text{area under line } y = x} = \frac{\int_0^1 [x - L(x)]\,dx}{\int_0^1 x\,dx}$$

$$= \frac{\int_0^1 [x - L(x)]\,dx}{[x^2/2]_0^1} = \frac{\int_0^1 [x - L(x)]\,dx}{1/2} = 2\int_0^1 [x - L(x)]\,dx$$

(b) $L(x) = \frac{5}{12}x^2 + \frac{7}{12}x \;\Rightarrow\; L(50\%) = L\left(\frac{1}{2}\right) = \frac{5}{48} + \frac{7}{24} = \frac{19}{48} = 0.3958\overline{3}$, so the bottom 50% of the households receive at most about 40% of the income. Using the result in part (a),

$$\text{coefficient of inequality} = 2\int_0^1 [x - L(x)]\,dx = 2\int_0^1 \left(x - \tfrac{5}{12}x^2 - \tfrac{7}{12}x\right)dx$$

$$= 2\int_0^1 \left(\tfrac{5}{12}x - \tfrac{5}{12}x^2\right)dx = 2\int_0^1 \tfrac{5}{12}\left(x - x^2\right)dx$$

$$= \tfrac{5}{6}\left[\tfrac{1}{2}x^2 - \tfrac{1}{3}x^3\right]_0^1 = \tfrac{5}{6}\left(\tfrac{1}{2} - \tfrac{1}{3}\right) = \tfrac{5}{6}\left(\tfrac{1}{6}\right) = \tfrac{5}{36}$$

65. $$\int_1^e \frac{x^2 + x + 1}{x}\,dx = \int_1^e \left(x + 1 + \frac{1}{x}\right)dx = \left[\tfrac{1}{2}x^2 + x + \ln|x|\right]_1^e$$

$$= \left(\tfrac{1}{2}e^2 + e + \ln e\right) - \left(\tfrac{1}{2} + 1 + \ln 1\right) = \tfrac{1}{2}e^2 + e - \tfrac{1}{2}$$

67. $$\int \left(x^2 + 1 + \frac{1}{x^2 + 1}\right)dx = \frac{x^3}{3} + x + \tan^{-1} x + C$$

The Substitution Rule

1. Let $u = 3x$. Then $du = 3\,dx$, so $dx = \frac{1}{3}du$. Thus,

$\int \cos 3x\,dx = \int \cos u\left(\frac{1}{3}\,du\right) = \frac{1}{3}\int \cos u\,du = \frac{1}{3}\sin u + C = \frac{1}{3}\sin 3x + C$. Don't forget that it is often very easy to check an indefinite integration by differentiating your answer. In this case,

$\dfrac{d}{dx}\left(\frac{1}{3}\sin 3x + C\right) = \frac{1}{3}(\cos 3x)\cdot 3 = \cos 3x$, the desired result.

3. Let $u = x^3 + 1$. Then $du = 3x^2\,dx$ and $x^2\,dx = \frac{1}{3}\,du$, so

$$\int x^2\sqrt{x^3 + 1}\,dx = \int \sqrt{u}\left(\tfrac{1}{3}\,du\right) = \frac{1}{3}\frac{u^{3/2}}{3/2} + C = \tfrac{1}{3}\cdot\tfrac{2}{3}u^{3/2} + C = \tfrac{2}{9}(x^3 + 1)^{3/2} + C.$$

5. Let $u = 1 + 2x$. Then $du = 2\,dx$ and $dx = \frac{1}{2}\,du$, so

$$\int \frac{4}{(1 + 2x)^3}\,dx = 4\int u^{-3}\left(\tfrac{1}{2}\,du\right) = 2\frac{u^{-2}}{-2} + C = -\frac{1}{u^2} + C = -\frac{1}{(1 + 2x)^2} + C.$$

7. Let $u = x^2 + 3$. Then $du = 2x\,dx$, so $\int 2x(x^2+3)^4\,dx = \int u^4\,du = \frac{1}{5}u^5 + C = \frac{1}{5}(x^2+3)^5 + C$.

9. Let $u = 3x - 2$. Then $du = 3\,dx$ and $dx = \frac{1}{3}\,du$, so
$\int (3x-2)^{20}\,dx = \int u^{20}\left(\frac{1}{3}\,du\right) = \frac{1}{3}\cdot\frac{1}{21}u^{21} + C = \frac{1}{63}(3x-2)^{21} + C$.

11. Let $u = 1 + x + 2x^2$. Then $du = (1+4x)\,dx$, so

$$\int \frac{1+4x}{\sqrt{1+x+2x^2}}\,dx = \int \frac{du}{\sqrt{u}} = \int u^{-1/2}\,du = \frac{u^{1/2}}{1/2} + C = 2\sqrt{1+x+2x^2} + C.$$

13. Let $u = 2y + 1$. Then $du = 2\,dy$ and $dy = \frac{1}{2}\,du$, so

$$\int \frac{3}{(2y+1)^5}\,dy = \int 3u^{-5}\left(\tfrac{1}{2}\,du\right) = \frac{3}{2}\cdot\frac{1}{-4}u^{-4} + C = \frac{-3}{8(2y+1)^4} + C.$$

15. Let $u = 4 - t$. Then $du = -dt$ and $dt = -du$, so
$\int \sqrt{4-t}\,dt = \int u^{1/2}\,(-du) = -\frac{2}{3}u^{3/2} + C = -\frac{2}{3}(4-t)^{3/2} + C$.

17. Let $u = \pi t$. Then $du = \pi\,dt$ and $dt = \frac{1}{\pi}\,du$, so
$\int \sin \pi t\,dt = \int \sin u\left(\frac{1}{\pi}\,du\right) = \frac{1}{\pi}(-\cos u) + C = -\frac{1}{\pi}\cos \pi t + C$.

19. Let $u = \sqrt{t}$. Then $du = \dfrac{dt}{2\sqrt{t}}$ and $\dfrac{1}{\sqrt{t}}\,dt = 2\,du$, so

$$\int \frac{\cos\sqrt{t}}{\sqrt{t}}\,dt = \int \cos u\,(2\,du) = 2\sin u + C = 2\sin\sqrt{t} + C.$$

21. Let $u = \sin\theta$. Then $du = \cos\theta\,d\theta$, so $\int \cos\theta\,\sin^6\theta\,d\theta = \int u^6\,du = \frac{1}{7}u^7 + C = \frac{1}{7}\sin^7\theta + C$.

23. Let $u = 1 + z^3$. Then $du = 3z^2\,dz$ and $z^2\,dz = \frac{1}{3}\,du$, so

$$\int \frac{z^2}{\sqrt[3]{1+z^3}}\,dz = \int u^{-1/3}\left(\tfrac{1}{3}\,du\right) = \tfrac{1}{3}\cdot\tfrac{3}{2}u^{2/3} + C = \tfrac{1}{2}(1+z^3)^{2/3} + C.$$

25. Let $u = \cot x$. Then $du = -\csc^2 x\,dx$ and $\csc^2 x\,dx = -du$, so

$$\int \sqrt{\cot x}\,\csc^2 x\,dx = \int \sqrt{u}\,(-du) = -\frac{u^{3/2}}{3/2} + C = -\tfrac{2}{3}(\cot x)^{3/2} + C.$$

27. Let $u = \sec x$. Then $du = \sec x\,\tan x\,dx$, so
$\int \sec^3 x\,\tan x\,dx = \int \sec^2 x\,(\sec x\,\tan x)\,dx = \int u^2\,du = \frac{1}{3}u^3 + C = \frac{1}{3}\sec^3 x + C$.

29. Let $u = b + cx^{a+1}$. Then $du = (a+1)cx^a\,dx$, so

$$\int x^a\sqrt{b+cx^{a+1}}\,dx = \int u^{1/2}\frac{1}{(a+1)c}\,du = \frac{1}{(a+1)c}\left(\tfrac{2}{3}u^{3/2}\right) + C = \frac{2}{3c(a+1)}\left(b+cx^{a+1}\right)^{3/2} + C.$$

31. Let $u = x + 2$. Then $du = dx$, so

$$\begin{aligned}\int \frac{x}{\sqrt[4]{x+2}}\,dx &= \int \frac{u-2}{\sqrt[4]{u}}\,du = \int \left(u^{3/4} - 2u^{-1/4}\right)du = \tfrac{4}{7}u^{7/4} - 2\cdot\tfrac{4}{3}u^{3/4} + C \\ &= \tfrac{4}{7}(x+2)^{7/4} - \tfrac{8}{3}(x+2)^{3/4} + C\end{aligned}$$

In Exercises 33–36, let $f(x)$ denote the integrand and $F(x)$ its antiderivative (with $C = 0$).

33. $f(x) = \dfrac{3x-1}{(3x^2-2x+1)^4}$.

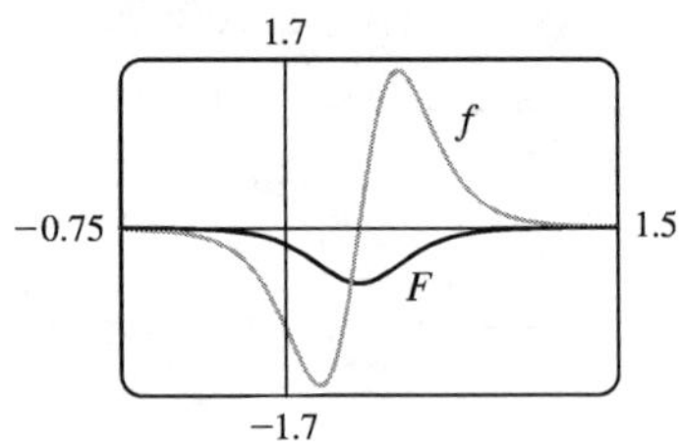

$u = 3x^2 - 2x + 1 \quad \Rightarrow \quad du = (6x-2)\,dx = 2(3x-1)\,dx$, so

$$\int \frac{3x-1}{(3x^2-2x+1)^4}\,dx = \int \frac{1}{u^4}\left(\frac{1}{2}\,du\right) = \frac{1}{2}\int u^{-4}\,du$$

$$= -\frac{1}{6}u^{-3} + C = -\frac{1}{6\,(3x^2-2x+1)^3} + C$$

Notice that at $x = \frac{1}{3}$, f changes from negative to positive, and F has a local minimum.

35. $f(x) = \sin^3 x\,\cos x$. $u = \sin x \quad \Rightarrow \quad du = \cos x\,dx$, so

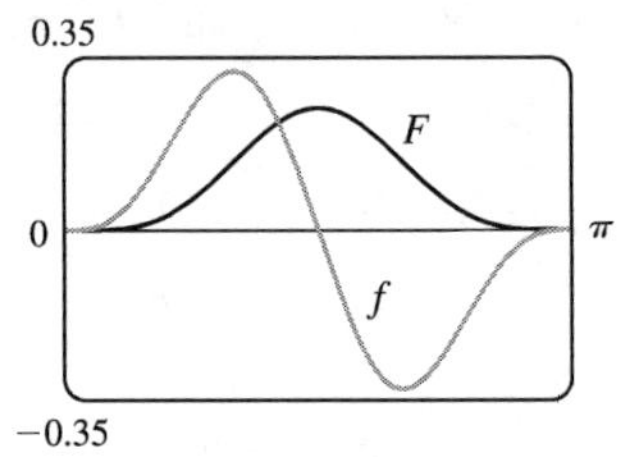

$$\textstyle\int \sin^3 x\,\cos x\,dx = \int u^3\,du = \frac{1}{4}u^4 + C = \frac{1}{4}\sin^4 x + C$$

Note that at $x = \frac{\pi}{2}$, f changes from positive to negative and F has a local maximum. Also, both f and F are periodic with period π, so at $x = 0$ and at $x = \pi$, f changes from negative to positive and F has local minima.

37. Let $u = x - 1$, so $du = dx$. When $x = 0$, $u = -1$; when $x = 2$, $u = 1$. Thus, $\int_0^2 (x-1)^{25}\,dx = \int_{-1}^1 u^{25}\,du = 0$ by Theorem 6(b), since $f(u) = u^{25}$ is an odd function.

39. Let $u = 1 + 2x^3$, so $du = 6x^2\,dx$. When $x = 0$, $u = 1$; when $x = 1$, $u = 3$. Thus,

$$\textstyle\int_0^1 x^2(1+2x^3)^5\,dx = \int_1^3 u^5\left(\frac{1}{6}\,du\right) = \frac{1}{6}\left[\frac{1}{6}u^6\right]_1^3 = \frac{1}{36}(3^6 - 1^6) = \frac{1}{36}(729-1) = \frac{728}{36} = \frac{182}{9}$$

41. Let $u = t/4$, so $du = \frac{1}{4}\,dt$. When $t = 0$, $u = 0$; when $t = \pi$, $u = \pi/4$. Thus,
$\int_0^\pi \sec^2(t/4)\,dt = \int_0^{\pi/4} \sec^2 u\,(4\,du) = 4\big[\tan u\big]_0^{\pi/4} = 4\big(\tan\frac{\pi}{4} - \tan 0\big) = 4(1-0) = 4.$

43. $\int_{-\pi/6}^{\pi/6} \tan^3\theta\,d\theta = 0$ by Theorem 6(b), since $f(\theta) = \tan^3\theta$ is an odd function.

45. Let $u = \cos\theta$, so $du = -\sin\theta\,d\theta$. When $\theta = 0$, $u = 1$; when $\theta = \frac{\pi}{3}$, $u = \frac{1}{2}$. Thus,

$$\int_0^{\pi/3} \frac{\sin\theta}{\cos^2\theta}\,d\theta = \int_1^{1/2} \frac{-du}{u^2} = \int_{1/2}^1 u^{-2}\,du = \left[-\frac{1}{u}\right]_{1/2}^1 = -1 - (-2) = 1.$$

47. Let $u = 1 + 2x$, so $du = 2\,dx$. When $x = 0$, $u = 1$; when $x = 13$, $u = 27$. Thus,

$$\int_0^{13} \frac{dx}{\sqrt[3]{(1+2x)^2}} = \int_1^{27} u^{-2/3}\left(\tfrac{1}{2}\,du\right) = \left[\tfrac{1}{2}\cdot 3u^{1/3}\right]_1^{27} = \tfrac{3}{2}(3-1) = 3.$$

49. Let $u = x - 1$, so $u + 1 = x$ and $du = dx$. When $x = 1$, $u = 0$; when $x = 2$, $u = 1$. Thus,
$\int_1^2 x\sqrt{x-1}\,dx = \int_0^1 (u+1)\sqrt{u}\,du = \int_0^1 (u^{3/2} + u^{1/2})\,du = \left[\frac{2}{5}u^{5/2} + \frac{2}{3}u^{3/2}\right]_0^1 = \frac{2}{5} + \frac{2}{3} = \frac{16}{15}.$

51. $\displaystyle\int_0^4 \frac{dx}{(x-2)^3}$ does not exist since $f(x) = \dfrac{1}{(x-2)^3}$ has an infinite discontinuity at $x = 2$.

53. Let $u = x^2 + a^2$, so $du = 2x\,dx$ and $x\,dx = \frac{1}{2}\,du$. When $x = 0$, $u = a^2$; when $x = a$, $u = 2a^2$. Thus,

$$\int_0^a x\sqrt{x^2+a^2}\,dx = \int_{a^2}^{2a^2} u^{1/2}\left(\tfrac{1}{2}\,du\right) = \tfrac{1}{2}\left[\tfrac{2}{3}u^{3/2}\right]_{a^2}^{2a^2} = \left[\tfrac{1}{3}u^{3/2}\right]_{a^2}^{2a^2}$$
$$= \tfrac{1}{3}\left[(2a^2)^{3/2} - (a^2)^{3/2}\right] = \tfrac{1}{3}\left(2\sqrt{2}-1\right)a^3$$

55. From the graph, it appears that the area under the curve is about $1 + \left(\text{a little more than } \frac{1}{2}\cdot 1\cdot 0.7\right)$, or about 1.4. The exact area is given by $A = \int_0^1 \sqrt{2x+1}\,dx$. Let $u = 2x+1$, so $du = 2\,dx$. The limits change to $2\cdot 0+1 = 1$ and $2\cdot 1+1 = 3$, and

$A = \int_1^3 \sqrt{u}\left(\frac{1}{2}\,du\right) = \frac{1}{2}\left[\frac{2}{3}u^{3/2}\right]_1^3 = \frac{1}{3}\left(3\sqrt{3}-1\right) = \sqrt{3} - \frac{1}{3} \approx 1.399.$

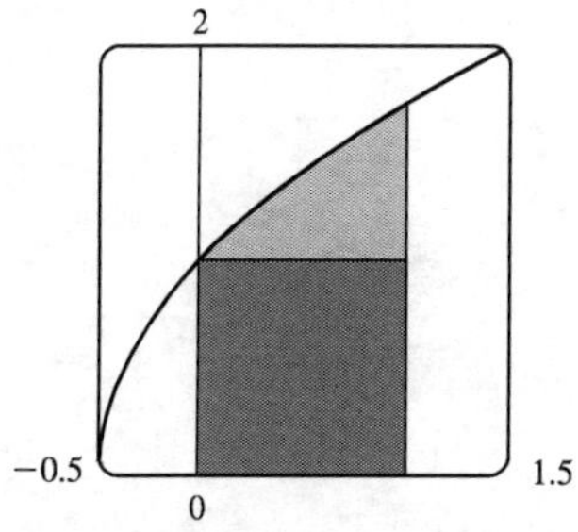

57. First write the integral as a sum of two integrals:
$I = \int_{-2}^2 (x+3)\sqrt{4-x^2}\,dx = I_1 + I_2 = \int_{-2}^2 x\sqrt{4-x^2}\,dx + \int_{-2}^2 3\sqrt{4-x^2}\,dx$. $I_1 = 0$ by Theorem 6(b), since $f(x) = x\sqrt{4-x^2}$ is an odd function and we are integrating from $x = -2$ to $x = 2$. We interpret I_2 as three times the area of a semicircle with radius 2, so $I = 0 + 3\cdot\frac{1}{2}\left(\pi\cdot 2^2\right) = 6\pi$.

59. The volume of inhaled air in the lungs at time t is

$$V(t) = \int_0^t f(u)\,du = \int_0^t \tfrac{1}{2}\sin\left(\tfrac{2\pi}{5}\,u\right)du = \int_0^{2\pi t/5} \tfrac{1}{2}\sin v\left(\tfrac{5}{2\pi}\,dv\right) \quad \left[\text{substitute } v = \tfrac{2\pi}{5}u,\ dv = \tfrac{2\pi}{5}\,du\right]$$
$$= \tfrac{5}{4\pi}\left[-\cos v\right]_0^{2\pi t/5} = \tfrac{5}{4\pi}\left[-\cos\left(\tfrac{2\pi}{5}t\right)+1\right] = \tfrac{5}{4\pi}\left[1-\cos\left(\tfrac{2\pi}{5}t\right)\right] \text{ liters}$$

61. Let $u = 2x$. Then $du = 2\,dx$, so $\int_0^2 f(2x)\,dx = \int_0^4 f(u)\left(\frac{1}{2}\,du\right) = \frac{1}{2}\int_0^4 f(u)\,du = \frac{1}{2}(10) = 5$.

63. (a) Let $u = -x$. Then $du = -dx$, so

$\int_a^b f(-x)\,dx = \int_{-a}^{-b} f(u)(-du) = \int_{-b}^{-a} f(u)\,du = \int_{-b}^{-a} f(x)\,dx$

From the diagram, we see that the equality follows from the fact that we are reflecting the graph of f, and the limits of integration, about the y-axis.

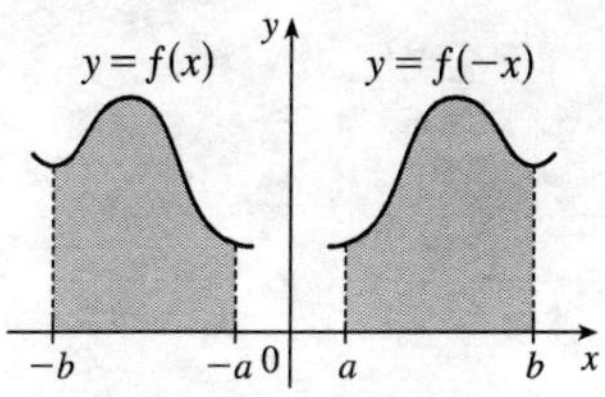

(b) Let $u = x + c$. Then $du = dx$, so

$\int_a^b f(x+c)\,dx = \int_{a+c}^{b+c} f(u)\,du = \int_{a+c}^{b+c} f(x)\,dx$

From the diagram, we see that the equality follows from the fact that we are translating the graph of f, and the limits of integration, by a distance c.

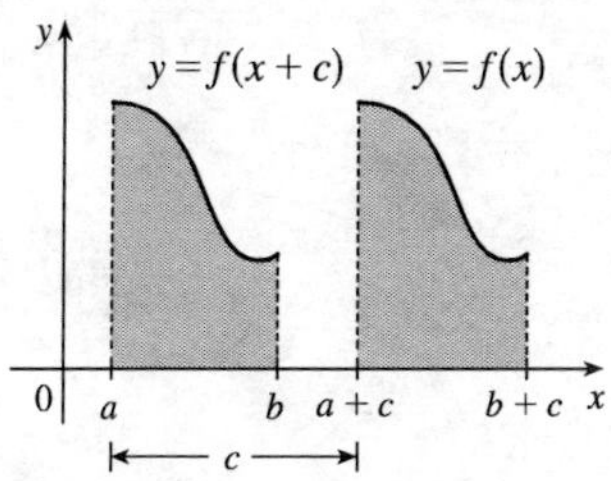

65. Let $u = 1 - x$. Then $x = 1 - u$ and $dx = -du$, so
$\int_0^1 x^a(1-x)^b\,dx = \int_1^0 (1-u)^a\,u^b(-du) = \int_0^1 u^b(1-u)^a\,du = \int_0^1 x^b(1-x)^a\,dx.$

67. Let $u = 5 - 3x$. Then $du = -3\,dx$ and $dx = -\frac{1}{3}\,du$, so

$$\int \frac{dx}{5-3x} = \int \frac{1}{u}\left(-\tfrac{1}{3}\,du\right) = -\tfrac{1}{3}\ln|u| + C = -\tfrac{1}{3}\ln|5-3x| + C.$$

69. Let $u = \ln x$. Then $du = \dfrac{dx}{x}$, so $\displaystyle\int \frac{(\ln x)^2}{x}\,dx = \int u^2\,du = \tfrac{1}{3}u^3 + C = \tfrac{1}{3}(\ln x)^3 + C$.

71. Let $u = 1 + e^x$. Then $du = e^x\,dx$, so $\int e^x\sqrt{1+e^x}\,dx = \int \sqrt{u}\,du = \frac{2}{3}u^{3/2} + C = \frac{2}{3}(1+e^x)^{3/2} + C$.
Or: Let $u = \sqrt{1+e^x}$. Then $u^2 = 1 + e^x$ and $2u\,du = e^x\,dx$, so
$\int e^x\sqrt{1+e^x}\,dx = \int u\cdot 2u\,du = \frac{2}{3}u^3 + C = \frac{2}{3}(1+e^x)^{3/2} + C$.

73. Let $u = \ln x$. Then $du = \dfrac{dx}{x}$, so $\displaystyle\int \frac{dx}{x\ln x} = \int \frac{du}{u} = \ln|u| + C = \ln|\ln x| + C$.

75. $\displaystyle\int \cot x\,dx = \int \frac{\cos x}{\sin x}\,dx$. Let $u = \sin x$. Then $du = \cos x\,dx$, so
$\int \cot x\,dx = \int \frac{1}{u}\,du = \ln|u| + C = \ln|\sin x| + C$.

77. Let $u = 1 + x^2$. Then $du = 2x\,dx$, so

$$\int \frac{1+x}{1+x^2}\,dx = \int \frac{1}{1+x^2}\,dx + \int \frac{x}{1+x^2}\,dx = \tan^{-1}x + \int \frac{\frac{1}{2}du}{u} = \tan^{-1}x + \tfrac{1}{2}\ln|u| + C$$
$$= \tan^{-1}x + \tfrac{1}{2}\ln\left|1+x^2\right| + C = \tan^{-1}x + \tfrac{1}{2}\ln\left(1+x^2\right) + C \text{ [since } 1+x^2 > 0\text{]}.$$

79. Let $u = 1/x$, so $du = -1/x^2\,dx$. When $x = 1$, $u = 1$; when $x = 2$, $u = \frac{1}{2}$. Thus,

$$\int_1^2 \frac{e^{1/x}}{x^2}\,dx = \int_1^{1/2} e^u\,(-du) = -\left[e^u\right]_1^{1/2} = -(e^{1/2} - e) = e - \sqrt{e}.$$

81. Let $u = \ln x$, so $du = \dfrac{dx}{x}$. When $x = e$, $u = 1$; when $x = e^4$; $u = 4$. Thus,

$$\int_e^{e^4} \frac{dx}{x\sqrt{\ln x}} = \int_1^4 u^{-1/2}\,du = 2\left[u^{1/2}\right]_1^4 = 2(2-1) = 2.$$

83. $\dfrac{x\sin x}{1+\cos^2 x} = x\cdot\dfrac{\sin x}{2-\sin^2 x} = xf(\sin x)$, where $f(t) = \dfrac{t}{2-t^2}$. By Exercise 66,

$$\int_0^\pi \frac{x\sin x}{1+\cos^2 x}\,dx = \int_0^\pi xf(\sin x)\,dx = \frac{\pi}{2}\int_0^\pi f(\sin x)\,dx = \frac{\pi}{2}\int_0^\pi \frac{\sin x}{1+\cos^2 x}\,dx$$

Let $u = \cos x$. Then $du = -\sin x\,dx$. When $x = \pi$, $u = -1$ and when $x = 0$, $u = 1$. So

$$\frac{\pi}{2}\int_0^\pi \frac{\sin x}{1+\cos^2 x}\,dx = -\frac{\pi}{2}\int_1^{-1} \frac{du}{1+u^2} = \frac{\pi}{2}\int_{-1}^1 \frac{du}{1+u^2} = \frac{\pi}{2}\left[\tan^{-1}u\right]_{-1}^1$$
$$= \frac{\pi}{2}\left[\tan^{-1}1 - \tan^{-1}(-1)\right] = \frac{\pi}{2}\left[\frac{\pi}{4} - \left(-\frac{\pi}{4}\right)\right] = \frac{\pi^2}{4}$$

The Mean Value Theorem

1. $f(x) = x^2 - 4x + 1$, $[0, 4]$. Since f is a polynomial, it is continuous and differentiable on $\mathbb{R}$, so it is continuous on $[0, 4]$ and differentiable on $(0, 4)$. Also, $f(0) = 1 = f(4)$. $f'(c) = 0 \Leftrightarrow 2c - 4 = 0 \Leftrightarrow c = 2$, which is in the open interval $(0, 4)$, so $c = 2$ satisfies the conclusion of Rolle's Theorem.

3. $f(x) = \sin 2\pi x$, $[-1, 1]$. f, being the composite of the sine function and the polynomial $2\pi x$, is continuous and differentiable on $\mathbb{R}$, so it is continuous on $[-1, 1]$ and differentiable on $(-1, 1)$. Also, $f(-1) = 0 = f(1)$. $f'(c) = 0 \Leftrightarrow 2\pi \cos 2\pi c = 0 \Leftrightarrow \cos 2\pi c = 0 \Leftrightarrow 2\pi c = \pm\frac{\pi}{2} + 2\pi n \Leftrightarrow c = \pm\frac{1}{4} + n$. If $n = 0$ or ± 1, then $c = \pm\frac{1}{4}, \pm\frac{3}{4}$ is in $(-1, 1)$.

5. $f(x) = 1 - x^{2/3}$. $f(-1) = 1 - (-1)^{2/3} = 1 - 1 = 0 = f(1)$. $f'(x) = -\frac{2}{3}x^{-1/3}$, so $f'(c) = 0$ has no solution. This does not contradict Rolle's Theorem, since $f'(0)$ does not exist, and so f is not differentiable on $(-1, 1)$.

7. $\dfrac{f(8) - f(0)}{8 - 0} = \dfrac{6 - 4}{8} = \dfrac{1}{4}$. The values of c which satisfy $f'(c) = \frac{1}{4}$ seem to be about $c = 0.8, 3.2, 4.4$, and 6.1.

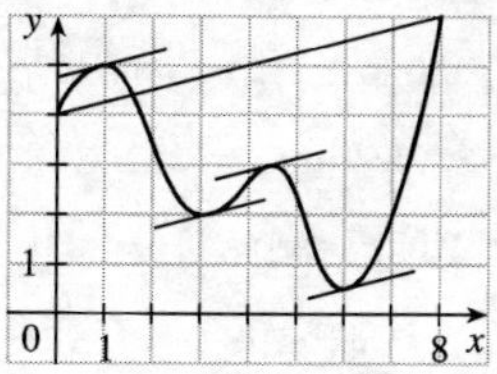

9. (a), (b) The equation of the secant line is

$$y - 5 = \frac{8.5 - 5}{8 - 1}(x - 1) \Leftrightarrow$$

$y = \frac{1}{2}x + \frac{9}{2}$.

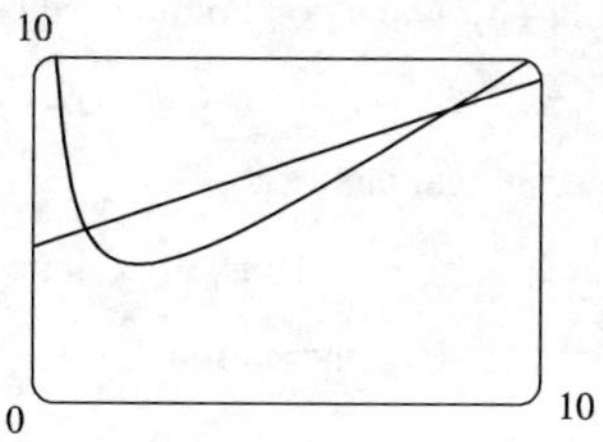

(c) $f(x) = x + 4/x \Rightarrow f'(x) = 1 - 4/x^2$.

So $f'(c) = \frac{1}{2} \Rightarrow c^2 = 8 \Rightarrow c = 2\sqrt{2}$, and

$f(c) = 2\sqrt{2} + \frac{4}{2\sqrt{2}} = 3\sqrt{2}$. Thus, an equation of the tangent line is $y - 3\sqrt{2} = \frac{1}{2}(x - 2\sqrt{2}) \Leftrightarrow$

$y = \frac{1}{2}x + 2\sqrt{2}$.

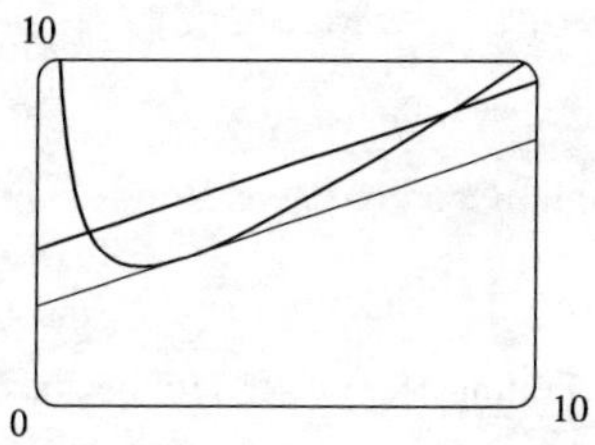

11. $f(x) = 3x^2 + 2x + 5$, $[-1, 1]$. f is continuous on $[-1, 1]$ and differentiable on $(-1, 1)$ since polynomials are continuous and differentiable on $\mathbb{R}$. $f'(c) = \dfrac{f(b) - f(a)}{b - a} \Leftrightarrow 6c + 2 = \dfrac{f(1) - f(-1)}{1 - (-1)} = \dfrac{10 - 6}{2} = 2 \Leftrightarrow$ $6c = 0 \Leftrightarrow c = 0$, which is in $(-1, 1)$.

13. $f(x) = \sqrt[3]{x}$, $[0, 1]$. f is continuous on $\mathbb{R}$ and differentiable on $(-\infty, 0) \cup (0, \infty)$, so f is continuous on $[0, 1]$ and differentiable on $(0, 1)$. $f'(c) = \dfrac{f(b) - f(a)}{b - a} \Leftrightarrow \dfrac{1}{3c^{2/3}} = \dfrac{f(1) - f(0)}{1 - 0} \Leftrightarrow \dfrac{1}{3c^{2/3}} = \dfrac{1 - 0}{1} \Leftrightarrow$ $3c^{2/3} = 1 \Leftrightarrow c^{2/3} = \frac{1}{3} \Leftrightarrow c^2 = \left(\frac{1}{3}\right)^3 = \frac{1}{27} \Leftrightarrow c = \pm\sqrt{\frac{1}{27}} = \pm\frac{\sqrt{3}}{9}$, but only $\frac{\sqrt{3}}{9}$ is in $(0, 1)$.

15. $f(x) = |x - 1|$. $f(3) - f(0) = |3 - 1| - |0 - 1| = 1$. Since $f'(c) = -1$ if $c < 1$ and $f'(c) = 1$ if $c > 1$, $f'(c)(3 - 0) = \pm 3$ and so is never equal to 1. This does not contradict the Mean Value Theorem since $f'(1)$ does not exist.

17. Let $f(x) = 1 + 2x + x^3 + 4x^5$. Then $f(-1) = -6 < 0$ and $f(0) = 1 > 0$. Since f is a polynomial, it is continuous, so the Intermediate Value Theorem says that there is a number c between -1 and 0 such that $f(c) = 0$. Thus, the given equation has a real root. Suppose the equation has distinct real roots a and b with $a < b$. Then $f(a) = f(b) = 0$. Since f is a polynomial, it is differentiable on (a, b) and continuous on $[a, b]$. By Rolle's Theorem, there is a number r in (a, b) such that $f'(r) = 0$. But $f'(x) = 2 + 3x^2 + 20x^4 \geq 2$ for all x, so $f'(x)$ can never be 0. This contradiction shows that the equation can't have two distinct real roots. Hence, it has exactly one real root.

19. Let $f(x) = x^3 - 15x + c$ for x in $[-2, 2]$. If f has two real roots a and b in $[-2, 2]$, with $a < b$, then $f(a) = f(b) = 0$. Since the polynomial f is continuous on $[a, b]$ and differentiable on (a, b), Rolle's Theorem implies that there is a number r in (a, b) such that $f'(r) = 0$. Now $f'(r) = 3r^2 - 15$. Since r is in (a, b), which is contained in $[-2, 2]$, we have $|r| < 2$, so $r^2 < 4$. It follows that $3r^2 - 15 < 3 \cdot 4 - 15 = -3 < 0$. This contradicts $f'(r) = 0$, so the given equation can't have two real roots in $[-2, 2]$. Hence, it has at most one real root in $[-2, 2]$.

21. (a) Suppose that a cubic polynomial $P(x)$ has roots $a_1 < a_2 < a_3 < a_4$, so $P(a_1) = P(a_2) = P(a_3) = P(a_4)$. By Rolle's Theorem there are numbers c_1, c_2, c_3 with $a_1 < c_1 < a_2$, $a_2 < c_2 < a_3$ and $a_3 < c_3 < a_4$ and $P'(c_1) = P'(c_2) = P'(c_3) = 0$. Thus, the second-degree polynomial $P'(x)$ has three distinct real roots, which is impossible.

(b) We prove by induction that a polynomial of degree n has at most n real roots. This is certainly true for $n = 1$. Suppose that the result is true for all polynomials of degree n and let $P(x)$ be a polynomial of degree $n + 1$. Suppose that $P(x)$ has more than $n + 1$ real roots, say $a_1 < a_2 < a_3 < \cdots < a_{n+1} < a_{n+2}$. Then $P(a_1) = P(a_2) = \cdots = P(a_{n+2}) = 0$. By Rolle's Theorem there are real numbers $c_1, \ldots, c_{n+1}$ with $a_1 < c_1 < a_2, \ldots, a_{n+1} < c_{n+1} < a_{n+2}$ and $P'(c_1) = \cdots = P'(c_{n+1}) = 0$. Thus, the nth degree polynomial $P'(x)$ has at least $n + 1$ roots. This contradiction shows that $P(x)$ has at most $n + 1$ real roots.

23. By the Mean Value Theorem, $f(4) - f(1) = f'(c)(4 - 1)$ for some $c \in (1, 4)$. But for every $c \in (1, 4)$ we have $f'(c) \geq 2$. Putting $f'(c) \geq 2$ into the above equation and substituting $f(1) = 10$, we get $f(4) = f(1) + f'(c)(4 - 1) = 10 + 3f'(c) \geq 10 + 3 \cdot 2 = 16$. So the smallest possible value of $f(4)$ is 16.

25. Suppose that such a function f exists. By the Mean Value Theorem there is a number $0 < c < 2$ with $f'(c) = \dfrac{f(2) - f(0)}{2 - 0} = \dfrac{5}{2}$. But this is impossible since $f'(x) \le 2 < \frac{5}{2}$ for all x, so no such function can exist.

27. We use Exercise 26 with $f(x) = \sqrt{1+x}$, $g(x) = 1 + \frac{1}{2}x$, and $a = 0$. Notice that $f(0) = 1 = g(0)$ and $f'(x) = \dfrac{1}{2\sqrt{1+x}} < \dfrac{1}{2} = g'(x)$ for $x > 0$. So by Exercise 26, $f(b) < g(b) \;\Rightarrow\; \sqrt{1+b} < 1 + \frac{1}{2}b$ for $b > 0$.

Another method: Apply the Mean Value Theorem directly to either $f(x) = 1 + \frac{1}{2}x - \sqrt{1+x}$ or $g(x) = \sqrt{1+x}$ on $[0, b]$.

29. Let $f(x) = \sin x$ and let $b < a$. Then $f(x)$ is continuous on $[b, a]$ and differentiable on (b, a). By the Mean Value Theorem, there is a number $c \in (b, a)$ with $\sin a - \sin b = f(a) - f(b) = f'(c)(a - b) = (\cos c)(a - b)$. Thus, $|\sin a - \sin b| \le |\cos c|\,|b - a| \le |a - b|$. If $a < b$, then $|\sin a - \sin b| = |\sin b - \sin a| \le |b - a| = |a - b|$. If $a = b$, both sides of the inequality are 0.

31. For $x > 0$, $f(x) = g(x)$, so $f'(x) = g'(x)$. For $x < 0$, $f'(x) = (1/x)' = -1/x^2$ and $g'(x) = (1 + 1/x)' = -1/x^2$, so again $f'(x) = g'(x)$. However, the domain of $g(x)$ is not an interval [it is $(-\infty, 0) \cup (0, \infty)$] so we cannot conclude that $f - g$ is constant (in fact it is not).

33. Let $g(t)$ and $h(t)$ be the position functions of the two runners and let $f(t) = g(t) - h(t)$. By hypothesis, $f(0) = g(0) - h(0) = 0$ and $f(b) = g(b) - h(b) = 0$, where b is the finishing time. Then by the Mean Value Theorem, there is a time c, with $0 < c < b$, such that $f'(c) = \dfrac{f(b) - f(0)}{b - 0}$. But $f(b) = f(0) = 0$, so $f'(c) = 0$. Since $f'(c) = g'(c) - h'(c) = 0$, we have $g'(c) = h'(c)$. So at time c, both runners have the same speed $g'(c) = h'(c)$.

The Definite Integral

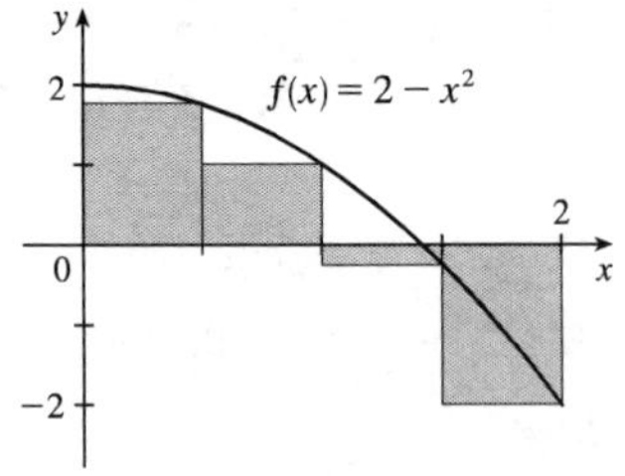

1. $R_4 = \sum\limits_{i=1}^{4} f(x_i)\,\Delta x \quad [x_i^* = x_i \text{ is a right endpoint and } \Delta x = 0.5]$

$= 0.5\,[f(0.5) + f(1) + f(1.5) + f(2)] \quad [f(x) = 2 - x^2]$

$= 0.5\,[1.75 + 1 + (-0.25) + (-2)]$

$= 0.5(0.5) = 0.25$

The Riemann sum represents the sum of the areas of the two rectangles above the x-axis minus the sum of the areas of the two rectangles below the x-axis; that is, the *net area* of the rectangles with respect to the x-axis.

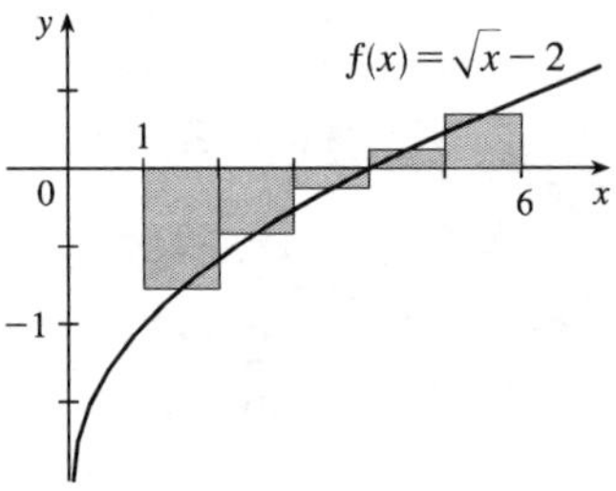

3. $M_5 = \sum\limits_{i=1}^{5} f(\overline{x}_i)\,\Delta x \quad [x_i^* = \overline{x}_i = \frac{1}{2}(x_{i-1} + x_i) \text{ is a midpoint and } \Delta x = 1]$

$= 1\,[f(1.5) + f(2.5) + f(3.5) + f(4.5) + f(5.5)] \quad [f(x) = \sqrt{x} - 2]$

≈ -0.856759

The Riemann sum represents the sum of the areas of the two rectangles above the x-axis minus the sum of the areas of the three rectangles below the x-axis.

5. $\Delta x = (b - a)/n = (8 - 0)/4 = 8/4 = 2.$

(a) Using the right endpoints to approximate $\int_0^8 f(x)\,dx$, we have

$$\sum_{i=1}^{4} f(x_i)\,\Delta x = 2[f(2) + f(4) + f(6) + f(8)] \approx 2[1 + 2 + (-2) + 1] = 4.$$

(b) Using the left endpoints to approximate $\int_0^8 f(x)\,dx$, we have

$$\sum_{i=1}^{4} f(x_{i-1})\,\Delta x = 2[f(0) + f(2) + f(4) + f(6)] \approx 2[2 + 1 + 2 + (-2)] = 6.$$

(c) Using the midpoint of each subinterval to approximate $\int_0^8 f(x)\,dx$, we have

$$\sum_{i=1}^{4} f(\overline{x}_i)\,\Delta x = 2[f(1) + f(3) + f(5) + f(7)] \approx 2[3 + 2 + 1 + (-1)] = 10.$$

7. Since f is increasing, $L_5 \le \int_0^{25} f(x)\,dx \le R_5$.

$$\text{Lower estimate} = L_5 = \sum_{i=1}^{5} f(x_{i-1})\,\Delta x = 5[f(0) + f(5) + f(10) + f(15) + f(20)]$$
$$= 5(-42 - 37 - 25 - 6 + 15) = 5(-95) = -475$$

$$\text{Upper estimate} = R_5 = \sum_{i=1}^{5} f(x_i)\,\Delta x = 5[f(5) + f(10) + f(15) + f(20) + f(25)]$$
$$= 5(-37 - 25 - 6 + 15 + 36) = 5(-17) = -85$$

9. $\Delta x = (10 - 2)/4 = 2$, so the endpoints are 2, 4, 6, 8, and 10, and the midpoints are 3, 5, 7, and 9. The Midpoint Rule gives $\int_2^{10} \sqrt{x^3 + 1}\,dx \approx \sum\limits_{i=1}^{4} f(\overline{x}_i)\,\Delta x = 2\left(\sqrt{3^3 + 1} + \sqrt{5^3 + 1} + \sqrt{7^3 + 1} + \sqrt{9^3 + 1}\right) \approx 124.1644.$

11. $\Delta x = (1-0)/5 = 0.2$, so the endpoints are 0, 0.2, 0.4, 0.6, 0.8, and 1, and the midpoints are 0.1, 0.3, 0.5, 0.7, and 0.9. The Midpoint Rule gives

$$\int_0^1 \sin(x^2)\,dx \approx \sum_{i=1}^{5} f(\overline{x}_i)\,\Delta x = 0.2\left[\sin(0.1)^2 + \sin(0.3)^2 + \sin(0.5)^2 + \sin(0.7)^2 + \sin(0.9)^2\right] \approx 0.3084.$$

13. In Maple, we use the command `with(student);` to load the sum and box commands, then `m:=middlesum(sin(x^2),x=0..1,5);` which gives us the sum in summation notation, then `M:=evalf(m);` which gives $M_5 \approx 0.30843908$, confirming the result of Exercise 11. The command `middlebox(sin(x^2),x=0..1,5)` generates the graph. Repeating for $n = 10$ and $n = 20$ gives $M_{10} \approx 0.30981629$ and $M_{20} \approx 0.31015563$.

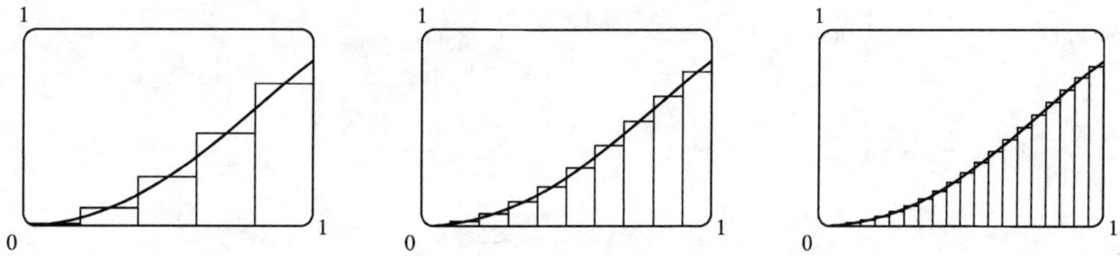

15. We'll create the table of values to approximate $\int_0^\pi \sin x\,dx$ by using the program in the solution to Exercise 5.1.7 with $Y_1 = \sin x$, Xmin $= 0$, Xmax $= \pi$, and $n = 5$, 10, 50, and 100.

The values of R_n appear to be approaching 2.

n	R_n
5	1.933766
10	1.983524
50	1.999342
100	1.999836

17. On $[0, \pi]$, $\lim_{n\to\infty} \sum_{i=1}^{n} x_i \sin x_i\,\Delta x = \int_0^\pi x \sin x\,dx$.

19. On $[1, 8]$, $\lim_{n\to\infty} \sum_{i=1}^{n} \sqrt{2x_i^* + (x_i^*)^2}\,\Delta x = \int_1^8 \sqrt{2x + x^2}\,dx$.

21. Note that $\Delta x = \dfrac{5-(-1)}{n} = \dfrac{6}{n}$ and $x_i = -1 + i\,\Delta x = -1 + \dfrac{6i}{n}$.

$$\int_{-1}^{5} (1+3x)\,dx = \lim_{n\to\infty} \sum_{i=1}^{n} f(x_i)\,\Delta x = \lim_{n\to\infty} \sum_{i=1}^{n} \left[1 + 3\left(-1 + \frac{6i}{n}\right)\right] \frac{6}{n}$$

$$= \lim_{n\to\infty} \frac{6}{n} \sum_{i=1}^{n} \left[-2 + \frac{18i}{n}\right] = \lim_{n\to\infty} \frac{6}{n} \left[\sum_{i=1}^{n}(-2) + \sum_{i=1}^{n} \frac{18i}{n}\right]$$

$$= \lim_{n\to\infty} \frac{6}{n} \left[-2n + \frac{18}{n} \sum_{i=1}^{n} i\right] = \lim_{n\to\infty} \frac{6}{n} \left[-2n + \frac{18}{n} \cdot \frac{n(n+1)}{2}\right]$$

$$= \lim_{n\to\infty} \left[-12 + \frac{108}{n^2} \cdot \frac{n(n+1)}{2}\right] = \lim_{n\to\infty} \left[-12 + 54\frac{n+1}{n}\right]$$

$$= \lim_{n\to\infty} \left[-12 + 54\left(1 + \frac{1}{n}\right)\right] = -12 + 54 \cdot 1 = 42$$

23. Note that $\Delta x = \dfrac{2-0}{n} = \dfrac{2}{n}$ and $x_i = 0 + i\,\Delta x = \dfrac{2i}{n}$.

$$\begin{aligned}\int_0^2 (2-x^2)\,dx &= \lim_{n\to\infty}\sum_{i=1}^n f(x_i)\,\Delta x = \lim_{n\to\infty}\sum_{i=1}^n \left(2 - \frac{4i^2}{n^2}\right)\left(\frac{2}{n}\right) \\ &= \lim_{n\to\infty}\frac{2}{n}\left[\sum_{i=1}^n 2 - \frac{4}{n^2}\sum_{i=1}^n i^2\right] = \lim_{n\to\infty}\frac{2}{n}\left(2n - \frac{4}{n^2}\sum_{i=1}^n i^2\right) \\ &= \lim_{n\to\infty}\left[4 - \frac{8}{n^3}\cdot\frac{n(n+1)(2n+1)}{6}\right] = \lim_{n\to\infty}\left(4 - \frac{4}{3}\cdot\frac{n+1}{n}\cdot\frac{2n+1}{n}\right) \\ &= \lim_{n\to\infty}\left[4 - \frac{4}{3}\left(1+\frac{1}{n}\right)\left(2+\frac{1}{n}\right)\right] = 4 - \tfrac{4}{3}\cdot 1\cdot 2 = \tfrac{4}{3}\end{aligned}$$

25. Note that $\Delta x = \dfrac{2-1}{n} = \dfrac{1}{n}$ and $x_i = 1 + i\,\Delta x = 1 + i(1/n) = 1 + i/n$.

$$\begin{aligned}\int_1^2 x^3\,dx &= \lim_{n\to\infty}\sum_{i=1}^n f(x_i)\,\Delta x = \lim_{n\to\infty}\sum_{i=1}^n\left(1+\frac{i}{n}\right)^3\left(\frac{1}{n}\right) = \lim_{n\to\infty}\frac{1}{n}\sum_{i=1}^n\left(\frac{n+i}{n}\right)^3 \\ &= \lim_{n\to\infty}\frac{1}{n^4}\sum_{i=1}^n\left(n^3 + 3n^2 i + 3ni^2 + i^3\right) = \lim_{n\to\infty}\frac{1}{n^4}\left[\sum_{i=1}^n n^3 + \sum_{i=1}^n 3n^2 i + \sum_{i=1}^n 3ni^2 + \sum_{i=1}^n i^3\right] \\ &= \lim_{n\to\infty}\frac{1}{n^4}\left[n\cdot n^3 + 3n^2\sum_{i=1}^n i + 3n\sum_{i=1}^n i^2 + \sum_{i=1}^n i^3\right] \\ &= \lim_{n\to\infty}\left[1 + \frac{3}{n^2}\cdot\frac{n(n+1)}{2} + \frac{3}{n^3}\cdot\frac{n(n+1)(2n+1)}{6} + \frac{1}{n^4}\cdot\frac{n^2(n+1)^2}{4}\right] \\ &= \lim_{n\to\infty}\left[1 + \frac{3}{2}\cdot\frac{n+1}{n} + \frac{1}{2}\cdot\frac{n+1}{n}\cdot\frac{2n+1}{n} + \frac{1}{4}\cdot\frac{(n+1)^2}{n^2}\right] \\ &= \lim_{n\to\infty}\left[1 + \frac{3}{2}\left(1+\frac{1}{n}\right) + \frac{1}{2}\left(1+\frac{1}{n}\right)\left(2+\frac{1}{n}\right) + \frac{1}{4}\left(1+\frac{1}{n}\right)^2\right] = 1 + \frac{3}{2} + \frac{1}{2}\cdot 2 + \frac{1}{4} = 3.75\end{aligned}$$

27.
$$\begin{aligned}\int_a^b x\,dx &= \lim_{n\to\infty}\frac{b-a}{n}\sum_{i=1}^n\left[a + \frac{b-a}{n}i\right] = \lim_{n\to\infty}\left[\frac{a(b-a)}{n}\sum_{i=1}^n 1 + \frac{(b-a)^2}{n^2}\sum_{i=1}^n i\right] \\ &= \lim_{n\to\infty}\left[\frac{a(b-a)}{n}n + \frac{(b-a)^2}{n^2}\cdot\frac{n(n+1)}{2}\right] = a(b-a) + \lim_{n\to\infty}\frac{(b-a)^2}{2}\left(1+\frac{1}{n}\right) \\ &= a(b-a) + \tfrac{1}{2}(b-a)^2 = (b-a)\left(a + \tfrac{1}{2}b - \tfrac{1}{2}a\right) = (b-a)\tfrac{1}{2}(b+a) = \tfrac{1}{2}\left(b^2 - a^2\right)\end{aligned}$$

29. $f(x) = \dfrac{x}{1+x^5}$, $a = 2$, $b = 6$, and $\Delta x = \dfrac{6-2}{n} = \dfrac{4}{n}$. Using Equation 3, we get $x_i^* = x_i = 2 + i\,\Delta x = 2 + \dfrac{4i}{n}$,

so $\displaystyle\int_2^6 \frac{x}{1+x^5}\,dx = \lim_{n\to\infty} R_n = \lim_{n\to\infty}\sum_{i=1}^n \frac{2+\dfrac{4i}{n}}{1+\left(2+\dfrac{4i}{n}\right)^5}\cdot\frac{4}{n}.$

31. $\Delta x = (\pi - 0)/n = \pi/n$ and $x_i^* = x_i = \pi i/n$.

$$\int_0^\pi \sin 5x\,dx = \lim_{n\to\infty}\sum_{i=1}^n(\sin 5x_i)\left(\frac{\pi}{n}\right) = \lim_{n\to\infty}\sum_{i=1}^n\left(\sin\frac{5\pi i}{n}\right)\frac{\pi}{n} \overset{\text{CAS}}{=} \pi\lim_{n\to\infty}\frac{1}{n}\cot\left(\frac{5\pi}{2n}\right) \overset{\text{CAS}}{=} \pi\left(\frac{2}{5\pi}\right) = \frac{2}{5}$$

33. (a) Think of $\int_0^2 f(x)\,dx$ as the area of a trapezoid with bases 1 and 3 and height 2. The area of a trapezoid is $A = \frac{1}{2}(b+B)h$, so $\int_0^2 f(x)\,dx = \frac{1}{2}(1+3)2 = 4$.

(b) $\int_0^5 f(x)\,dx = \int_0^2 f(x)\,dx + \int_2^3 f(x)\,dx + \int_3^5 f(x)\,dx$

trapezoid rectangle triangle

$= \frac{1}{2}(1+3)2 + 3\cdot 1 + \frac{1}{2}\cdot 2\cdot 3 = 4+3+3 = 10$

(c) $\int_5^7 f(x)\,dx$ is the negative of the area of the triangle with base 2 and height 3. $\int_5^7 f(x)\,dx = -\frac{1}{2}\cdot 2\cdot 3 = -3$.

(d) $\int_7^9 f(x)\,dx$ is the negative of the area of a trapezoid with bases 3 and 2 and height 2, so it equals $-\frac{1}{2}(B+b)h = -\frac{1}{2}(3+2)2 = -5$. Thus,
$\int_0^9 f(x)\,dx = \int_0^5 f(x)\,dx + \int_5^7 f(x)\,dx + \int_7^9 f(x)\,dx = 10 + (-3) + (-5) = 2.$

35. $\int_0^3 \left(\frac{1}{2}x - 1\right) dx$ can be interpreted as the area of the triangle above the x-axis minus the area of the triangle below the x-axis; that is,

$\frac{1}{2}(1)\left(\frac{1}{2}\right) - \frac{1}{2}(2)(1) = \frac{1}{4} - 1 = -\frac{3}{4}$.

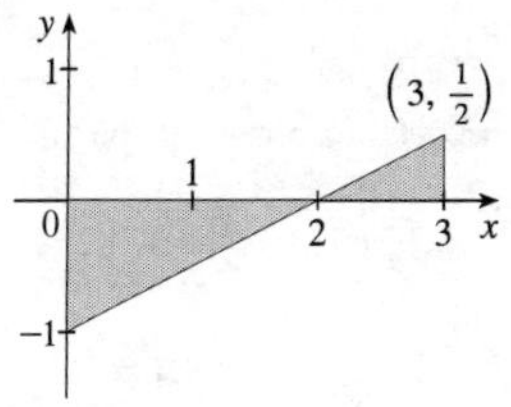

37. $\int_{-3}^0 \left(1+\sqrt{9-x^2}\,\right) dx$ can be interpreted as the area under the graph of $f(x) = 1+\sqrt{9-x^2}$ between $x=-3$ and $x=0$. This is equal to one-quarter the area of the circle with radius 3, plus the area of the rectangle, so

$\int_{-3}^0 \left(1+\sqrt{9-x^2}\,\right) dx = \frac{1}{4}\pi\cdot 3^2 + 1\cdot 3 = 3 + \frac{9}{4}\pi$.

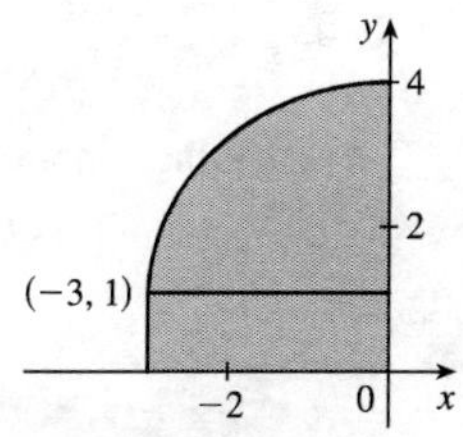

39. $\int_{-1}^2 |x|\,dx$ can be interpreted as the sum of the areas of the two shaded triangles; that is, $\frac{1}{2}(1)(1) + \frac{1}{2}(2)(2) = \frac{1}{2} + \frac{4}{2} = \frac{5}{2}$.

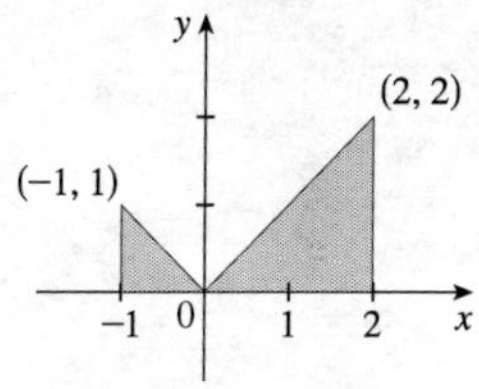

41. $\int_9^4 \sqrt{t}\,dt = -\int_4^9 \sqrt{t}\,dt$ [because we reversed the limits of integration]

$= -\int_4^9 \sqrt{x}\,dx$ [we can use any letter without changing the value of the integral]

$= -\frac{38}{3}$

43. $\int_0^1 (5-6x^2)\,dx = \int_0^1 5\,dx - 6\int_0^1 x^2\,dx = 5(1-0) - 6\left(\frac{1}{3}\right) = 5-2 = 3$

45. $\int_1^4 (2x^2-3x+1)\,dx = 2\int_1^4 x^2\,dx - 3\int_1^4 x\,dx + \int_1^4 1\,dx$

$= 2\cdot\frac{1}{3}(4^3-1^3) - 3\cdot\frac{1}{2}(4^2-1^2) + 1(4-1) = \frac{45}{2} = 22.5$

47. $\int_{-2}^2 f(x)\,dx + \int_2^5 f(x)\,dx - \int_{-2}^{-1} f(x)\,dx = \int_{-2}^5 f(x)\,dx + \int_{-1}^{-2} f(x)\,dx$ [by Property 5 and reversing limits]

$= \int_{-1}^5 f(x)\,dx$ [Property 5]

49. $\int_0^9 [2f(x)+3g(x)]\,dx = 2\int_0^9 f(x)\,dx + 3\int_0^9 g(x)\,dx = 2(37) + 3(16) = 122$

51. $0 \le \sin x < 1$ on $\left[0, \frac{\pi}{4}\right]$, so $\sin^3 x \le \sin^2 x$ on $\left[0, \frac{\pi}{4}\right]$. Hence, $\int_0^{\pi/4} \sin^3 x\,dx \le \int_0^{\pi/4} \sin^2 x\,dx$ (Property 7).

53. If $-1 \le x \le 1$, then $0 \le x^2 \le 1$ and $1 \le 1 + x^2 \le 2$, so $1 \le \sqrt{1+x^2} \le \sqrt{2}$ and $1[1-(-1)] \le \int_{-1}^{1} \sqrt{1+x^2}\,dx \le \sqrt{2}\,[1-(-1)]$ [Property 8]; that is, $2 \le \int_{-1}^{1} \sqrt{1+x^2}\,dx \le 2\sqrt{2}$.

55. If $1 \le x \le 2$, then $\frac{1}{2} \le \frac{1}{x} \le 1$, so $\frac{1}{2}(2-1) \le \int_1^2 \frac{1}{x}\,dx \le 1(2-1)$ or $\frac{1}{2} \le \int_1^2 \frac{1}{x}\,dx \le 1$.

57. If $\frac{\pi}{4} \le x \le \frac{\pi}{3}$, then $1 \le \tan x \le \sqrt{3}$, so $1\left(\frac{\pi}{3} - \frac{\pi}{4}\right) \le \int_{\pi/4}^{\pi/3} \tan x\,dx \le \sqrt{3}\left(\frac{\pi}{3} - \frac{\pi}{4}\right)$ or

$\frac{\pi}{12} \le \int_{\pi/4}^{\pi/3} \tan x\,dx \le \frac{\pi}{12}\sqrt{3}$.

59. For $-1 \le x \le 1$, $0 \le x^4 \le 1$ and $1 \le \sqrt{1+x^4} \le \sqrt{2}$, so $1[1-(-1)] \le \int_{-1}^{1} \sqrt{1+x^4}\,dx \le \sqrt{2}\,[1-(-1)]$ or

$2 \le \int_{-1}^{1} \sqrt{1+x^4}\,dx \le 2\sqrt{2}$.

61. $\sqrt{x^4+1} \ge \sqrt{x^4} = x^2$, so $\int_1^3 \sqrt{x^4+1}\,dx \ge \int_1^3 x^2\,dx = \frac{1}{3}\left(3^3 - 1^3\right) = \frac{26}{3}$.

63. Using a regular partition and right endpoints as in the proof of Property 2, we calculate

$$\int_a^b cf(x)\,dx = \lim_{n\to\infty} \sum_{i=1}^{n} cf(x_i)\,\Delta x_i = \lim_{n\to\infty} c \sum_{i=1}^{n} f(x_i)\,\Delta x_i = c \lim_{n\to\infty} \sum_{i=1}^{n} f(x_i)\,\Delta x_i = c\int_a^b f(x)\,dx.$$

65. Since $-\left|f(x)\right| \le f(x) \le \left|f(x)\right|$, it follows from Property 7 that

$$-\int_a^b |f(x)|\,dx \le \int_a^b f(x)\,dx \le \int_a^b |f(x)|\,dx \quad\Rightarrow\quad \left|\int_a^b f(x)\,dx\right| \le \int_a^b |f(x)|\,dx$$

Note that the definite integral is a real number, and so the following property applies: $-a \le b \le a \;\Rightarrow\; |b| \le a$ for all real numbers b and nonnegative numbers a.

67. $\displaystyle\lim_{n\to\infty} \sum_{i=1}^{n} \frac{i^4}{n^5} = \lim_{n\to\infty} \frac{1}{n} \sum_{i=1}^{n} \left(\frac{i}{n}\right)^4 = \int_0^1 x^4\,dx$

69. Choose $x_i = 1 + \dfrac{i}{n}$ and $x_i^* = \sqrt{x_{i-1}x_i} = \sqrt{\left(1 + \dfrac{i-1}{n}\right)\left(1 + \dfrac{i}{n}\right)}$. Then

$$\begin{aligned}
\int_1^2 x^{-2}dx &= \lim_{n\to\infty} \frac{1}{n} \sum_{i=1}^{n} \frac{1}{\left(1+\frac{i-1}{n}\right)\left(1+\frac{i}{n}\right)} \\
&= \lim_{n\to\infty} n \sum_{i=1}^{n} \frac{1}{(n+i-1)(n+i)} \\
&= \lim_{n\to\infty} n \sum_{i=1}^{n} \left(\frac{1}{n+i-1} - \frac{1}{n+i}\right) \qquad \text{[by the hint]} \\
&= \lim_{n\to\infty} n \left(\sum_{i=0}^{n-1} \frac{1}{n+i} - \sum_{i=1}^{n} \frac{1}{n+i}\right) \\
&= \lim_{n\to\infty} n \left(\left[\frac{1}{n} + \frac{1}{n+1} + \cdots + \frac{1}{2n-1}\right] - \left[\frac{1}{n+1} + \cdots + \frac{1}{2n-1} + \frac{1}{2n}\right]\right) \\
&= \lim_{n\to\infty} n \left(\frac{1}{n} - \frac{1}{2n}\right) = \lim_{n\to\infty} \left(1 - \tfrac{1}{2}\right) = \tfrac{1}{2}
\end{aligned}$$

Review

CONCEPT CHECK

1. (a) $\sum_{i=1}^{n} f(x_i^*)\,\Delta x$ is an expression for a Riemann sum of a function f.
x_i^* is a point in the ith subinterval $[x_{i-1}, x_i]$ and Δx is the length of the subintervals.

(b) See Figure 1 in Section 5.2.

(c) In Section 5.2, see Figure 3 and the paragraph beside it.

2. (a) See Definition 5.2.2.

(b) See Figure 2 in Section 5.2.

(c) In Section 5.2, see Figure 4 and the paragraph above it.

3. See the Fundamental Theorem of Calculus after Example 8 in Section 5.3.

4. (a) See the Net Change Theorem after Example 5 in Section 5.4.

(b) $\int_{t_1}^{t_2} r(t)\,dt$ represents the change in the amount of water in the reservoir between time t_1 and time t_2.

5. (a) $\int_{60}^{120} v(t)\,dt$ represents the change in position of the particle from $t = 60$ to $t = 120$ seconds.

(b) $\int_{60}^{120} |v(t)|\,dt$ represents the total distance traveled by the particle from $t = 60$ to 120 seconds.

(c) $\int_{60}^{120} a(t)\,dt$ represents the change in the velocity of the particle from $t = 60$ to $t = 120$ seconds.

6. (a) $\int f(x)\,dx$ is the family of functions $\{F \mid F' = f\}$. Any two such functions differ by a constant.

(b) The connection is given by the Evaluation Theorem: $\int_a^b f(x)\,dx = \left[\int f(x)\,dx\right]_a^b$ if f is continuous.

7. The precise version of this statement is given by the Fundamental Theorem of Calculus. See the statement of this theorem and the paragraph that follows it at the end of Section 5.3.

8. See the Substitution Rule (5.5.4). This says that it is permissible to operate with the dx after an integral sign as if it were a differential.

TRUE-FALSE QUIZ

1. True by Property 2 of the Integral in Section 5.2.

3. True by Property 3 of the Integral in Section 5.2.

5. False. For example, let $f(x) = x^2$. Then $\int_0^1 \sqrt{x^2}\,dx = \int_0^1 x\,dx = \frac{1}{2}$, but $\sqrt{\int_0^1 x^2\,dx} = \sqrt{\frac{1}{3}} = \frac{1}{\sqrt{3}}$.

7. True by Comparison Property 7 of the Integral in Section 5.2.

9. True. The integrand is an odd function that is continuous on $[-1, 1]$, so the result follows from Theorem 5.5.6(b).

11. False. The function $f(x) = 1/x^4$ is not bounded on the interval $[-2, 1]$. It has an infinite discontinuity at $x = 0$, so it is not integrable on the interval. (If the integral were to exist, a positive value would be expected, by Comparison Property 6 of Integrals.)

13. False. For example, the function $y = |x|$ is continuous on $\mathbb{R}$, but has no derivative at $x = 0$.

EXERCISES

1. (a)

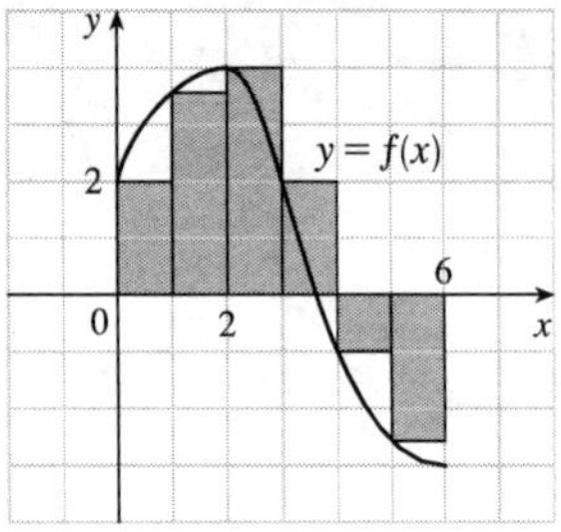

$$L_6 = \sum_{i=1}^{6} f(x_{i-1})\,\Delta x \quad [\Delta x = \tfrac{6-0}{6} = 1]$$
$$= f(x_0)\cdot 1 + f(x_1)\cdot 1 + f(x_2)\cdot 1 + f(x_3)\cdot 1 + f(x_4)\cdot 1 + f(x_5)\cdot 1$$
$$\approx 2 + 3.5 + 4 + 2 + (-1) + (-2.5) = 8$$

The Riemann sum represents the sum of the areas of the four rectangles above the x-axis minus the sum of the areas of the two rectangles below the x-axis.

(b)

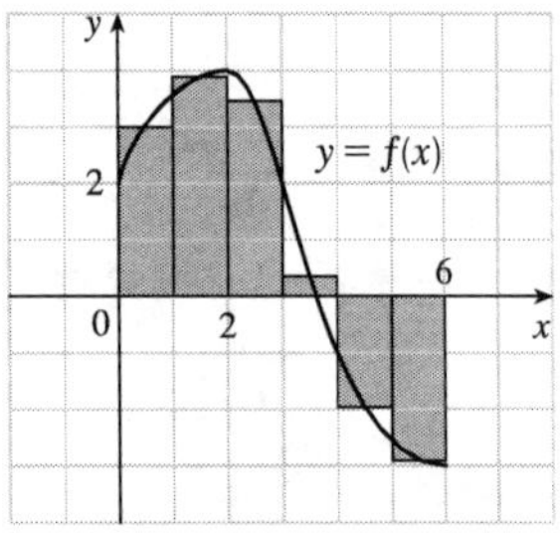

$$M_6 = \sum_{i=1}^{6} f(\overline{x}_i)\,\Delta x \quad [\Delta x = \tfrac{6-0}{6} = 1]$$
$$= f(\overline{x}_1)\cdot 1 + f(\overline{x}_2)\cdot 1 + f(\overline{x}_3)\cdot 1 + f(\overline{x}_4)\cdot 1 + f(\overline{x}_5)\cdot 1 + f(\overline{x}_6)\cdot 1$$
$$= f(0.5) + f(1.5) + f(2.5) + f(3.5) + f(4.5) + f(5.5)$$
$$\approx 3 + 3.9 + 3.4 + 0.3 + (-2) + (-2.9) = 5.7$$

The Riemann sum represents the sum of the areas of the four rectangles above the x-axis minus the sum of the areas of the two rectangles below the x-axis.

3. $\int_0^1 (x + \sqrt{1-x^2}\,)\,dx = \int_0^1 x\,dx + \int_0^1 \sqrt{1-x^2}\,dx = I_1 + I_2.$

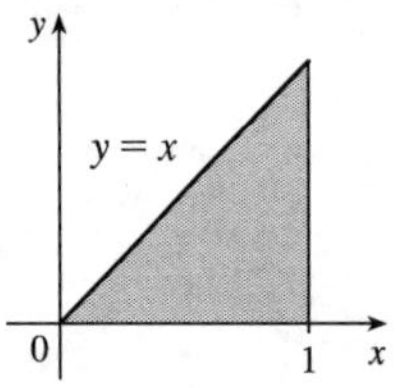

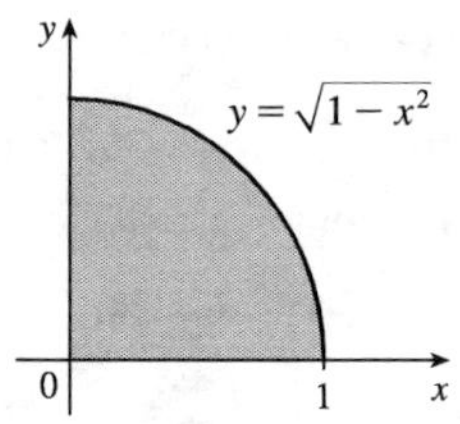

I_1 can be interpreted as the area of the triangle shown in the figure and I_2 can be interpreted as the area of the quarter-circle. Area $= \frac{1}{2}(1)(1) + \frac{1}{4}(\pi)(1)^2 = \frac{1}{2} + \frac{\pi}{4}$.

5. $\int_0^6 f(x)\,dx = \int_0^4 f(x)\,dx + \int_4^6 f(x)\,dx \;\Rightarrow\; 10 = 7 + \int_4^6 f(x)\,dx \;\Rightarrow\; \int_4^6 f(x)\,dx = 10 - 7 = 3$

7. First note that either a or b must be the graph of $\int_0^x f(t)\,dt$, since $\int_0^0 f(t)\,dt = 0$, and $c(0) \neq 0$. Now notice that $b > 0$ when c is increasing, and that $c > 0$ when a is increasing. It follows that c is the graph of $f(x)$, b is the graph of $f'(x)$, and a is the graph of $\int_0^x f(t)\,dt$.

9. $\int_1^2 (8x^3 + 3x^2)\,dx = \left[8\cdot\frac{1}{4}x^4 + 3\cdot\frac{1}{3}x^3\right]_1^2 = \left[2x^4 + x^3\right]_1^2 = (2\cdot 2^4 + 2^3) - (2+1) = 40 - 3 = 37$

11. $\int_0^1 (1 - x^9)\,dx = \left[x - \frac{1}{10}x^{10}\right]_0^1 = \left(1 - \frac{1}{10}\right) - 0 = \frac{9}{10}$

13. $\displaystyle\int_1^9 \frac{\sqrt{u} - 2u^2}{u}\,du = \int_1^9 (u^{-1/2} - 2u)\,du = \left[2u^{1/2} - u^2\right]_1^9 = (6 - 81) - (2 - 1) = -76$

15. Let $u = y^2 + 1$, so $du = 2y\,dy$ and $y\,dy = \frac{1}{2}\,du$. When $y = 0$, $u = 1$; when $y = 1$, $u = 2$. Thus,

$\int_0^1 y(y^2+1)^5\,dy = \int_1^2 u^5\left(\frac{1}{2}\,du\right) = \frac{1}{2}\left[\frac{1}{6}u^6\right]_1^2 = \frac{1}{12}(64-1) = \frac{63}{12} = \frac{21}{4}$.

17. $\displaystyle\int_1^5 \frac{dt}{(t-4)^2}$ does not exist because the function $f(t) = \dfrac{1}{(t-4)^2}$ has an infinite discontinuity at $t = 4$;

that is, f is discontinuous on the interval $[1, 5]$.

19. Let $u = v^3$, so $du = 3v^2\,dv$. When $v = 0$, $u = 0$; when $v = 1$, $u = 1$. Thus,

$\int_0^1 v^2\cos(v^3)\,dv = \int_0^1 \cos u\left(\frac{1}{3}\,du\right) = \frac{1}{3}\left[\sin u\right]_0^1 = \frac{1}{3}(\sin 1 - 0) = \frac{1}{3}\sin 1$.

21. Let $u = x^2 + 4x$. Then $du = (2x+4)\,dx = 2(x+2)\,dx$, so

$\displaystyle\int \frac{x+2}{\sqrt{x^2+4x}}\,dx = \int u^{-1/2}\left(\tfrac{1}{2}\,du\right) = \tfrac{1}{2}\cdot 2u^{1/2} + C = \sqrt{u} + C = \sqrt{x^2+4x} + C.$

23. Let $u = \sin \pi t$. Then $du = \pi\,\cos\pi t\,dt$, so $\int \sin\pi t\,\cos\pi t\,dt = \int u\left(\frac{1}{\pi}\,du\right) = \frac{1}{\pi}\cdot\frac{1}{2}u^2 + C = \frac{1}{2\pi}(\sin\pi t)^2 + C$.

25. Let $u = 2\theta$. Then $du = 2\,d\theta$, so

$$\begin{aligned}\int_0^{\pi/8} \sec 2\theta \tan 2\theta\,d\theta &= \int_0^{\pi/4} \sec u \tan u\left(\tfrac{1}{2}\,du\right) = \tfrac{1}{2}[\sec u]_0^{\pi/4} = \tfrac{1}{2}\left(\sec\tfrac{\pi}{4} - \sec 0\right)\\ &= \tfrac{1}{2}\left(\sqrt{2}-1\right) = \tfrac{1}{2}\sqrt{2} - \tfrac{1}{2}\end{aligned}$$

27. Since $x^2 - 4 < 0$ for $0 \le x < 2$ and $x^2 - 4 > 0$ for $2 < x \le 3$, we have $\left|x^2-4\right| = -(x^2-4) = 4 - x^2$ for $0 \le x < 2$ and $\left|x^2-4\right| = x^2 - 4$ for $2 < x \le 3$. Thus,

$$\begin{aligned}\int_0^3 \left|x^2-4\right|dx &= \int_0^2 (4-x^2)\,dx + \int_2^3 (x^2-4)\,dx = \left[4x - \frac{x^3}{3}\right]_0^2 + \left[\frac{x^3}{3} - 4x\right]_2^3\\ &= \left(8 - \tfrac{8}{3}\right) - 0 + (9-12) - \left(\tfrac{8}{3} - 8\right) = \tfrac{16}{3} - 3 + \tfrac{16}{3} = \tfrac{32}{3} - \tfrac{9}{3} = \tfrac{23}{3}\end{aligned}$$

In Exercises 29 and 30, let $f(x)$ denote the integrand and $F(x)$ its antiderivative (with $C = 0$).

29. Let $u = 1 + \sin x$. Then $du = \cos x\,dx$, so

$\displaystyle\int \frac{\cos x\,dx}{\sqrt{1+\sin x}} = \int u^{-1/2}\,du = 2u^{1/2} + C = 2\sqrt{1+\sin x} + C.$

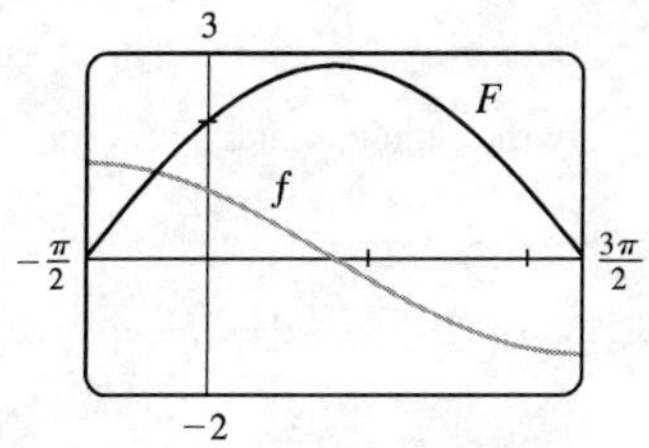

31. From the graph, it appears that the area under the curve $y = x\sqrt{x}$ between $x = 0$ and $x = 4$ is somewhat less than half the area of an 8×4 rectangle, so perhaps about 13 or 14. To find the exact value, we evaluate

$\int_0^4 x\sqrt{x}\,dx = \int_0^4 x^{3/2}\,dx = \left[\frac{2}{5}x^{5/2}\right]_0^4 = \frac{2}{5}(4)^{5/2} = \frac{64}{5} = 12.8.$

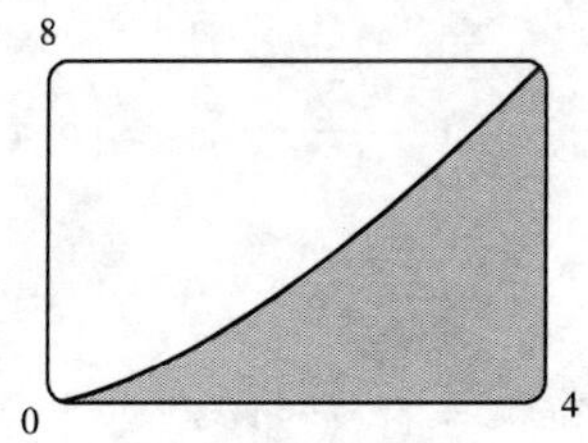

33. By FTC1, $F(x) = \int_1^x \sqrt{1+t^4}\,dt \;\Rightarrow\; F'(x) = \sqrt{1+x^4}$.

35. $g(x) = \int_0^{x^3} \frac{t\,dt}{\sqrt{1+t^3}}$. Let $y = g(u)$ and $u = x^3$.

Then $g'(x) = \frac{dy}{dx} = \frac{dy}{du}\frac{du}{dx} = \frac{u}{\sqrt{1+u^3}}3x^2 = \frac{x^3}{\sqrt{1+x^9}}3x^2 = \frac{3x^5}{\sqrt{1+x^9}}$.

37. $y = \int_{\sqrt{x}}^{x} \frac{\cos\theta}{\theta}\,d\theta = \int_{\sqrt{x}}^{1} \frac{\cos\theta}{\theta}\,d\theta + \int_{1}^{x} \frac{\cos\theta}{\theta}\,d\theta = \int_{1}^{x} \frac{\cos\theta}{\theta}\,d\theta - \int_{1}^{\sqrt{x}} \frac{\cos\theta}{\theta}\,d\theta \Rightarrow$

$y' = \frac{\cos x}{x} - \frac{\cos\sqrt{x}}{\sqrt{x}}\,\frac{1}{2\sqrt{x}} = \frac{2\cos x - \cos\sqrt{x}}{2x}$

39. If $1 \le x \le 3$, then $\sqrt{1^2+3} \le \sqrt{x^2+3} \le \sqrt{3^2+3} \Rightarrow 2 \le \sqrt{x^2+3} \le 2\sqrt{3}$, so $2(3-1) \le \int_1^3 \sqrt{x^2+3}\,dx \le 2\sqrt{3}(3-1)$; that is, $4 \le \int_1^3 \sqrt{x^2+3}\,dx \le 4\sqrt{3}$.

41. $0 \le x \le 1 \Rightarrow 0 \le \cos x \le 1 \Rightarrow x^2\cos x \le x^2 \Rightarrow$

$\int_0^1 x^2\cos x\,dx \le \int_0^1 x^2\,dx = \frac{1}{3}\left[x^3\right]_0^1 = \frac{1}{3}$ [Property 7].

43. Let $f(x) = \sqrt{1+x^3}$ on $[0,1]$. The Midpoint Rule with $n = 5$ gives

$$\begin{aligned}\int_0^1 \sqrt{1+x^3}\,dx &\approx \tfrac{1}{5}[f(0.1)+f(0.3)+f(0.5)+f(0.7)+f(0.9)]\\ &= \tfrac{1}{5}\left[\sqrt{1+(0.1)^3}+\sqrt{1+(0.3)^3}+\cdots+\sqrt{1+(0.9)^3}\right] \approx 1.110\end{aligned}$$

45. Note that $r(t) = b'(t)$, where $b(t) =$ the number of barrels of oil consumed up to time t. So, by the Net Change Theorem, $\int_0^3 r(t)\,dt = b(3) - b(0)$ represents the number of barrels of oil consumed from Jan. 1, 2000, through Jan. 1, 2003.

47. We use the Midpoint Rule with $n = 6$ and $\Delta t = \frac{24-0}{6} = 4$. The increase in the bee population was

$$\begin{aligned}\int_0^{24} r(t)\,dt \approx M_6 &= 4[r(2)+r(6)+r(10)+r(14)+r(18)+r(22)]\\ &\approx 4[50+1000+7000+8550+1350+150]\\ &= 4(18{,}100) = 72{,}400\end{aligned}$$

49. By the Fundamental Theorem of Calculus, we know that $F(x) = \int_a^x t^2\sin(t^2)\,dt$ is an antiderivative of $f(x) = x^2\sin(x^2)$. This integral cannot be expressed in any simpler form. Since $\int_a^a f\,dt = 0$ for any a, we can take $a = 1$, and then $F(1) = 0$, as required. So $F(x) = \int_1^x t^2\sin(t^2)\,dt$ is the desired function.

51. $\int_0^x f(t)\,dt = x\sin x + \int_0^x \frac{f(t)}{1+t^2}\,dt \Rightarrow f(x) = x\cos x + \sin x + \frac{f(x)}{1+x^2}$ (by differentiation) $\Rightarrow$

$f(x)\left(1 - \frac{1}{1+x^2}\right) = x\cos x + \sin x \Rightarrow f(x)\left(\frac{x^2}{1+x^2}\right) = x\cos x + \sin x \Rightarrow$

$f(x) = \frac{1+x^2}{x^2}(x\cos x + \sin x)$

53. Let $u = f(x)$ and $du = f'(x)\,dx$. So $2\int_a^b f(x)f'(x)\,dx = 2\int_{f(a)}^{f(b)} u\,du = \left[u^2\right]_{f(a)}^{f(b)} = [f(b)]^2 - [f(a)]^2$.

55. Let $u = 1 - x$. Then $du = -dx$, so $\int_0^1 f(1-x)\,dx = \int_1^0 f(u)(-du) = \int_0^1 f(u)\,du = \int_0^1 f(x)\,dx$.

APPLICATIONS OF INTEGRATION

Areas between Curves

1. $A = \int_{x=0}^{x=4} (y_T - y_B)\,dx = \int_0^4 \left[(5x - x^2) - x\right] dx = \int_0^4 (4x - x^2)\,dx$

$= \left[2x^2 - \frac{1}{3}x^3\right]_0^4 = \left(32 - \frac{64}{3}\right) - (0) = \frac{32}{3}$

3. $A = \int_{y=0}^{y=1} (x_R - x_L)\,dy = \int_0^1 \left[\sqrt{y} - (y^2 - 1)\right] dy = \int_0^1 \left(y^{1/2} - y^2 + 1\right) dy$

$= \left[\frac{2}{3}y^{3/2} - \frac{1}{3}y^3 + y\right]_0^1 = \left(\frac{2}{3} - \frac{1}{3} + 1\right) - (0) = \frac{4}{3}$

5. $A = \int_{-1}^{2} \left[(9 - x^2) - (x + 1)\right] dx$

$= \int_{-1}^{2} (8 - x - x^2)\,dx$

$= \left[8x - \frac{x^2}{2} - \frac{x^3}{3}\right]_{-1}^{2}$

$= \left(16 - 2 - \frac{8}{3}\right) - \left(-8 - \frac{1}{2} + \frac{1}{3}\right)$

$= 22 - 3 + \frac{1}{2} = \frac{39}{2}$

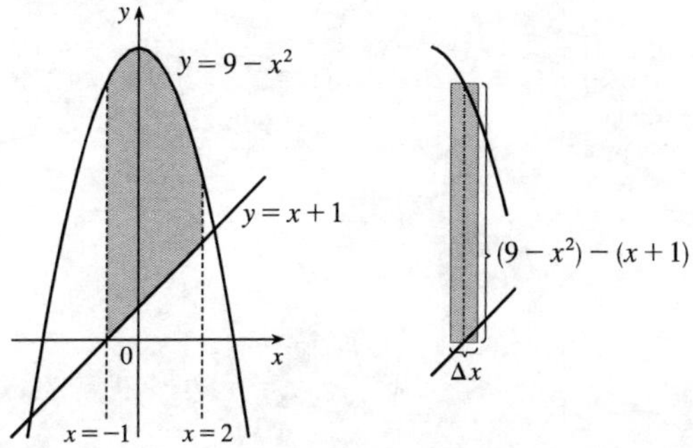

7. The curves intersect when $x = x^2 \;\Rightarrow\; x^2 - x = 0 \;\Leftrightarrow\; x(x - 1) = 0 \;\Leftrightarrow\; x = 0, 1.$

$A = \int_0^1 (x - x^2)\,dx$

$= \left[\frac{1}{2}x^2 - \frac{1}{3}x^3\right]_0^1$

$= \frac{1}{2} - \frac{1}{3}$

$= \frac{1}{6}$

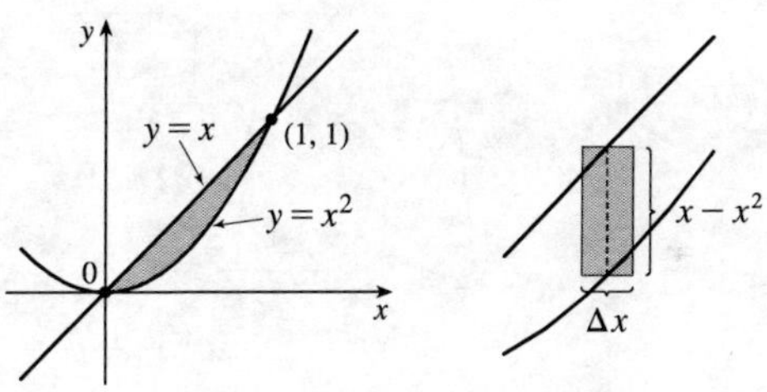

9. First find the points of intersection: $\sqrt{x+3} = \dfrac{x+3}{2} \;\Rightarrow\; \left(\sqrt{x+3}\right)^2 = \left(\dfrac{x+3}{2}\right)^2 \;\Rightarrow\; x + 3 = \frac{1}{4}(x+3)^2$

$\Rightarrow\; 4(x+3) - (x+3)^2 = 0 \;\Rightarrow\; (x+3)[4 - (x+3)] = 0 \;\Rightarrow\; (x+3)(1 - x) = 0 \;\Rightarrow\; x = -3$ or 1. So

$$A = \int_{-3}^{1} \left(\sqrt{x+3} - \frac{x+3}{2}\right) dx = \left[\frac{2}{3}(x+3)^{3/2} - \frac{(x+3)^2}{4}\right]_{-3}^{1} = \left(\frac{16}{3} - 4\right) - (0 - 0) = \frac{4}{3}$$

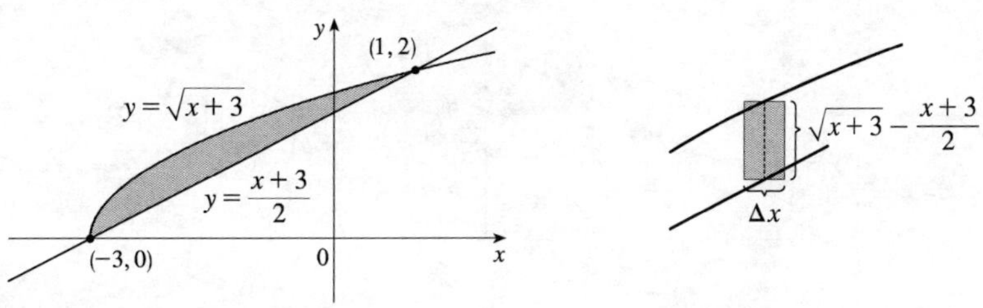

11. $A = \int_0^1 \left(\sqrt{x} - x^2\right) dx$

$= \left[\frac{2}{3}x^{3/2} - \frac{1}{3}x^3\right]_0^1$

$= \frac{2}{3} - \frac{1}{3}$

$= \frac{1}{3}$

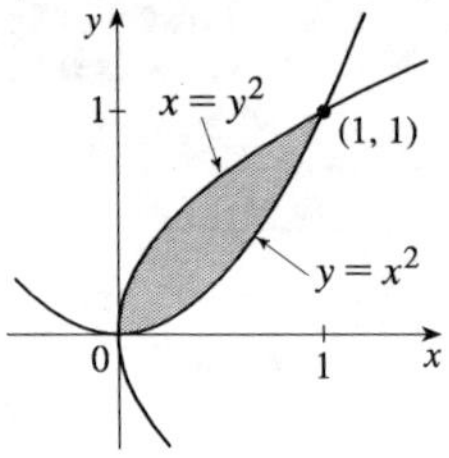

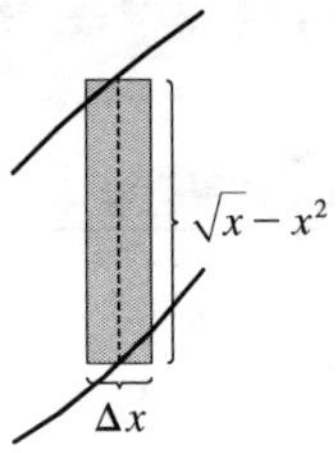

13. $12 - x^2 = x^2 - 6 \quad \Leftrightarrow \quad 2x^2 = 18 \quad \Leftrightarrow \quad x^2 = 9 \quad \Leftrightarrow \quad x = \pm 3$, so

$A = \int_{-3}^{3} \left[\left(12 - x^2\right) - \left(x^2 - 6\right)\right] dx = 2\int_0^3 \left(18 - 2x^2\right) dx$ [by symmetry]

$= 2\left[18x - \frac{2}{3}x^3\right]_0^3 = 2\left[(54 - 18) - 0\right] = 2(36) = 72$

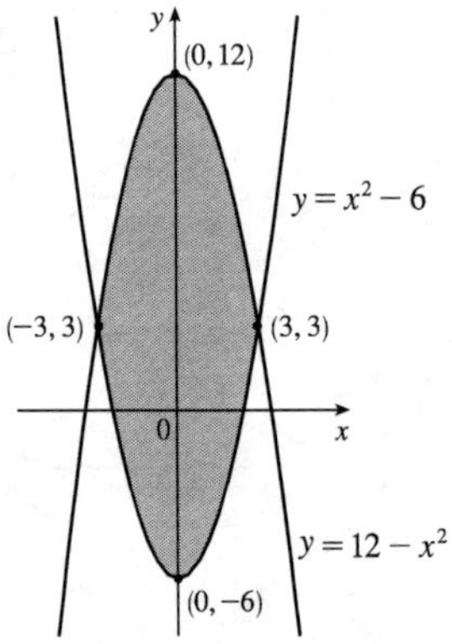

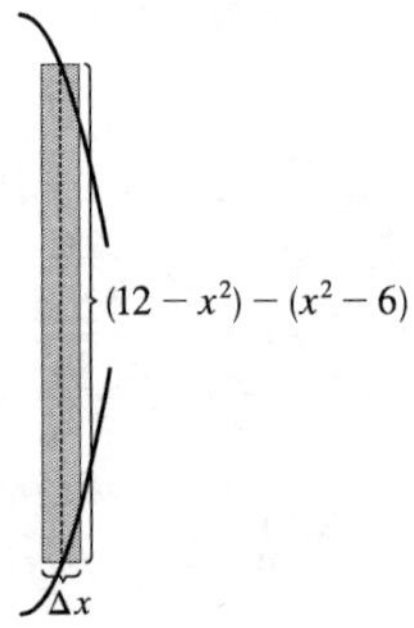

15. $\frac{1}{2}x = \sqrt{x} \quad \Rightarrow \quad \frac{1}{4}x^2 = x \quad \Rightarrow \quad x^2 - 4x = 0 \quad \Rightarrow \quad x(x - 4) = 0 \quad \Rightarrow \quad x = 0$ or 4, so

$A = \int_0^4 \left(\sqrt{x} - \frac{1}{2}x\right) dx + \int_4^9 \left(\frac{1}{2}x - \sqrt{x}\right) dx = \left[\frac{2}{3}x^{3/2} - \frac{1}{4}x^2\right]_0^4 + \left[\frac{1}{4}x^2 - \frac{2}{3}x^{3/2}\right]_4^9$

$= \left[\left(\frac{16}{3} - 4\right) - 0\right] + \left[\left(\frac{81}{4} - 18\right) - \left(4 - \frac{16}{3}\right)\right] = \frac{81}{4} + \frac{32}{3} - 26 = \frac{59}{12}$

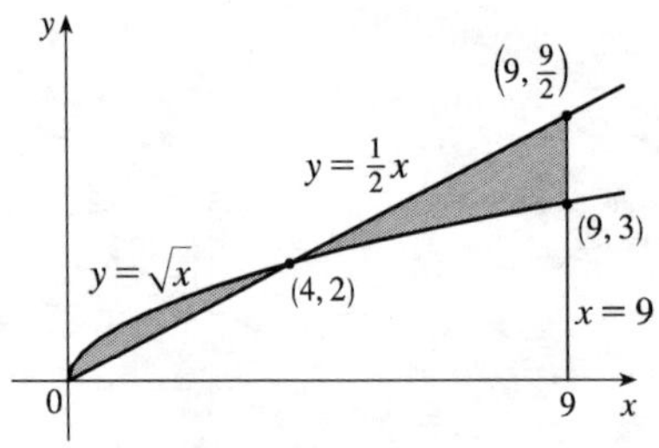

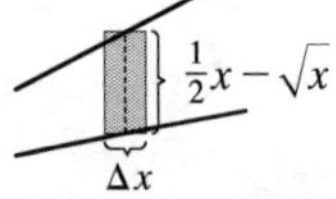

For $4 < x < 9$

17. $2y^2 = 1 - y \;\Leftrightarrow\; 2y^2 + y - 1 = 0 \;\Leftrightarrow\; (2y-1)(y+1) = 0 \;\Leftrightarrow\; y = \frac{1}{2}$ or -1, so $x = \frac{1}{2}$ or 2 and
$A = \int_{-1}^{1/2} \left[(1-y) - 2y^2\right] dy = \int_{-1}^{1/2} \left(1 - y - 2y^2\right) dy = \left[y - \frac{1}{2}y^2 - \frac{2}{3}y^3\right]_{-1}^{1/2}$

$= \left(\frac{1}{2} - \frac{1}{8} - \frac{1}{12}\right) - \left(-1 - \frac{1}{2} + \frac{2}{3}\right) = \frac{7}{24} - \left(-\frac{5}{6}\right) = \frac{7}{24} + \frac{20}{24} = \frac{27}{24} = \frac{9}{8}$

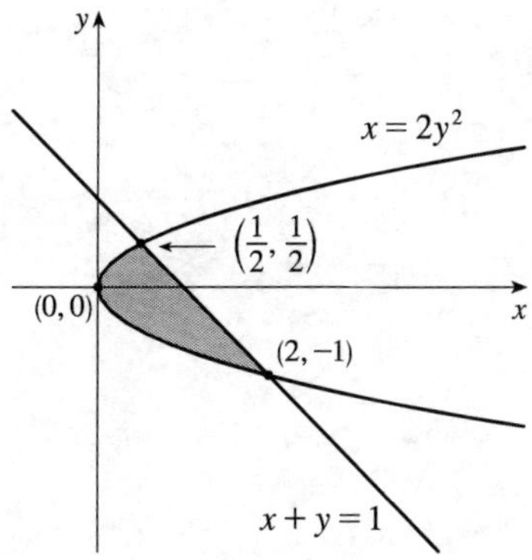

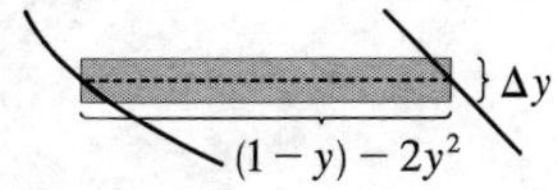

19. The curves intersect when $1 - y^2 = y^2 - 1 \;\Leftrightarrow\; 2 = 2y^2 \;\Leftrightarrow\; y^2 = 1 \;\Leftrightarrow\; y = \pm 1$.

$$A = \int_{-1}^{1} \left[(1-y^2) - (y^2 - 1)\right] dy$$
$$= \int_{-1}^{1} 2(1-y^2)\, dy$$
$$= 2 \cdot 2 \int_{0}^{1} (1-y^2)\, dy$$
$$= 4\left[y - \tfrac{1}{3}y^3\right]_0^1 = 4\left(1 - \tfrac{1}{3}\right) = \tfrac{8}{3}$$

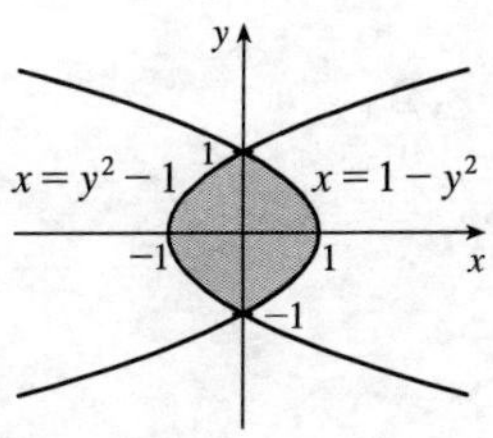

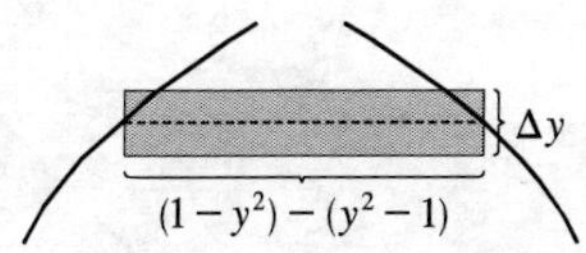

21. Notice that $\cos x = \sin 2x = 2\sin x \cos x \;\Leftrightarrow$

$2\sin x \cos x - \cos x = 0 \;\Leftrightarrow\; \cos x\,(2\sin x - 1) = 0 \;\Leftrightarrow$

$2\sin x = 1$ or $\cos x = 0 \;\Leftrightarrow\; x = \frac{\pi}{6}$ or $\frac{\pi}{2}$.

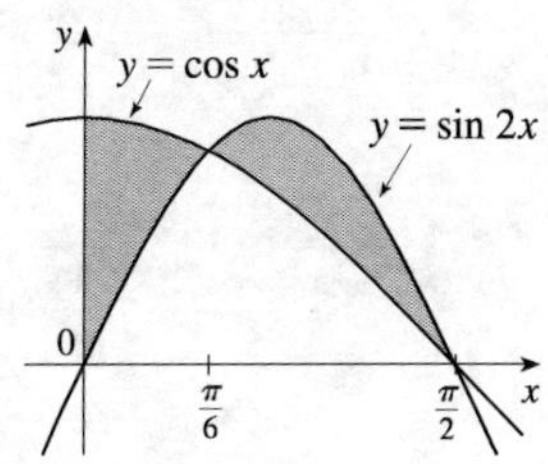

$A = \int_0^{\pi/6} (\cos x - \sin 2x)\, dx + \int_{\pi/6}^{\pi/2} (\sin 2x - \cos x)\, dx$

$= \left[\sin x + \frac{1}{2}\cos 2x\right]_0^{\pi/6} + \left[-\frac{1}{2}\cos 2x - \sin x\right]_{\pi/6}^{\pi/2}$

$= \frac{1}{2} + \frac{1}{2} \cdot \frac{1}{2} - \left(0 + \frac{1}{2} \cdot 1\right) + \left(\frac{1}{2} - 1\right) - \left(-\frac{1}{2} \cdot \frac{1}{2} - \frac{1}{2}\right) = \frac{1}{2}$

23. From the graph, we see that the curves intersect at $x=0$, $x=\frac{\pi}{2}$, and $x=\pi$. By symmetry,

$$A=\int_0^{\pi}\left|\cos x-\left(1-\frac{2x}{\pi}\right)\right|dx=2\int_0^{\pi/2}\left[\cos x-\left(1-\frac{2x}{\pi}\right)\right]dx=2\int_0^{\pi/2}\left(\cos x-1+\frac{2x}{\pi}\right)dx$$

$$=2\left[\sin x-x+\tfrac{1}{\pi}x^2\right]_0^{\pi/2}=2\left[\left(1-\tfrac{\pi}{2}+\tfrac{1}{\pi}\cdot\tfrac{\pi^2}{4}\right)-0\right]=2\left(1-\tfrac{\pi}{2}+\tfrac{\pi}{4}\right)=2-\tfrac{\pi}{2}$$

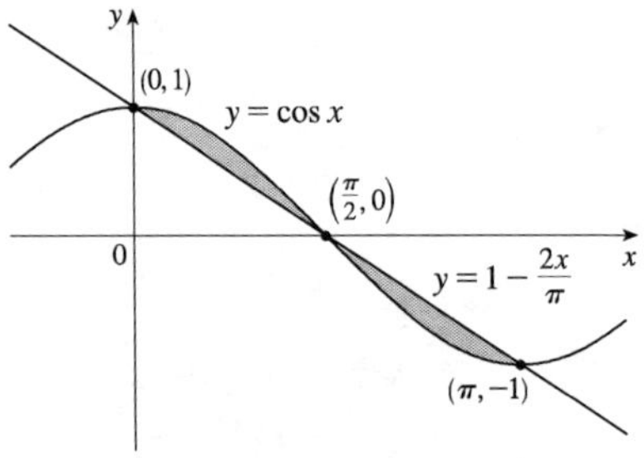

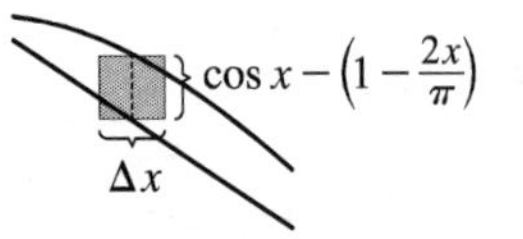

25. Graph the three functions $y=x$, $y=-\frac{1}{2}x$, and $y=-2x+3$; then determine the points of intersection: $(0,0)$, $(1,1)$, and $(2,-1)$.

$$A=\int_0^1\left[x-\left(-\tfrac{1}{2}x\right)\right]dx+\int_1^2\left[(3-2x)-\left(-\tfrac{1}{2}x\right)\right]dx=\int_0^1\tfrac{3}{2}x\,dx+\int_1^2\left(3-\tfrac{3}{2}x\right)dx$$

$$=\left[\tfrac{3}{4}x^2\right]_0^1+\left[3x-\tfrac{3}{4}x^2\right]_1^2=\left(\tfrac{3}{4}-0\right)+(6-3)-\left(3-\tfrac{3}{4}\right)=\tfrac{3}{2}$$

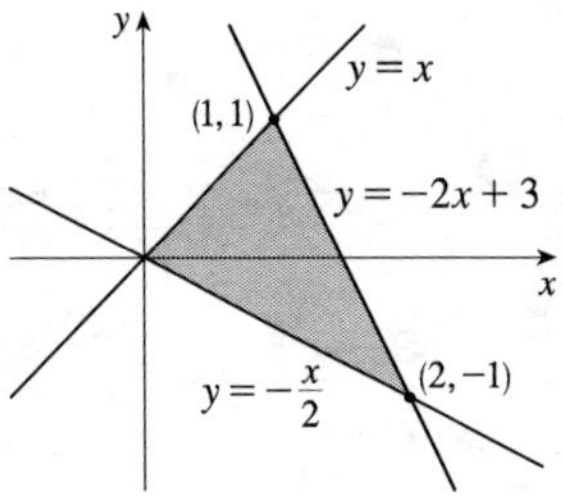

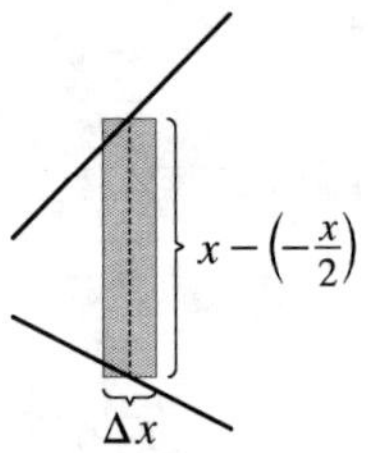

27. An equation of the line through $(0,0)$ and $(2,1)$ is $y=\frac{1}{2}x$; through $(0,0)$ and $(-1,6)$ is $y=-6x$; through $(2,1)$ and $(-1,6)$ is $y=-\frac{5}{3}x+\frac{13}{3}$.

$$A=\int_{-1}^0\left[\left(-\tfrac{5}{3}x+\tfrac{13}{3}\right)-(-6x)\right]dx+\int_0^2\left[\left(-\tfrac{5}{3}x+\tfrac{13}{3}\right)-\tfrac{1}{2}x\right]dx$$

$$=\int_{-1}^0\left(\tfrac{13}{3}x+\tfrac{13}{3}\right)dx+\int_0^2\left(-\tfrac{13}{6}x+\tfrac{13}{3}\right)dx$$

$$=\tfrac{13}{3}\int_{-1}^0(x+1)\,dx+\tfrac{13}{3}\int_0^2\left(-\tfrac{1}{2}x+1\right)dx$$

$$=\tfrac{13}{3}\left[\tfrac{1}{2}x^2+x\right]_{-1}^0+\tfrac{13}{3}\left[-\tfrac{1}{4}x^2+x\right]_0^2$$

$$=\tfrac{13}{3}\left[0-\left(\tfrac{1}{2}-1\right)\right]+\tfrac{13}{3}\left[(-1+2)-0\right]=\tfrac{13}{3}\cdot\tfrac{1}{2}+\tfrac{13}{3}\cdot 1=\tfrac{13}{2}$$

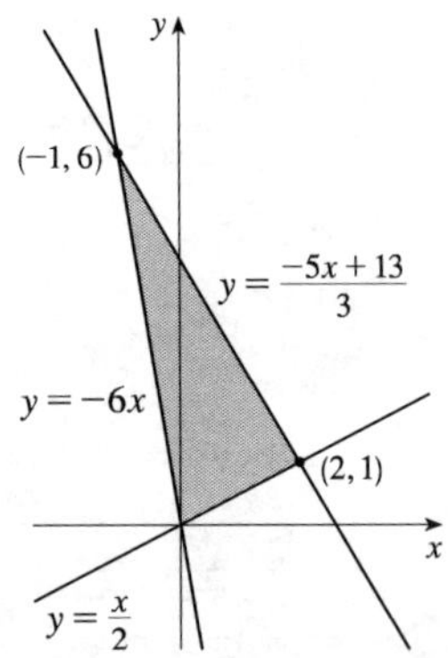

29. $A = \int_{-1}^{1} \left| x^3 - x \right| dx$

$= 2 \int_{0}^{1} (x - x^3)\, dx$ [by symmetry]

$= 2\left[\frac{1}{2}x^2 - \frac{1}{4}x^4\right]_0^1$

$= 2\left(\frac{1}{2} - \frac{1}{4}\right) = \frac{1}{2}$

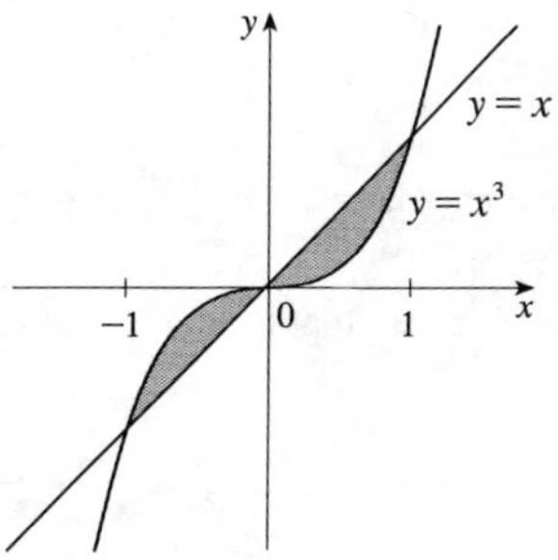

31. Let $f(x) = \cos^2\left(\frac{\pi x}{4}\right) - \sin^2\left(\frac{\pi x}{4}\right)$ and $\Delta x = \frac{1-0}{4}$.

The shaded area is given by

$A = \int_0^1 f(x)\,dx \approx M_4$

$= \frac{1}{4}\left[f\left(\frac{1}{8}\right) + f\left(\frac{3}{8}\right) + f\left(\frac{5}{8}\right) + f\left(\frac{7}{8}\right)\right]$

≈ 0.6407

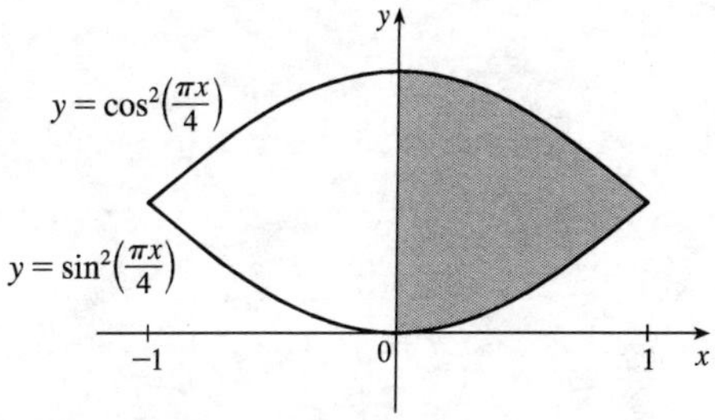

33.

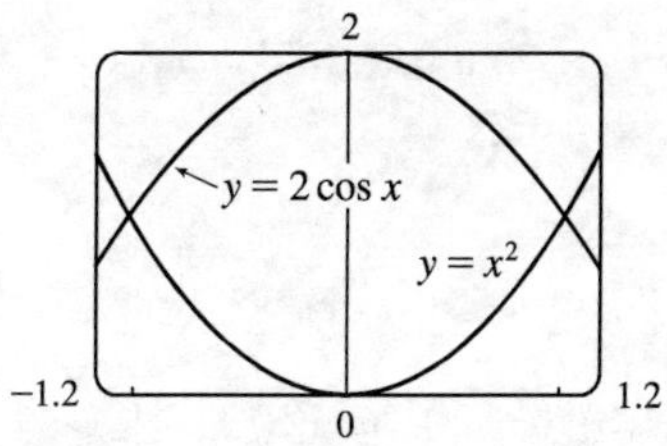

From the graph, we see that the curves intersect at $x = \pm a \approx \pm 1.02$, with $2\cos x > x^2$ on $(-a, a)$. So the area of the region bounded by the curves is

$$A = \int_{-a}^{a} \left(2\cos x - x^2\right) dx = 2\int_0^a \left(2\cos x - x^2\right) dx$$

$$= 2\left[2\sin x - \tfrac{1}{3}x^3\right]_0^a \approx 2.70$$

35.

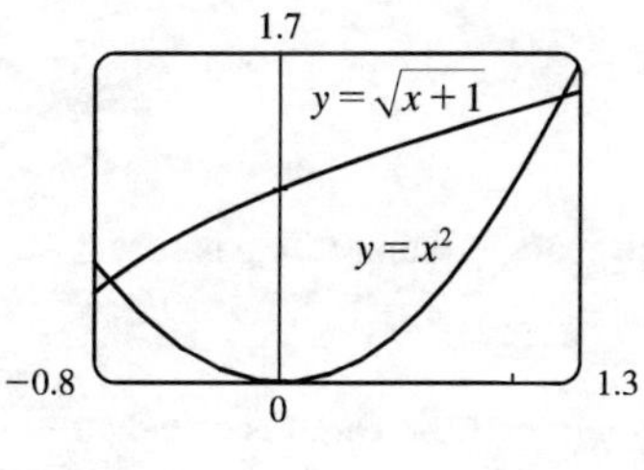

From the graph, we see that the curves intersect at $x = a \approx -0.72$ and at $x = b \approx 1.22$, with $\sqrt{x+1} > x^2$ on (a, b). So the area between the curves is

$$A = \int_a^b \left(\sqrt{x+1} - x^2\right) dx = \left[\tfrac{2}{3}(x+1)^{3/2} - \tfrac{1}{3}x^3\right]_a^b$$

$$\approx 1.38$$

37. As the figure illustrates, the curves $y = x$ and $y = x^5 - 6x^3 + 4x$ enclose a four-part region symmetric about the origin (since $x^5 - 6x^3 + 4x$ and x are odd functions of x). The curves intersect at values of x where $x^5 - 6x^3 + 4x = x$; that is, where $x(x^4 - 6x^2 + 3) = 0$. That happens at $x = 0$ and where $x^2 = \dfrac{6 \pm \sqrt{36 - 12}}{2} = 3 \pm \sqrt{6}$; that is, at $x = -\sqrt{3 + \sqrt{6}}, -\sqrt{3 - \sqrt{6}}, 0, \sqrt{3 - \sqrt{6}}$, and $\sqrt{3 + \sqrt{6}}$.

The exact area is

$$\begin{aligned} 2\int_0^{\sqrt{3+\sqrt{6}}} \left|(x^5 - 6x^3 + 4x) - x\right| dx &= 2\int_0^{\sqrt{3+\sqrt{6}}} \left|x^5 - 6x^3 + 3x\right| dx \\ &= 2\int_0^{\sqrt{3-\sqrt{6}}} (x^5 - 6x^3 + 3x)\, dx + 2\int_{\sqrt{3-\sqrt{6}}}^{\sqrt{3+\sqrt{6}}} (-x^5 + 6x^3 - 3x)\, dx \\ &\overset{\text{CAS}}{=} 12\sqrt{6} - 9 \end{aligned}$$

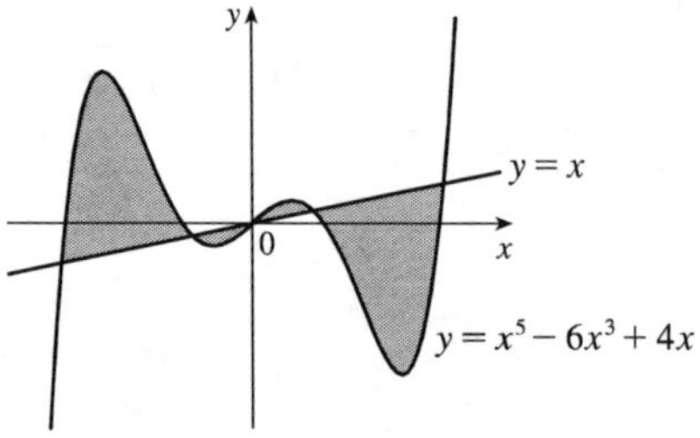

39. 1 second $= \frac{1}{3600}$ hour, so 10 s $= \frac{1}{360}$ h. With the given data, we can take $n = 5$ to use the Midpoint Rule.

$\Delta t = \frac{1/360 - 0}{5} = \frac{1}{1800}$, so

$$\begin{aligned} \text{distance}_{\text{Kelly}} - \text{distance}_{\text{Chris}} &= \int_0^{1/360} v_K\, dt - \int_0^{1/360} v_C\, dt = \int_0^{1/360} (v_K - v_C)\, dt \\ &\approx M_5 = \tfrac{1}{1800}\left[(v_K - v_C)(1) + (v_K - v_C)(3) + (v_K - v_C)(5)\right. \\ &\qquad\qquad \left. + (v_K - v_C)(7) + (v_K - v_C)(9)\right] \\ &= \tfrac{1}{1800}\left[(22 - 20) + (52 - 46) + (71 - 62) + (86 - 75) + (98 - 86)\right] \\ &= \tfrac{1}{1800}(2 + 6 + 9 + 11 + 12) = \tfrac{1}{1800}(40) = \tfrac{1}{45} \text{ mile, or } 117\tfrac{1}{3} \text{ feet} \end{aligned}$$

41. We know that the area under curve A between $t = 0$ and $t = x$ is $\int_0^x v_A(t)\, dt = s_A(x)$, where $v_A(t)$ is the velocity of car A and s_A is its displacement. Similarly, the area under curve B between $t = 0$ and $t = x$ is $\int_0^x v_B(t)\, dt = s_B(x)$.

(a) After one minute, the area under curve A is greater than the area under curve B. So car A is ahead after one minute.

(b) The area of the shaded region has numerical value $s_A(1) - s_B(1)$, which is the distance by which A is ahead of B after 1 minute.

(c) After two minutes, car B is traveling faster than car A and has gained some ground, but the area under curve A from $t = 0$ to $t = 2$ is still greater than the corresponding area for curve B, so car A is still ahead.

(d) From the graph, it appears that the area between curves A and B for $0 \le t \le 1$ (when car A is going faster), which corresponds to the distance by which car A is ahead, seems to be about 3 squares. Therefore, the cars will be side by side at the time x where the area between the curves for $1 \le t \le x$ (when car B is going faster) is the same as the area for $0 \le t \le 1$. From the graph, it appears that this time is $x \approx 2.2$. So the cars are side by side when $t \approx 2.2$ minutes.

43.

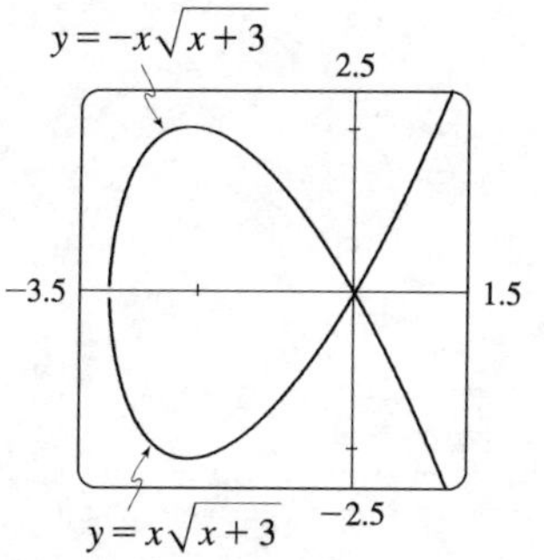

To graph this function, we must first express it as a combination of explicit functions of y; namely, $y = \pm x\sqrt{x+3}$. We can see from the graph that the loop extends from $x = -3$ to $x = 0$, and that by symmetry, the area we seek is just twice the area under the top half of the curve on this interval, the equation of the top half being $y = -x\sqrt{x+3}$. So the area is $A = 2\int_{-3}^{0}\left(-x\sqrt{x+3}\right)dx$. We substitute $u = x + 3$, so $du = dx$ and the limits change to 0 and 3, and we get

$$A = -2\int_0^3 \left[(u-3)\sqrt{u}\right]du = -2\int_0^3\left(u^{3/2} - 3u^{1/2}\right)du$$

$$= -2\left[\tfrac{2}{5}u^{5/2} - 2u^{3/2}\right]_0^3 = -2\left[\tfrac{2}{5}\left(3^2\sqrt{3}\right) - 2\left(3\sqrt{3}\right)\right] = \tfrac{24}{5}\sqrt{3}$$

45.

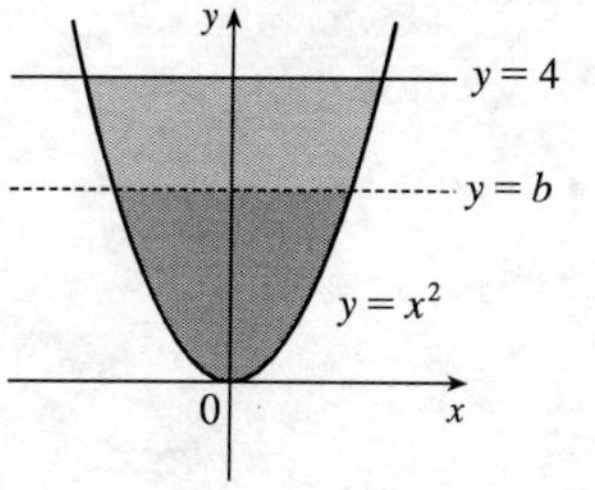

By the symmetry of the problem, we consider only the first quadrant, where $y = x^2 \Rightarrow x = \sqrt{y}$. We are looking for a number b such that $\int_0^b \sqrt{y}\,dy = \int_b^4 \sqrt{y}\,dy \Rightarrow \tfrac{2}{3}\left[y^{3/2}\right]_0^b = \tfrac{2}{3}\left[y^{3/2}\right]_b^4 \Rightarrow$ $b^{3/2} = 4^{3/2} - b^{3/2} \Rightarrow 2b^{3/2} = 8 \Rightarrow b^{3/2} = 4 \Rightarrow$ $b = 4^{2/3} \approx 2.52$.

47. We first assume that $c > 0$, since c can be replaced by $-c$ in both equations without changing the graphs, and if $c = 0$ the curves do not enclose a region. We see from the graph that the enclosed area A lies between $x = -c$ and $x = c$, and by symmetry, it is equal to four times the area in the first quadrant.

The enclosed area is

$$A = 4\int_0^c \left(c^2 - x^2\right)dx = 4\left[c^2x - \tfrac{1}{3}x^3\right]_0^c$$

$$= 4\left(c^3 - \tfrac{1}{3}c^3\right) = 4\left(\tfrac{2}{3}c^3\right) = \tfrac{8}{3}c^3$$

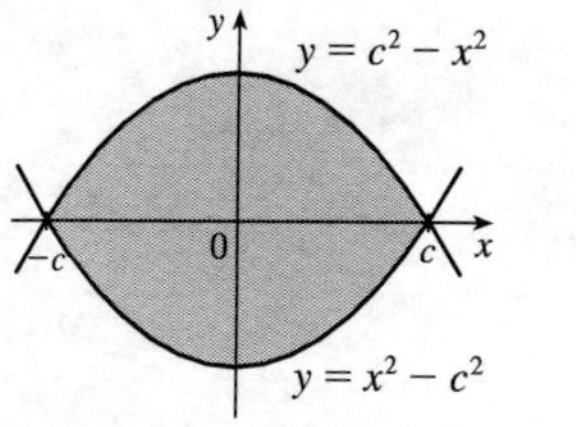

So $A = 576 \Leftrightarrow \tfrac{8}{3}c^3 = 576 \Leftrightarrow c^3 = 216 \Leftrightarrow c = \sqrt[3]{216} = 6$.

Note that $c = -6$ is another solution, since the graphs are the same.

49. $A = \int_1^2 \left(\frac{1}{x} - \frac{1}{x^2}\right) dx = \left[\ln x + \frac{1}{x}\right]_1^2$

$= \left(\ln 2 + \frac{1}{2}\right) - (\ln 1 + 1)$

$= \ln 2 - \frac{1}{2} \approx 0.19$

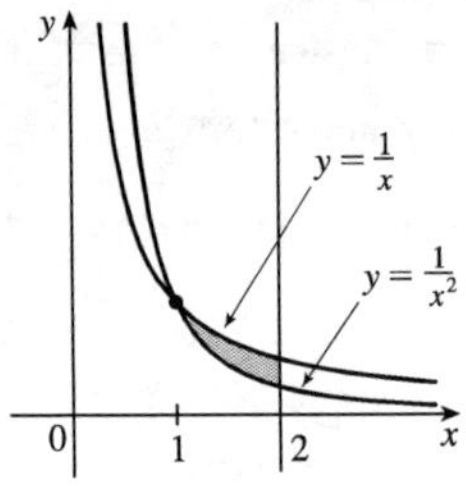

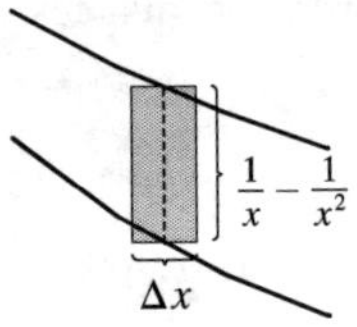

51. The curves intersect when $x^2 = \frac{2}{x^2+1} \Leftrightarrow x^4 + x^2 = 2 \Leftrightarrow x^4 + x^2 - 2 = 0 \Leftrightarrow$

$(x^2+2)(x^2-1) = 0 \Leftrightarrow x^2 = 1 \Leftrightarrow x = \pm 1.$

$A = \int_{-1}^1 \left(\frac{2}{x^2+1} - x^2\right) dx$

$= 2\int_0^1 \left(\frac{2}{x^2+1} - x^2\right) dx$

$= 2\left[2\tan^{-1} x - \frac{1}{3}x^3\right]_0^1 = 2\left(2 \cdot \frac{\pi}{4} - \frac{1}{3}\right)$

$= \pi - \frac{2}{3} \approx 2.47$

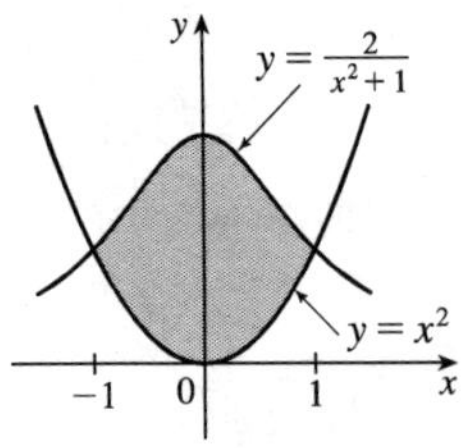

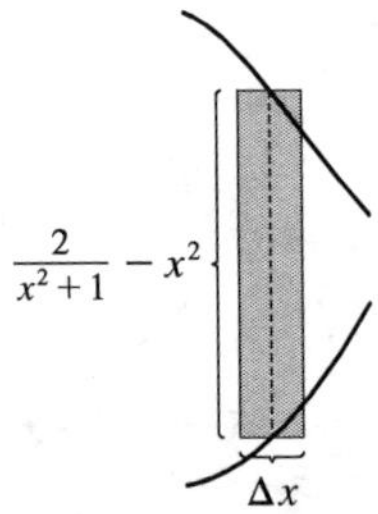

Work

1. By Equation 2, $W = Fd = (900)(8) = 7200$ J.

3. $$W = \int_a^b f(x)\,dx = \int_0^9 \frac{10}{(1+x)^2}\,dx = 10\int_1^{10} \frac{1}{u^2}\,du \qquad [u = 1+x,\ du = dx]$$
$$= 10\left[-\frac{1}{u}\right]_1^{10} = 10\left(-\tfrac{1}{10}+1\right) = 9 \text{ ft-lb}$$

5. The force function is given by $F(x)$ (in newtons) and the work (in joules) is the area under the curve, given by $\int_0^8 F(x)\,dx = \int_0^4 F(x)\,dx + \int_4^8 F(x)\,dx = \frac{1}{2}(4)(30) + (4)(30) = 180$ J.

7. $10 = f(x) = kx = \frac{1}{3}k$ [4 inches $= \frac{1}{3}$ foot], so $k = 30$ lb/ft and $f(x) = 30x$. Now 6 inches $= \frac{1}{2}$ foot, so $W = \int_0^{1/2} 30x\,dx = \left[15x^2\right]_0^{1/2} = \frac{15}{4}$ ft-lb.

9. If $\int_0^{0.12} kx\,dx = 2$ J, then $2 = \left[\frac{1}{2}kx^2\right]_0^{0.12} = \frac{1}{2}k(0.0144) = 0.0072k$ and $k = \frac{2}{0.0072} = \frac{2500}{9} \approx 277.78$ N/m. Thus, the work needed to stretch the spring from 35 cm to 40 cm is
$\int_{0.05}^{0.10} \frac{2500}{9}x\,dx = \left[\frac{1250}{9}x^2\right]_{1/20}^{1/10} = \frac{1250}{9}\left(\frac{1}{100} - \frac{1}{400}\right) = \frac{25}{24} \approx 1.04$ J.

11. $f(x) = kx$, so $30 = \frac{2500}{9}x$ and $x = \frac{270}{2500}$ m $= 10.8$ cm

In Exercises 13–20, n is the number of subintervals of length Δx, and x_i^* is a sample point in the ith subinterval $[x_{i-1}, x_i]$.

13. (a) The portion of the rope from x ft to $(x + \Delta x)$ ft below the top of the building weighs $\frac{1}{2}\,\Delta x$ lb and must be lifted x_i^* ft, so its contribution to the total work is $\frac{1}{2}x_i^*\,\Delta x$ ft-lb. The total work is
$$W = \lim_{n\to\infty} \sum_{i=1}^{n} \tfrac{1}{2}x_i^*\,\Delta x = \int_0^{50} \tfrac{1}{2}x\,dx = \left[\tfrac{1}{4}x^2\right]_0^{50} = \tfrac{2500}{4} = 625 \text{ ft-lb}$$
Notice that the exact height of the building does not matter (as long as it is more than 50 ft).

(b) When half the rope is pulled to the top of the building, the work to lift the top half of the rope is $W_1 = \int_0^{25} \frac{1}{2}x\,dx = \left[\frac{1}{4}x^2\right]_0^{25} = \frac{625}{4}$ ft-lb. The bottom half of the rope is lifted 25 ft and the work needed to accomplish that is $W_2 = \int_{25}^{50} \frac{1}{2}\cdot 25\,dx = \frac{25}{2}\left[x\right]_{25}^{50} = \frac{625}{2}$ ft-lb. The total work done in pulling half the rope to the top of the building is $W = W_1 + W_2 = \frac{625}{2} + \frac{625}{4} = \frac{3}{4}\cdot 625 = \frac{1875}{4}$ ft-lb.

15. The work needed to lift the cable is $\lim\limits_{n\to\infty} \sum_{i=1}^{n} 2x_i^*\,\Delta x = \int_0^{500} 2x\,dx = \left[x^2\right]_0^{500} = 250{,}000$ ft-lb. The work needed to lift the coal is 800 lb $\cdot$ 500 ft $= 400{,}000$ ft-lb. Thus, the total work required is $250{,}000 + 400{,}000 = 650{,}000$ ft-lb.

17. At a height of x meters ($0 \leq x \leq 12$), the mass of the rope is $(0.8 \text{ kg/m})(12 - x \text{ m}) = (9.6 - 0.8x)$ kg and the mass of the water is $\left(\frac{36}{12} \text{ kg/m}\right)(12 - x \text{ m}) = (36 - 3x)$ kg. The mass of the bucket is 10 kg, so the total mass is $(9.6 - 0.8x) + (36 - 3x) + 10 = (55.6 - 3.8x)$ kg, and hence, the total force is $9.8(55.6 - 3.8x)$ N.

The work needed to lift the bucket Δx m through the ith subinterval of $[0, 12]$ is $9.8(55.6 - 3.8x_i^*)\Delta x$, so the total work is

$$W = \lim_{n\to\infty} \sum_{i=1}^{n} 9.8(55.6 - 3.8x_i^*)\,\Delta x = \int_0^{12} (9.8)(55.6 - 3.8x)\,dx = 9.8\left[55.6x - 1.9x^2\right]_0^{12}$$

$$= 9.8(393.6) \approx 3857 \text{ J}$$

19. A "slice" of water Δx m thick and lying at a depth of x_i^* m (where $0 \leq x_i^* \leq \frac{1}{2}$) has volume $(2 \times 1 \times \Delta x)$ m^3, a mass of $2000\,\Delta x$ kg, weighs about $(9.8)(2000\,\Delta x) = 19{,}600\,\Delta x$ N, and thus requires about $19{,}600x_i^*\,\Delta x$ J of work for its removal. So $W = \lim_{n\to\infty} \sum_{i=1}^{n} 19{,}600x_i^*\,\Delta x = \int_0^{1/2} 19{,}600x\,dx = \left[9800x^2\right]_0^{1/2} = 2450$ J.

21. A rectangular "slice" of water Δx m thick and lying x ft above the bottom has width x ft and volume $8x\,\Delta x$ m^3. It weighs about $(9.8 \times 1000)(8x\,\Delta x)$ N, and must be lifted $(5 - x)$ m by the pump, so the work needed is about $(9.8 \times 10^3)(5 - x)(8x\,\Delta x)$J. The total work required is

$$W \approx \int_0^3 (9.8 \times 10^3)(5 - x)8x\,dx = (9.8 \times 10^3)\int_0^3 (40x - 8x^2)\,dx = (9.8 \times 10^3)\left[20x^2 - \tfrac{8}{3}x^3\right]_0^3$$

$$= (9.8 \times 10^3)(180 - 72) = (9.8 \times 10^3)(108) = 1058.4 \times 10^3 \approx 1.06 \times 10^6 \text{ J}$$

23. Measure depth x downward from the flat top of the tank, so that $0 \leq x \leq 2$ ft. Then $\Delta W = (62.5)\left(2\sqrt{4 - x^2}\right)(8\,\Delta x)(x + 1)$ ft-lb, so

$$W \approx (62.5)(16)\int_0^2 (x + 1)\sqrt{4 - x^2}\,dx = 1000\left(\int_0^2 x\sqrt{4 - x^2}\,dx + \int_0^2 \sqrt{4 - x^2}\,dx\right)$$

$$= 1000\left[\int_0^4 u^{1/2}\left(\tfrac{1}{2}\,du\right) + \tfrac{1}{4}\pi\left(2^2\right)\right] \qquad \text{[Put } u = 4 - x^2\text{, so } du = -2x\,dx\text{]}$$

$$= 1000\left(\left[\tfrac{1}{2} \cdot \tfrac{2}{3}u^{3/2}\right]_0^4 + \pi\right) = 1000\left(\tfrac{8}{3} + \pi\right) \approx 5.8 \times 10^3 \text{ ft-lb}$$

Note: The second integral represents the area of a quarter-circle of radius 2.

25. If only 4.7×10^5 J of work is done, then only the water above a certain level (call it h) will be pumped out. So we use the same formula as in Exercise 21, except that the work is fixed, and we are trying to find the lower limit of integration: $4.7 \times 10^5 \approx \int_h^3 (9.8 \times 10^3)(5 - x)8x\,dx = (9.8 \times 10^3)\left[20x^2 - \tfrac{8}{3}x^3\right]_h^3 \Leftrightarrow$ $\frac{4.7}{9.8} \times 10^2 \approx 48 = \left(20 \cdot 3^2 - \frac{8}{3} \cdot 3^3\right) - \left(20h^2 - \frac{8}{3}h^3\right) \Leftrightarrow$ $2h^3 - 15h^2 + 45 = 0$. To find the solution of this equation, we plot $2h^3 - 15h^2 + 45$ between $h = 0$ and $h = 3$. We see that the equation is satisfied for $h \approx 2.0$. So the depth of water remaining in the tank is about 2.0 m.

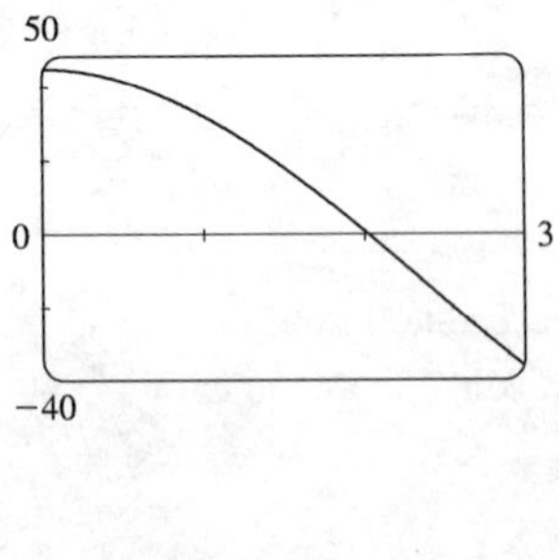

27. $V = \pi r^2 x$, so V is a function of x and P can also be regarded as a function of x. If $V_1 = \pi r^2 x_1$ and $V_2 = \pi r^2 x_2$, then

$$\begin{aligned} W &= \int_{x_1}^{x_2} F(x)\,dx = \int_{x_1}^{x_2} \pi r^2 P(V(x))\,dx \\ &= \int_{x_1}^{x_2} P(V(x))\,dV(x) \qquad [\text{Let } V(x) = \pi r^2 x, \text{ so } dV(x) = \pi r^2\,dx.] \\ &= \int_{V_1}^{V_2} P(V)\,dV \quad \text{by the Substitution Rule.} \end{aligned}$$

29. $W = \displaystyle\int_a^b F(r)\,dr = \int_a^b G\frac{m_1 m_2}{r^2}\,dr = Gm_1m_2\left[\frac{-1}{r}\right]_a^b = Gm_1m_2\left(\frac{1}{a} - \frac{1}{b}\right)$

Average Value of a Function

1. $f_{\text{ave}} = \dfrac{1}{b-a}\displaystyle\int_a^b f(x)\,dx = \frac{1}{1-(-1)}\int_{-1}^{1} x^2\,dx = \tfrac{1}{2}\cdot 2\int_0^1 x^2\,dx = \left[\tfrac{1}{3}x^3\right]_0^1 = \tfrac{1}{3}$

3. $g_{\text{ave}} = \frac{1}{\frac{\pi}{2}-0}\int_0^{\pi/2} \cos x\,dx = \frac{2}{\pi}[\sin x]_0^{\pi/2} = \frac{2}{\pi}(1-0) = \frac{2}{\pi}$

5. $f_{\text{ave}} = \frac{1}{5-0}\int_0^5 t\sqrt{1+t^2}\,dt = \frac{1}{5}\int_1^{26}\sqrt{u}\left(\frac{1}{2}\,du\right) \qquad [u = 1+t^2,\ du = 2t\,dt]$

$= \frac{1}{10}\int_1^{26} u^{1/2}\,du = \frac{1}{10}\cdot\frac{2}{3}\left[u^{3/2}\right]_1^{26} = \frac{1}{15}\left(26^{3/2}-1\right)$

7. $h_{\text{ave}} = \frac{1}{\pi-0}\int_0^{\pi}\cos^4 x\,\sin x\,dx = \frac{1}{\pi}\int_1^{-1} u^4(-du) \quad [u = \cos x,\ du = -\sin x\,dx]$

$= \frac{1}{\pi}\int_{-1}^{1} u^4\,du = \frac{1}{\pi}\cdot 2\int_0^1 u^4 du = \frac{2}{\pi}\left[\frac{1}{5}u^5\right]_0^1 = \frac{2}{5\pi}$

9. (a) $f_{\text{ave}} = \dfrac{1}{5-2}\displaystyle\int_2^5 (x-3)^2\,dx = \tfrac{1}{3}\left[\tfrac{1}{3}(x-3)^3\right]_2^5$

$= \frac{1}{9}\left[2^3 - (-1)^3\right] = \frac{1}{9}(8+1) = 1$

(b) $f(c) = f_{\text{ave}} \Leftrightarrow (c-3)^2 = 1 \Leftrightarrow c - 3 = \pm 1$

$\Leftrightarrow c = 2$ or 4

(c)

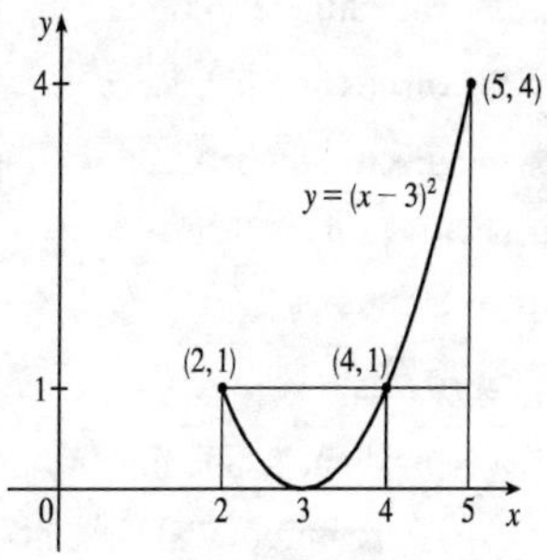

11. (a) $f_{\text{ave}} = \dfrac{1}{\pi-0}\displaystyle\int_0^{\pi} (2\sin x - \sin 2x)\,dx$

$= \frac{1}{\pi}\left[-2\cos x + \frac{1}{2}\cos 2x\right]_0^{\pi}$

$= \frac{1}{\pi}\left[\left(2+\frac{1}{2}\right) - \left(-2+\frac{1}{2}\right)\right] = \frac{4}{\pi}$

(b) $f(c) = f_{\text{ave}} \Leftrightarrow 2\sin c - \sin 2c = \frac{4}{\pi} \Leftrightarrow$

$c_1 \approx 1.238$ or $c_2 \approx 2.808$

(c)

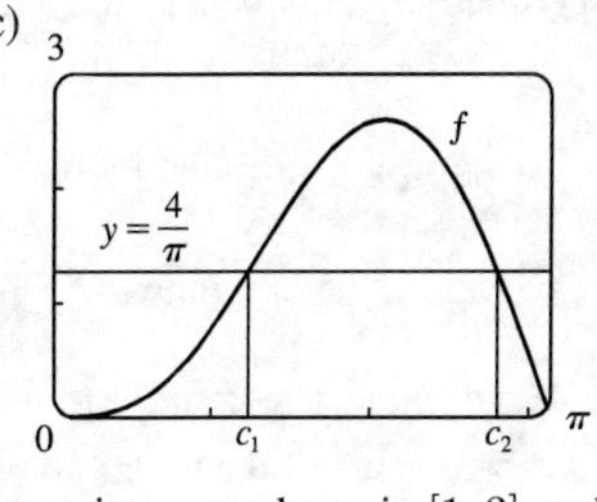

13. f is continuous on $[1, 3]$, so by the Mean Value Theorem for Integrals there exists a number c in $[1, 3]$ such that $\int_1^3 f(x)\,dx = f(c)(3-1) \Rightarrow 8 = 2f(c)$; that is, there is a number c such that $f(c) = \frac{8}{2} = 4$.

15. $f_{\text{ave}} = \dfrac{1}{50-20}\displaystyle\int_{20}^{50} f(x)\,dx \approx \tfrac{1}{30}M_3 = \dfrac{1}{30}\cdot\dfrac{50-20}{3}[f(25)+f(35)+f(45)]$

$= \tfrac{1}{3}(38+29+48) = \tfrac{115}{3} = 38\tfrac{1}{3}$

17. Let $t = 0$ and $t = 12$ correspond to 9 A.M. and 9 P.M., respectively.

$T_{\text{ave}} = \frac{1}{12-0}\int_0^{12}\left[50 + 14\sin\frac{1}{12}\pi t\right]dt = \frac{1}{12}\left[50t - 14\cdot\frac{12}{\pi}\cos\frac{1}{12}\pi t\right]_0^{12}$

$= \frac{1}{12}\left[50\cdot 12 + 14\cdot\frac{12}{\pi} + 14\cdot\frac{12}{\pi}\right] = \left(50 + \frac{28}{\pi}\right)°\text{F} \approx 59°\text{F}$

19. $\rho_{\text{ave}} = \dfrac{1}{8}\displaystyle\int_0^8 \dfrac{12}{\sqrt{x+1}}\,dx = \tfrac{3}{2}\displaystyle\int_0^8 (x+1)^{-1/2}\,dx = \left[3\sqrt{x+1}\right]_0^8 = 9 - 3 = 6\ \text{kg/m}$

21. $V_{\text{ave}} = \frac{1}{5}\int_0^5 V(t)\,dt = \frac{1}{5}\int_0^5 \frac{5}{4\pi}\left[1-\cos\left(\frac{2}{5}\pi t\right)\right]dt = \frac{1}{4\pi}\int_0^5\left[1-\cos\left(\frac{2}{5}\pi t\right)\right]dt$

$= \frac{1}{4\pi}\left[t - \frac{5}{2\pi}\sin\left(\frac{2}{5}\pi t\right)\right]_0^5 = \frac{1}{4\pi}\left[(5-0)-0\right] = \frac{5}{4\pi} \approx 0.4\ \text{L}$

23. Let $F(x) = \int_a^x f(t)\,dt$ for x in $[a,b]$. Then F is continuous on $[a,b]$ and differentiable on (a,b), so by the Mean Value Theorem there is a number c in (a,b) such that $F(b) - F(a) = F'(c)(b-a)$. But $F'(x) = f(x)$ by the Fundamental Theorem of Calculus. Therefore, $\int_a^b f(t)\,dt - 0 = f(c)(b-a)$.

Review

CONCEPT CHECK

1. (a) See Section 6.1, Figure 2 and Equations 6.1.1 and 6.1.2.

(b) Instead of using "top minus bottom" and integrating from left to right, we use "right minus left" and integrate from bottom to top. See Figures 11 and 12 in Section 6.1.

2. The numerical value of the area represents the number of meters by which Sue is ahead of Kathy after 1 minute.

3. (a) See the discussion in Section 6.2, near Figures 2 and 3, ending in the Definition of Volume.

(b) See the discussion between Examples 5 and 6 in Section 6.2. If the cross-section is a disk, find the radius in terms of x or y and use $A = \pi(\text{radius})^2$. If the cross-section is a washer, find the inner radius r_{in} and outer radius r_{out} and use $A = \pi\left(r_{\text{out}}^2\right) - \pi\left(r_{\text{in}}^2\right)$.

4. (a) $V = 2\pi rh\,\Delta r = (\text{circumference})\,(\text{height})\,(\text{thickness})$

(b) For a typical shell, find the circumference and height in terms of x or y and calculate $V = \int_a^b (\text{circumference})\,(\text{height})\,(dx \text{ or } dy)$, where a and b are the limits on x or y.

(c) Sometimes slicing produces washers or disks whose radii are difficult (or impossible) to find explicitly. On other occasions, the cylindrical shell method leads to an easier integral than slicing does.

5. $\int_0^6 f(x)\,dx$ represents the amount of work done. Its units are newton-meters, or joules.

6. (a) The average value of a function f on an interval $[a,b]$ is $f_{\text{ave}} = \dfrac{1}{b-a}\displaystyle\int_a^b f(x)\,dx$.

(b) The Mean Value Theorem for Integrals says that there is a number c at which the value of f is exactly equal to the average value of the function, that is, $f(c) = f_{\text{ave}}$. For a geometric interpretation of the Mean Value Theorem for Integrals, see Figure 2 in Section 6.5 and the discussion that accompanies it.

EXERCISES

1. $0 = x^2 - x - 6 = (x-3)(x+2) \Leftrightarrow x = 3$ or -2. So

$$A = \int_{-2}^{3} \left[0 - (x^2 - x - 6)\right] dx = \int_{-2}^{3} (-x^2 + x + 6)\, dx$$
$$= \left[-\tfrac{1}{3}x^3 + \tfrac{1}{2}x^2 + 6x\right]_{-2}^{3}$$
$$= \left(-9 + \tfrac{9}{2} + 18\right) - \left(\tfrac{8}{3} + 2 - 12\right)$$
$$= \tfrac{125}{6}$$

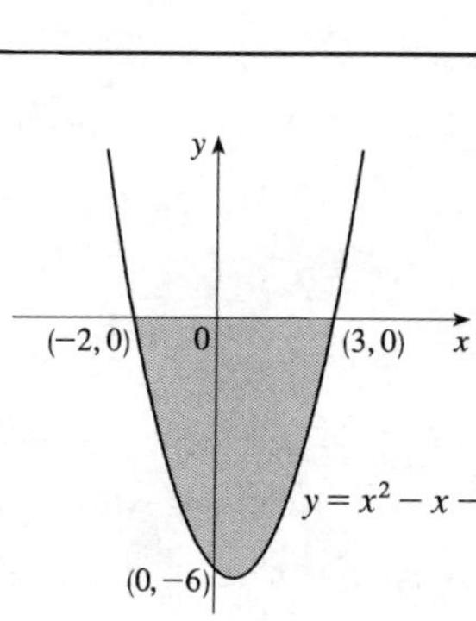

3. $1 - x^2 = 1 - \sqrt{x} \Leftrightarrow -x^2 = -\sqrt{x} \Leftrightarrow x^2 = \sqrt{x} \Rightarrow x^4 = x$
$\Rightarrow x^4 - x = 0 \Rightarrow x(x^3 - 1) = 0 \Rightarrow$
$x(x-1)(x^2 + x + 1) = 0 \Rightarrow x = 0$ or 1. So

$$A = \int_0^1 \left[(1 - x^2) - (1 - \sqrt{x})\right] dx = \int_0^1 \left(\sqrt{x} - x^2\right) dx$$
$$= \left[\tfrac{2}{3}x^{3/2} - \tfrac{1}{3}x^3\right]_0^1 = \tfrac{2}{3} - \tfrac{1}{3} = \tfrac{1}{3}$$

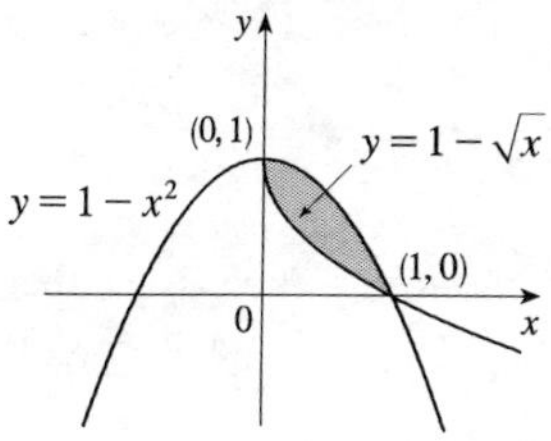

5. $$A = \int_0^2 \left[\sin\left(\tfrac{\pi x}{2}\right) - (x^2 - 2x)\right] dx$$
$$= \left[-\tfrac{2}{\pi}\cos\left(\tfrac{\pi x}{2}\right) - \tfrac{1}{3}x^3 + x^2\right]_0^2$$
$$= \left(\tfrac{2}{\pi} - \tfrac{8}{3} + 4\right) - \left(-\tfrac{2}{\pi} - 0 + 0\right) = \tfrac{4}{3} + \tfrac{4}{\pi}$$

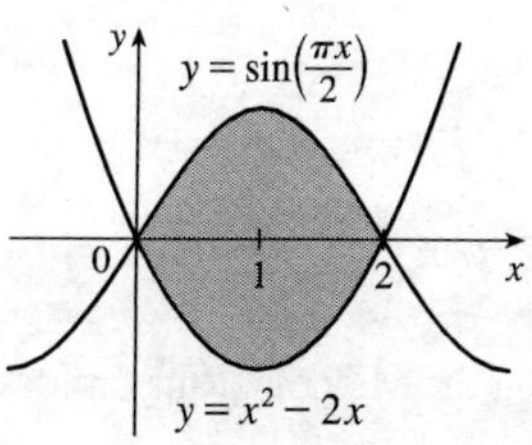

7. Using washers with inner radius x^2 and outer radius $2x$, we have

$$V = \pi \int_0^2 \left[(2x)^2 - \left(x^2\right)^2\right] dx = \pi \int_0^2 \left(4x^2 - x^4\right) dx$$
$$= \pi\left[\tfrac{4}{3}x^3 - \tfrac{1}{5}x^5\right]_0^2 = \pi\left(\tfrac{32}{3} - \tfrac{32}{5}\right) = 32\pi \cdot \tfrac{2}{15}$$
$$= \tfrac{64\pi}{15}$$

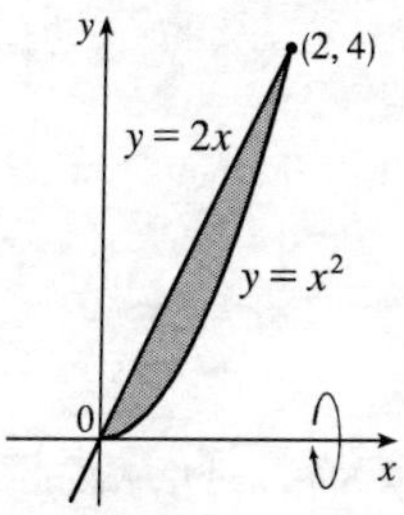

9. $$V = \pi \int_{-3}^{3} \left\{\left[(9 - y^2) - (-1)\right]^2 - \left[0 - (-1)\right]^2\right\} dy$$
$$= 2\pi \int_0^3 \left[(10 - y^2)^2 - 1\right] dy$$
$$= 2\pi \int_0^3 (100 - 20y^2 + y^4 - 1)\, dy$$
$$= 2\pi \int_0^3 (99 - 20y^2 + y^4)\, dy = 2\pi\left[99y - \tfrac{20}{3}y^3 + \tfrac{1}{5}y^5\right]_0^3$$
$$= 2\pi\left(297 - 180 + \tfrac{243}{5}\right) = \tfrac{1656\pi}{5}$$

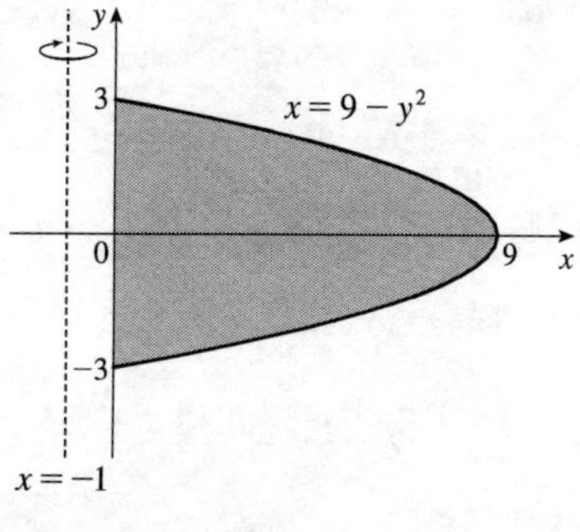

11. The graph of $x^2 - y^2 = a^2$ is a hyperbola with right and left branches. Solving for y gives us $y^2 = x^2 - a^2 \Rightarrow$ $y = \pm\sqrt{x^2 - a^2}$. We'll use shells and the height of each shell is $\sqrt{x^2 - a^2} - \left(-\sqrt{x^2 - a^2}\right) = 2\sqrt{x^2 - a^2}$.

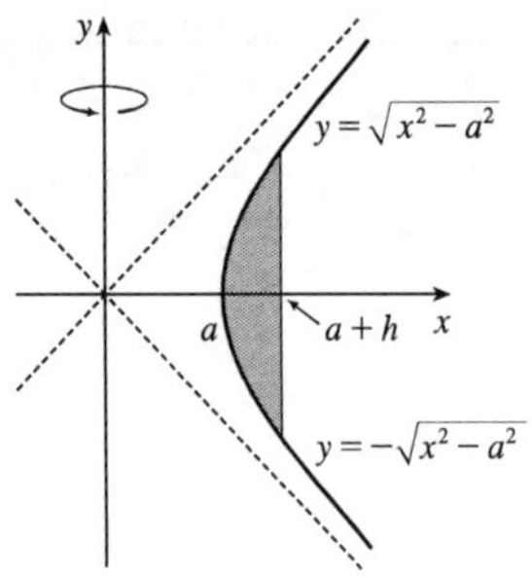

The volume is $V = \int_a^{a+h} 2\pi x \cdot 2\sqrt{x^2 - a^2}\,dx$. To evaluate, let $u = x^2 - a^2$, so $du = 2x\,dx$ and $x\,dx = \frac{1}{2}\,du$. When $x = a$, $u = 0$, and when $x = a + h$, $u = (a+h)^2 - a^2 = a^2 + 2ah + h^2 - a^2 = 2ah + h^2$.

Thus, $V = 4\pi \int_0^{2ah+h^2} \sqrt{u}\left(\frac{1}{2}\,du\right) = 2\pi\left[\frac{2}{3}u^{3/2}\right]_0^{2ah+h^2} = \frac{4}{3}\pi\left(2ah + h^2\right)^{3/2}$.

13. $V = \int_0^1 \pi\left[\left(1 - x^3\right)^2 - \left(1 - x^2\right)^2\right] dx$

15. (a) A cross-section is a washer with inner radius x^2 and outer radius x.

$V = \int_0^1 \pi\left[(x)^2 - \left(x^2\right)^2\right] dx = \int_0^1 \pi\left(x^2 - x^4\right) dx = \pi\left[\frac{1}{3}x^3 - \frac{1}{5}x^5\right]_0^1 = \pi\left[\frac{1}{3} - \frac{1}{5}\right] = \frac{2\pi}{15}$

(b) A cross-section is a washer with inner radius y and outer radius $\sqrt{y}$.

$V = \int_0^1 \pi\left[\left(\sqrt{y}\right)^2 - y^2\right] dy = \int_0^1 \pi\left(y - y^2\right) dy = \pi\left[\frac{1}{2}y^2 - \frac{1}{3}y^3\right]_0^1 = \pi\left[\frac{1}{2} - \frac{1}{3}\right] = \frac{\pi}{6}$

(c) A cross-section is a washer with inner radius $2 - x$ and outer radius $2 - x^2$.

$V = \int_0^1 \pi\left[\left(2 - x^2\right)^2 - (2 - x)^2\right] dx = \int_0^1 \pi\left(x^4 - 5x^2 + 4x\right) dx = \pi\left[\frac{1}{5}x^5 - \frac{5}{3}x^3 + 2x^2\right]_0^1$
$= \pi\left[\frac{1}{5} - \frac{5}{3} + 2\right] = \frac{8\pi}{15}$

17. (a) Using the Midpoint Rule on $[0, 1]$ with $f(x) = \tan(x^2)$ and $n = 4$, we estimate

$$A = \int_0^1 \tan(x^2)\,dx \approx \frac{1}{4}\left[\tan\left(\left(\tfrac{1}{8}\right)^2\right) + \tan\left(\left(\tfrac{3}{8}\right)^2\right) + \tan\left(\left(\tfrac{5}{8}\right)^2\right) + \tan\left(\left(\tfrac{7}{8}\right)^2\right)\right] \approx \tfrac{1}{4}(1.53) \approx 0.38$$

(b) Using the Midpoint Rule on $[0, 1]$ with $f(x) = \pi\tan^2(x^2)$ (for disks) and $n = 4$, we estimate

$$V = \int_0^1 f(x)\,dx \approx \frac{1}{4}\pi\left[\tan^2\left(\left(\tfrac{1}{8}\right)^2\right) + \tan^2\left(\left(\tfrac{3}{8}\right)^2\right) + \tan^2\left(\left(\tfrac{5}{8}\right)^2\right) + \tan^2\left(\left(\tfrac{7}{8}\right)^2\right)\right] \approx \tfrac{\pi}{4}(1.114) \approx 0.87$$

19. The solid is obtained by rotating the region $\mathcal{R} = \left\{(x, y) \mid 0 \le x \le \frac{\pi}{2}, 0 \le y \le \cos x\right\}$ about the y-axis.

21. The solid is obtained by rotating the region $\mathcal{R} = \left\{(x, y) \mid 0 \le y \le 2, 0 \le x \le 4 - y^2\right\}$ about the x-axis.

23. Take the base to be the disk $x^2 + y^2 \le 9$. Then $V = \int_{-3}^{3} A(x)\,dx$, where $A(x_0)$ is the area of the isosceles right triangle whose hypotenuse lies along the line $x = x_0$ in the xy-plane. The length of the hypotenuse is $2\sqrt{9 - x^2}$ and the length of each leg is $\sqrt{2}\sqrt{9 - x^2}$. $A(x) = \frac{1}{2}\left(\sqrt{2}\sqrt{9 - x^2}\right)^2 = 9 - x^2$, so

$V = 2\int_0^3 A(x)\,dx = 2\int_0^3 \left(9 - x^2\right) dx = 2\left[9x - \frac{1}{3}x^3\right]_0^3 = 2(27 - 9) = 36.$

25. Equilateral triangles with sides measuring $\frac{1}{4}x$ meters have height $\frac{1}{4}x\sin 60^\circ = \frac{\sqrt{3}}{8}x$. Therefore,

$$A(x) = \tfrac{1}{2}\cdot\tfrac{1}{4}x\cdot\tfrac{\sqrt{3}}{8}x = \tfrac{\sqrt{3}}{64}x^2.\ \ V = \int_0^{20} A(x)\,dx = \tfrac{\sqrt{3}}{64}\int_0^{20} x^2\,dx = \tfrac{\sqrt{3}}{64}\left[\tfrac{1}{3}x^3\right]_0^{20} = \tfrac{8000\sqrt{3}}{64\cdot 3} = \tfrac{125\sqrt{3}}{3}\text{ m}^3.$$

27. $f(x) = kx \;\Rightarrow\; 30\text{ N} = k(15-12)\text{ cm} \;\Rightarrow\; k = 10\text{ N/cm} = 1000\text{ N/m}.\ 20\text{ cm} - 12\text{ cm} = 0.08\text{ m} \;\Rightarrow$

$$W = \int_0^{0.08} kx\,dx = 1000\int_0^{0.08} x\,dx = 500\left[x^2\right]_0^{0.08} = 500(0.08)^2 = 3.2\text{ N-m} = 3.2\text{ J}.$$

29. (a) The parabola has equation $y = ax^2$ with vertex at the origin and passing through $(4,4)$. $4 = a\cdot 4^2 \;\Rightarrow\; a = \frac{1}{4} \;\Rightarrow\; y = \frac{1}{4}x^2 \;\Rightarrow\; x^2 = 4y$ $\Rightarrow\; x = 2\sqrt{y}$. Each circular disk has radius $2\sqrt{y}$ and is moved $4-y$ ft.

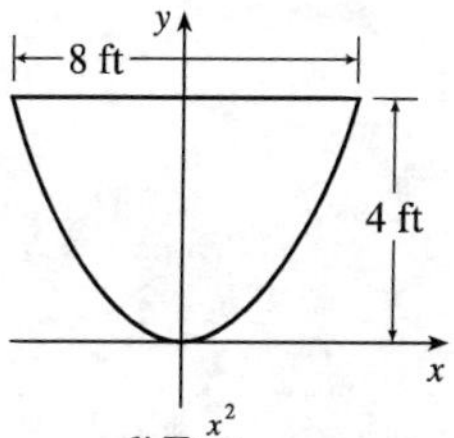

$$W = \int_0^4 \pi\left(2\sqrt{y}\right)^2 62.5\,(4-y)\,dy = 250\pi\int_0^4 y(4-y)\,dy$$

$$= 250\pi\left[2y^2 - \tfrac{1}{3}y^3\right]_0^4 = 250\pi\left(32 - \tfrac{64}{3}\right) = \tfrac{8000\pi}{3} \approx 8378\text{ ft-lb}$$

(b) In part (a) we knew the final water level (0) but not the amount of work done. Here we use the same equation, except with the work fixed, and the lower limit of integration (that is, the final water level — call it h) unknown: $W = 4000 \;\Leftrightarrow\; 250\pi\left[2y^2 - \frac{1}{3}y^3\right]_h^4 = 4000 \;\Leftrightarrow$

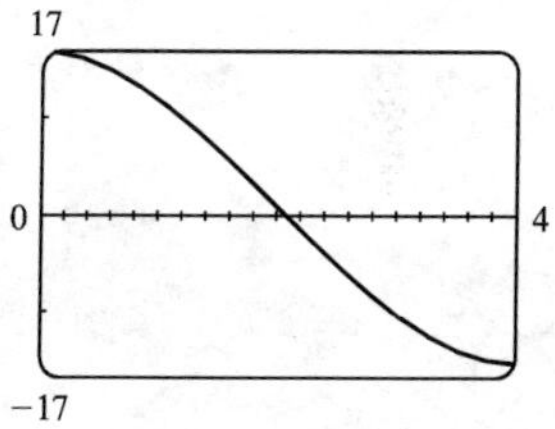

$\frac{16}{\pi} = \left[\left(32 - \frac{64}{3}\right) - \left(2h^2 - \frac{1}{3}h^3\right)\right] \;\Leftrightarrow\; h^3 - 6h^2 + 32 - \frac{48}{\pi} = 0.$

We graph the function $f(h) = h^3 - 6h^2 + 32 - \frac{48}{\pi}$ on the interval $[0,4]$ to see where it is 0. From the graph, $f(h) = 0$ for $h \approx 2.1$. So the depth of water remaining is about 2.1 ft.

31. $\displaystyle\lim_{h\to 0} f_{\text{ave}} = \lim_{h\to 0}\frac{1}{(x+h)-x}\int_x^{x+h} f(t)\,dt = \lim_{h\to 0}\frac{F(x+h)-F(x)}{h}$, where $F(x) = \int_a^x f(t)\,dt$. But we recognize this limit as being $F'(x)$ by the definition of a derivative. Therefore, $\displaystyle\lim_{h\to 0} f_{\text{ave}} = F'(x) = f(x)$ by FTC1.

The Fundamental Theorem of Calculus

1. The precise version of this statement is given by the Fundamental Theorem of Calculus. See the statement of this theorem and the paragraph that follows it at the end of Section 5.3.

3. (a) $g(x) = \int_0^x f(t)\,dt$.

$$g(0) = \int_0^0 f(t)\,dt = 0$$

$$g(1) = \int_0^1 f(t)\,dt = 1 \cdot 2 = 2 \text{ [rectangle]},$$

$$g(2) = \int_0^2 f(t)\,dt = \int_0^1 f(t)\,dt + \int_1^2 f(t)\,dt = g(1) + \int_1^2 f(t)\,dt$$

$$= 2 + 1 \cdot 2 + \tfrac{1}{2} \cdot 1 \cdot 2 = 5 \text{ [rectangle plus triangle]},$$

$$g(3) = \int_0^3 f(t)\,dt = g(2) + \int_2^3 f(t)\,dt = 5 + \tfrac{1}{2} \cdot 1 \cdot 4 = 7,$$

$$g(6) = g(3) + \int_3^6 f(t)\,dt \text{ [the integral is negative since } f \text{ lies under the } x\text{-axis]}$$

$$= 7 + \left[-\left(\tfrac{1}{2} \cdot 2 \cdot 2 + 1 \cdot 2\right)\right] = 7 - 4 = 3$$

(d)

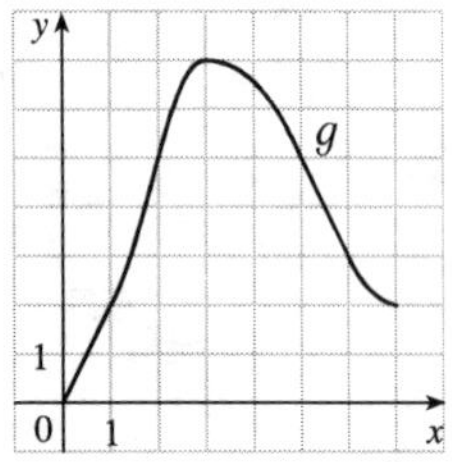

(b) g is increasing on $(0, 3)$ because as x increases from 0 to 3, we keep adding more area.

(c) g has a maximum value when we start subtracting area; that is, at $x = 3$.

5.

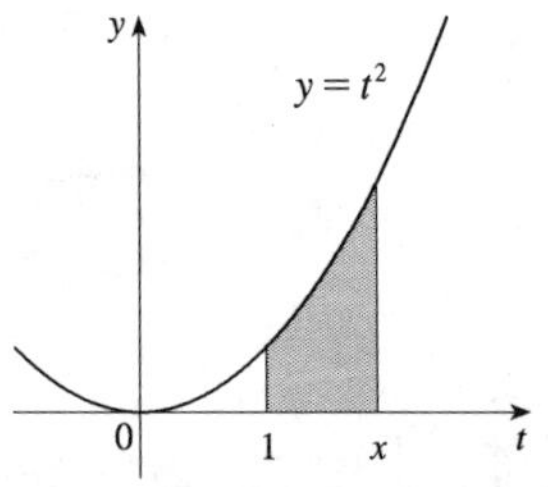

(a) By FTC1 with $f(t) = t^2$ and $a = 1$, $g(x) = \int_1^x t^2\,dt \Rightarrow$
$g'(x) = f(x) = x^2$.

(b) Using FTC2, $g(x) = \int_1^x t^2\,dt = \left[\tfrac{1}{3}t^3\right]_1^x = \tfrac{1}{3}x^3 - \tfrac{1}{3} \Rightarrow$
$g'(x) = x^2$.

7. $f(t) = \sqrt{1 + 2t}$ and $g(x) = \int_0^x \sqrt{1 + 2t}\,dt$, so by FTC1, $g'(x) = f(x) = \sqrt{1 + 2x}$.

9. $f(t) = t^2 \sin t$ and $g(y) = \int_2^y t^2 \sin t\,dt$, so by FTC1, $g'(y) = f(y) = y^2 \sin y$.

11. $F(x) = \int_x^2 \cos(t^2)\,dt = -\int_2^x \cos(t^2)\,dt \Rightarrow F'(x) = -\cos(x^2)$

13. Let $u = \dfrac{1}{x}$. Then $\dfrac{du}{dx} = -\dfrac{1}{x^2}$. Also, $\dfrac{dh}{dx} = \dfrac{dh}{du}\dfrac{du}{dx}$, so

$$h'(x) = \frac{d}{dx}\int_2^{1/x} \sin^4 t\,dt = \frac{d}{du}\int_2^u \sin^4 t\,dt \cdot \frac{du}{dx} = \sin^4 u\,\frac{du}{dx} = \frac{-\sin^4(1/x)}{x^2}.$$

15. Let $u = \sqrt{x}$. Then $\dfrac{du}{dx} = \dfrac{1}{2\sqrt{x}}$. Also, $\dfrac{dy}{dx} = \dfrac{dy}{du}\dfrac{du}{dx}$, so

$$y' = \frac{d}{dx}\int_3^{\sqrt{x}} \frac{\cos t}{t}\,dt = \frac{d}{du}\int_3^u \frac{\cos t}{t}\,dt \cdot \frac{du}{dx} = \frac{\cos u}{u} \cdot \frac{1}{2\sqrt{x}} = \frac{\cos\sqrt{x}}{\sqrt{x}} \cdot \frac{1}{2\sqrt{x}} = \frac{\cos\sqrt{x}}{2x}.$$

17. Let $w = 1 - 3x$. Then $\dfrac{dw}{dx} = -3$. Also, $\dfrac{dy}{dx} = \dfrac{dy}{dw}\dfrac{dw}{dx}$, so

$$y' = \frac{d}{dx}\int_{1-3x}^1 \frac{u^3}{1 + u^2}\,du = \frac{d}{dw}\int_w^1 \frac{u^3}{1 + u^2}\,du \cdot \frac{dw}{dx}$$

$$= -\frac{d}{dw}\int_1^w \frac{u^3}{1 + u^2}\,du \cdot \frac{dw}{dx} = -\frac{w^3}{1 + w^2}(-3) = \frac{3(1 - 3x)^3}{1 + (1 - 3x)^2}$$

19. $\displaystyle\int_{-1}^{3} x^5\,dx = \left[\frac{x^6}{6}\right]_{-1}^{3} = \frac{3^6}{6} - \frac{(-1)^6}{6} = \frac{729-1}{6} = \frac{364}{3}$

21. $\int_2^8 (4x+3)\,dx = \left[\frac{4}{2}x^2 + 3x\right]_2^8 = \left(2\cdot 8^2 + 3\cdot 8\right) - \left(2\cdot 2^2 + 3\cdot 2\right) = 152 - 14 = 138$

23. $\int_0^1 x^{4/5}\,dx = \left[\frac{5}{9}x^{9/5}\right]_0^1 = \frac{5}{9} - 0 = \frac{5}{9}$

25. $\displaystyle\int_1^2 \frac{3}{t^4}\,dt = 3\int_1^2 t^{-4}\,dt = 3\left[\frac{t^{-3}}{-3}\right]_1^2 = \frac{3}{-3}\left[\frac{1}{t^3}\right]_1^2 = -1\left(\frac{1}{8} - 1\right) = \frac{7}{8}$

27. $\displaystyle\int_{-5}^{5} \frac{2}{x^3}\,dx$ does not exist because the function $f(x) = \dfrac{2}{x^3}$ has an infinite discontinuity at $x = 0$; that is, f is discontinuous on the interval $[-5, 5]$.

29. $\int_0^2 x(2+x^5)\,dx = \int_0^2 (2x + x^6)\,dx = \left[x^2 + \frac{1}{7}x^7\right]_0^2 = \left(4 + \frac{128}{7}\right) - (0+0) = \frac{156}{7}$

31. $\int_0^{\pi/4} \sec^2 t\,dt = \left[\tan t\right]_0^{\pi/4} = \tan\frac{\pi}{4} - \tan 0 = 1 - 0 = 1$

33. $\int_\pi^{2\pi} \csc^2\theta\,d\theta$ does not exist because the function $f(\theta) = \csc^2\theta$ has infinite discontinuities at $\theta = \pi$ and $\theta = 2\pi$; that is, f is discontinuous on the interval $[\pi, 2\pi]$.

35. $\int_0^2 f(x)\,dx = \int_0^1 x^4\,dx + \int_1^2 x^5\,dx = \left[\frac{1}{5}x^5\right]_0^1 + \left[\frac{1}{6}x^6\right]_1^2 = \left(\frac{1}{5} - 0\right) + \left(\frac{64}{6} - \frac{1}{6}\right) = 10.7$

37. From the graph, it appears that the area is about 60. The actual area is $\int_0^{27} x^{1/3}dx = \left[\frac{3}{4}x^{4/3}\right]_0^{27} = \frac{3}{4}\cdot 81 - 0 = \frac{243}{4} = 60.75$. This is $\frac{3}{4}$ of the area of the viewing rectangle.

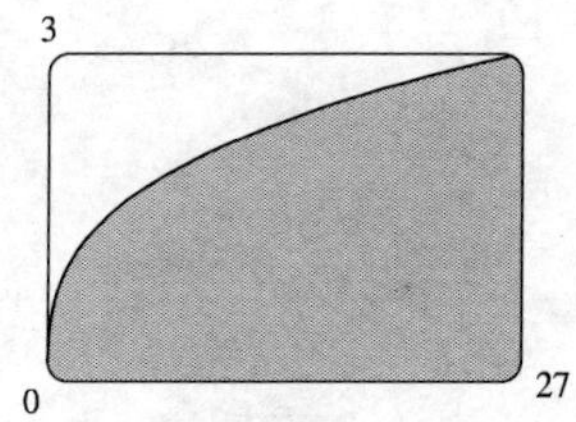

39. It appears that the area under the graph is about $\frac{2}{3}$ of the area of the viewing rectangle, or about $\frac{2}{3}\pi \approx 2.1$. The actual area is $\int_0^\pi \sin x\,dx = [-\cos x]_0^\pi = (-\cos\pi) - (-\cos 0) = -(-1) + 1 = 2$.

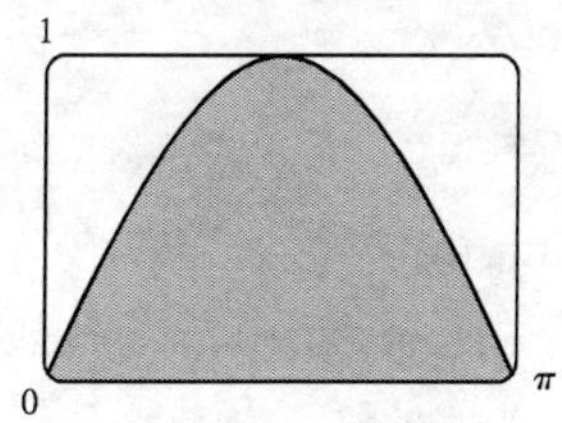

41. $\int_{-1}^2 x^3\,dx = \left[\frac{1}{4}x^4\right]_{-1}^2 = 4 - \frac{1}{4} = \frac{15}{4} = 3.75$

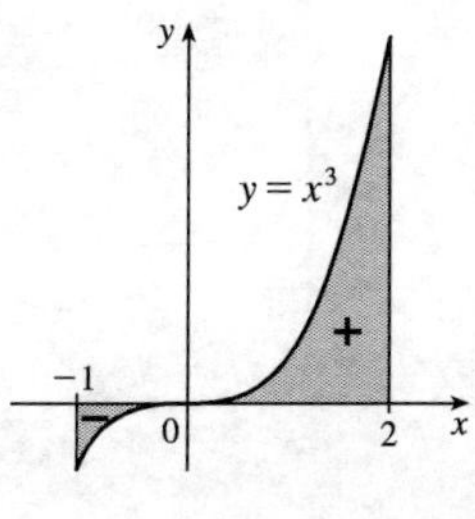

43. $g(x) = \int_{2x}^{3x} \frac{u^2-1}{u^2+1}\,du = \int_{2x}^{0} \frac{u^2-1}{u^2+1}\,du + \int_{0}^{3x} \frac{u^2-1}{u^2+1}\,du = -\int_{0}^{2x} \frac{u^2-1}{u^2+1}\,du + \int_{0}^{3x} \frac{u^2-1}{u^2+1}\,du \;\Rightarrow$

$$g'(x) = -\frac{(2x)^2-1}{(2x)^2+1} \cdot \frac{d}{dx}(2x) + \frac{(3x)^2-1}{(3x)^2+1} \cdot \frac{d}{dx}(3x) = -2 \cdot \frac{4x^2-1}{4x^2+1} + 3 \cdot \frac{9x^2-1}{9x^2+1}$$

45. $y = \int_{\sqrt{x}}^{x^3} \sqrt{t}\sin t\,dt = \int_{\sqrt{x}}^{1} \sqrt{t}\sin t\,dt + \int_{1}^{x^3} \sqrt{t}\sin t\,dt = -\int_{1}^{\sqrt{x}} \sqrt{t}\sin t\,dt + \int_{1}^{x^3} \sqrt{t}\sin t\,dt \;\Rightarrow$

$$y' = -\sqrt[4]{x}\,(\sin\sqrt{x}) \cdot \frac{d}{dx}(\sqrt{x}) + x^{3/2}\sin(x^3) \cdot \frac{d}{dx}(x^3) = -\frac{\sqrt[4]{x}\sin\sqrt{x}}{2\sqrt{x}} + x^{3/2}\sin(x^3)\,(3x^2)$$

$$= 3x^{7/2}\sin(x^3) - \frac{\sin\sqrt{x}}{2\sqrt[4]{x}}$$

47. $F(x) = \int_1^x f(t)\,dt \;\Rightarrow\; F'(x) = f(x) = \int_1^{x^2} \frac{\sqrt{1+u^4}}{u}\,du \left[\text{since } f(t) = \int_1^{t^2} \frac{\sqrt{1+u^4}}{u}\,du\right] \;\Rightarrow$

$$F''(x) = f'(x) = \frac{\sqrt{1+(x^2)^4}}{x^2} \cdot \frac{d}{dx}(x^2) = \frac{\sqrt{1+x^8}}{x^2} \cdot 2x = \frac{2\sqrt{1+x^8}}{x}.$$ So $F''(2) = \sqrt{1+2^8} = \sqrt{257}$.

49. (a) The Fresnel function $S(x) = \int_0^x \sin(\frac{\pi}{2}t^2)\,dt$ has local maximum values where $0 = S'(x) = \sin(\frac{\pi}{2}x^2)$ and S' changes from positive to negative. For $x > 0$, this happens when $\frac{\pi}{2}x^2 = (2n-1)\pi$ [odd multiples of π] $\Leftrightarrow$ $x^2 = 2(2n-1) \;\Leftrightarrow\; x = \sqrt{4n-2}$, n any positive integer. For $x < 0$, S' changes from positive to negative where $\frac{\pi}{2}x^2 = 2n\pi$ [even multiples of π] $\Leftrightarrow\; x^2 = 4n \;\Leftrightarrow\; x = -2\sqrt{n}$. S' does not change sign at $x = 0$.

(b) S is concave upward on those intervals where $S''(x) > 0$. Differentiating our expression for $S'(x)$, we get $S''(x) = \cos(\frac{\pi}{2}x^2)(2\frac{\pi}{2}x) = \pi x\cos(\frac{\pi}{2}x^2)$. For $x > 0$, $S''(x) > 0$ where $\cos(\frac{\pi}{2}x^2) > 0 \;\Leftrightarrow\; 0 < \frac{\pi}{2}x^2 < \frac{\pi}{2}$ or $(2n-\frac{1}{2})\pi < \frac{\pi}{2}x^2 < (2n+\frac{1}{2})\pi$, n any integer $\Leftrightarrow\; 0 < x < 1$ or $\sqrt{4n-1} < x < \sqrt{4n+1}$, n any positive integer. For $x < 0$, $S''(x) > 0$ where $\cos(\frac{\pi}{2}x^2) < 0 \;\Leftrightarrow\; (2n-\frac{3}{2})\pi < \frac{\pi}{2}x^2 < (2n-\frac{1}{2})\pi$, n any integer $\Leftrightarrow\; 4n-3 < x^2 < 4n-1 \;\Leftrightarrow\; \sqrt{4n-3} < |x| < \sqrt{4n-1} \;\Rightarrow\; \sqrt{4n-3} < -x < \sqrt{4n-1}$ $\Rightarrow\; -\sqrt{4n-3} > x > -\sqrt{4n-1}$, so the intervals of upward concavity for $x < 0$ are $(-\sqrt{4n-1}, -\sqrt{4n-3})$, n any positive integer. To summarize: S is concave upward on the intervals $(0,1)$, $(-\sqrt{3},-1)$, $(\sqrt{3},\sqrt{5})$, $(-\sqrt{7},-\sqrt{5})$, $(\sqrt{7},3)$, ...

(c) In Maple, we use `plot({int(sin(Pi*t^2/2),t=0..x),0.2},x=0..2);`. Note that Maple recognizes the Fresnel function, calling it `FresnelS(x)`. In Mathematica, we use `Plot[{Integrate[Sin[Pi*t^2/2],{t,0,x}],0.2},{x,0,2}]`. In Derive, we load the utility file `FRESNEL` and plot `FRESNEL_SIN(x)`. From the graphs, we see that $\int_0^x \sin(\frac{\pi}{2}t^2)\,dt = 0.2$ at $x \approx 0.74$.

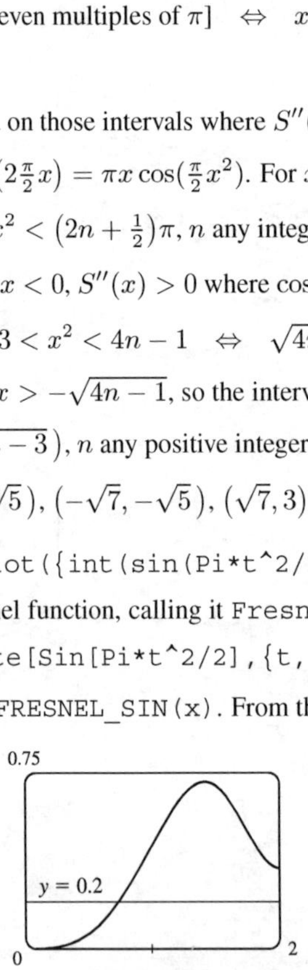

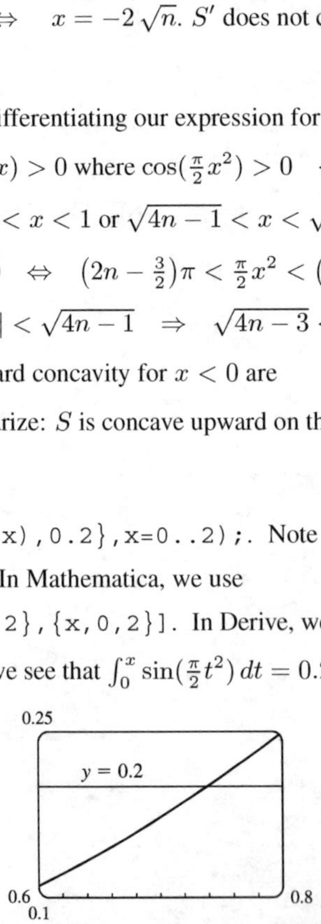

APPENDIXES

A Numbers, Inequalities, and Absolute Values

1. $|5-23| = |-18| = 18$

3. $|-\pi| = \pi$ because $\pi > 0$.

5. $\left|\sqrt{5}-5\right| = -\left(\sqrt{5}-5\right) = 5-\sqrt{5}$ because $\sqrt{5}-5<0$.

7. For $x<2$, $x-2<0$, so $|x-2| = -(x-2) = 2-x$.

9. $|x+1| = \begin{cases} x+1 & \text{for } x+1 \geq 0 \quad \Leftrightarrow \quad x \geq -1 \\ -(x+1) & \text{for } x+1<0 \quad \Leftrightarrow \quad x<-1 \end{cases}$

11. $\left|x^2+1\right| = x^2+1$ (since $x^2+1 \geq 0$ for all x).

13. $2x+7>3 \quad \Leftrightarrow \quad 2x>-4 \quad \Leftrightarrow \quad x>-2$, so $x \in (-2,\infty)$.

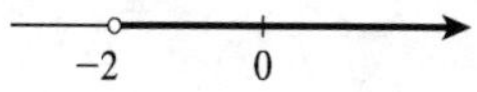

15. $1-x \leq 2 \quad \Leftrightarrow \quad -x \leq 1 \quad \Leftrightarrow \quad x \geq -1$, so $x \in [-1,\infty)$.

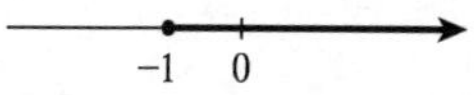

17. $2x+1<5x-8 \quad \Leftrightarrow \quad 9<3x \quad \Leftrightarrow \quad 3<x$, so $x \in (3,\infty)$.

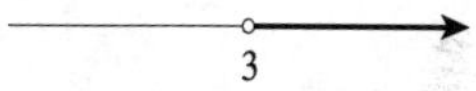

19. $-1<2x-5<7 \quad \Leftrightarrow \quad 4<2x<12 \quad \Leftrightarrow \quad 2<x<6$, so $x \in (2,6)$.

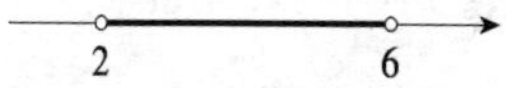

21. $0 \leq 1-x<1 \quad \Leftrightarrow \quad -1 \leq -x<0 \quad \Leftrightarrow \quad 1 \geq x>0$, so $x \in (0,1]$.

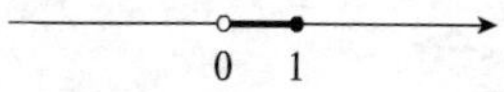

23. $4x<2x+1 \leq 3x+2$. So $4x<2x+1 \quad \Leftrightarrow \quad 2x<1 \quad \Leftrightarrow \quad x<\frac{1}{2}$, and $2x+1 \leq 3x+2 \quad \Leftrightarrow \quad -1 \leq x$. Thus, $x \in \left[-1,\frac{1}{2}\right)$.

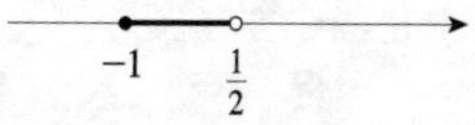

25. $(x-1)(x-2)>0$. *Case 1:* (both factors are positive, so their product is positive)

$x-1>0 \quad \Leftrightarrow \quad x>1$, and $x-2>0 \quad \Leftrightarrow \quad x>2$, so $x \in (2,\infty)$.

Case 2: (both factors are negative, so their product is positive)

$x-1<0 \quad \Leftrightarrow \quad x<1$, and $x-2<0 \quad \Leftrightarrow \quad x<2$, so $x \in (-\infty,1)$.

Thus, the solution set is $(-\infty,1) \cup (2,\infty)$.

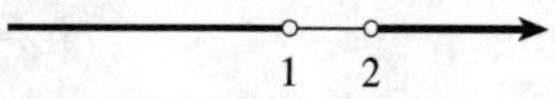

27. $2x^2+x \leq 1 \quad \Leftrightarrow \quad 2x^2+x-1 \leq 0 \quad \Leftrightarrow \quad (2x-1)(x+1) \leq 0$. *Case 1:* $2x-1 \geq 0 \quad \Leftrightarrow \quad x \geq \frac{1}{2}$, and $x+1 \leq 0 \quad \Leftrightarrow \quad x \leq -1$, which is an impossible combination. *Case 2:* $2x-1 \leq 0 \quad \Leftrightarrow \quad x \leq \frac{1}{2}$, and $x+1 \geq 0$ $\Leftrightarrow \quad x \geq -1$, so $x \in \left[-1,\frac{1}{2}\right]$. Thus, the solution set is $\left[-1,\frac{1}{2}\right]$.

29. $x^2+x+1>0 \quad \Leftrightarrow \quad x^2+x+\frac{1}{4}+\frac{3}{4}>0 \quad \Leftrightarrow \quad \left(x+\frac{1}{2}\right)^2+\frac{3}{4}>0$. But since $\left(x+\frac{1}{2}\right)^2 \geq 0$ for every real x, the original inequality will be true for all real x as well. Thus, the solution set is $(-\infty,\infty)$.

31. $x^2 < 3 \iff x^2 - 3 < 0 \iff (x - \sqrt{3})(x + \sqrt{3}) < 0$. *Case 1:* $x > \sqrt{3}$ and $x < -\sqrt{3}$, which is impossible. *Case 2:* $x < \sqrt{3}$ and $x > -\sqrt{3}$. Thus, the solution set is $(-\sqrt{3}, \sqrt{3})$.

Another method: $x^2 < 3 \iff |x| < \sqrt{3} \iff -\sqrt{3} < x < \sqrt{3}$.

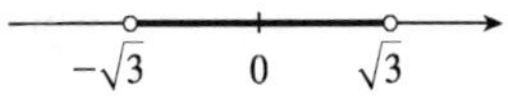

33. $x^3 - x^2 \leq 0 \iff x^2(x - 1) \leq 0$. Since $x^2 \geq 0$ for all x, the inequality is satisfied when $x - 1 \leq 0 \iff x \leq 1$. Thus, the solution set is $(-\infty, 1]$.

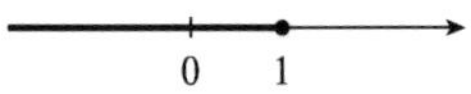

35. $x^3 > x \iff x^3 - x > 0 \iff x(x^2 - 1) > 0 \iff x(x - 1)(x + 1) > 0$. Constructing a table:

Interval	x	$x - 1$	$x + 1$	$x(x-1)(x+1)$
$x < -1$	$-$	$-$	$-$	$-$
$-1 < x < 0$	$-$	$-$	$+$	$+$
$0 < x < 1$	$+$	$-$	$+$	$-$
$x > 1$	$+$	$+$	$+$	$+$

Since $x^3 > x$ when the last column is positive, the solution set is $(-1, 0) \cup (1, \infty)$.

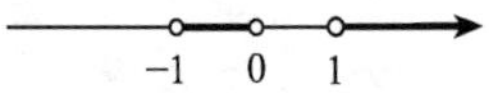

37. $1/x < 4$. This is clearly true for $x < 0$. So suppose $x > 0$. then $1/x < 4 \iff 1 < 4x \iff \frac{1}{4} < x$. Thus, the solution set is $(-\infty, 0) \cup (\frac{1}{4}, \infty)$.

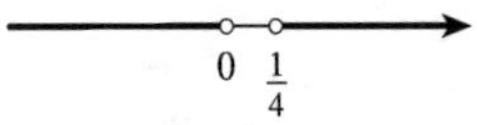

39. $C = \frac{5}{9}(F - 32) \Rightarrow F = \frac{9}{5}C + 32$. So $50 \leq F \leq 95 \Rightarrow 50 \leq \frac{9}{5}C + 32 \leq 95 \Rightarrow 18 \leq \frac{9}{5}C \leq 63 \Rightarrow 10 \leq C \leq 35$. So the interval is $[10, 35]$.

41. (a) Let T represent the temperature in degrees Celsius and h the height in km. $T = 20$ when $h = 0$ and T decreases by 10 °C for every km (1 °C for each 100-m rise). Thus, $T = 20 - 10h$ when $0 \leq h \leq 12$.

(b) From part (a), $T = 20 - 10h \Rightarrow 10h = 20 - T \Rightarrow h = 2 - T/10$. So $0 \leq h \leq 5 \Rightarrow 0 \leq 2 - T/10 \leq 5 \Rightarrow -2 \leq -T/10 \leq 3 \Rightarrow -20 \leq -T \leq 30 \Rightarrow 20 \geq T \geq -30 \Rightarrow -30 \leq T \leq 20$. Thus, the range of temperatures (in °C) to be expected is $[-30, 20]$.

43. $|2x| = 3 \iff$ either $2x = 3$ or $2x = -3 \iff x = \frac{3}{2}$ or $x = -\frac{3}{2}$.

45. $|x + 3| = |2x + 1| \iff$ either $x + 3 = 2x + 1$ or $x + 3 = -(2x + 1)$. In the first case, $x = 2$, and in the second case, $x + 3 = -2x - 1 \iff 3x = -4 \iff x = -\frac{4}{3}$. So the solutions are $-\frac{4}{3}$ and 2.

47. By Property 5 of absolute values, $|x| < 3 \iff -3 < x < 3$, so $x \in (-3, 3)$.

49. $|x - 4| < 1 \iff -1 < x - 4 < 1 \iff 3 < x < 5$, so $x \in (3, 5)$.

51. $|x + 5| \geq 2 \iff x + 5 \geq 2$ or $x + 5 \leq -2 \iff x \geq -3$ or $x \leq -7$, so $x \in (-\infty, -7] \cup [-3, \infty)$.

53. $|2x - 3| \leq 0.4 \iff -0.4 \leq 2x - 3 \leq 0.4 \iff 2.6 \leq 2x \leq 3.4 \iff 1.3 \leq x \leq 1.7$, so $x \in [1.3, 1.7]$.

55. $1 \leq |x| \leq 4$. So either $1 \leq x \leq 4$ or $1 \leq -x \leq 4 \iff -1 \geq x \geq -4$. Thus, $x \in [-4, -1] \cup [1, 4]$.

57. $a(bx - c) \geq bc \iff bx - c \geq \dfrac{bc}{a} \iff bx \geq \dfrac{bc}{a} + c = \dfrac{bc + ac}{a} \iff x \geq \dfrac{bc + ac}{ab}$

59. $ax + b < c \quad \Leftrightarrow \quad ax < c - b \quad \Leftrightarrow \quad x > \dfrac{c-b}{a}$ (since $a < 0$)

61. $|(x+y) - 5| = |(x-2) + (y-3)| \leq |x-2| + |y-3| < 0.01 + 0.04 = 0.05$

63. If $a < b$ then $a + a < a + b$ and $a + b < b + b$. So $2a < a + b < 2b$. Dividing by 2, we get $a < \frac{1}{2}(a+b) < b$.

65. $|ab| = \sqrt{(ab)^2} = \sqrt{a^2b^2} = \sqrt{a^2}\,\sqrt{b^2} = |a|\,|b|$

67. If $0 < a < b$, then $a \cdot a < a \cdot b$ and $a \cdot b < b \cdot b$ [using Rule 3 of Inequalities]. So $a^2 < ab < b^2$ and hence $a^2 < b^2$.

69. Observe that the sum, difference and product of two integers is always an integer. Let the rational numbers be represented by $r = m/n$ and $s = p/q$ (where m, n, p and q are integers with $n \neq 0$, $q \neq 0$). Now $r + s = \dfrac{m}{n} + \dfrac{p}{q} = \dfrac{mq + pn}{nq}$, but $mq + pn$ and nq are both integers, so $\dfrac{mq+pn}{nq} = r + s$ is a rational number by definition. Similarly, $r - s = \dfrac{m}{n} - \dfrac{p}{q} = \dfrac{mq - pn}{nq}$ is a rational number. Finally, $r \cdot s = \dfrac{m}{n} \cdot \dfrac{p}{q} = \dfrac{mp}{nq}$ but mp and nq are both integers, so $\dfrac{mp}{nq} = r \cdot s$ is a rational number by definition.

B Coordinate Geometry and Lines

1. From the Distance Formula with $x_1 = 1$, $x_2 = 4$, $y_1 = 1$, $y_2 = 5$, we find the distance from $(1,1)$ to $(4,5)$ to be $\sqrt{(4-1)^2 + (5-1)^2} = \sqrt{3^2 + 4^2} = \sqrt{25} = 5$.

3. $\sqrt{(-1-6)^2 + [3-(-2)]^2} = \sqrt{(-7)^2 + 5^2} = \sqrt{74}$

5. $\sqrt{(4-2)^2 + (-7-5)^2} = \sqrt{2^2 + (-12)^2} = \sqrt{148} = 2\sqrt{37}$

7. From (2), the slope is $\dfrac{11-5}{4-1} = \dfrac{6}{3} = 2$.

9. With $P(-3,3)$ and $Q(-1,-6)$, the slope m of the line through P and Q is $m = \dfrac{-6-3}{-1-(-3)} = -\dfrac{9}{2}$.

11. Since $|AC| = \sqrt{(-4-0)^2 + (3-2)^2} = \sqrt{(-4)^2 + 1^2} = \sqrt{17}$ and $|BC| = \sqrt{[-4-(-3)]^2 + [3-(-1)]^2} = \sqrt{(-1)^2 + 4^2} = \sqrt{17}$, the triangle has two sides of equal length, and so is isosceles.

13. Using $A(-2,9)$, $B(4,6)$, $C(1,0)$, and $D(-5,3)$, we have

$|AB| = \sqrt{[4-(-2)]^2 + (6-9)^2} = \sqrt{6^2 + (-3)^2} = \sqrt{45} = \sqrt{9}\,\sqrt{5} = 3\,\sqrt{5}$,

$|BC| = \sqrt{(1-4)^2 + (0-6)^2} = \sqrt{(-3)^2 + (-6)^2} = \sqrt{45} = \sqrt{9}\,\sqrt{5} = 3\,\sqrt{5}$,

$|CD| = \sqrt{(-5-1)^2 + (3-0)^2} = \sqrt{(-6)^2 + 3^2} = \sqrt{45} = \sqrt{9}\,\sqrt{5} = 3\,\sqrt{5}$, and

$|DA| = \sqrt{[-2-(-5)]^2 + (9-3)^2} = \sqrt{3^2 + 6^2} = \sqrt{45} = \sqrt{9}\,\sqrt{5} = 3\,\sqrt{5}$. So all sides are of equal length and we have a rhombus. Moreover, $m_{AB} = \dfrac{6-9}{4-(-2)} = -\dfrac{1}{2}$, $m_{BC} = \dfrac{0-6}{1-4} = 2$, $m_{CD} = \dfrac{3-0}{-5-1} = -\dfrac{1}{2}$, and $m_{DA} = \dfrac{9-3}{-2-(-5)} = 2$, so the sides are perpendicular. Thus, A, B, C, and D are vertices of a square.

15. The slope of the line segment AB is $\frac{4-1}{7-1}=\frac{1}{2}$, the slope of CD is $\frac{7-10}{-1-5}=\frac{1}{2}$, the slope of BC is $\frac{10-4}{5-7}=-3$, and the slope of DA is $\frac{1-7}{1-(-1)}=-3$. So AB is parallel to CD and BC is parallel to DA. Hence $ABCD$ is a parallelogram.

17. $x=3$

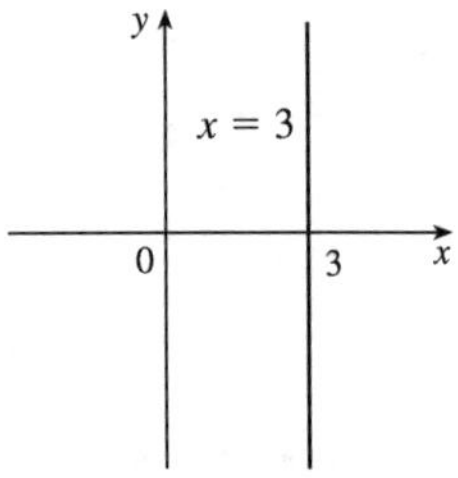

19. $xy=0 \quad \Leftrightarrow \quad x=0$ or $y=0$. The graph consists of the coordinate axes.

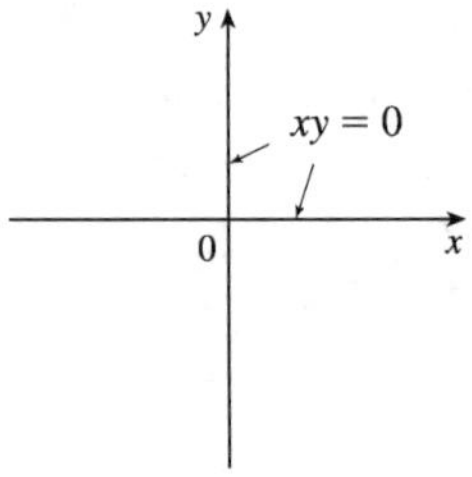

21. By the point-slope form of the equation of a line, an equation of the line through $(2,-3)$ with slope 6 is $y-(-3)=6(x-2)$ or $y=6x-15$.

23. $y-7=\frac{2}{3}(x-1)$ or $y=\frac{2}{3}x+\frac{19}{3}$

25. The slope of the line through $(2,1)$ and $(1,6)$ is $m=\frac{6-1}{1-2}=-5$, so an equation of the line is $y-1=-5(x-2)$ or $y=-5x+11$.

27. By the slope-intercept form of the equation of a line, an equation of the line is $y=3x-2$.

29. Since the line passes through $(1,0)$ and $(0,-3)$, its slope is $m=\frac{-3-0}{0-1}=3$, so an equation is $y=3x-3$.

Another method: From Exercise 61, $\frac{x}{1}+\frac{y}{-3}=1 \quad \Rightarrow \quad -3x+y=-3 \quad \Rightarrow \quad y=3x-3$.

31. Since $m=0$, $y-5=0(x-4)$ or $y=5$.

33. Putting the line $x+2y=6$ into its slope-intercept form gives us $y=-\frac{1}{2}x+3$, so we see that this line has slope $-\frac{1}{2}$. Thus, we want the line of slope $-\frac{1}{2}$ that passes through the point $(1,-6)$: $y-(-6)=-\frac{1}{2}(x-1) \quad \Leftrightarrow$ $y=-\frac{1}{2}x-\frac{11}{2}$.

35. $2x+5y+8=0 \quad \Leftrightarrow \quad y=-\frac{2}{5}x-\frac{8}{5}$. Since this line has slope $-\frac{2}{5}$, a line perpendicular to it would have slope $\frac{5}{2}$, so the required line is $y-(-2)=\frac{5}{2}[x-(-1)] \quad \Leftrightarrow \quad y=\frac{5}{2}x+\frac{1}{2}$.

37. $x+3y=0 \quad \Leftrightarrow \quad y=-\frac{1}{3}x$, so the slope is $-\frac{1}{3}$ and the y-intercept is 0.

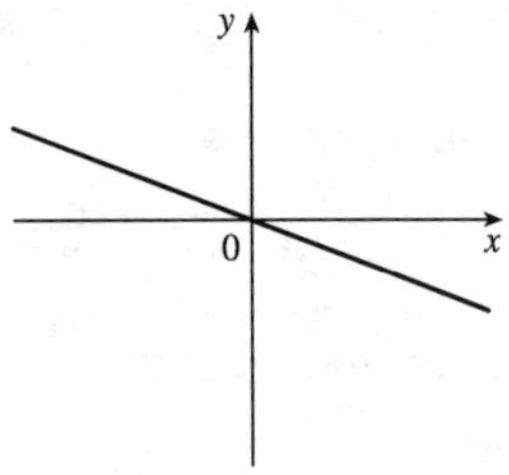

39. $y=-2$ is a horizontal line with slope 0 and y-intercept -2.

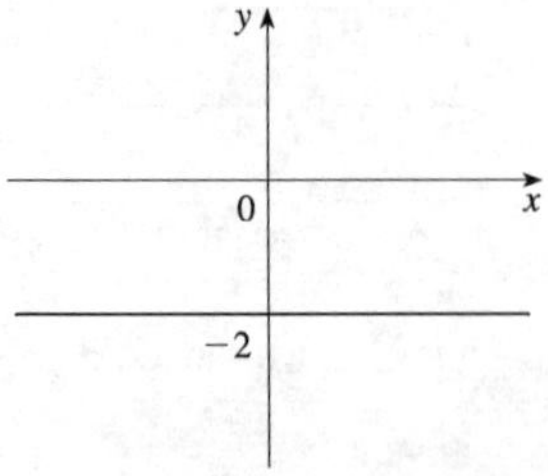

41. $3x - 4y = 12 \Leftrightarrow y = \frac{3}{4}x - 3$, so the slope is $\frac{3}{4}$ and the y-intercept is -3.

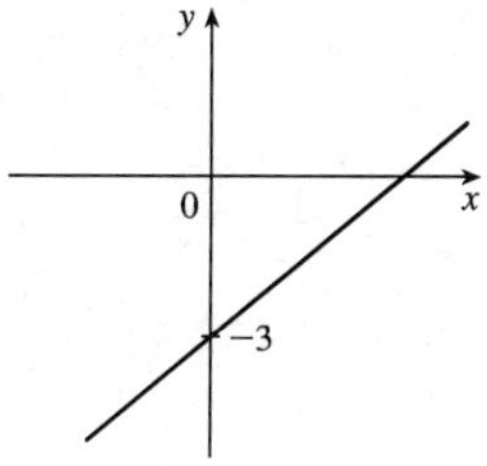

43. $\{(x, y) \mid x < 0\}$

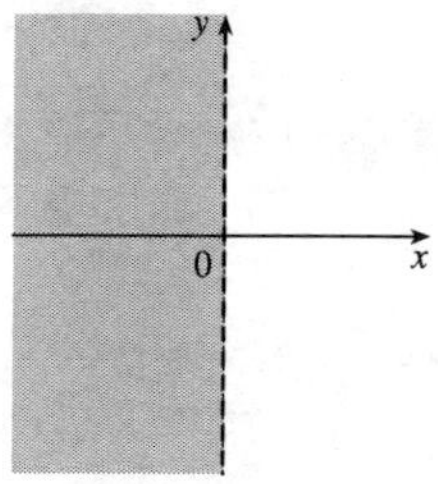

45. $\{(x, y) \mid xy < 0\} = \{(x, y) \mid x < 0 \text{ and } y > 0\}$
$\cup \{(x, y) \mid x > 0 \text{ and } y < 0\}$

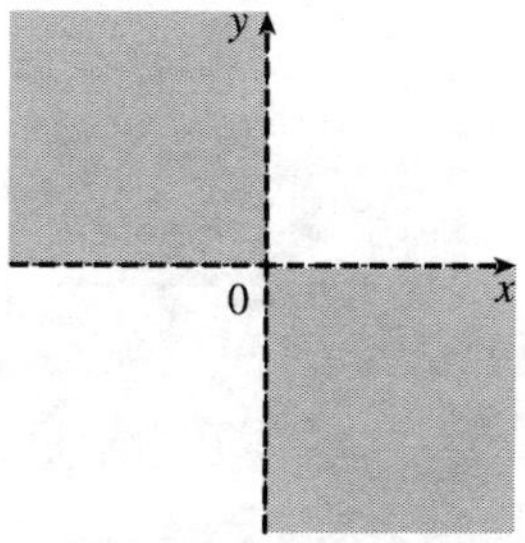

47. $\left\{(x, y) \,\middle|\, |x| \leq 2\right\} = \{(x, y) \mid -2 \leq x \leq 2\}$

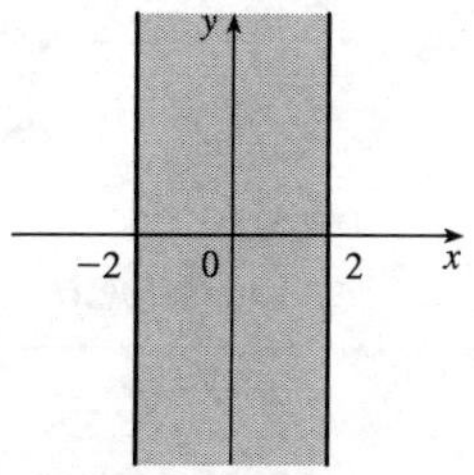

49. $\{(x, y) \mid 0 \leq y \leq 4, x \leq 2\}$

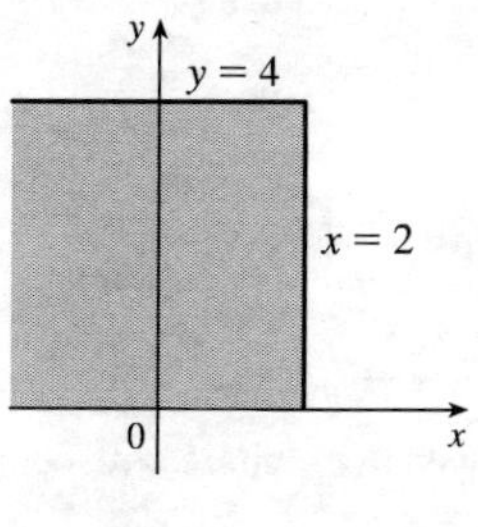

51. $\{(x, y) \mid 1 + x \leq y \leq 1 - 2x\}$

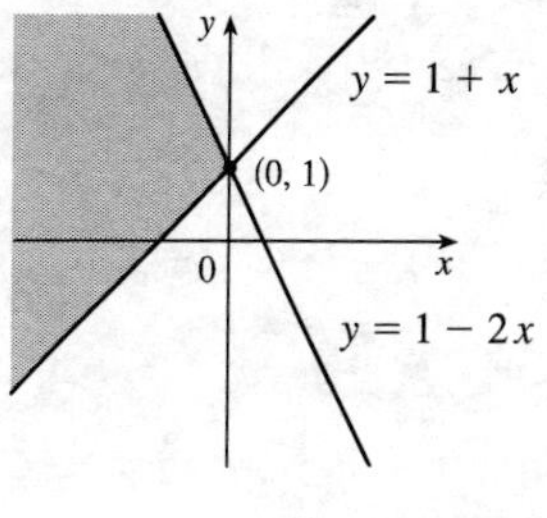

53. Let $P(0, y)$ be a point on the y-axis. The distance from P to $(5, -5)$ is $\sqrt{(5-0)^2 + (-5-y)^2} = \sqrt{5^2 + (y+5)^2}$. The distance from P to $(1, 1)$ is $\sqrt{(1-0)^2 + (1-y)^2} = \sqrt{1^2 + (y-1)^2}$. We want these distances to be equal:
$\sqrt{5^2 + (y+5)^2} = \sqrt{1^2 + (y-1)^2} \Leftrightarrow 5^2 + (y+5)^2 = 1^2 + (y-1)^2 \Leftrightarrow$
$25 + (y^2 + 10y + 25) = 1 + (y^2 - 2y + 1) \Leftrightarrow 12y = -48 \Leftrightarrow y = -4$. So the desired point is $(0, -4)$.

55. (a) Using the midpoint formula from Exercise 54 with $(1, 3)$ and $(7, 15)$, we get $\left(\frac{1+7}{2}, \frac{3+15}{2}\right) = (4, 9)$.

(b) Using the same formula, we get $\left(\frac{-1+8}{2}, \frac{6-12}{2}\right) = \left(\frac{7}{2}, -3\right)$.

57. $2x - y = 4 \;\Leftrightarrow\; y = 2x - 4 \;\Rightarrow\; m_1 = 2$ and $6x - 2y = 10 \;\Leftrightarrow\; 2y = 6x - 10 \;\Leftrightarrow\; y = 3x - 5 \;\Rightarrow$ $m_2 = 3$. Since $m_1 \neq m_2$, the two lines are not parallel. To find the point of intersection: $2x - 4 = 3x - 5 \;\Leftrightarrow$ $x = 1 \;\Rightarrow\; y = -2$. Thus, the point of intersection is $(1, -2)$.

59. With $A(1, 4)$ and $B(7, -2)$, the slope of segment AB is $\frac{-2-4}{7-1} = -1$, so its perpendicular bisector has slope 1. The midpoint of AB is $\left(\frac{1+7}{2}, \frac{4+(-2)}{2}\right) = (4, 1)$, so an equation of the perpendicular bisector is $y - 1 = 1(x - 4)$ or $y = x - 3$.

61. (a) Since the x-intercept is a, the point $(a, 0)$ is on the line, and similarly since the y-intercept is b, $(0, b)$ is on the line. Hence, the slope of the line is $m = \dfrac{b-0}{0-a} = -\dfrac{b}{a}$. Substituting into $y = mx + b$ gives $y = -\dfrac{b}{a}x + b \;\Leftrightarrow$ $\dfrac{b}{a}x + y = b \;\Leftrightarrow\; \dfrac{x}{a} + \dfrac{y}{b} = 1$.

(b) Letting $a = 6$ and $b = -8$ gives $\dfrac{x}{6} + \dfrac{y}{-8} = 1 \;\Leftrightarrow\; -8x + 6y = -48$ [multiply by -48] $\Leftrightarrow$ $6y = 8x - 48 \;\Leftrightarrow\; 3y = 4x - 24 \;\Leftrightarrow\; y = \frac{4}{3}x - 8$.

C Graphs of Second-Degree Equations

1. An equation of the circle with center $(3, -1)$ and radius 5 is $(x - 3)^2 + (y + 1)^2 = 5^2 = 25$.

3. The equation has the form $x^2 + y^2 = r^2$. Since $(4, 7)$ lies on the circle, we have $4^2 + 7^2 = r^2 \;\Rightarrow\; r^2 = 65$. So the required equation is $x^2 + y^2 = 65$.

5. $x^2 + y^2 - 4x + 10y + 13 = 0 \;\Leftrightarrow\; x^2 - 4x + y^2 + 10y = -13 \;\Leftrightarrow$ $(x^2 - 4x + 4) + (y^2 + 10y + 25) = -13 + 4 + 25 = 16 \;\Leftrightarrow\; (x - 2)^2 + (y + 5)^2 = 4^2$. Thus, we have a circle with center $(2, -5)$ and radius 4.

7. $x^2 + y^2 + x = 0 \;\Leftrightarrow\; \left(x^2 + x + \frac{1}{4}\right) + y^2 = \frac{1}{4} \;\Leftrightarrow\; \left(x + \frac{1}{2}\right)^2 + y^2 = \left(\frac{1}{2}\right)^2$. Thus, we have a circle with center $\left(-\frac{1}{2}, 0\right)$ and radius $\frac{1}{2}$.

9. $2x^2 + 2y^2 - x + y = 1 \;\Leftrightarrow\; 2\left(x^2 - \frac{1}{2}x + \frac{1}{16}\right) + 2\left(y^2 + \frac{1}{2}y + \frac{1}{16}\right) = 1 + \frac{1}{8} + \frac{1}{8} \;\Leftrightarrow$ $2\left(x - \frac{1}{4}\right)^2 + 2\left(y + \frac{1}{4}\right)^2 = \frac{5}{4} \;\Leftrightarrow\; \left(x - \frac{1}{4}\right)^2 + \left(y + \frac{1}{4}\right)^2 = \frac{5}{8}$. Thus, we have a circle with center $\left(\frac{1}{4}, -\frac{1}{4}\right)$ and radius $\frac{\sqrt{5}}{2\sqrt{2}} = \frac{\sqrt{10}}{4}$.

11. $y = -x^2$. Parabola

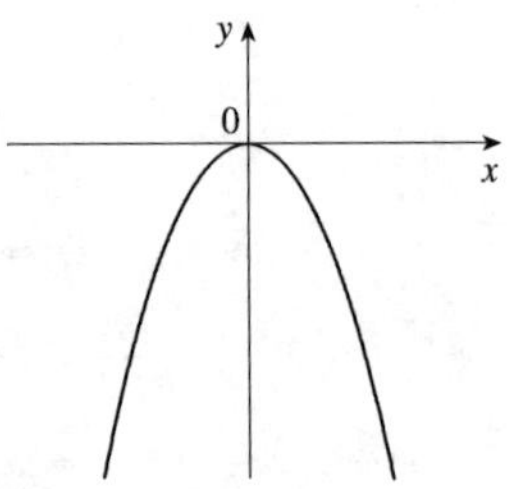

13. $x^2 + 4y^2 = 16 \;\Leftrightarrow\; \dfrac{x^2}{16} + \dfrac{y^2}{4} = 1$. Ellipse

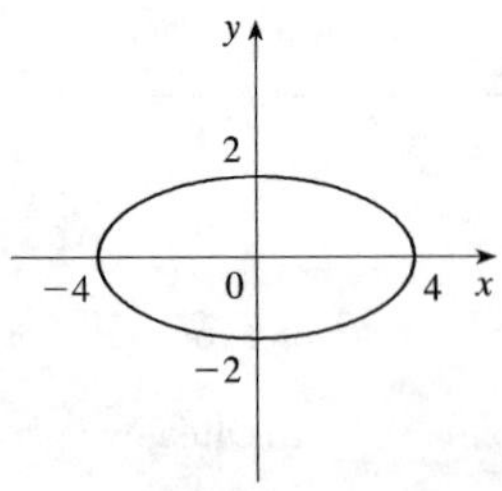

15. $16x^2 - 25y^2 = 400 \quad \Leftrightarrow \quad \dfrac{x^2}{25} - \dfrac{y^2}{16} = 1.$

Hyperbola

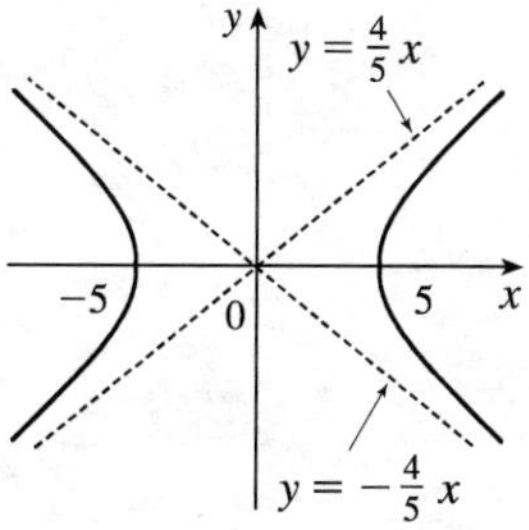

17. $4x^2 + y^2 = 1 \quad \Leftrightarrow \quad \dfrac{x^2}{1/4} + y^2 = 1.$ Ellipse

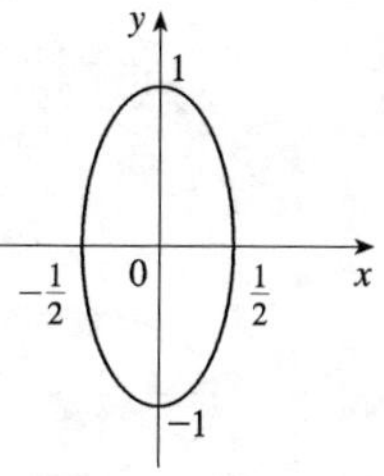

19. $x = y^2 - 1.$ Parabola with vertex at $(-1, 0)$

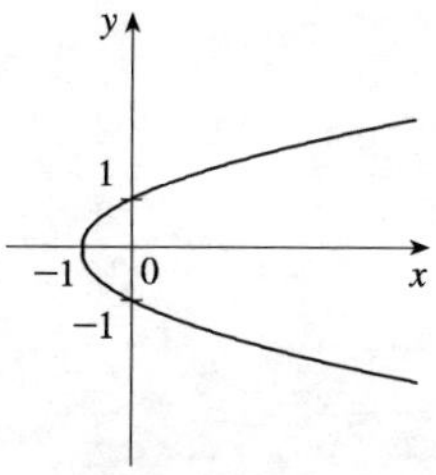

21. $9y^2 - x^2 = 9 \quad \Leftrightarrow \quad y^2 - \dfrac{x^2}{9} = 1.$ Hyperbola

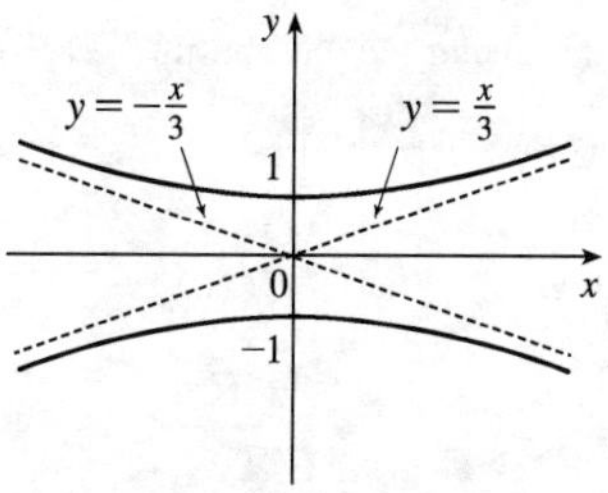

23. $xy = 4.$ Hyperbola

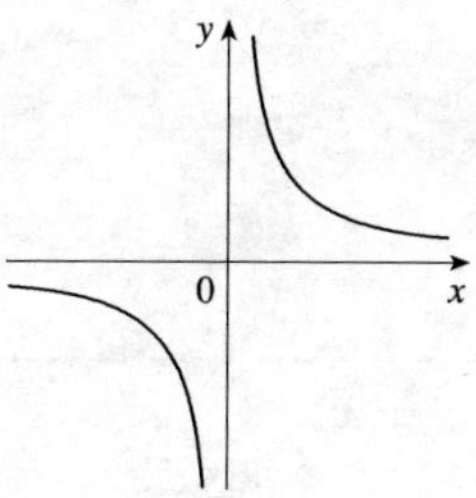

25. $9(x-1)^2 + 4(y-2)^2 = 36 \quad \Leftrightarrow$

$\dfrac{(x-1)^2}{4} + \dfrac{(y-2)^2}{9} = 1.$ Ellipse centered at $(1, 2)$

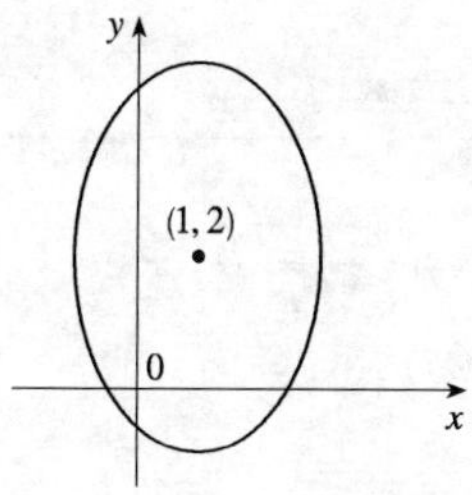

27. $y = x^2 - 6x + 13 = (x^2 - 6x + 9) + 4 =$ $(x-3)^2 + 4.$ Parabola with vertex at $(3, 4)$

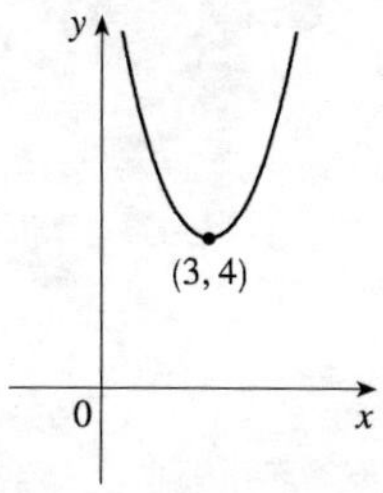

29. $x = -y^2 + 4.$ Parabola with vertex at $(4, 0)$

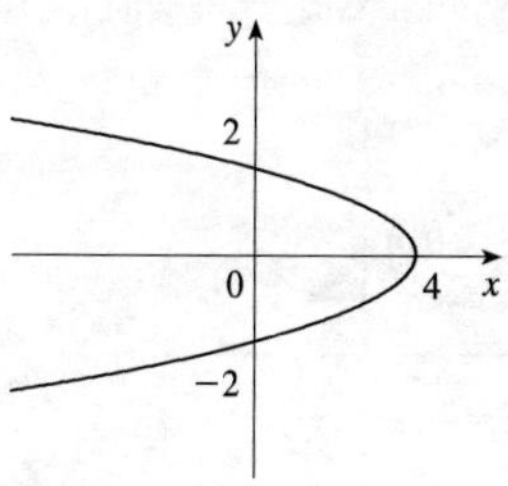

31. $x^2 + 4y^2 - 6x + 5 = 0 \quad \Leftrightarrow$
$(x^2 - 6x + 9) + 4y^2 = -5 + 9 = 4 \quad \Leftrightarrow$
$\dfrac{(x-3)^2}{4} + y^2 = 1$. Ellipse centered at $(3, 0)$

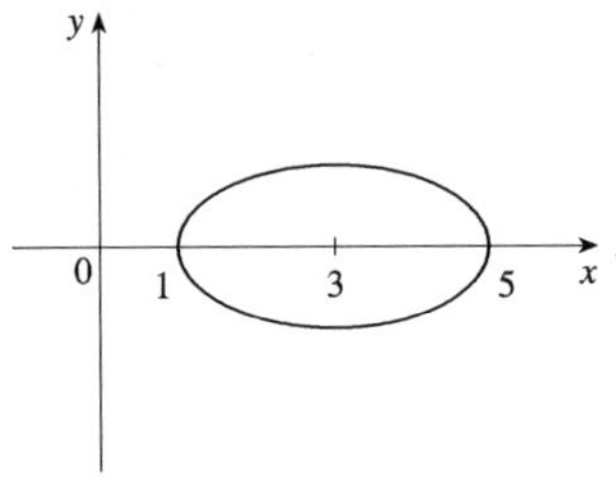

33. $y = 3x$ and $y = x^2$ intersect where $3x = x^2 \quad \Leftrightarrow$
$0 = x^2 - 3x = x(x-3)$, that is, at $(0, 0)$ and $(3, 9)$.

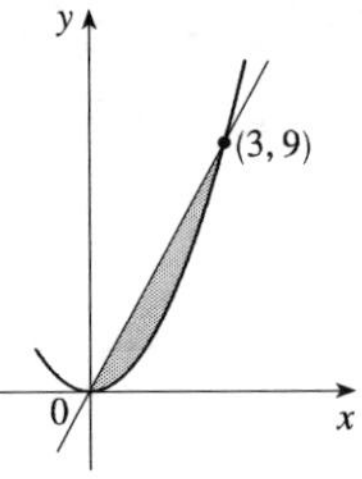

35. The parabola must have an equation of the form $y = a(x-1)^2 - 1$. Substituting $x = 3$ and $y = 3$ into the equation gives $3 = a(3-1)^2 - 1$, so $a = 1$, and the equation is $y = (x-1)^2 - 1 = x^2 - 2x$.
Note that using the other point $(-1, 3)$ would have given the same value for a, and hence the same equation.

37. $\{(x, y) \mid x^2 + y^2 \leq 1\}$

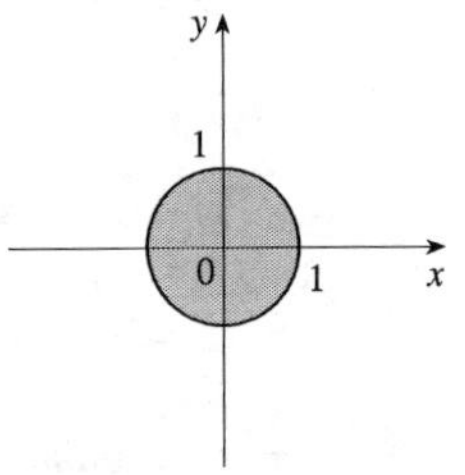

39. $\{(x, y) \mid y \geq x^2 - 1\}$

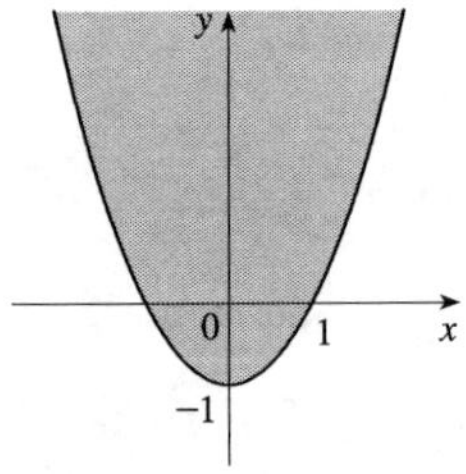

D Trigonometry

1. $210° = 210\left(\frac{\pi}{180}\right) = \frac{7\pi}{6}$ rad

3. $9° = 9\left(\frac{\pi}{180}\right) = \frac{\pi}{20}$ rad

5. $900° = 900\left(\frac{\pi}{180}\right) = 5\pi$ rad

7. 4π rad $= 4\pi\left(\frac{180}{\pi}\right) = 720°$

9. $\frac{5\pi}{12}$ rad $= \frac{5\pi}{12}\left(\frac{180}{\pi}\right) = 75°$

11. $-\frac{3\pi}{8}$ rad $= -\frac{3\pi}{8}\left(\frac{180}{\pi}\right) = -67.5°$

13. Using Formula 3, $a = r\theta = 36 \cdot \frac{\pi}{12} = 3\pi$ cm.

15. Using Formula 3, $\theta = a/r = \frac{1}{1.5} = \frac{2}{3}$ rad $= \frac{2}{3}\left(\frac{180}{\pi}\right) = \left(\frac{120}{\pi}\right)° \approx 38.2°$.

17.

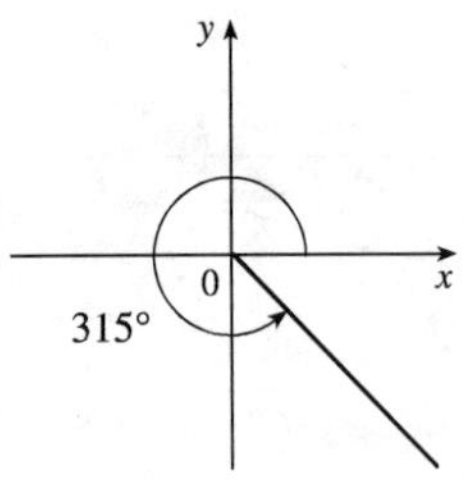

19.

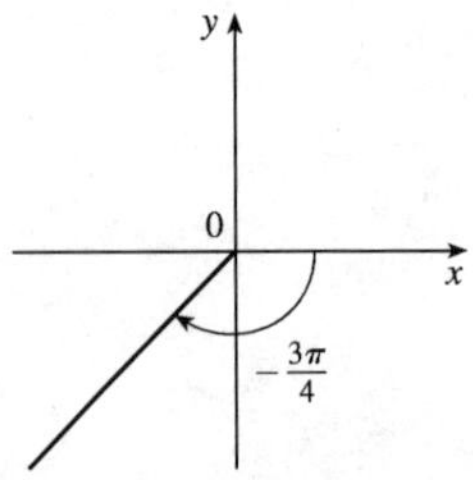

21.

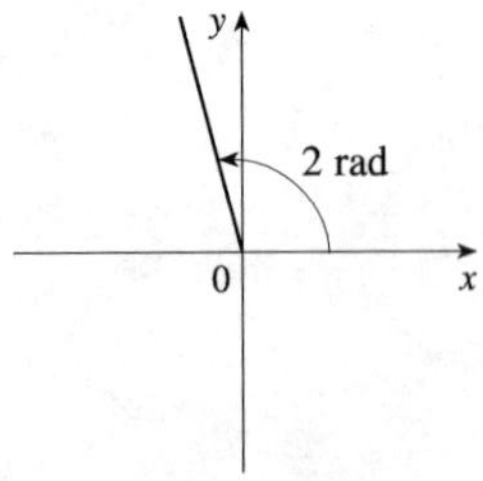

23. 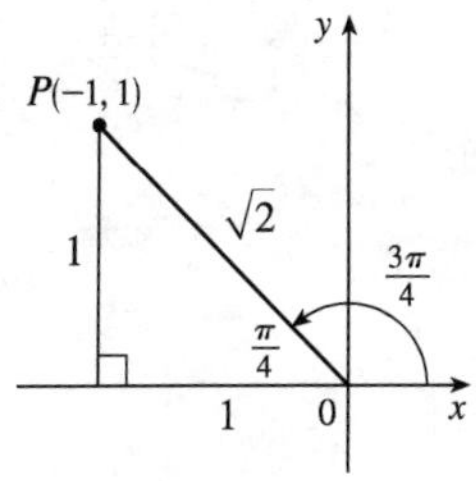

From the diagram we see that a point on the terminal side is $P(-1,1)$. Therefore, taking $x=-1$, $y=1$, $r=\sqrt{2}$ in the definitions of the trigonometric ratios, we have $\sin\frac{3\pi}{4}=\frac{1}{\sqrt{2}}$, $\cos\frac{3\pi}{4}=-\frac{1}{\sqrt{2}}$, $\tan\frac{3\pi}{4}=-1$, $\csc\frac{3\pi}{4}=\sqrt{2}$, $\sec\frac{3\pi}{4}=-\sqrt{2}$, and $\cot\frac{3\pi}{4}=-1$.

25. 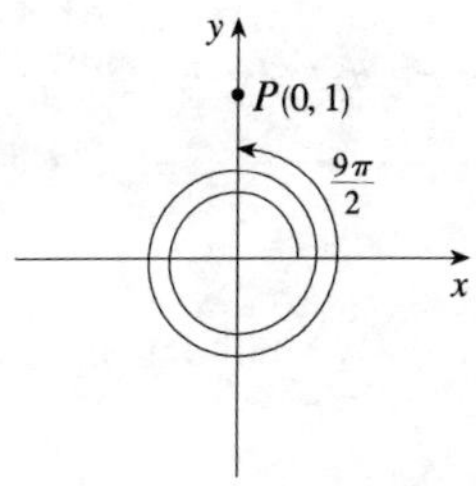

From the diagram we see that a point on the terminal line is $P\,(0,1)$. Therefore taking $x=0$, $y=1$, $r=1$ in the definitions of the trigonometric ratios, we have $\sin\frac{9\pi}{2}=1$, $\cos\frac{9\pi}{2}=0$, $\tan\frac{9\pi}{2}=y/x$ is undefined since $x=0$, $\csc\frac{9\pi}{2}=1$, $\sec\frac{9\pi}{2}=r/x$ is undefined since $x=0$, and $\cot\frac{9\pi}{2}=0$.

27. 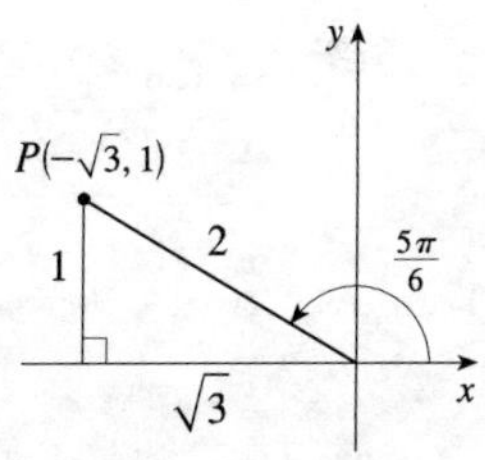

Using Figure 8 we see that a point on the terminal line is $P\left(-\sqrt{3},1\right)$. Therefore taking $x=-\sqrt{3}$, $y=1$, $r=2$ in the definitions of the trigonometric ratios, we have $\sin\frac{5\pi}{6}=\frac{1}{2}$, $\cos\frac{5\pi}{6}=-\frac{\sqrt{3}}{2}$, $\tan\frac{5\pi}{6}=-\frac{1}{\sqrt{3}}$, $\csc\frac{5\pi}{6}=2$, $\sec\frac{5\pi}{6}=-\frac{2}{\sqrt{3}}$, and $\cot\frac{5\pi}{6}=-\sqrt{3}$.

29. $\sin\theta=y/r=\frac{3}{5}\;\Rightarrow\; y=3$, $r=5$, and $x=\sqrt{r^2-y^2}=4$ (since $0<\theta<\frac{\pi}{2}$). Therefore taking $x=4$, $y=3$, $r=5$ in the definitions of the trigonometric ratios, we have $\cos\theta=\frac{4}{5}$, $\tan\theta=\frac{3}{4}$, $\csc\theta=\frac{5}{3}$, $\sec\theta=\frac{5}{4}$, and $\cot\theta=\frac{4}{3}$.

31. $\frac{\pi}{2}<\phi<\pi\;\Rightarrow\;\phi$ is in the second quadrant, where x is negative and y is positive. Therefore $\sec\phi=r/x=-1.5=-\frac{3}{2}\;\Rightarrow\; r=3$, $x=-2$, and $y=\sqrt{r^2-x^2}=\sqrt{5}$. Taking $x=-2$, $y=\sqrt{5}$, and $r=3$ in the definitions of the trigonometric ratios, we have $\sin\phi=\frac{\sqrt{5}}{3}$, $\cos\phi=-\frac{2}{3}$, $\tan\phi=-\frac{\sqrt{5}}{2}$, $\csc\phi=\frac{3}{\sqrt{5}}$, and $\cot\theta=-\frac{2}{\sqrt{5}}$.

33. $\pi<\beta<2\pi$ means that β is in the third or fourth quadrant where y is negative. Also since $\cot\beta=x/y=3$ which is positive, x must also be negative. Therefore $\cot\beta=x/y=\frac{3}{1}\;\Rightarrow\; x=-3$, $y=-1$, and $r=\sqrt{x^2+y^2}=\sqrt{10}$. Taking $x=-3$, $y=-1$ and $r=\sqrt{10}$ in the definitions of the trigonometric ratios, w have $\sin\beta=-\frac{1}{\sqrt{10}}$, $\cos\beta=-\frac{3}{\sqrt{10}}$, $\tan\beta=\frac{1}{3}$, $\csc\beta=-\sqrt{10}$, and $\sec\beta=-\frac{\sqrt{10}}{3}$.

35. $\sin 35^\circ=\dfrac{x}{10}\;\Rightarrow\; x=10\sin 35^\circ\approx 5.73576$ cm

37. $\tan\frac{2\pi}{5}=\dfrac{x}{8}\;\Rightarrow\; x=8\tan\frac{2\pi}{5}\approx 24.62147$ cm

39. 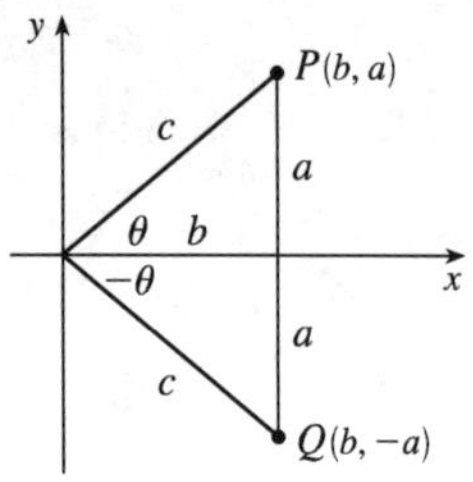

(a) From the diagram we see that $\sin\theta = \frac{y}{r} = \frac{a}{c}$, and

$$\sin(-\theta) = \frac{-a}{c} = -\frac{a}{c} = -\sin\theta.$$

(b) Again from the diagram we see that $\cos\theta = \frac{x}{r} = \frac{b}{c} = \cos(-\theta)$.

41. (a) Using (12a) and (13a), we have

$$\tfrac{1}{2}\left[\sin(x+y) + \sin(x-y)\right] = \tfrac{1}{2}\left[\sin x\cos y + \cos x\sin y + \sin x\cos y - \cos x\sin y\right]$$
$$= \tfrac{1}{2}\left(2\sin x\cos y\right) = \sin x\cos y$$

(b) This time, using (12b) and (13b), we have

$$\tfrac{1}{2}\left[\cos(x+y) + \cos(x-y)\right] = \tfrac{1}{2}\left[\cos x\cos y - \sin x\sin y + \cos x\cos y + \sin x\sin y\right]$$
$$= \tfrac{1}{2}\left(2\cos x\cos y\right) = \cos x\cos y$$

(c) Again using (12b) and (13b), we have

$$\tfrac{1}{2}\left[\cos(x-y) - \cos(x+y)\right] = \tfrac{1}{2}\left[\cos x\cos y + \sin x\sin y - \cos x\cos y + \sin x\sin y\right]$$
$$= \tfrac{1}{2}\left(2\sin x\sin y\right) = \sin x\sin y$$

43. Using (12a), we have $\sin\left(\frac{\pi}{2} + x\right) = \sin\frac{\pi}{2}\cos x + \cos\frac{\pi}{2}\sin x = 1\cdot\cos x + 0\cdot\sin x = \cos x$.

45. Using (6), we have $\sin\theta\cot\theta = \sin\theta\cdot\dfrac{\cos\theta}{\sin\theta} = \cos\theta$.

47. $\sec y - \cos y = \dfrac{1}{\cos y} - \cos y$ [by (6)] $= \dfrac{1-\cos^2 y}{\cos y} = \dfrac{\sin^2 y}{\cos y}$ [by (7)] $= \dfrac{\sin y}{\cos y}\sin y = \tan y\sin y$ [by (6)]

49. $\cot^2\theta + \sec^2\theta = \dfrac{\cos^2\theta}{\sin^2\theta} + \dfrac{1}{\cos^2\theta}$ [by (6)] $= \dfrac{\cos^2\theta\cos^2\theta + \sin^2\theta}{\sin^2\theta\cos^2\theta}$

$= \dfrac{\left(1-\sin^2\theta\right)\left(1-\sin^2\theta\right) + \sin^2\theta}{\sin^2\theta\cos^2\theta}$ [by (7)] $= \dfrac{1-\sin^2\theta+\sin^4\theta}{\sin^2\theta\cos^2\theta}$

$= \dfrac{\cos^2\theta+\sin^4\theta}{\sin^2\theta\cos^2\theta}$ [by (7)] $= \dfrac{1}{\sin^2\theta} + \dfrac{\sin^2\theta}{\cos^2\theta} = \csc^2\theta + \tan^2\theta$ [by (6)]

51. Using (14a), we have $\tan 2\theta = \tan(\theta+\theta) = \dfrac{\tan\theta+\tan\theta}{1-\tan\theta\tan\theta} = \dfrac{2\tan\theta}{1-\tan^2\theta}$.

53. Using (15a) and (16a),

$$\sin x\sin 2x + \cos x\cos 2x = \sin x\left(2\sin x\cos x\right) + \cos x\left(2\cos^2 x - 1\right) = 2\sin^2 x\cos x + 2\cos^3 x - \cos x$$
$$= 2\left(1-\cos^2 x\right)\cos x + 2\cos^3 x - \cos x \text{ [by (7)]}$$
$$= 2\cos x - 2\cos^3 x + 2\cos^3 x - \cos x = \cos x$$

Or: $\sin x\sin 2x + \cos x\cos 2x = \cos(2x - x)$ [by 13(b)] $= \cos x$

55. $\dfrac{\sin\phi}{1-\cos\phi} = \dfrac{\sin\phi}{1-\cos\phi}\cdot\dfrac{1+\cos\phi}{1+\cos\phi} = \dfrac{\sin\phi\,(1+\cos\phi)}{1-\cos^2\phi} = \dfrac{\sin\phi\,(1+\cos\phi)}{\sin^2\phi}$ [by (7)]

$= \dfrac{1+\cos\phi}{\sin\phi} = \dfrac{1}{\sin\phi} + \dfrac{\cos\phi}{\sin\phi} = \csc\phi + \cot\phi$ [by (6)]

57. Using (12a),

$$\begin{aligned}\sin 3\theta + \sin\theta &= \sin(2\theta+\theta) + \sin\theta = \sin 2\theta\cos\theta + \cos 2\theta\sin\theta + \sin\theta \\ &= \sin 2\theta\cos\theta + \left(2\cos^2\theta - 1\right)\sin\theta + \sin\theta \quad \text{[by (16a)]} \\ &= \sin 2\theta\cos\theta + 2\cos^2\theta\sin\theta - \sin\theta + \sin\theta = \sin 2\theta\cos\theta + \sin 2\theta\cos\theta \quad \text{[by (15a)]} \\ &= 2\sin 2\theta\cos\theta\end{aligned}$$

59. Since $\sin x = \frac{1}{3}$ we can label the opposite side as having length 1, the hypotenuse as having length 3, and use the Pythagorean Theorem to get that the adjacent side has length $\sqrt{8}$. Then, from the diagram, $\cos x = \frac{\sqrt{8}}{3}$. Similarly we have that $\sin y = \frac{3}{5}$. Now use (12a):

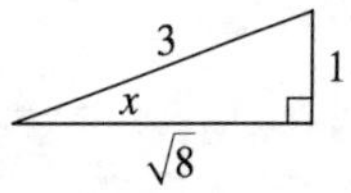

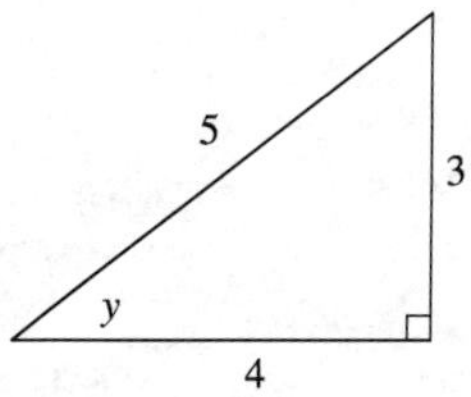

$\sin(x+y) = \sin x\cos y + \cos x\,\sin y = \frac{1}{3}\cdot\frac{4}{5} + \frac{\sqrt{8}}{3}\cdot\frac{3}{5} = \frac{4}{15} + \frac{3\sqrt{8}}{15} = \frac{4+6\sqrt{2}}{15}$.

61. Using (13b) and the values for $\cos x$ and $\sin y$ obtained in Exercise 59, we have

$$\cos(x-y) = \cos x\cos y + \sin x\sin y = \frac{\sqrt{8}}{3}\cdot\frac{4}{5} + \frac{1}{3}\cdot\frac{3}{5} = \frac{8\sqrt{2}+3}{15}$$

63. Using (15a) and the values for $\sin y$ and $\cos y$ obtained in Exercise 59, we have

$$\sin 2y = 2\sin y\cos y = 2\cdot\frac{3}{5}\cdot\frac{4}{5} = \frac{24}{25}$$

65. $2\cos x - 1 = 0 \quad\Leftrightarrow\quad \cos x = \frac{1}{2} \quad\Rightarrow\quad x = \frac{\pi}{3}, \frac{5\pi}{3}$ for $x \in [0, 2\pi]$.

67. $2\sin^2 x = 1 \quad\Leftrightarrow\quad \sin^2 x = \frac{1}{2} \quad\Leftrightarrow\quad \sin x = \pm\frac{1}{\sqrt{2}} \quad\Rightarrow\quad x = \frac{\pi}{4}, \frac{3\pi}{4}, \frac{5\pi}{4}, \frac{7\pi}{4}$.

69. Using (15a), we have $\sin 2x = \cos x \quad\Leftrightarrow\quad 2\sin x\cos x - \cos x = 0 \quad\Leftrightarrow\quad \cos x(2\sin x - 1) = 0 \quad\Leftrightarrow$ $\cos x = 0$ or $2\sin x - 1 = 0 \quad\Rightarrow\quad x = \frac{\pi}{2}, \frac{3\pi}{2}$ or $\sin x = \frac{1}{2} \quad\Rightarrow\quad x = \frac{\pi}{6}$ or $\frac{5\pi}{6}$. Therefore, the solutions are $x = \frac{\pi}{6}, \frac{\pi}{2}, \frac{5\pi}{6}, \frac{3\pi}{2}$.

71. $\sin x = \tan x \quad\Leftrightarrow\quad \sin x - \tan x = 0 \quad\Leftrightarrow\quad \sin x - \dfrac{\sin x}{\cos x} = 0 \quad\Leftrightarrow\quad \sin x\left(1 - \dfrac{1}{\cos x}\right) = 0 \quad\Leftrightarrow\quad \sin x = 0$ or $1 - \dfrac{1}{\cos x} = 0 \quad\Rightarrow\quad x = 0, \pi, 2\pi$ or $1 = \dfrac{1}{\cos x} \quad\Rightarrow\quad \cos x = 1 \quad\Rightarrow\quad x = 0, 2\pi$. Therefore the solutions are $x = 0, \pi, 2\pi$.

73. We know that $\sin x = \frac{1}{2}$ when $x = \frac{\pi}{6}$ or $\frac{5\pi}{6}$, and from Figure 13(a), we see that $\sin x \le \frac{1}{2} \quad\Rightarrow\quad 0 \le x \le \frac{\pi}{6}$ or $\frac{5\pi}{6} \le x \le 2\pi$ for $x \in [0, 2\pi]$.

75. $\tan x = -1$ when $x = \frac{3\pi}{4}, \frac{7\pi}{4}$, and $\tan x = 1$ when $x = \frac{\pi}{4}$ or $\frac{5\pi}{4}$. From Figure 14 we see that $-1 < \tan x < 1$ $\Rightarrow\quad 0 \le x < \frac{\pi}{4}, \frac{3\pi}{4} < x < \frac{5\pi}{4}$, and $\frac{7\pi}{4} < x \le 2\pi$.

77. $y = \cos\left(x - \frac{\pi}{3}\right)$. We start with the graph of $y = \cos x$ and shift it $\frac{\pi}{3}$ units to the right.

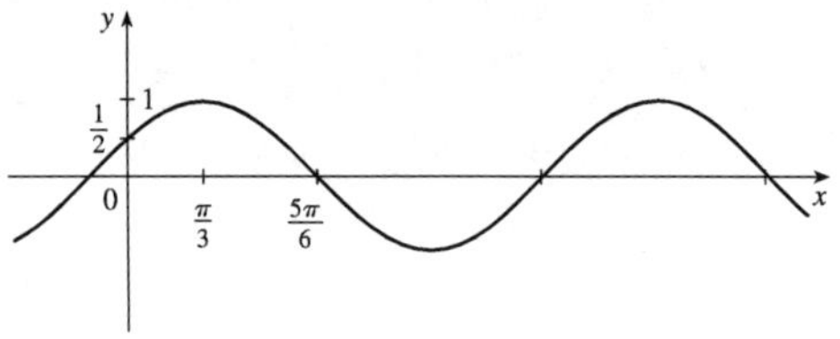

79. $y = \frac{1}{3}\tan\left(x - \frac{\pi}{2}\right)$. We start with the graph of $y = \tan x$, shift it $\frac{\pi}{2}$ units to the right and compress it to $\frac{1}{3}$ of its original vertical size.

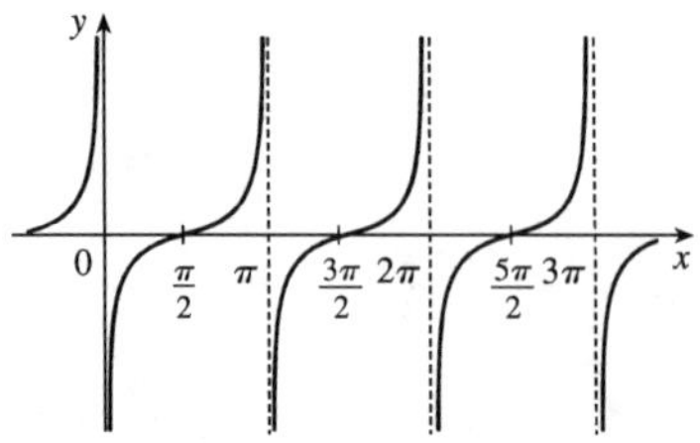

81. $y = |\sin x|$. We start with the graph of $y = \sin x$ and reflect the parts below the x-axis about the x-axis.

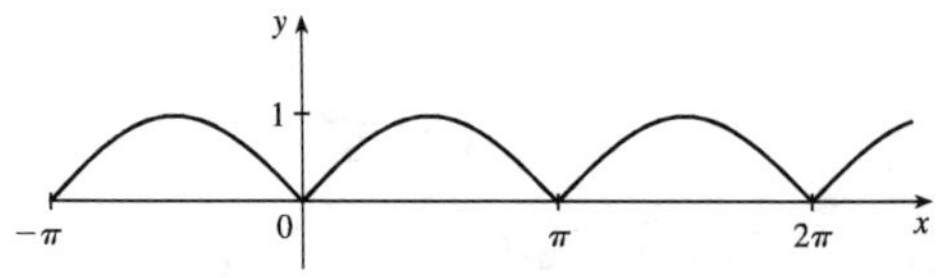

83. From the figure in the text, we see that $x = b\cos\theta$, $y = b\sin\theta$, and from the distance formula we have that the distance c from (x, y) to $(a, 0)$ is $c = \sqrt{(x-a)^2 + (y-0)^2} \;\Rightarrow$

$$c^2 = (b\cos\theta - a)^2 + (b\sin\theta)^2 = b^2\cos^2\theta - 2ab\cos\theta + a^2 + b^2\sin^2\theta$$

$$= a^2 + b^2\left(\cos^2\theta + \sin^2\theta\right) - 2ab\cos\theta = a^2 + b^2 - 2ab\cos\theta \quad \text{[by (7)]}$$

85. Using the Law of Cosines, we have $c^2 = 1^2 + 1^2 - 2(1)(1)\cos(\alpha - \beta) = 2\left[1 - \cos(\alpha - \beta)\right]$. Now, using the distance formula, $c^2 = |AB|^2 = (\cos\alpha - \cos\beta)^2 + (\sin\alpha - \sin\beta)^2$. Equating these two expressions for c^2, we get $2\left[1 - \cos(\alpha - \beta)\right] = \cos^2\alpha + \sin^2\alpha + \cos^2\beta + \sin^2\beta - 2\cos\alpha\cos\beta - 2\sin\alpha\sin\beta \;\Rightarrow$ $1 - \cos(\alpha - \beta) = 1 - \cos\alpha\cos\beta - \sin\alpha\sin\beta \;\Rightarrow\; \cos(\alpha - \beta) = \cos\alpha\cos\beta + \sin\alpha\sin\beta$.

87. In Exercise 86 we used the subtraction formula for cosine to prove the addition formula for cosine. Using that formula with $x = \frac{\pi}{2} - \alpha$, $y = \beta$, we get $\cos\left[\left(\frac{\pi}{2} - \alpha\right) + \beta\right] = \cos\left(\frac{\pi}{2} - \alpha\right)\cos\beta - \sin\left(\frac{\pi}{2} - \alpha\right)\sin\beta \;\Rightarrow$ $\cos\left[\frac{\pi}{2} - (\alpha - \beta)\right] = \cos\left(\frac{\pi}{2} - \alpha\right)\cos\beta - \sin\left(\frac{\pi}{2} - \alpha\right)\sin\beta$. Now we use the identities given in the problem, $\cos\left(\frac{\pi}{2} - \theta\right) = \sin\theta$ and $\sin\left(\frac{\pi}{2} - \theta\right) = \cos\theta$, to get $\sin(\alpha - \beta) = \sin\alpha\cos\beta - \cos\alpha\sin\beta$.

89. Using the formula from part (a), the area of the triangle is $\frac{1}{2}(10)(3)\sin 107° \approx 14.34457\ \text{cm}^2$.

E Sigma Notation

1. $\sum_{i=1}^{5} \sqrt{i} = \sqrt{1} + \sqrt{2} + \sqrt{3} + \sqrt{4} + \sqrt{5}$

3. $\sum_{i=4}^{6} 3^i = 3^4 + 3^5 + 3^6$

5. $\sum_{k=0}^{4} \frac{2k-1}{2k+1} = -1 + \frac{1}{3} + \frac{3}{5} + \frac{5}{7} + \frac{7}{9}$

7. $\sum_{i=1}^{n} i^{10} = 1^{10} + 2^{10} + 3^{10} + \cdots + n^{10}$

9. $\sum_{j=0}^{n-1} (-1)^j = 1 - 1 + 1 - 1 + \cdots + (-1)^{n-1}$

11. $1 + 2 + 3 + 4 + \cdots + 10 = \sum_{i=1}^{10} i$

13. $\frac{1}{2} + \frac{2}{3} + \frac{3}{4} + \frac{4}{5} + \cdots + \frac{19}{20} = \sum_{i=1}^{19} \frac{i}{i+1}$

15. $2 + 4 + 6 + 8 + \cdots + 2n = \sum_{i=1}^{n} 2i$

17. $1 + 2 + 4 + 8 + 16 + 32 = \sum_{i=0}^{5} 2^i$

19. $x + x^2 + x^3 + \cdots + x^n = \sum_{i=1}^{n} x^i$

21. $\sum_{i=4}^{8} (3i-2) = [3(4)-2] + [3(5)-2] + [3(6)-2] + [3(7)-2] + [3(8)-2] = 10 + 13 + 16 + 19 + 22 = 80$

23. $\sum_{j=1}^{6} 3^{j+1} = 3^2 + 3^3 + 3^4 + 3^5 + 3^6 + 3^7 = 9 + 27 + 81 + 243 + 729 + 2187 = 3276$

(For a more general method, see Exercise 47.)

25. $\sum_{n=1}^{20} (-1)^n = -1 + 1 - 1 + 1 - 1 + 1 - 1 + 1 - 1 + 1 - 1 + 1 - 1 + 1 - 1 + 1 - 1 + 1 - 1 + 1 = 0$

27. $\sum_{i=0}^{4} \left(2^i + i^2\right) = (1+0) + (2+1) + (4+4) + (8+9) + (16+16) = 61$

29. $\sum_{i=1}^{n} 2i = 2\sum_{i=1}^{n} i = 2 \cdot \frac{n(n+1)}{2}$ [by Theorem 3(c)] $= n(n+1)$

31.
$$\begin{aligned} \sum_{i=1}^{n} (i^2 + 3i + 4) &= \sum_{i=1}^{n} i^2 + 3\sum_{i=1}^{n} i + \sum_{i=1}^{n} 4 = \frac{n(n+1)(2n+1)}{6} + \frac{3n(n+1)}{2} + 4n \\ &= \tfrac{1}{6}\left[(2n^3 + 3n^2 + n) + (9n^2 + 9n) + 24n\right] = \tfrac{1}{6}\left(2n^3 + 12n^2 + 34n\right) \\ &= \tfrac{1}{3}n(n^2 + 6n + 17) \end{aligned}$$

33.
$$\begin{aligned} \sum_{i=1}^{n} (i+1)(i+2) &= \sum_{i=1}^{n} (i^2 + 3i + 2) = \sum_{i=1}^{n} i^2 + 3\sum_{i=1}^{n} i + \sum_{i=1}^{n} 2 \\ &= \frac{n(n+1)(2n+1)}{6} + \frac{3n(n+1)}{2} + 2n = \frac{n(n+1)}{6}[(2n+1) + 9] + 2n \\ &= \frac{n(n+1)}{3}(n+5) + 2n = \frac{n}{3}[(n+1)(n+5) + 6] = \frac{n}{3}\left(n^2 + 6n + 11\right) \end{aligned}$$

35. $\sum_{i=1}^{n}(i^3 - i - 2) = \sum_{i=1}^{n} i^3 - \sum_{i=1}^{n} i - \sum_{i=1}^{n} 2 = \left[\frac{n(n+1)}{2}\right]^2 - \frac{n(n+1)}{2} - 2n$

$= \frac{1}{4}n(n+1)\left[n(n+1) - 2\right] - 2n = \frac{1}{4}n(n+1)(n+2)(n-1) - 2n$

$= \frac{1}{4}n\left[(n+1)(n-1)(n+2) - 8\right] = \frac{1}{4}n\left[(n^2-1)(n+2) - 8\right] = \frac{1}{4}n(n^3 + 2n^2 - n - 10)$

37. By Theorem 2(a) and Example 3, $\sum_{i=1}^{n} c = c\sum_{i=1}^{n} 1 = cn.$

39. $\sum_{i=1}^{n}\left[(i+1)^4 - i^4\right] = (2^4 - 1^4) + (3^4 - 2^4) + (4^4 - 3^4) + \cdots + \left[(n+1)^4 - n^4\right]$

$= (n+1)^4 - 1^4 = n^4 + 4n^3 + 6n^2 + 4n$

On the other hand,

$$\sum_{i=1}^{n}\left[(i+1)^4 - i^4\right] = \sum_{i=1}^{n}(4i^3 + 6i^2 + 4i + 1) = 4\sum_{i=1}^{n} i^3 + 6\sum_{i=1}^{n} i^2 + 4\sum_{i=1}^{n} i + \sum_{i=1}^{n} 1$$

$$= 4S + n(n+1)(2n+1) + 2n(n+1) + n \quad \left[\text{where } S = \sum_{i=1}^{n} i^3\right]$$

$$= 4S + 2n^3 + 3n^2 + n + 2n^2 + 2n + n = 4S + 2n^3 + 5n^2 + 4n$$

Thus, $n^4 + 4n^3 + 6n^2 + 4n = 4S + 2n^3 + 5n^2 + 4n$, from which it follows that

$4S = n^4 + 2n^3 + n^2 = n^2(n^2 + 2n + 1) = n^2(n+1)^2$ and $S = \left[\frac{n(n+1)}{2}\right]^2$.

41. (a) $\sum_{i=1}^{n}\left[i^4 - (i-1)^4\right] = (1^4 - 0^4) + (2^4 - 1^4) + (3^4 - 2^4) + \cdots + \left[n^4 - (n-1)^4\right] = n^4 - 0 = n^4$

(b) $\sum_{i=1}^{100}\left(5^i - 5^{i-1}\right) = (5^1 - 5^0) + (5^2 - 5^1) + (5^3 - 5^2) + \cdots + (5^{100} - 5^{99}) = 5^{100} - 5^0 = 5^{100} - 1$

(c) $\sum_{i=3}^{99}\left(\frac{1}{i} - \frac{1}{i+1}\right) = \left(\frac{1}{3} - \frac{1}{4}\right) + \left(\frac{1}{4} - \frac{1}{5}\right) + \left(\frac{1}{5} - \frac{1}{6}\right) + \cdots + \left(\frac{1}{99} - \frac{1}{100}\right) = \frac{1}{3} - \frac{1}{100} = \frac{97}{300}$

(d) $\sum_{i=1}^{n}(a_i - a_{i-1}) = (a_1 - a_0) + (a_2 - a_1) + (a_3 - a_2) + \cdots + (a_n - a_{n-1}) = a_n - a_0$

43. $\lim_{n\to\infty}\sum_{i=1}^{n}\frac{1}{n}\left(\frac{i}{n}\right)^2 = \lim_{n\to\infty}\frac{1}{n^3}\sum_{i=1}^{n} i^2 = \lim_{n\to\infty}\frac{1}{n^3}\frac{n(n+1)(2n+1)}{6} = \lim_{n\to\infty}\frac{1}{6}\left(1+\frac{1}{n}\right)\left(2+\frac{1}{n}\right)$

$= \frac{1}{6}(1)(2) = \frac{1}{3}$

45. $\lim_{n\to\infty}\sum_{i=1}^{n}\frac{2}{n}\left[\left(\frac{2i}{n}\right)^3 + 5\left(\frac{2i}{n}\right)\right] = \lim_{n\to\infty}\sum_{i=1}^{n}\left[\frac{16}{n^4}i^3 + \frac{20}{n^2}i\right] = \lim_{n\to\infty}\left[\frac{16}{n^4}\sum_{i=1}^{n} i^3 + \frac{20}{n^2}\sum_{i=1}^{n} i\right]$

$= \lim_{n\to\infty}\left[\frac{16}{n^4}\frac{n^2(n+1)^2}{4} + \frac{20}{n^2}\frac{n(n+1)}{2}\right] = \lim_{n\to\infty}\left[\frac{4(n+1)^2}{n^2} + \frac{10n(n+1)}{n^2}\right]$

$= \lim_{n\to\infty}\left[4\left(1+\frac{1}{n}\right)^2 + 10\left(1+\frac{1}{n}\right)\right] = 4\cdot 1 + 10\cdot 1 = 14$

47. Let $S = \sum_{i=1}^{n} ar^{i-1} = a + ar + ar^2 + \cdots + ar^{n-1}$. Multiplying both sides by r gives us $rS = ar + ar^2 + \cdots + ar^{n-1} + ar^n$. Subtracting the first equation from the second, we find $(r-1)S = ar^n - a = a(r^n - 1)$, so $S = \dfrac{a(r^n - 1)}{r - 1}$ (since $r \neq 1$).

49. $\sum_{i=1}^{n} \left(2i + 2^i\right) = 2\sum_{i=1}^{n} i + \sum_{i=1}^{n} 2 \cdot 2^{i-1} = 2\dfrac{n(n+1)}{2} + \dfrac{2(2^n - 1)}{2 - 1} = 2^{n+1} + n^2 + n - 2$.
For the first sum we have used Theorem 3(c), and for the second, Exercise 47 with $a = r = 2$.